MW00843692

Energy Systems

Series Editor:

Panos M. Pardalos, University of Florida, USA

Josef Kallrath • Panos M. Pardalos
Steffen Rebennack • Max Scheidt
Editors

Optimization in the Energy Industry

Editors

Prof. Dr. Josef Kallrath
Am Mahlstein 8
67273 Weisenheim
Germany
josef.kallrath@t-online.de

Steffen Rebennack
University of Florida
Department of Industrial & Systems Engineering
303 Weil Hall, P.O.Box 116595
Gainesville FL 32611-6595
USA
steffen@ufl.edu

Prof. Panos M. Pardalos
University of Florida
Department of Industrial & Systems Engineering
303 Weil Hall, P.O.Box 116595
Gainesville FL 32611-6595
USA
pardalos@ufl.edu

Dr. Max Scheidt
ProCom GmbH
Luisenstraße 41
52070 Aachen
Germany
max.scheidt@procom.de

Cover art entitled "WATER AND SOIL, THE PRIMEVAL ELEMENTS TO CREATION" is designed by Elias Tyligadas.

ISBN: 978-3-540-88964-9 e-ISBN: 978-3-540-88965-6

Library of Congress Control Number: 2008942066

Cover design: WMX Design GmbH, Heidelberg

Printed on acid-free paper

springer.com

To our families.

Preface

ΚΑΙ ΟΤΙ ΟΥΤΕ ΕΛΑΧΙΣΤΟΝ ΕΣΤΙΝ ΕΝ ΤΑΙΣ ΑΡΧΑΙΣ ΟΥΤΕ ΤΟ ΜΕΓΙΣΤΟΝ. ΕΙ ΓΑΡ ΠΑΝ ΕΝ ΠΑΝΤΙ ΚΑΙ ΠΑΝ ΕΚ ΠΑΝΤΟΣ ΕΚΚΡΙΝΕΤΑΙ, ΚΑΙ ΑΠΟ ΤΟΥ ΕΛΑΧΙΣΤΟΥ ΔΟΚΟΥΝΤΟΣ ΕΚΡΙΘΗΣΣΕΤΑΙ ΤΙ ΕΛΛΑΣΣΟΝ ΕΚΕΙΝΟΥ ΚΑΙ ΤΟ ΜΕΓΙΣΤΟΝ ΔΟΚΟΥΝ ΑΠΟ ΤΙΝΟΣ ΕΞΕΚΡΙΘΗ ΕΑΥΤΟΥ ΜΕΙΖΟΝΟΣ.

Anaxagoras of Clazomenae (499BC - 428BC)

There is no smallest among the small
and no largest among the large.
But always something still smaller
and something still larger.

Today, the optimization of production planning processes by means of IT and quantitative methods is a de-facto standard in the energy industry. Franch et al. in Chapter 1 and Ikenouye in Chapter 2 give an introduction, overview, and reasons for this. Furthermore, the energy problem now is not only a challenging one but also one of the most important issues in the world from the political and economical points of view. In every country, the government is faced with the problem of how to adopt the system of 'Cap and Trade.' Especially energy consuming industries, such as steel, power, oil and chemicals, are seriously confronted with this problem.

This is also the reason why the German Operations Research Society (GOR) and one of its working groups, held a symposium with the title "Stochastic Optimization in the Energy Industry." During the 78th meeting of the GOR working group "Praxis der Mathematischen Optimierung/Real World Optimization" in Aachen at Procom GmbH on April 21/22, 2007, the speakers with an application background explained their requirements for stochastic optimization solutions based on practical experiences. The speakers from the research side and the software system suppliers examined different aspects of the whole subject – from the integration of wind energy, the chain of errors in nuclear power plants and the scheduling of hydroelectric power stations, and the risk assessment in trading activities to the various software systems which support stochastic optimization methods.

The symposium offered an interesting overview which reflected the requirements, possibilities and restrictions of "Stochastic Optimization in the Energy Industry." As the speakers came from all over the world (Brazil, USA, The Netherlands, Norway, Switzerland and Germany) it was also an ideal platform to exchange ideas across countries in the energy sector and beyond.

This book is partly based on the contributions the speakers made to the workshop, but also contains chapters provided by other colleagues. The chapters of the first part of the book give a general introduction to the field. The second part contains deterministic models, while the third part provides methods and applications involving uncertain data. The fourth part includes contributions which focus on pricing.

After opening the European markets for electricity, the energy supply companies expect both new risks and new chances. The ex-ante uncertain market price increasingly determines the amount of their self-generated energy. While the classic unit scheduling objective is the cost-optimal production plan, in liberalized energy markets a holistic examination of the power-station and trading portfolio results in multiple chances to improve the profit situation.

Borisovksy et al. in Chapter 3 consider the problem of constructing trading hubs in the structure of electricity wholesale markets. The nodes of a trading hub are used to calculate a reference price that can be employed by the market participants for different types of hedging. The need for such a reference price is the considerable variability of energy prices at different nodes of the electricity grid at different periods of time. Hub construction is viewed as a mathematical programming problem.

These changes in electric network infrastructure and government policies have created opportunities for the employment of distributed generation to achieve a variety of benefits. Fidalgo et al. in Chapter 4 propose a decisions support system to assess some of the technical benefits, namely, voltage profile improvement, power loss reduction, and network capacity investment deferral, brought through branch congestion reduction.

Bulatov discusses in Chapter 5 three special energy problems which can be solved in polynomial time, exploiting their convexity. These problems are:

Minimal shutdown during power shortages in a power supply system, search for optimal states in thermodynamic systems and optimal allocation of water resources.

The book covers several optimization issues for power plants. Kusiak & Song discuss in Chapter 6 the improvement of combustion processes with application in boiler performance. The modeling of nonlinear processes in nuclear power plant cores is discussed by Yatsenko et al. in Chapter 7. Design optimization of polygeneration energy systems are modeled via mixed-integer nonlinear programs by Liu et al. in Chapter 8 and also by Jüdes et al. in Chapter 9. Mathematical modeling of biomass-based power plants are discussed by Bruglieri & Liberti in Chapter 10 and by Lai et al. in Chapter 11. Electric power systems are considered by Woolley et al. in Chapter 12 and by Chiang et al. in Chapter 13.

Software systems geared to today's market requirements are able to represent the whole portfolio consisting of both generating and trading components. This increases the transparency of the whole planning process. At the same time, risks become apparent and have to be supervised and validated.

Due to increased cost pressure on power generation and trading companies, caused by operating under market conditions, a cost efficient management of the risks becomes more important. As a result of the liberalization of the markets for electrical energy, companies are exposed to higher uncertainties in power generation and trading planning, e.g., the volatility of the prices for electrical energy and for primary energies, especially natural gas. Risks and uncertainties are normally not yet explicitly considered by today's commercial optimization systems. In a deterministic approach, all information is considered to be certain. Actually, there are relative uncertainties in different exogenous factors, e.g., the prices in spot and futures trading, in load forecast, the expected input of wind energy, the water supply and the power stations' availability. However, in the academic world there are a lot of activities on that topic. The contributions of Eichhorn et al. in Chapter 14, Epe et al. in Chapter 15, Heitmann & Hamacher in Chapter 16, Bläsig & Haubrich in Chapter 17, Radziukynas & Radziukyniene in Chapter 18, and Weber et al. in Chapter 19 are all related to risk minimization and stochastic programming.

To derive robust decisions, stochastic optimization operations are suitable for mid- and long-term calculations although they generally take a long time for the computing work. In the electricity industry the observed increases of electricity price dynamics combined with the characteristic periodicity of related decision processes have motivated the use of multistage stochastic programming in recent years to provide flexible models for practical applications in the sector. Especially in power generation and trading, the planning process must obey highly complex interrelations between manifold influences. They range from short term price fluctuations as observed in spot markets to long term changes of fundamental influences. Not only changes in the electric supply system itself must be considered, but also the related availability and costs of required fuels. This is outlined by Frauendorfer & Güssow in Chapter 20.

Another example is the valuation of electricity swing option by Steinbach & Vollbrecht in Chapter 21. The optimization and subsequent hedging of reservoir discharges for a hydropower producer is discussed by Fleten & Wallace in Chapter 22.

This book can be read linearly, from beginning to end. This will give a good overview of how rich the world of energy is for mathematical optimization and especially optimization under uncertainty. The book covers a wide range of techniques and algorithms. Those readers already familiar with the topic are encouraged to visit directly the topics of their interest but we are sure they will also detect many facets of a field which will have a large impact on the future of mankind.

We would like to take this opportunity to thank the authors for their contributions, the referees, and the publisher for helping to produce this book.

June 2008

Josef Kallrath
Panos M. Pardalos
Steffen Rebennack
Max Scheidt

Contents

List of Contributors

Rüdiger Barth
Institute for Energy Economics
and the Rational Use of Energy
University Stuttgart
70565 Stuttgart
Germany
ruediger.barth@ier.
uni-stuttgart.de

Boris Blaesig
Institute of Power Systems
and Power Economics
Schinkelstrasse 6, 52056 Aachen
Germany
boris@blaesig.org

Pavel A. Borisovsky
Omsk State Technical University
11 Prospect Mira, 644050 Omsk
Russia
borisovski@mail.ru

Heike Brand
Institute for Energy Economics
and the Rational Use of Energy
University Stuttgart
70565 Stuttgart
Germany
heike.brand@ier.
uni-stuttgart.de

Maurizio Bruglieri
INDACO, Politecnico di Milano
Via Durando 38/a, 20158 Milano
Italy
maurizio.bruglieri@polimi.it

Valerian P. Bulatov
Melentiev Energy Systems
Institute of SB RAS 130
Lermontov Strasse
Irkutsk, 664033
Russia
bulatov@isem.sei.irk.ru

Hsiao-Dong Chiang
School of Electrical and Computer
Engineering
Cornell University, Ithaca
NY 14853
USA
chiang@ece.cornell.edu

William P. Clarke
School of Engineering
The University of Queensland
Brisbane, Qld 4067
Australia
B.Clarke@eng.uq.edu.au

Andreas Eichhorn
Humboldt-University Berlin
Department of Mathematics
10099 Berlin
Germany
http://www.math.hu-berlin.de/~eichhorn
eichhorn@math.hu-berlin.de

Alexa Epe
Ruhr-Universität Bochum
Universitätsstraße 150
44801 Bochum
Germany
epe@lee.rub.de

Anton V. Eremeev
Omsk Branch of Sobolev
Institute of Mathematics
SB RAS
13 Pevtsov St., 644099 Omsk
Russia
eremeev@ofim.oscsbras.ru

J.N. Fidalgo
INESC Porto and Faculdade
de Engenharia da
Universidade do Porto
Rua Dr. Roberto Frias
4200-465 Porto
Portugal
jfidalgo@inescporto.pt

Stein-Erik Fleten
Norwegian University of Science
and Technology
Department of Industrial
Economics and Technology
Management, Alfred Getz v. 1
7491 Trondheim
Norway
Stein-Erik.Fleten@iot.ntnu.no

Dalila B.M.M. Fontes
LIAAD - INESC Porto L.A.
and Faculdade de Economia da
Universidade do Porto
Rua Dr. Roberto Frias
4200-464 Porto
Portugal
fontes@fep.up.pt

Torben Franch
ProCom GmbH
Luisenstr. 41, 52070 Aachen
Germany
http://www.procom.de
Torben.Franch@procom.de

Karl Frauendorfer
Institute for Operations Research
and Computational Finance
University of St. Gallen
Switzerland
karl.frauendorfer@unisg.ch

Egor B. Grinkevich
Administrator of Trade System
for United Energy System of Russia
12 Krasnopresnenskaya
Naberezhnaya, 123610 Moscow
Russia
geb@rosenergo.com

Jens Güssow
Institute for Operations Research
and Computational Finance
University of St. Gallen
Switzerland
jens.guessow@unisg.ch

Thomas Hamacher
Max-Planck-Institut für
Plasmaphysik, Gruppe für
Energie und Systemstudien
Boltzmannstrasse 2 Garching
Germany
hamacher@ipp.mpg.de

Hans-Jürgen Haubrich
Institute of Power Systems
and Power Economics
Schinkelstrasse 6, 52056 Aachen
Germany
haubrich@iaew.rwth-aachen.de

Nina Heitmann
Max-Planck-Institut für
Plasmaphysik, Gruppe für
Energie und Systemstudien
Boltzmannstrasse 2
85748 Garching
Germany
nina.heitmann@ipp.mpg.de

Holger Heitsch
Humboldt-University Berlin
Department of Mathematics
10099 Berlin
Germany
http://www.math.hu-berlin.de/~heitsch
heitsch@math.hu-berlin.de

Susumu Ikenouye
Ike Ltd.
112-0012, 6-12-2-304, Otsuka
Bunkyoku, Tokyo
Japan
susumu.ikenouye@nifty.com

Quan-Yuan Jiang
School of Electrical Engineering
Zhejiang University, Hangzhou
P.R. China
jqy@zju.edu.cn

Marc Jüdes
Institute for Energy Engineering,
Technische Universität Berlin
Marchstrasse 18, 10587 Berlin
Germany
juedes@iet.tu-berlin.de

Sergey A. Klokov
Omsk Branch of Sobolev Institute
of Mathematics SB RAS
13 Pevtsov St., 644099 Omsk
Russia
klokov@ofim.oscsbras.ru

Abhay K. Koppar
Department of Agricultural
and Biological Engineering
University of Florida
Gainesville, FL 32607
USA
kopparak@ufl.edu

Christian Küchler
Humboldt–Universität zu Berlin
Unter den Linden 6, 10099 Berlin
Germany
ckuechler@math.hu-berlin.de

Andrew Kusiak
The University of Iowa
Department of Mechanical
and Industrial Engineering
3131 Seamans Center, Iowa City
IA 52242-1527
USA
andrew-kusiak@uiowa.edu

Takwai E. Lai
School of Engineering
The University of Queensland
Brisbane, Qld 4067
Australia
edsterlai@yahoo.com.au

Leo Liberti
LIX, Ecole Polytechnique
F-91128 Palaiseau
France
liberti@lix.polytechnique.fr

Pei Liu
Centre for Process Systems Engineering
Department of Chemical Engineering
Imperial College London
London SW7 2AZ
UK
pei.liu@imperial.ac.uk

Peter Meibom
Risø National Laboratory for Sustainable Energy
Technical University of Denmark
Roskilde
Denmark
peter.meibom@risoe.dk

Anna Nagurney
Department of Finance and Operations Management
Isenberg School of Management
University of Massachusetts
Amherst, MA, 01003
USA
nagurney@gbfin.umass.edu

Panos M. Pardalos
Department of Industrial and Systems Engineering
Center for Applied Optimization
University of Florida, Gainesville
FL 32611, USA
pardalos@ufl.edu

Efstratios N. Pistikopoulos
Centre for Process Systems Engineering
Department of Chemical Engineering
Imperial College London, London
SW7 2AZ
UK
e.pistikopoulos@imperial.ac.uk

Pratap C. Pullammanappallil
Department of Agricultural and Biological Engineering
University of Florida
Gainesville, FL 32607
USA
pcpratap@ufl.edu

Virginijus Radziukynas
Lithuanian Energy Institute
Laboratory of Systems Control and Automation
Lithuania
virginijus@mail.lei.lt

Ingrida Radziukyniene
Vytautas Magnus University
Faculty of Informatics
Lithuania
i.radziukyniene@if.vdu.lt

Steffen Rebennack
Department of Industrial and Systems Engineering
Center for Applied Optimization
University of Florida, Gainesville
FL 32611
USA
steffen@ufl.edu

Werner Römisch
Humboldt-University Berlin
Department of Mathematics
10099 Berlin
Germany
http://www.math.hu-berlin.de/~romisch
romisch@math.hu-berlin.de

Max Scheidt
ProCom GmbH
Luisenstrasse 41, 52070 Aachen
Germany
http://www.procom.de
Max.Scheidt@procom.de

Susana Silva
ALERT - Life Sciences Computing S.A.
Rua Antnio Bessa Leite
1430, 2° 4150-074 Porto
Portugal
susana.silva@alert.pt

Zhe Song
The University of Iowa
Department of Mechanical and Industrial Engineering
3131 Seamans Center, Iowa City
IA 52242-1527
USA
zhe-song@uiowa.edu

Marc C. Steinbach
Leibniz Universität Hannover
IfAM Welfengarten 1
30167 Hannover
Germany
www.ifam.uni-hannover.de/~steinbach
steinbach@ifam.uni-hannover.de

Günter Stock
Meischenfeld 11, 52076 Aachen
Germany
guenter.stock@t-online.de

John Stranlund
Department of Resource Economics
College of Natural Resources and the Environment
University of Massachusetts
Amherst, MA 01003
USA
stranlund@resecon.umass.edu

George Tsatsaronis
Institute for Energy Engineering
Technische Universität Berlin
Marchstrasse 18, 10587 Berlin
Germany
tsatsaronis@iet.tu-berlin.de

Stefan Vigerske
Humboldt–Universität zu Berlin
Unter den Linden 6, 10099 Berlin
Germany
http://www.math.hu-berlin.de/~stefan
stefan@math.hu-berlin.de

Andrey V. Vinnikov
Administrator of Trade System for United Energy System of Russia Joint Institute for Nuclear Research
12 Krasnopresnenskaya Naberezhnaya, 123610 Moscow
Russia
vinnikov@rosenergo.com

Hans-Joachim Vollbrecht
Fachhochschule Vorarlberg
FZ PPE Sägerstrasse 4
6850 Dornbirn
Austria
www.staff.fh-vorarlberg.ac.at/hvhans-joachim.vollbrecht@fhv.at

Hermann-Josef Wagner
Ruhr-Universität Bochum
Universitätsstraße 150
44801 Bochum
Germany
lee@lee.rub.de

Stein W. Wallace
Chinese University of Hong Kong
Shatin NT, Hong Kong
China and Molde University College
P.O. Box 2110
6402 Molde
Norway
Stein.W.Wallace@hiMolde.no

Bin Wang
School of Electrical
and Computer Engineering
Cornell University
Ithaca, NY 14853
USA
bw297@cornell.edu

Christoph Weber
Management Sciences
and Energy
Economics Universität
Duisburg-Essen
Universitätsstraße
2, 45141 Essen
Germany
Christoph_Weber@uni-duisburg-essen.de

Oliver Woll
Universität Duisburg-Essen
Universitätsstraße 2, 45141 Essen
Germany
Oliver.Woll@uni-duisburg-essen.de

Trisha Woolley
Department of Finance and
Operations Management
Isenberg School of Management
University of Massachusetts
Amherst, MA, 01003
USA
twoolley@som.umass.edu

Vitaliy A. Yatsenko
Space Research Institute NASU
and NSAU 40 Prospect Academica
Glushkova 03680 Kyiv
Ukraine
vyatsenko@gmail.com

Conventions and Abbreviations

The following table contains in alphabetic order abbreviations used in at least two chapters of the book.

Abbreviation	*Meaning*
cf.	Confer (compare)
CHP	Combined heat and power
CVaR	Conditional value-at-risk
e.g.	Exempli gratia (for example)
EEX	European energy exchange
GHG	GreenHouse gas
HRSG	Heat recovery steam generator
i.e.	Id est (that is)
ISO	Independent system operator
LP	Linear programming
MIP	Mixed integer (linear) programming
MINLP	Mixed integer nonlinear programming
NLP	Nonlinear programming
OPF	Optimal power flow
PSO	Particle swarm optimization
s.t.	Subject to
SLP	Successive linear programming
SQP	Successive quadratic programming

Part I

Challenges and Perspectives of Optimization in the Energy Industry

1

Current and Future Challenges for Production Planning Systems

Torben Franch, Max Scheidt, and Günter Stock

Summary. This article elaborates on the coming challenges production planning departments in utilities are facing in the near and remote future. Firstly, we will motivate the complexity of production planning, followed by a general solution approach to this task. The development of a new generation of energy management tools seems necessary to fulfill the need to handle uncertainty and eventually cover stochastic processes in energy planning. These new energy management systems have to include complex workflows and different methods and tools into the planning process.

Key words: Energy management, Uncertainty in energy planning

1.1 Introduction

Energy planning can be complicated. Due to its techno-economic nature it was already complex in monopolistic times and has gone from 'complex' to 'very complex' thereafter.

First of all, it is important to explain what production planning in the energy industry or energy planning, respectively, means. Production planning is the commercial and technical organization that uses power plants to generate income. It is the key organizational function that translates production capacity into commercial value. In a nutshell, this means that without production planning, power plants are not generating any income.

The objective for production planning is clearly to maximize the profits that can be created by running power plants. As power plants inherently produce more than electricity, the maximization of profits is typically subject to a number of restrictions. These restrictions are particularly heat supply but also technical restrictions and ancillary service commitments. Experience shows that production planning becomes very complex as soon as power plants produce more than just straight power.

1.2 Production Planning – History and Present

A good example for how complex production planning really is and what significant commercial impact it can have is depicted in Fig. 1.1. The producer's every day production capacity of his power plants is offered to the Nord Pool exchange. When it is profitable, production is sold. The set of assets consists of a number of smaller and larger production units using different fuels. Furthermore, heat is supplied to a stretched-out heat grid and different steam grids. This example of production planning shows very clearly that even small improvements in performance can have a significant impact on results. Moreover, small planning mistakes can have very serious commercial and operational consequences.

In Fig. 1.1 actual hourly production in December 2004 is depicted. At first glance, it can be difficult to understand how this can be an optimal production plan. However, there are some good explanations. The variation in production is a function of many factors such as weekend stops, ancillary services delivery, and commercial production. In the chart, one can see the 'coal-minimum' and the 'oil-minimum' situations where reserves are delivered automatically and manually. On closer examination, it is even possible to see that different on-duty crews have different views of what is maximum and minimum production capacity.

The deregulation of energy markets has had a very significant impact on production planning: Firstly, the purpose of planning has changed from minimizing cost of delivery to maximizing profits. Secondly, new markets have emerged, like spot power, gas, and CO_2. Thirdly, the roles of market

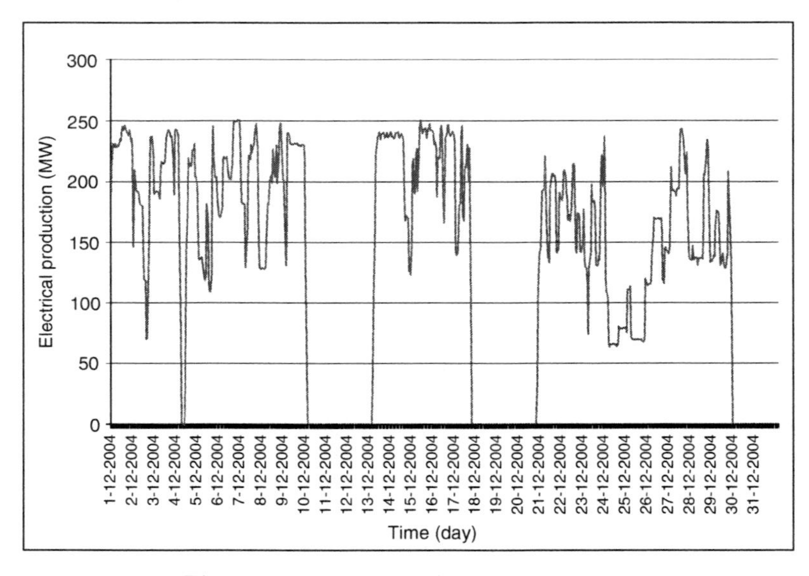

Fig. 1.1. Production planning in practice

participants have changed. Consequently, as a result of this, the production planning workflow has changed as well.

In order to understand where production planning and production planning tools are today, it makes sense to look at the historical framework. The European energy markets have been deregulated in the past 10 years and this had a considerable impact on how energy companies behave in the market and organize themselves, see [1]. Firstly, deregulation meant that the purpose of an energy company changed. Today, companies very much strive to make profits for their owners whereas prior to deregulation, the objective was to minimize delivery costs to consumers. In the past, very often the result of a year was decided when the annual budget was drawn up. Secondly, deregulation has opened new markets. Today, it is possible to trade spot power and gas, imbalances and CO_2 emission rights – all products that were not even known a few years ago. Lastly, deregulation changed the roles of market participants. In some countries, this led to new players entering the markets, yet in other countries, this resulted in the emergence of a few and very large energy giants.

To illustrate how much all these factors have influenced production planning, taking a look at an illustration of production planning work processes prior to deregulation makes sense.

Prior to deregulation, production planning consisted of the forecasting of load and later the computation of the optimal production plan, see Fig. 1.2. While this looks like a relatively simple task, it can be a difficult calculation, especially if the production system is complex. Previously, the focus of attention was mostly on technical power plant availability and how to meet production requirements. In those days fuel prices were relatively stable and hence there was no need for daily calculations. Instead, calculations were made weekly or even less frequently. For shorter periods, a prioritization of production units was sufficient. Deregulation and the emergence of new markets changed all this radically.

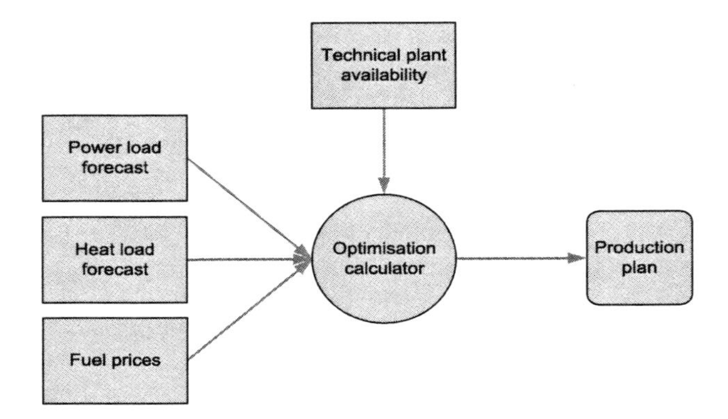

Fig. 1.2. Production planning before deregulation

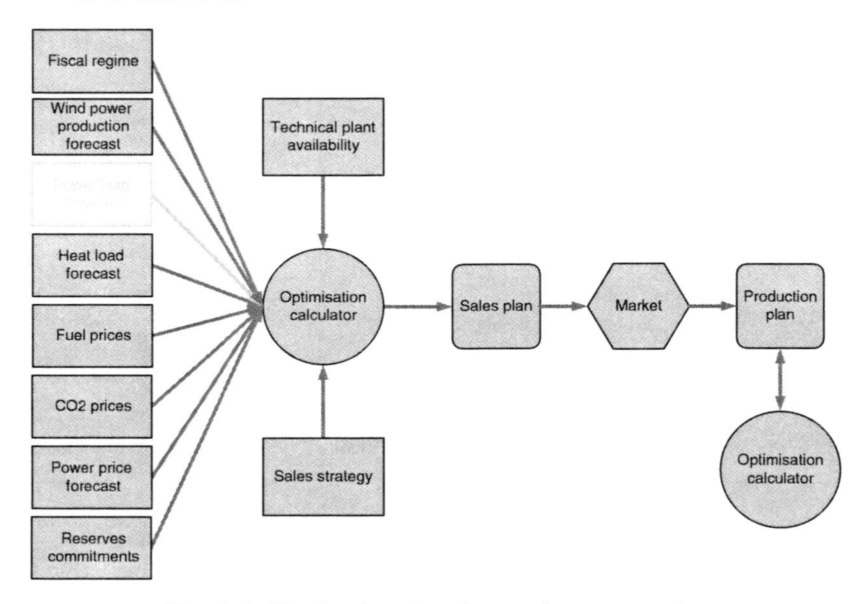

Fig. 1.3. Production planning work process today

Today, however, the amount of input data is not only much larger but inputs are also much more volatile, see Fig. 1.3. This means that production planners have to work very efficiently day in and day out to compile information, do the necessary analysis and planning and then submit these to the exchanges before noon. That means they have complex workflows, many methods, lots of data and less time for it all. At the same time, the new deregulated environment called for the development of new systems for effective data management and shorter calculation time for optimization. The good news is that power load forecasting is no longer a task for production planning. Today, this is the task of the retail manager. Furthermore, there are now several new trading platforms, like exchanges, over the counter trading, cross border trading and intraday trading. This is why sales strategies play an important role. All in all, nowadays, production planning has very much become a task of optimizing sales in an environment of volatile power and fuel prices.

1.3 The Coming Challenge: Handling Uncertainty

"It's hard to predict, especially the future". This well-known saying attributed to Winston Churchill proves to be valid in production planning as well. In fact, production planning is very much exposed to risks and uncertainties, although not much attention has been paid to this aspect for quite some time. One of the most volatile commodities in the world is power, even more volatile than fuel oil prices. As a comparison, in the period April 2006-March 2007, the fuel oil price has varied from US $50 to 75 per barrel, while the Nord Pool price showed much larger variations and German EEX prices have been even more volatile. This makes it very difficult to predict power prices a day ahead.

Fig. 1.4 depicts the base load prices for 2006 in the Nord Pool area DK2.

But it is even harder to predict hourly prices and profiles, which is shown in Fig. 1.5 for Nord Pool DK2.

While hourly spot prices are so difficult to predict, they are one of the most important parameters in a production plan. Wrong forecasts of spot prices can lead to wrong decisions. If you base heat planning on a wrong spot price profile, you could end up with power production in low price hours and heat production in high price hours. Generally, you have to optimize the combined heat, steam and power production portfolio regarding your forecasts

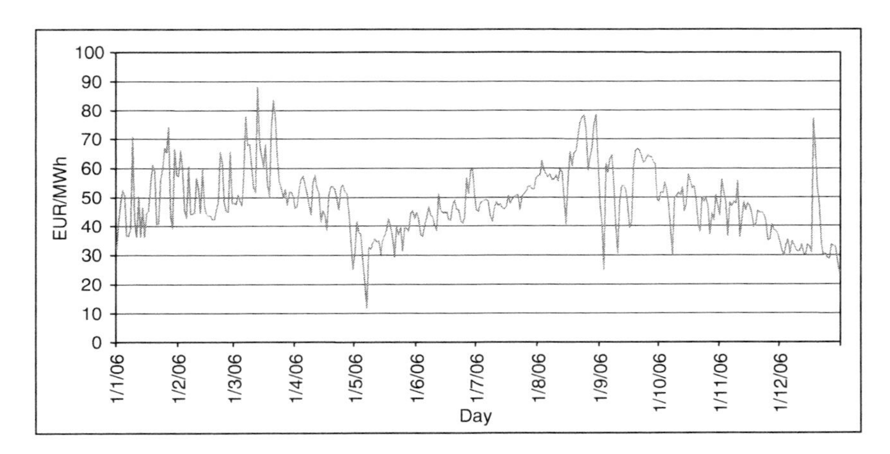

Fig. 1.4. Spot prices (Nord Pool DK2)

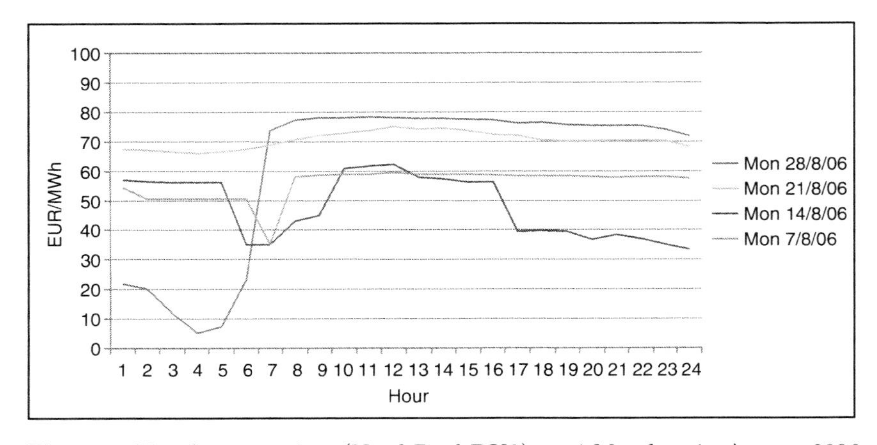

Fig. 1.5. Hourly spot prices (Nord Pool DK2) on 4 Mondays in August 2006

of district heating, steam production and spot prices. This is naturally always prone to errors resulting in imbalances between your day-ahead planning and the required and delivered customer load.

While it is yet impossible to forecast exact values, in fact sometimes it is possible to forecast the direction of imbalances. One example can be found in the field of wind power forecasting.

The graph in Fig. 1.6 shows the forecasting of wind power production at a Baltic Sea wind farm and the actual production curve. It shows that the prediction for wind power production a day-ahead is very accurate.

However, the problem is that predictions are not always as good. As can be seen in Fig. 1.7, which shows said wind farm on another day. This time, the forecast results in notable imbalances which are priced with different imbalance costs for each hour. The graph illustrates also the commercial risk attached with such a wrong prediction regarding the exact time of the wind load curve.

Forecasts of power prices and wind power production are by far not the only sources of uncertainty and of commercial risks. There is uncertainty in heat load forecasts, fuel prices, unit failures and many more. Basically, uncertainty cannot be avoided. Uncertainty about input parameters leads to imbalances – and even wrong decisions. This is especially true for virtual power plants, see [5]. Also, one can forecast some effects in a short time horizon. The key to this problem is handling the risks effectively. This is important because the commercial implications can be very substantial. So, how do you do production planning under uncertainty? One approach is to ignore it, because

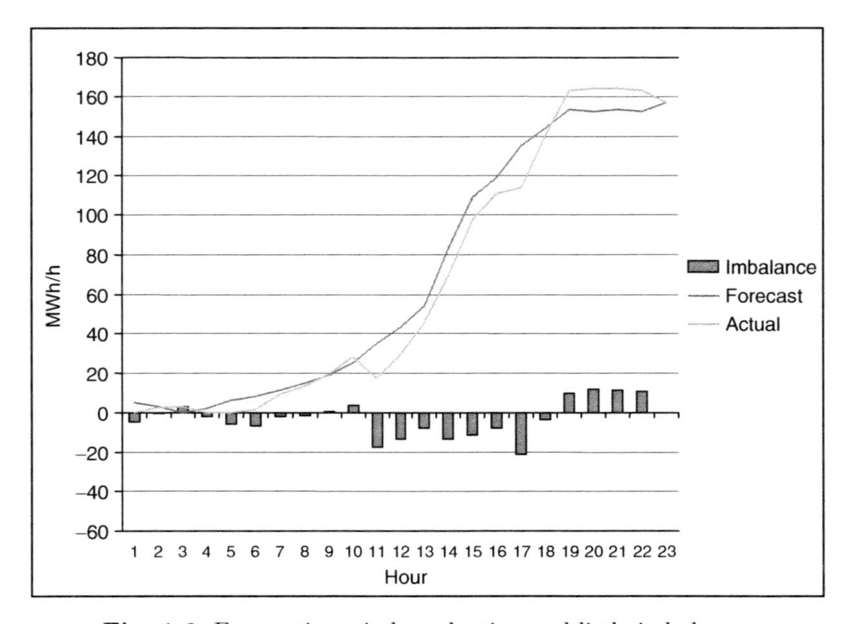

Fig. 1.6. Forecasting wind production and little imbalances

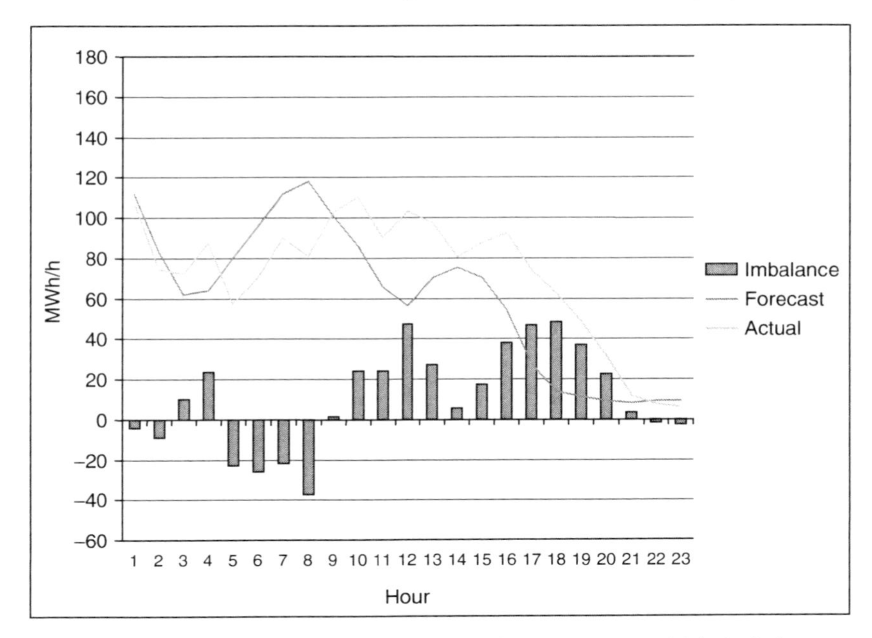

Fig. 1.7. Forecasting wind production with time error and big imbalances

production planning is complex enough, already. Another approach is to deal with it. This means to start acknowledging that input parameters to production planning are uncertain. Rather than avoiding it, it makes sense to accept it, work with it and even to exploit the opportunities it brings. The good thing is that sometimes being wrong does not have serious consequences. It also means to acknowledge that input parameters are not always symmetrical and that it is sometimes possible to predict the shape of the distributions.

There are many reasons why markets will become even more volatile in the future. One reason is the increasing share of renewable production capacity. Moreover, the deregulation of the gas markets will be another source of uncertainty. The effect of global warming will lead to shortages of cooling water and cause additional volatility in the market. How politicians will respond to this also causes concern. Furthermore, CO_2 quotas are predicted to come in short supply.

1.4 Requirements for Future Production Planning Systems

Overall, the energy management systems as they exist today form a strong basis. The last 10 years have shown great achievements: Despite enormous changes in the market environment, the industry has been able to adapt without market failures leading to blackouts. The market participants have been able to cope with very large changes in the commercial, legal and regulatory environment. User-friendly tools for modeling power plant systems and for solving complex optimization problems have been developed. Production planning has developed from being a technical activity to being a commercial core competence.

To exemplify the latter, Fig. 1.8 depicts the BoFiT modeling environment. BoFiT is a production planning solution suite widely used in German and European utilities, e.g. Vattenfall Europe [3] and Stadtwerke Munich [6]. It features among other things a graphical user interface that facilitates the development of the features of a model and explain its results within teams. It also helps to explain the results to the business staff using their own language.

Now, it is time to face the next challenge: Efficient handling of uncertainty and automation of time-consuming business processes. In future, energy management tools will have to be developed further, much in the way that risk management systems have developed with a far stronger focus on strategies and trading opportunities.

With the deregulation of energy markets, uncertainty became a key feature of the commercial management in many energy companies, like risk management and hedging, financial trading portfolios, new end-user products with fixed price components. Production planning is very much exposed to risks now, however, for some reason this had received less attention in the field of multi-commodity systems.

So what could a future production planning solution look like?

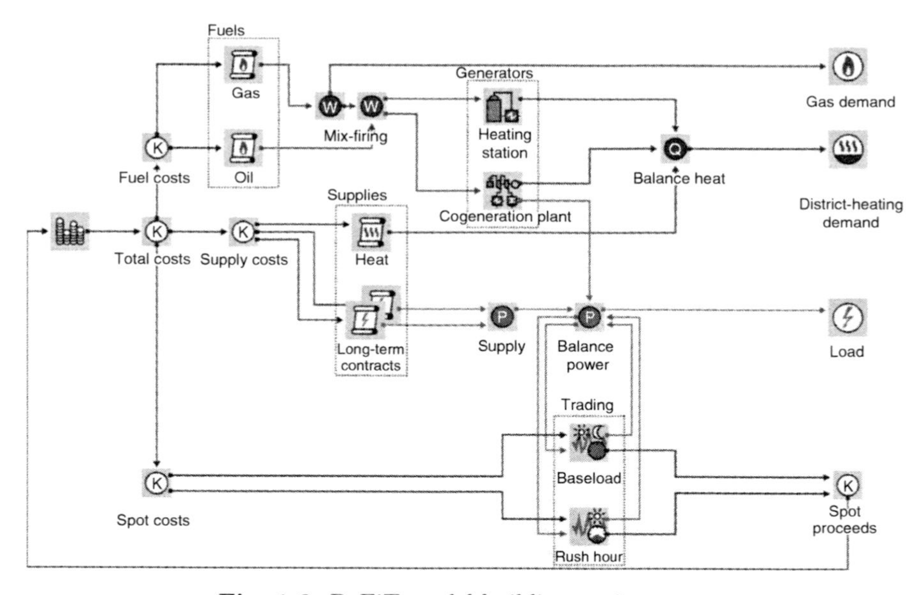

Fig. 1.8. BoFiT model building environment

First of all, planning tools should be bridges between mathematical methods and the user. Hence, they should provide a comfortable way to use a graphical modeling environment. Furthermore, planning tools should be easy to integrate and to adapt in an IT system environment. Planning tools should support relevant user decisions. They should deliver reliable results and offer a quick response time. Classical decision horizons are long-term, medium-term, day-ahead, and intraday. They should be supported by the planning tool see Fig. 1.9. This overall frame covers the period from 15 min to 60 months. Long-term covers several years to one or two decades. Medium-term reaches from 1 week to 60 months. Day-ahead covers the period from 24 h to 1 week and intraday concerns 15 min to 30 h. Optimizing these planning horizons requires corresponding grid load forecasts, sales forecasts, market forecasts, demand forecasts of clients and client groups.

On top of the above-mentioned requirements, new production planning systems need to support a different approach of choosing a market strategy. Figure 1.10 shows the basic modules of future production planning systems. There exist various input parameters which are put in order of decreasing volatility. Hence, the most volatile parameter is the "imbalance price forecast" and the least volatile are "reserves commitments". The input to the planning are not just single-point forecasts but some form of uncertain or stochastic data. These inputs enter into a trading strategy analysis module where it is possible to evaluate different strategies with different combinations of input data. Part of this calculation can be an optimization calculation that is integrated in the trading strategy analysis tool. The result of the trading strategy analysis is a sales plan which in turn leads to a production plan.

The benefits of such a new type of energy management system are very obvious: The user is now choosing a market strategy that reflects the uncertainty in the market and which is optimized to exploit possibilities of spikes as well as to minimize expected imbalance costs. The question is whether this type of system is simple to create. The consensus is that more work has to be

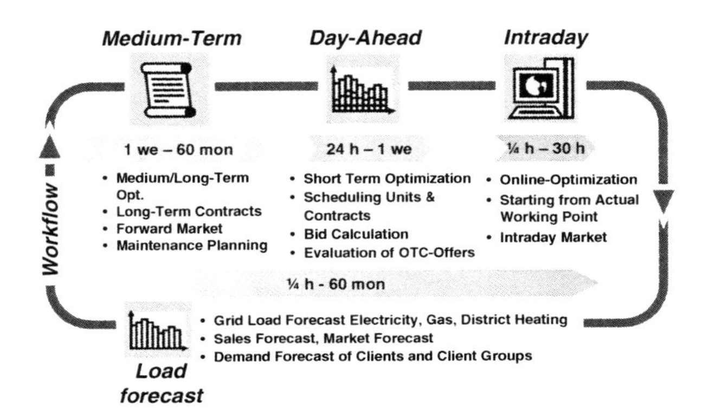

Fig. 1.9. Decision horizons and results in production planning

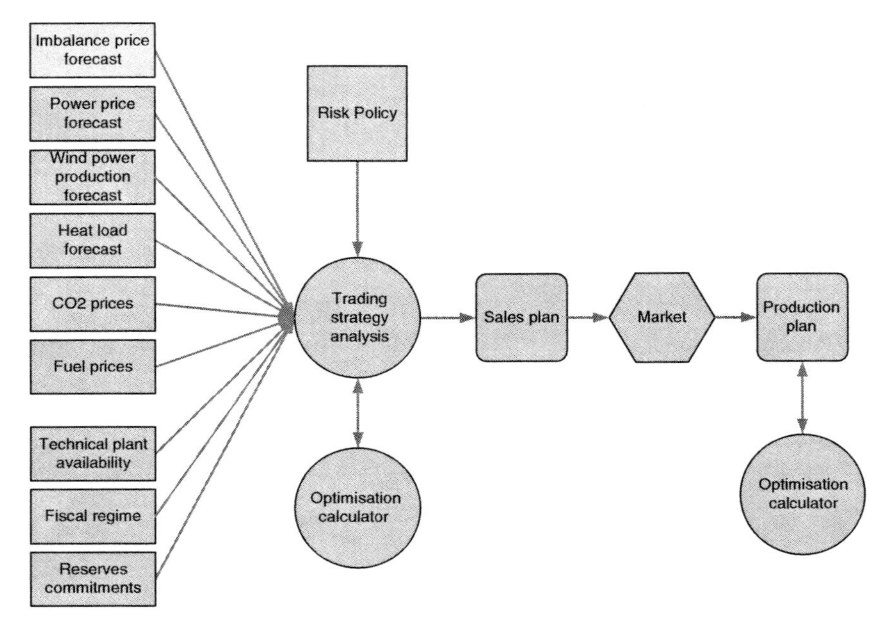

Fig. 1.10. Production planning system of tomorrow

done and this will involve stronger cooperation between research institutions and solution providers, e.g. see [2, 4].

Handling uncertainty implies a need to include stochasticity. Evaluation of different strategies leads to handling a multitude of calculations and scenarios, which ultimately requires an automation environment and extensive data management system to support the users efficiently.

From an IT perspective, integration aspects call for the application of a modern service-oriented architecture, the principles of which are exemplified in Fig. 1.11. It facilitates the different phases in the life cycle of a production planning solution, being process configuration, process execution, and process control. Major benefits of the SOA are its flexibility in deployment and its readiness to add new services e.g. stochastic optimization kernels or Monte Carlo simulations.

The SOA facilitates the definition and automatic execution of workflows. This is shown in Fig. 1.12. Following the detailed analysis of the business processes these are orchestrated in a graphical user interface. Once approved, the workflows are executed automatically at certain times or manually. They are controlled by showing the actual parts of the workflow being successfully or unsuccessfully executed. The services and the data inputs are combined and executed in the order of this workflow. The results are stored in a time series management system and can be visualized in user-defined reports.

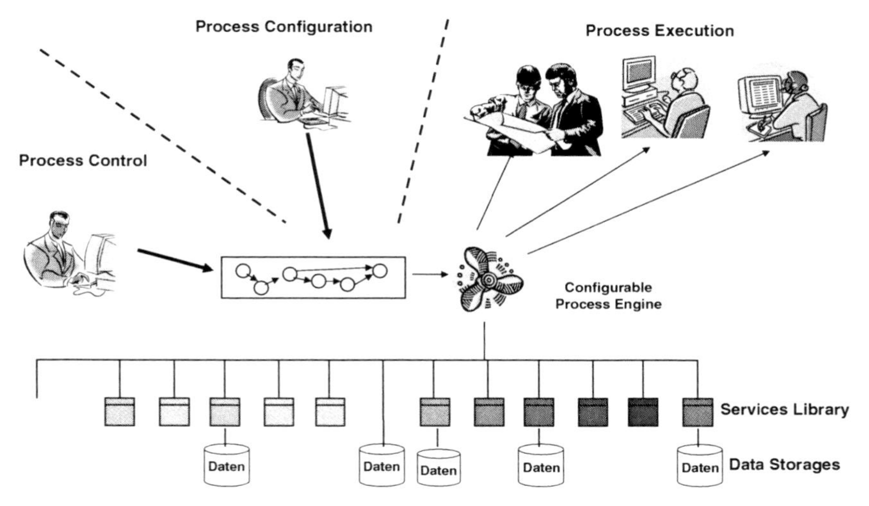

Fig. 1.11. Principle of a Service Oriented Architecture (SOA)

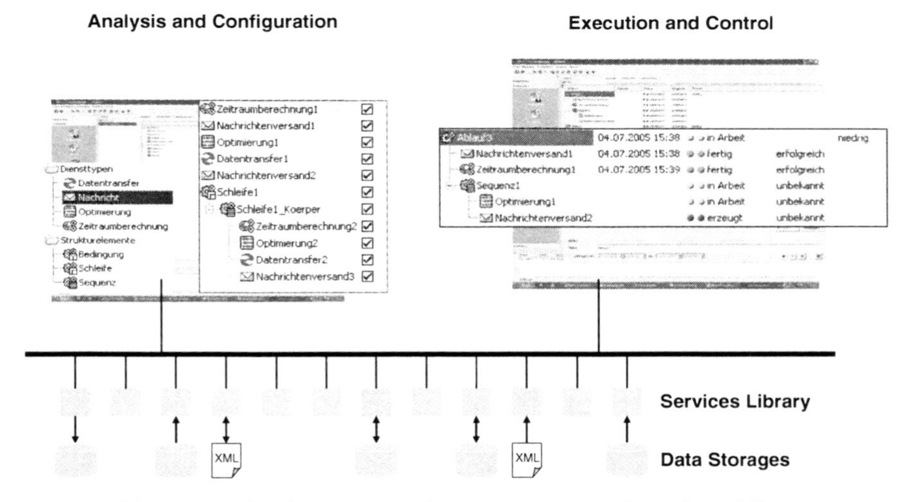

Fig. 1.12. Configuration and automatic execution of workflows

While the requirements from business and IT are fairly clear today, there is still a good deal of research to be done on the core issue of handling uncertainty. It is of pre-eminent importance to find a meaningful way how to describe and represent uncertain input. Unless a very simple and systematic way to estimate uncertain input parameters can be found for production planners, there is little chance that such a system with stochastic optimization tasks will be used by the clients in multi-commodity production planning.

Furthermore, the need for quick response times in real planning and bidding situations has to be fulfilled. However, there is a growing need to enhance energy management systems to deal with uncertain (stochastic) input because of the requirements of the planning process as shown above.

Today's planning systems for co-generation of thermal and electrical production are in general not equipped to deal with uncertain input. Nevertheless, the data models used must not be so different from stochastic models because the fundamental efficiency curves of power plants or the maximum or minimum power production capacity of the plants are not stochastic. Therefore, there is a possibility to migrate existing deterministic models by using stochastic input distributions and scenario tree techniques. The future will have to prove the benefits of those approaches in real production planning processes.

Apart from the technical challenges, there are also educational and organisational issues which need to be addressed rather sooner than later. It is necessary to educate production planners to deal with uncertainty. On top of that it is necessary to educate planners and traders in each other's 'languages'. On one hand, production planners will have to become more familiar with the complicated language of financial traders who juggle terms like delta-hedging or spread options and many more. On the other hand, traders need to become more familiar with the technical characteristics and physical limitations of power plants and co-generation units respectively. The fact that the economic implications of production planning decisions are coming more into the focal point of planning, leads to the question, whether production planning should be executed by the trading companies or the power plant owners. There is good reason for both choices and a lot of internal struggles upon the right answer to the question is currently ongoing in many European utilities. Depending on the final decision, it will be necessary to check and afterwards adjust the business processes around production planning.

1.5 Conclusion

Production planning has come a long way over the past 10 years. A number of methods and tools have been developed which make it possible to operate in new markets and new environments. So far, major focus has been placed on developing tools that can support production planning in a situation where uncertainty is ignored. Nevertheless, risk management and handling uncertainty is an area that still needs to be improved. As the future is most likely to bring more volatility, the next step forward is to start finding a way to efficiently manage risk and uncertainties and especially to be ready to exploit the opportunities this brings. Finally, this integration should be linked with process and workflow automation systems. This enables the automation of those very complex calculations which are going to integrate a number of different tools and methods to achieve certain goals under tight time schedules.

All technical improvements need to be accompanied by corresponding organisational and educational measures to ensure an outmost exploitation of the business improving potential which the improved planning systems offer. This is the challenge energy companies have to master!

References

1. L. Ilie, A. Horobet, and C. Popescu. Liberalization and regulation in the EU energy market. Technical Report 6419, Munich Personal RePEc Archive, October 2007. http://mpra.ub.uni-muenchen.de/6419/
2. M. Scheidt. Agent-based simulation of liberalized electricity markets. *Proceedings of The second Asia-Pacific Conference of International Agent Technology (IAT01)*, pages 505–510, September 2001
3. M. Scheidt, T. Jung, and P. Malinowski. Integrated power station operation optimization – BoFiT and Vattenfall Europe case study. *Proceedings of International Conference The European Electricity Market EEM-04*, September 2004
4. M. Scheidt and B. Kozlowski. Risikoorientierte Optimierung: Die Suche nach dem effizienten Portfolio. *e|m|w*, (5):43–48, 2004. (in German)
5. G. Stock and M. Henle. Integration "Virtueller Kraftwerke" in Querverbundsysteme. *Euroheat and Power, Fernwärme international*, 31(3):58–63, 2002. (in German)
6. G. Stock, H. Kohlmeier, and A. Ressenig. Kostentransparenz durch Energiemanagement: Stadtwerke München optimieren Energieerzeugung. *BWK*, 55(3):32–36, 2003. (in German)

2

The Earth Warming Problem: Practical Modeling in Industrial Enterprises

Susumu Ikenouye

Summary. The earth warming problem will be one of the most difficult problems for industrial enterprises in the world. Heavily energy consuming industries, i.e., steel, power, refinery and chemical, have to establish a powerful management system to deal with the Earth warming problem. The core of this management system is the planning function. The planner should take more complicated criteria into consideration than before. Some of the criteria conflict with each other. At the same time, surroundings of the planning work will be continuously unstable because of political and economical changes in the world. We have to make an effort to implement a planning tool to help planners facing uncertain problems under multi criteria. The idea of modeling is the first step to accomplish a practical planning tool for ordinary planning persons for daily decision making work processes. Mathematical programming approaches are very promising to develop this kind of planning tool.

2.1 Introduction

The earth warming problem has been studied scientifically for many years [3]. Now, this challenging problem is one of the most important issues in the world from both the political and economical point of view. In all countries, governments are faced with the problem how to adopt the system of "Cap and Trade." Especially, energy consuming industries, e.g., steel, power, oil and chemical, are seriously confronted with this problem.

Zoning of the earth warming problem is shown in Fig. 2.1. Obviously, the *earth*, *country* and *enterprise* are basic zones to be modeled. Furthermore, the *complex* of industrial companies is very important in the discussion of emission control. Close connection between factories by fuel/product pipelines and by power lines will make a strong contribution to save energy and to reduce GreenHouse Gas (GHG) in a entire complex.

Management procedures for GHG emissions in each zone should have good simulation functions to estimate how much quantity of GHG will be generated. It is desirable that this simulator embeds optimization techniques. Practical procedures for GHG emission control have to be continuously and robust.

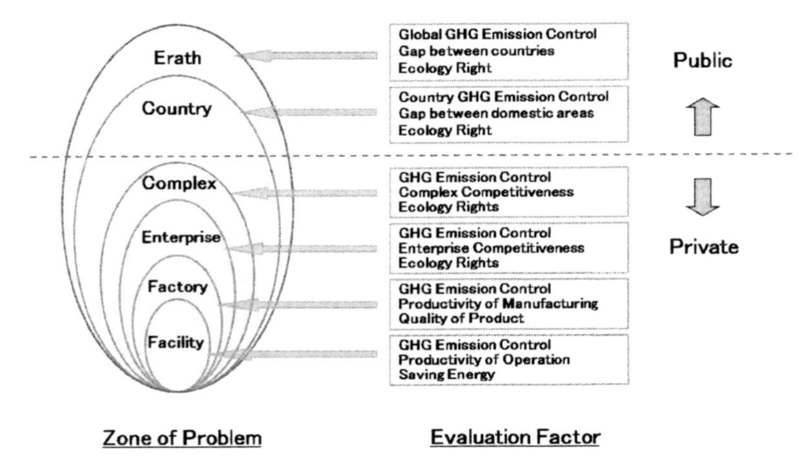

Fig. 2.1. Boundaries of the Earth warming problem

The simulation function has specific evaluation items depending on the character of each zone. Every industrial company has to have simulation functions containing economical metrics and GHG emission metrics. The simulation for production, capital investment and purchase of carbon credit has to be done simultaneously.

The quality of the product is naturally very important for the competitiveness of an industrial enterprise. Until now, there is no good estimation to compare these metrics in simulation and optimization. Good approaches and methods of quality evaluation are expected for a more reasonable simulation.

2.2 Management: What Changes will Affect the Planning Work?

GHG emission control in industrial companies can be done as a management cycle of PDCA (Plan-Do-See and Check) like a financial budget control. A planning tool in phase P should have enough ability to make an optimal plan. The planner has to asses a plan by GHG emission besides economical and technological points of view. In some cases, there will be severe conflicts between economical metrics and GHG emission metrics.

The strongest impact of the change is illustrated in Fig. 2.2. We have to think how to design a new tool of planning in this confliction. In general, operations research (OR) technology offers *multicriteria programming* and *goal programming*, [4, 6]. However, until now, practical applications of both methods cannot be found in real management systems of industrial companies.

A table of objective criteria will contain the following crucial factors:

- *Economics*: sales, income, cost, depreciation expenses, capital investment, debt, return on asset (ROA)

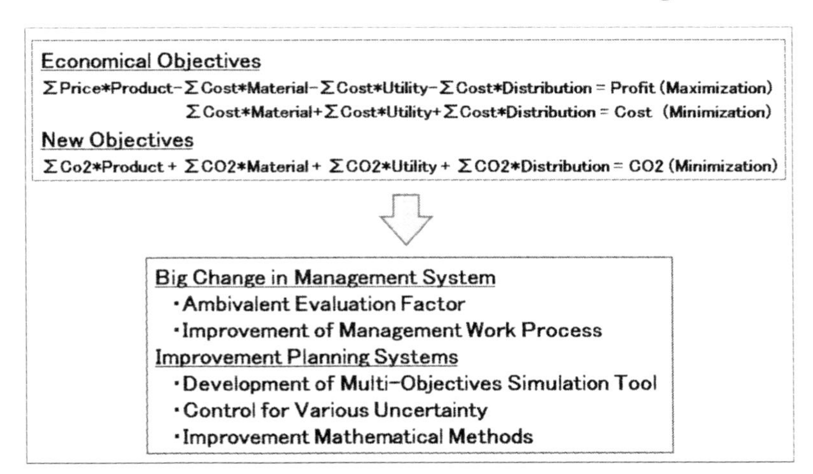

Fig. 2.2. Big change in management system

- *Environment*: GHG, CO_2, carbon credit
- *Technology*: production effectiveness, quality of product

In any way, through real work of planning, the planner has pay attention to all metrics above mentioned. We have to try to find a good method to include these metrics as objectives for planning.

2.3 Modeling: How to Make a Practical Model for the Earth Warming Problem?

2.3.1 Structure of the Model for the Industry

Heavily energy consuming industries, such as, steel, power, oil and chemical have specific models for mathematical calculations. In general, this models are a combination of process flow models and network models. A long term model is likely to be of multi periods.

Criteria of such a model contains metrics as mentioned before as possible. From the point of mathematical programming, all of these metrics introduced are target constraints. In each case study, one of the constraints will be the objective. In some case, a set of constraints will form multiobjectives.

The model we discussed is an abstract one and it will be divided into several models to be solved by methods of OR. AS a whole, the model will be a complex of sub models and methodologies.

2.3.2 The Model Type for Planning Work of an Industrial Company

The following three types of models are very effective for practical planning work:

1. Enterprise-wide model of single term (single-period model)
2. Enterprise-wide model of road map (multiperiod model)
3. Process and network model as a social model

All models are for long-term planning, annual planning and longer time scale. Shorter time scale plans, such as, monthly plan, production scheduling and process control plan, provably have other aspects in technology and engineering points of view.

Enterprise-Wide Model of Single Term

The first model is applied for enterprise-wide planning in a single term. The planner will use this model in the case study of an annual business plan and a production plan including judgment on investments for facility and purchasing carbon credits. This model contains the selection problem. Integer variables should be used for the selection of capital investment and purchase of carbon credits.

Enterprise-Wide Road Map Model

The second model covers several time periods. A Road Map Plan of GHG emission control as the Kyoto Protocol in 1997 [2, 7] has been discussed for several years. This model is almost the same as a connected single-term model of Sect. 2.3.2. The decision problem which investment should be selected and when it will be done can be modeled as a mixed-integer linear programming (MIP) problem. However, it will be very difficult to solve a single-term enterprise-wide model as one monolithic model. In every time period, the

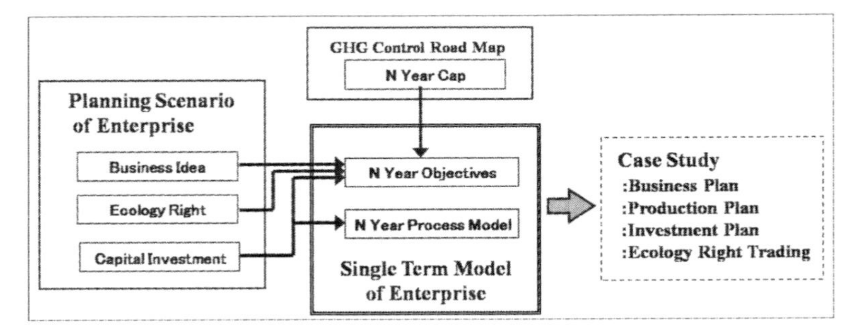

Fig. 2.3. Single term planning including GHG emission control

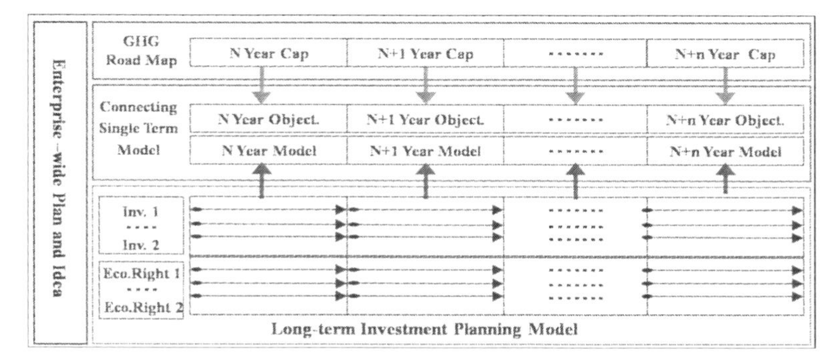

Fig. 2.4. Configuration of enterprise road map model

production process model has to be modified by adding all candidates of investments for the process flow. We have to find another idea to solve this complicated large-scale problem.

The planning work to deal with Road Map Plans like the Kyoto Protocol should consider a forecast for the coming 5 years or more. The planner has to face very strong uncertainty in any situation. So, this model shall be modified very often. We need good remodeling functions to perform the planning work smoothly.

Process and Network Models as Social Models

This model is a combination of a process flow model and a network model. Process flow models are very popular as refinery models like PIMS of Aspentech [1]. Network models are just like logistic models. They show power transmission lines, fuel, steam, water and other utilities. The structure of a process and network model is good for an industrial complex to simulate and to control GHG emission. Usually, a typical complex is composed out of power plant, refinery, steal and petrochemical. All these industries are consuming a lot of energy and are generating huge GHG.

Process and network models are composed by a set of elements connected in a network. Each element shows one company or one factory. This element is a production process flow model that can be solved as standalone mathematical model with multicriteria objectives.

Every element in a process and network model is an independent company. This model is able to simulate in detail the cooperation of companies as one independent company. This ability is very useful to evaluate competitiveness of a specific area or country.

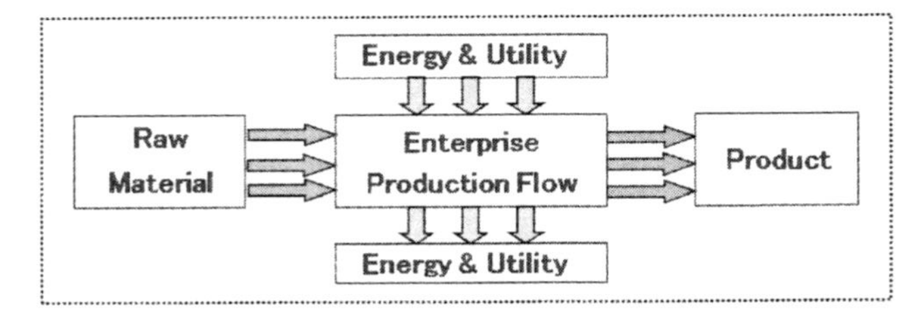

Fig. 2.5. Element (enterprise) of process and network model

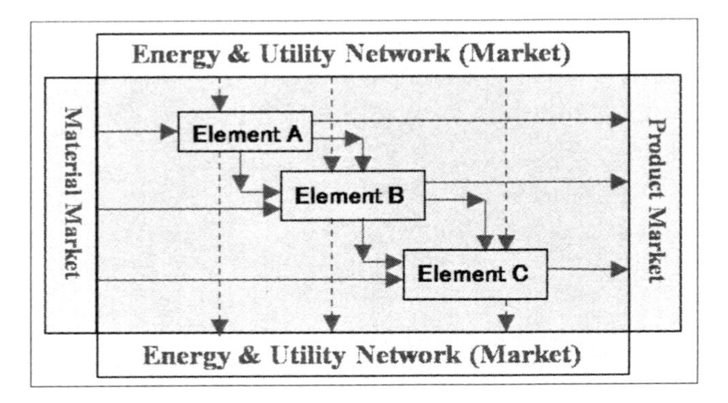

Fig. 2.6. Process and network model

2.4 Problems When Applying to Real World

2.4.1 Practical Multipurpose Programming

Large and complicated models like a Road Map Plan of GHG emission control is not so easy to apply in ordinary planning of practical management work. Planning work processes cannot be covered by any IT system and by any OR methods completely. The problem of earth warming is not explained by scientific approach enough. So, many points remain unsolved for the coming years. Most processes of decision making will be done by planner. As mentioned before, the mathematical model that we discuss has several submodels that could be solved by a steady mathematical method like linear programs (LPs) or MIPs.

Multicriteria optimization models for GHG emission control is a new idea. There is no deep experience of application in real work. For the time being, practical solution for planning work of GHG control is still heuristic way supported by OR methods partially.

2.4.2 Effort in OR

Decomposition methods will have a large influence to produce practical solutions. The planner can easily understand what happens in the calculation processes. Visual modeling tools are also helpful to illustrate and interpret the model. The planner should judge by adopting heavy criteria and a clear understanding of interdepending relationships between submodels and each criteria.

Mathematical effort to solve models having contradiction and uncertainty is very important and essential. Multicriteria programming, goal programming and stochastic programming [5, 8] are expected to be more easily to use in ordinary work.

From the point of view that practical solutions for planning work are still heuristic, decomposition of how to solve this problem should be considered carefully. Mathematical programming, connected with other methods like constraint programming and rule base system or metaheuristics, may yield efficient hybrid method, able to solve large-scale real world problems.

2.5 Conclusion

Our understanding of the Earth warming problem will change continuously from now on. As a consequence, in the work of enterprise management, the planner has to prepare basic and natural methods to cope with the situation changing in the world. Although, there a clear ideas and methodologies for solving multicriteria optimization problems with conflicting goals, there are no off-the-shelf models and solvers available. The very first, important step is to develop a reasonable model. Nature and characters of the problem must be analyzed to find a way for appropriate modeling and solving.

References

1. Aspen Tech. Inc. *Users PIMS Manual.* Aspen Tech. Inc., Cambridge, 1995
2. Giulio A. De Leo, Luca Rizzi, Andrea Caizzi, and Marino Gatto. Carbon emissions: The economic benefits of the Kyoto Protocol. *Nature*, 413:478–479, 2001
3. John Houghton. *Global Warming: The Complete Briefing.* 3rd edition, Cambridge University Press, Cambridge, UK, 2004
4. Josef Kallrath and John M. Wilson. *Business Optimisation Using Mathematical Programming.* MacMillian Business, London, 1997
5. David Morton. Overview of Stochastic Programming Applications. Dash Optimization, 29 May 2002; http://www.dashoptimization.com/home/downloads/pdf/StochasticApplications.pdf
6. P. M. Pardalos, Y. Siskos, and C. Zopounidis, editors. *Advances in Multicriteria Analysis.* Nonconvex Optimization and Its Applications. Springer, Berlin, 1995
7. United Nations. Kyoto Protocol to the United Nartions Framework Convention on Climate Change. http://unfccc.int/resource/docs/convkp/kpeng.pdf, 1998
8. S. Uryasev and P. M. Pardalos, editors. *Stochastic Optimization: Algorithms and Applications*, volume 54. Applied Optimization. Springer, Berlin, 2001

Part II

Deterministic Methods

3

Trading Hubs Construction for Electricity Markets

Pavel A. Borisovsky, Anton V. Eremeev, Egor B. Grinkevich, Sergey A. Klokov, and Andrey V. Vinnikov

Summary. In this chapter, we consider a problem of constructing trading hubs in the structure of the electricity wholesale markets. The nodes of a trading hub are used to calculate a reference price that can be employed by the market participants for different types of hedging. The need for such a reference price is due to considerable variability of energy prices at different nodes of the electricity grid at different periods of time. The hubs construction is viewed as a mathematical programming problem here. We discuss its connections with clustering problems, consider the heuristic algorithms of solution and indicate some complexity issues. The performance of algorithms is illustrated on the real-life data.

3.1 Introduction

In the modern electricity spot markets the price is not unique, it varies from one node of the power grid to another and it also depends on time. The market participants in this situation are interested in one or several reference prices to hedge the price risks and to settle the forward contracts. These reference prices can be calculated by taking an average of the energy prices in a number of nodes with the most typical price dynamics in the given region. A set of such nodes with a specific formula for computing the average is called a *trading hub*. For short, in what follows, we will use the term *hub*.

Large electricity markets, such as PJM Interconnection (USA), Midwest ISO (USA), United Energy System (Russia) and others, provide a number of hubs. In this case, each buyer or seller prefers the hub approximating the most closely the nodal price of this participant. The hubs in electricity markets have some similarity with the hubs in oil and gas markets, but each of these commodities has unique features which require relevant trading instruments [2].

Successfully functioning hubs contribute to emergence of *derivatives*, the financial instruments (contracts) that do not represent ownership rights in any asset but, rather, derive their value from the value of the underlying commodity. The derivatives may serve as efficient tools for isolating financial risk and hedging to reduce exposure to risk [6]. The hubs also contribute to the success of electronic trading systems providing the aggregated data on price dynamics over the system. In the case of electricity markets, the hub price is usually defined as a simple or a weighted average of the nodal prices over the nodes comprising the hub. Due to this reason the hub price is less volatile than the prices at individual nodes. This feature is of particular importance because the liquidity of futures contracts depends significantly on predictability of the price of the underlying commodity (liquidity here means large volume of trade operations and easiness to find a contracting party).

To define a hub, it is sufficient to select a set of nodes and the weights to average the prices over these nodes. In the present chapter, we will consider this task as an optimization problem, keeping in mind that usually the optimization is just one of the steps in the decision making process of designing a hub. This process in practice involves a lot of negotiations between the market participants and administration, so that the human expertise often plays an essential role. In some cases it has been proposed to define the hubs without any optimization, e.g. the hubs may consist exclusively of generation nodes, grouped on the regional basis, with weights equal to the historical volume of the generation or the installed capacity, but such approach is not always applicable. One of the promising statistical approaches to hubs construction is based on the principal component analysis [3] but detailed presentation of this method is outside of the scope of this chapter.

The trading hubs construction problems considered below have similar terminology to the *hubs location problems* [4], however, these classes of problems are different. The hubs location problems are mainly motivated by the applications where certain elements of a system are actually connected via hubs, while in our case the trading hubs are purely virtual constructions and no physical connections are associated with them. Also, in the hubs location problems there is no equivalent of the hub price, which plays an important role in our case.

The remaining part of the chapter is organized as follows. In Sect. 3.2, we discuss the motivation for the hubs construction problems, the ways of using the hubs in the electricity markets and the properties demanded from them. Here we also provide a brief review of hubs implemented in some electricity markets. The criteria and constraints, formulated in Sect. 3.2, are converted into mathematical programming problems in Sect. 3.3 and some basic properties of these problems are discussed here. The hub construction problems often turn out to be large-scale non-convex optimization problems, which makes it relevant to look for appropriate heuristics to solve them. Some of these heuristics are presented and evaluated in Sects. 3.4 and 3.5. Section 3.6 contains the conclusions.

3.2 Hedging in the Electricity Markets and Hubs Usage

3.2.1 Price Volatility

The trading hub construction problems appear in the context of the modern electricity markets based on the *Locational Marginal Pricing* (LMP). LMP is a mechanism for using market-based pricing for managing transmission congestion and thermal losses in the electricity grid. The energy prices at different locations vary due to transmission congestion, which prevents relatively low-price generation from meeting the loads beyond a certain neighborhood. If not the congestion and transmission losses, the energy price would be uniform all over transmission grid. The market clearing LMP price is determined by an *Independent System Operator* (ISO) on the basis of solution to a mathematical programming problem, known as the *economic dispatch problem*. The LMP price at any node is taken to be a Lagrange multiplier of the power flow balance constraint associated with this node. The details of this approach can be found, e.g. in [5, 17, 26, 29]. Computation of LMP prices requires that all market players submit to an ISO their bids for generation and load. If the price and dispatch schedule computation takes place a day before dispatch, this is called the *day-ahead market*. Additionally, an ISO may support other similar markets scheduled at later time, e.g. an hour-ahead market and the real-time market.

Due to the difficulty to store electric energy for significant time and due to high variability of demand for this good during a day, the LMP price of electricity is highly volatile. An example of price behaviour can be seen in Fig. 3.1. This figure contains the Real-Time data of 13 Feb 2007 obtained from the web site of PJM system operator `http://www.pjm.com`. The LMP prices of many energy markets have a strong dependence on the geographical location and the grid topology [27]. This is why in many cases it is important to establish regional hubs, defining reference prices that closely approximate the cost of energy in the area and may be used for hedging (compare the graphs of PJM Eastern hub and node CARKSVI in Fig. 3.1).

3.2.2 Basic Hedging Strategies and Hubs Usage

Hedging by Means of Futures Contracts

Deliveries in the futures market are organized in physical or financial form. The first, physical delivery assumes that the seller at the maturity must hold the specified in the contract quantity of good at the specified warehouse. The seller then sends delivery call to the buyer who transfers money at the price specified in the contract and the seller transfers the right of possession to the buyer. The essence of the contract is its price. Fixing the price when signing the contract allows both the buyer and the seller to secure their cash flow for the future.

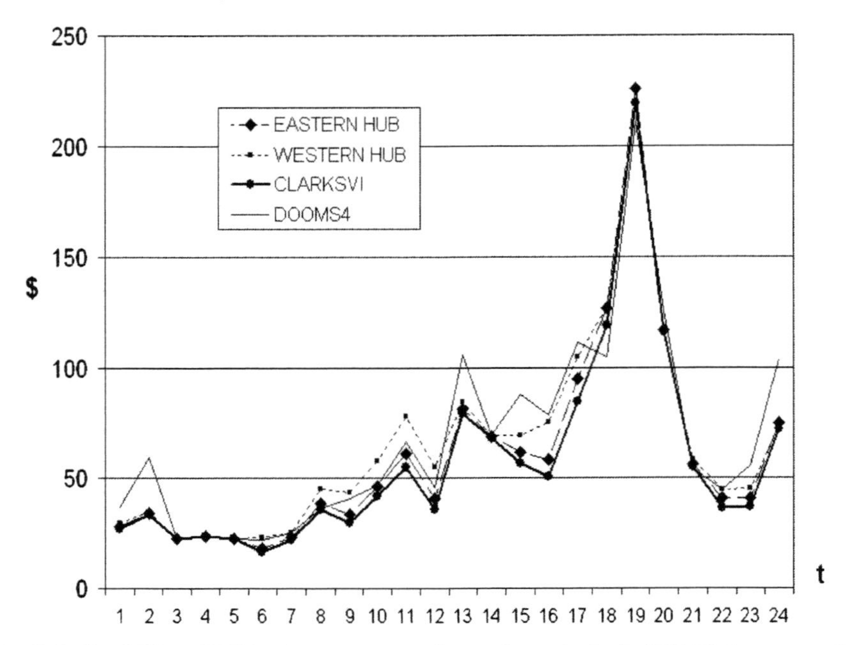

Fig. 3.1. Real-Time LMP prices at two nodes and two hubs in PJM Interconnection as of 13 Feb 2007

The seller of the contract can either deliver the good to the warehouse or buy the good in the warehouse from an agent (third party) at the spot price, while receiving from the buyer the price set in the contract. This observation motivates the second, financial form of delivery.

The financial form of delivery implies that entering into a futures contract at price C_f would yield for the seller the amount of money M_s depending on the spot price C:

$$M_s = C_f - C\,. \tag{3.1}$$

For the buyer, the result M_b is just the opposite:

$$M_b = C - C_f\,. \tag{3.2}$$

So, financially the futures contract results for the seller in:

- Receiving the difference between the contract and the spot prices when the contract price is higher than the spot price
- Paying the difference between the spot and the contract prices when the contract price is lower than the spot price

Since electricity can not be stored, the physical settlement would imply that the seller physically delivers electricity at the rate specified in the contract while the buyer transfers money to the seller's account at the specified price. Nevertheless, in the case of electricity, the usage of common electric grid is unavoidable and, therefore, the physical delivery of electricity involves a lot more detailed coordination of actions of all participants compared to physical delivery of other products. For this reason, the more flexible financial form of delivery is widely accepted in the electricity markets. The delivery period is usually 1 month, the amount is 1 MWh each hour. Also, contracts of 1-week, 1-quarter and 1-year duration are traded at some markets (Nord Pool, EEX).

Hedging by Means of Financial Transmission Rights (FTRs)

The FTRs are hedging instruments, designed for compensation of price difference between the nodes separated by congested transmission lines in the electricity spot markets with locational marginal pricing [15, 17, 26] (e.g. day ahead or real time markets). The FTR contracts can be defined from any node to any other node. If the hub price is defined as the simple average or a weighted average of a set of nodal prices, then the FTRs may be defined between any node and a hub, or between two hubs. The market of node-to-node FTRs may be quite illiquid due to a large number of nodes in the system. Liquidity can be increased by usage of hubs because an FTR from node 1 to node 2 is decomposable into two FTRs: from node 1 to hub and from hub to node 2 [17]. Traders may obtain the FTRs to cover the basis risk between their own node and the hub or between two different hubs. If the nodal price of a participant is sufficiently close to the price of some hub, the basis risk from node to hub may be quite small with most of the basis risk being between hubs.

In some markets, organized according with the Standard Market Design principles, the rules of the day-ahead market allow to post *virtual* demand/supply bids, i.e. the bids for buying or selling the energy, not supported by real generation capacity or load (e.g. New England Pool in the USA). In such a case, a significant amount of virtual demand/supply may be concentrated in a hub and the latter can be viewed as a trading point with risk-hedging potential [21].

The spot price in the electricity markets is significantly volatile even on the daily scale, therefore, when hubs are designed to be used for FTR transactions or for virtual bidding, it is important to take into account the hourly prices in historical data. Alternatively, if the hub is designed only to be used for 1-week, 1-month or annual futures, then the statistical data may be averaged over these time periods to simplify the analysis and to design the hub more adequate to the market tools in use.

The Hedge Ratio

Here we describe the most widely used method of hedging by standard contracts utilizing the so-called *hedge ratio*. This method is well-known among risk managers and it is based on easily calculable and understandable quadratic distance measure between the prices.

Consider a producer selling each hour 1 MWh of electricity on the spot market at price c_{it}, where t denotes hour and i denotes the node of the electric grid to which the producer belongs. The value c_{it} is a random number.

Since the producer sells 1 MWh at price c_{it}, the amount of money $M_{it}^{(s)}$ he gets from the spot market in that hour is

$$M_{it}^{(s)} = c_{it} \,. \tag{3.3}$$

The producer is going to hedge the position in the spot market by entering into a financially settled futures contract at price C_f. The settlement of the futures contract is defined relatively to the spot price c_{Ht} in the hub H meaning that according to (3.1) his participation in the futures market results in the value

$$M_{it}^{(f)} = C_f - c_{Ht} \tag{3.4}$$

per each 1 MWh of the delivery. Note that c_{Ht} is a random value as well.

Suppose the producer located in node i sells h_i futures contracts (h_i may be greater or less than 1). In this case the producer receives a total amount M_{it}

$$M_{it} = M_{it}^{(s)} + h_i M_{it}^{(f)} = c_{it} + h_i \left(C_f - c_{Ht}\right). \tag{3.5}$$

For efficient hedging the producer aims to minimize the variance of M_{it} [31]:

$$\text{Min} \;\; \sigma^2(M_{it}). \tag{3.6}$$

The only possible parameter the hedger can change is the amount h of futures contracts that he sells to hedge a unit of sold good in the spot market. Hence, (3.6) is transformed into

$$\frac{d\sigma^2(M_{it})}{dh_i} = 0. \tag{3.7}$$

Expanding the variance of the sum and respecting that C_f is not random we obtain

$$\frac{d}{dh_i}\left(\sigma^2(c_{it}) + h_i^2\sigma^2(c_{Ht}) - 2\rho(c_{it}, c_{Ht})h_i\sigma(c_{it})\sigma(c_{Ht})\right) = 0, \tag{3.8}$$

$$h_i\sigma(c_{Ht}) - \rho(c_{it}, c_{Ht})\sigma(c_{it}) = 0, \tag{3.9}$$

$$h_i = \rho(c_{it}, c_{Ht})\frac{\sigma(c_{it})}{\sigma(c_{Ht})}, \tag{3.10}$$

where $\rho(\cdot,\cdot)$ denotes the correlation.

Substituting the value of h_i (known as the *hedge ratio*) into (3.5) and calculating its variance we have:

$$\sigma^2(M_{i,t}) = \left(1 - \rho^2(c_{it}, c_{Ht})\right)\sigma^2(c_{it}), \tag{3.11}$$

i.e. an optimal hedge ratio decreases the variance of the initial position by factor of $1 - \rho^2(c_{it}, c_{Ht})$. Hence the standard criterion of selecting a good hub for a given node would be to take the one with the maximal correlation with the nodal price.

Unfortunately, reliable estimation of correlation and variance for a given node and given hour is practically impossible due to volatile behaviour of electricity prices [31]. One may take the historical prices data in order to calculate the hedge ratio as discussed above, but it will be only a rough estimate.

For the futures contracts with delivery periods of 1 month duration, the index t in the above relations could be replaced by the month index τ. In such a case, if μ_τ denotes the set of hours belonging to month τ, instead of the values c_{it} one would use

$$c_{i\tau} = \frac{\sum\limits_{t\in\mu_\tau} c_{it}A_{it}}{\sum\limits_{t\in\mu_\tau} A_{it}}, \tag{3.12}$$

where A_{it} is the amount of electricity sold in the node i at hour t. A discussion of positive and negative factors of averaging with time-dependent weights A_{it}, as compared to the simple average, can be found in [7]. In any case, replacement of the hourly prices by average prices over certain time periods reduces the amount of input data for the decision support system, simplifying the analysis. At the same time, this approach reduces the amount of useful information at the input, e.g. the distinct behaviour of on-peak and off-peak prices can not be seen after such an aggregation. With this in mind, we will usually talk about time t indexed in hours, unless otherwise is stated.

It is clear that a large number of hubs would allow to find an appropriate hub for any node. However, large number of futures can not be liquid simultaneously. At most of the exchanges, the number of different liquid futures is small. Usually there exists one lead contract attracting most of liquidity and 3–5 supplementary contracts. Unfortunately, it is difficult to evaluate quantitatively the liquidity of each hub in a given collection of hubs in advance, because this property depends on many organizational factors and strategies of market participants. Hence, the upper bound on the number of hubs is often used as a simplified liquidity requirement.

The necessary number of hubs m may be evaluated by means of the Principal Component Analysis [3]. Usage of this method in hubs design is established on the basis of assumption that each node i is hedging in all existing hubs with the hedge ratios minimizing (3.6) with respect to each hub. Each eigenvector (principal component) of the sample correlation matrix between all nodal prices can, theoretically, define a hub. The greater its eigenvalue, the greater amount of the total variance of the nodal prices it carries. A decision about the number of principal components, that are practically significant, can be made e.g. using Kaiser's or Cattell's criteria [34].

Another constraint, which is also connected with hub liquidity is a lower bound on the number of nodes in a hub, when the hub price is computed as the simple average of the nodal prices. This constraint ensures stability of the hub price under minor modifications of the grid (permanent exclusion or temporary outage of nodes due to planned repair or unexpected breakdown). In general, taking average over a larger number of nodes usually decreases the variance of hub price, which makes it more predictable for the traders, increasing the liquidity. The lower bound on the number of hub nodes, equal to 50 or 100 appears to be sufficient (see the examples in Sect. 3.2.3).

Note that when the hub price is computed as a weighted average, the nodal weight may be chosen arbitrary close to zero, even though, formally, this node is used for calculation of the hub price. This makes it meaningless to impose the lower bound on the number of nodes when the hub price is computed as a weighted average. There are some ways to modify this condition for the weighted case as well, but for simplicity we will consider only the unweighted case here.

To sum up, informally, the Hubs Construction Problem consists in finding a sufficiently small number of large hubs which would explain as much as possible the price dynamics in most of the nodes of the electric grid. The ways to formally state this problem will be addressed in Sect. 3.3.

3.2.3 Hubs Design of Some Existing Markets

In this section, we briefly survey several cases of hubs design in large-scale electricity markets based on locational marginal pricing.

Midwest ISO

The system operator Midwest ISO manages one of the largest electricity markets in the USA. The grid of Midwest ISO consists of more than 30,000 nodes and the LMPs are computed for about 1,500 nodes. The overall installed capacity of Midwest ISO generation is near 150 GW. The electricity market is organized according to the Standard Market Design principles [8]. There are four trading hubs in this market: Cinergy, Michigan, Illinois and Minnesota consisting of about 330, 260, 150 and 170 nodes, respectively. These hubs were chosen by LECG, LLC and the Midwest ISO in conjunction with the Trading Hubs Task Force in year 2003 [13].

One of the main requirements in the design of the Midwest ISO hubs was that the hub price should move consistently with the prices in the corresponding target region and most of locations in the target region are "close" to the hub in terms of price difference. Also, it was required that it should be unlikely for a significant portion of the trading hub to be lost from service. The volatility of hub price had to be low, implying that plausible patterns of transmission congestion and individual transmission outages should not cause the trading hub price to substantially diverge from prices in the target region. The trading hub definitions had to be fixed i.e. once a hub was defined, the set of hub nodes and their weights are not changed.

In view of these requirements, the optimization model with given number of hubs was applied (see Sect. 3.3.1 below). The input data consisted of the nodal prices for each 2-h period of the year in 1,290 nodes. A straightforward solution of this problem by means of commercial optimization packages was impossible because of its high dimensionality. For this reason the problem formulation was simplified by setting the weights of all nodes equal to 1. This simplification allowed to apply the standard statistical clustering algorithm H-means (see Sect. 3.4 below) to form 30 clusters of nodes. Inspection of these clusters showed several sufficiently large candidates with relatively low distance between the hub price and the nodal prices in their target regions. Some of the outliers were manually excluded from these clusters on the basis of the scatter plot analysis and evaluation of the experts. The subsequent analysis consisted in comparison of the hub prices with the electricity price in 37 load area zones of major market participants to determine the cluster best fit for each of them.

PJM Interconnection

operates one of the largest wholesale electricity markets in the world. The overall installed capacity of Midwest ISO generation is near 160 GW, the number of market participants is more than 350. The market structure complies with the Standard Market Design. There are two actively traded hubs: Western (near 110 nodes) and Eastern (near 240 nodes), eight localized hubs: AEP Gen, AEP Dayton, Chicago Gen, Chicago, Domnion, Northern Illinois, New Jersey, Ohio and an interface hub Western Interface.

The hub price is computed as a weighted average of the real-time LMP prices with a fixed set of the nodal weights. The largest volume of trade is concentrated in PJM-Western hub, due to its stability to the influence of system constraints and its location between large load areas and areas of generation. The PJM-Western monthly futures are traded at NYMEX stock exchange for on-peak and off-peak hours (see `http://www.nymex.com/JM\_desc.aspx`).

To determine the composition of the PJM-Western Hub, the nodal prices were analysed under various historic transmission constrained conditions [25]. The standard cluster analysis tools were used to determine candidate clusters of nodes that respond in a similar way under many different transmission constraints. For each of these clusters, an optimization problem was solved to determine the node weights that minimize the distance between the hub price and the energy price in the subregion for which the hub is targeted. Originally, in year 1998 PJM-Western Hub consisted of nodes selected from PEPCO, BGE, Penelec and MetEd zones. Later this hub went through some changes, with addition of nodes from APS, ComEd, AEP, Dayton, Dominion, and Duquesne and RECO zones.

ISO New England

system operator is responsible for New England's bulk power generation and transmission system with an installed capacity of 32 GW and more than 200 market participants. The market is organized according with the Standard Market Design. The power grid of ISO New England has only one hub,

NEPOOL Hub, allocated between the areas of prevailing generation at North and West and the areas of prevailing consumption at Connecticut and Northeast Massachusetts/Boston. The hub price is the simple average of the nodal prices at 32 nodes. The choice of these nodes was based on statistical analysis, using simulated nodal prices [14].

Electricity Market of United Energy System of Russia

Administrator of Trade System for United Energy System of Russia operates a wholesale electricity market with an overall installed capacity near 200 GW and 200 market participants, more than 100 of them are generators. The day-ahead market is based on the locational marginal pricing, where the hourly prices on electricity are computed for more than 7,000. The mathematical models of the current market, which is functioning since September 2006, and its two-sector predecessor are described in [5]. There are four hubs in the European zone of Russia: Center-Europe, Center-West, Volga and Urals. The other zone is located in Siberia, it has two hubs: Kuznetsk Basin and Krasnoyarsk – see `http://www.np-ats.ru`. The sizes of smaller hubs are close to 50 nodes and the size of the largest hub Center-West is over 300 nodes. The hub prices are computed as simple average of the day-ahead locational marginal prices. All hubs consist of high voltage nodes (not less than 220 kV), which ensures that local congestions and grid modifications do not influence the hub price a lot. The sets of hub nodes in the European zone were chosen using the H-means clustering algorithm with subsequent expertise. The Principal Component Analysis, applied to the European zone indicates that the largest eigenvalue corresponds to the average electricity price in this zone. The first eigenvalue greatly exceeds all other eigenvalues and there are four other principle components of significant value, which is consistent with the number of existing hubs.

3.3 Problem Formulations

In this section, we discuss the mathematical formulations of the Hub Construction Problems, taking into account the criteria and constraints considered in Sect. 3.2.2 above.

As it was mentioned before, the trading hubs should be constructed, taking into account the historical data of the nodal prices, preferably, covering a whole preceding year or several years. If the hubs are designed for monthly futures, without separation of on-peak and off-peak futures, then the historical data may be aggregated into 1-month elementary periods. If the hubs are also aimed to be used for different types of futures, for virtual bidding at the day-ahead market or for the FTR contracts, then such time aggregation is inappropriate. It is important that the input data represent the price dynamics in all seasons and, if the hourly prices are not averaged over 1-week or

1-month period, it is also important that the data reflects different modes of the system: with congestions and without them, on-peak and off-peak hours, working days and weekends. For terminological convenience, we will usually call the elementary historical time intervals "hours" and denote them by t. The number of elementary time intervals in the historical data will be denoted by T.

Here we assume that the distance measure between the hub price and the nodal price is computed as the sum of squared differences over all hours t, $t = 1, \ldots, T$, i.e. the squared Euclidean distance in T-dimensional space. If appropriate, the Euclidean distance may be substituted by some other standard metric, or by the observed variance or observed correlation (the latter should be maximized) as it was discussed in Sect. 3.2.2. The sum of squared differences appears to be the most widely used criterion and many clustering methods are well suited to it.

In what follows, n will denote the number of nodes, where the LMP prices are computed, and c_{it} will be the LMP price in node $i = 1, \ldots, n$ at hour $t = 1, \ldots, T$. To allow different nodes to have different significance at different time, the weighting factors $w_{i,t}$ can be introduced. One of the standard approaches to weighting is to take the weights equal to the traded volumes $w_{i,t} = A_{i,t}$. Alternatively, one can assign a set of constant weights equal to the installed capacity in the nodes or equal to the annual average traded volume.

We will say that a node i is assigned to hub j, if the market participant located at node i uses hub j (and only this hub) for hedging. The set of nodes, assigned to a hub j will be called the *target region* of the hub j. In practice, a market participant may hold a set of nodes of the grid and trade the electricity in all of these nodes with certain proportion of the traded volumes in the nodes.

3.3.1 Construction of a Given Number of Hubs

The problem formulation considered here is based on the assumption that a given number of hubs m, $m < n$ is sufficiently small to ensure sufficient liquidity of hubs. The goal is to minimize the total (weighted) deviation D of the hub prices from the nodal prices in the target regions of the hubs. In effect, this means that for simplicity we assume that each node belongs to one market participant and each participant holds one node. The mathematical formulation of this problem is as follows:

$$\text{Min } D = \sum_{i=1}^{n} \sum_{j=1}^{m} \delta_i^j \sum_{t=1}^{T} (c_{it} - c_t^j)^2 w_{it} \tag{3.13}$$

s.t.

$$c_t^j = \sum_{i=1}^{n} \alpha_i^j c_{it}, \; j = 1, \ldots, m, \; t = 1, \ldots, T, \tag{3.14}$$

$$\sum_{i=1}^{n} \alpha_i^j = 1, \ j = 1, \dots, m, \tag{3.15}$$

$$\sum_{j=1}^{m} \delta_i^j = 1, \ i = 1, \dots, n, \tag{3.16}$$

$$\delta_i^j \in \{0, 1\}, \quad \alpha_i^j \geq 0, \ i = 1, \dots, n, \ j = 1, \dots, m. \tag{3.17}$$

Here the variables δ_i^j, $i = 1, \dots, n$, $j = 1, \dots, m$ define which nodes are assigned to each hub, the variables α_i^j, $i = 1, \dots, n$ define the set of weights within the hub j and the variables c_t^j, $t = 1, \dots, T$ give the price of hub j, $j = 1, \dots, m$ at each hour t. Equation (3.14) gives the hub price calculation, while (3.16) ensures that each node is assigned to exactly one of the hubs. The constraint (3.15) serves for normalization of the hub price. Although the model would be meaningful without this constraint, in certain conditions it plays the role of a cut, as it will be seen in the proof of Proposition 1 below.

The Boolean variables δ_i^j may be substituted by real-valued variables ranging from 0 to 1. Although this relaxation of problem (3.13)–(3.17) allows each node to be assigned to several hubs simultaneously, it is easy to see that the relaxed formulation always has an optimal solution with Boolean values of all δ_i^j. (For each i one can assign $\delta_i^j = 1$ for a single hub j which minimizes $\sum_{t=1}^{T}(c_{it} - c_t^j)^2 w_{it}$.) This problem in the relaxed version was originally formulated by W. Hogan [16] for the case of two hubs and extended to optional number of hubs in [13], Appendix A.

As it is noted in [16], the relaxed formulation belongs to the class of nonconvex optimization problems, thus it is impossible to apply directly the efficient optimization techniques developed in convex optimization. However, once the set of all variables δ_i^j is fixed, the remaining variables may be found by solving a convex optimization problem; sometimes they may be assigned explicitly as we will see in the proof of Proposition 1 below. Alternatively, if the set of all variables α_i^j is given, the complementary assignment of the variables δ_i^j is straightforward. These properties may be exploited in the nonconvex optimization algorithms [19], if they are tailored for this problem.

Note that in a feasible solution one or several hubs may have empty target regions, i.e. for these hubs j holds

$$\sum_{i=1}^{n} \delta_i^j = 0.$$

We will call such assignments *degenerate*. Note that it is possible to eliminate the empty target regions, not increasing the objective function value. This can be done iteratively by finding a node k with the maximal value of

$$\sum_{j=1}^{m} \delta_k^j \sum_{t=1}^{T} (c_{kt} - c_t^j)^2 w_{kt}$$

and assigning it to a hub with an empty target region. The new assignment, coupled with the available set of real-valued variables α_i^j and c_t^j, gives a feasible solution and does not increase the previously found value of objective function (in fact this holds for any choice of k).

The following proposition is aimed at finding the best-possible set of real-valued variables, complementing a non-degenerate assignment of nodes.

Proposition 1. *Suppose, $w_{it} = w_i$ does not depend on t for all nodes i and a feasible non-degenerate assignment $\{\delta_i^j\}$ is given. Then the optimal price in hub j, $1 \leq j \leq m$ is calculated as the weighted average of prices in the assigned nodes:*

$$c_t^j = \sum_{i:\ \delta_i^j = 1} w_i c_{it} \Big/ \sum_{i:\ \delta_i^j = 1} w_i. \tag{3.18}$$

Proof. Denote by F_{jt} the deviation of nodal prices in hub j at hour t.

$$F_{jt} = \sum_{i=1}^{n} \delta_i^j (c_{it} - c_t^j)^2 w_i.$$

To find hub price c_t^j minimizing F_{jt} we differentiate it over c_t^j:

$$\frac{\partial F_{jt}}{\partial c_t^j} = -\sum_{i=1}^{n} \delta_i^j 2(c_{it} - c_t^j) w_i = 0.$$

Solving this equation we obtain

$$c_t^j = \sum_{i=1}^{n} \delta_i^j w_i c_{it} \Big/ \sum_{i=1}^{n} \delta_i^j w_i.$$

Denote $\alpha_i^j = w_i / \sum_{k=1}^{n} \delta_k^j w_k$ if $\delta_i^j = 1$ and $\alpha_i^j = 0$ otherwise. It is easy to check that $\sum_{i=1}^{n} \alpha_i^j = 1$, so the obtained solution is feasible and hence it is optimal. □

In conditions of Proposition 1, hub j, $1 \leq j \leq m$ is completely defined by its target region

$$H_j = \{i : \delta_i^j = 1,\ i = 1, \ldots, n\},$$

since the coefficients α_i^j are given by

$$\alpha_i^j = w_i \Big/ \sum_{k:\ \delta_k^j = 1} w_k.$$

In the special case where the weights w_i are all identical, the problem turns into the classical minimum sum-of-squares clustering problem: find a partition of a given finite set of vectors in Euclidean space into several disjoint sets (clusters), minimizing sum of squared distances from each element to the centroid of its cluster. Here centroid means the simple average of vectors in a cluster. Currently the complexity status of this problem is open, in spite of a number of attempts to prove that this problem is NP-hard (see the survey [1]). This problem has been deeply studied during the last 50 years and a number of exact and heuristic approaches to its solution have been developed (see the survey in [11]). Some of them will be discussed in Sect. 3.4.

In the case of identical weights w_i, the optimal hub price (3.18) equals to the simple average of the nodal prices, which makes it appropriate to impose a lower bound $n_{\min}$, $n_{\min} \leq n/m$ on the number of nodes in each hub:

$$n_{\min} \leq \sum_{i=1}^{n} \delta_i^j, \quad j = 1, \dots, m. \tag{3.19}$$

This modification of the problem is not studied as much as the minimum sum-of-squares clustering problem and its solution may require some modification of the well-known clustering methods or application of general-purpose optimization tools.

Let us consider what modification of the objective function (3.13) is required in order to minimize the total observed variance (3.6), assuming that participants use the hedge ratio approach described above. With simplifying assumption that the prices are stationary distributed, the estimated variance $\hat{\sigma}_i^2$ of price c_{it} in node i, as well as the estimated correlation $\hat{\rho}_{ij}$ between the nodal price c_{it} and the hub price c_t^j, can be expressed on the basis of the historical data. Then (3.11) leads to the following criterion:

$$\text{Min} \quad \sum_{i=1}^{n} \sum_{j=1}^{m} \delta_i^j \hat{\sigma}_i^2 \left(1 - \hat{\rho}_{ij}^2\right), \tag{3.20}$$

where

$$\hat{\sigma}_i^2 = \sum_{t=1}^{T} \left(c_{it} - \frac{1}{T} \sum_{t=1}^{T} c_{it} \right)^2 \Big/ (T-1),$$

$$\hat{\rho}_{ij}^2 = \frac{\left(T \sum_{t=1}^{T} c_{it} c_t^j - \sum_{t=1}^{T} c_{it} \sum_{t=1}^{T} c_t^j\right)^2}{\left(T \sum_{t=1}^{T} c_{it}^2 - \left(\sum_{t=1}^{T} c_{it}\right)^2\right)\left(T \sum_{t=1}^{T} (c_t^j)^2 - \left(\sum_{t=1}^{T} c_t^j\right)^2\right)}$$

for $i = 1, \dots, n, \; j = 1, \dots, m$.

3.3.2 Single Hub Selection

The purpose for formulation of the Single Hub Selection Problem in this section is to refine a set of m preliminary hubs by selecting a refined hub within each of them. The set of preliminary hubs may be a result of selecting a given number of hubs, or it may describe an existing set of hubs or zones in the electricity market.

We can assume without loss of generality that the set of nodes of the preliminary hub is $\{1,\ldots,N\}$, where $N \leq n$. In what follows, talking about the Single Hub Selection Problem we will use the term "hub" only for the refined hub, while the preliminary hub will be referred to as *a given set of nodes* or *cluster*.

We will assume that the hub is chosen with respect to the locational energy prices of the market participants situated in the target region of the hub. Let p_{rt} denote the energy price of participant r, $r = 1,\ldots,R$ at hour t. A particular definition of the price of participant does not matter. In case a participant r has the injection/withdrawal of energy within a single node i of the grid, the energy price of this participant usually equals c_{it}. If the injection/withdrawal of a participant is spread over a number of nodes, then the price p_{rt} may be calculated as a weighted average of the nodal prices according to some market rules.

Suppose the hub price is always computed as an average price over all included nodes, and require that the hub contains at least $n_{\min}$ nodes. Then the Single Hub Selection Problem consists in minimizing the sum of squared differences of the prices of participants from the hub price with respect to a given set of weights of market participants W_{rt}, $r = 1,\ldots,R$:

$$\text{Min } f = \sum_{t=1}^{T}\sum_{r=1}^{R}(c_t - p_{rt})^2 W_{rt} \tag{3.21}$$

s.t.

$$c_t = \frac{1}{L}\sum_{i=1}^{N} x_i c_{it},\ t = 1,\ldots,T, \tag{3.22}$$

$$\sum_{i=1}^{N} x_i = L, \tag{3.23}$$

$$L \geq n_{\min}, \tag{3.24}$$

$$x_i \in \{0,1\},\ i = 1,\ldots,N,\quad c_t \geq 0,\ t = 1,\ldots,T. \tag{3.25}$$

Here the binary variables x_i turn into 1 whenever node i is included into the hub. The variables c_t define the hub price at time t, $t = 1,\ldots,T$. The complexity status if this problem in the special case, when each participant is located in its own node, is established by the following proposition.

Proposition 2. *The Single Hub Selection Problem (3.21)–(3.25) is NP-hard even when* $R = N$, $p_{it} = c_{it}$ *for all* $i = 1,\ldots,N$, $t = 1,\ldots,T$ *and* $T = 2$.

The proof of Proposition 2, provided in the Appendix, is based on a transformation from an NP-complete Partition problem.

3.4 Heuristics for Construction of Given Number of Hubs

In this section, we discuss two well-known clustering heuristics in the context of the Hubs Construction Problem. We assume that all nodes are given constant weights w_i. According to Proposition 1, it is sufficient to partition the set of nodes $\{1, \ldots, n\}$ into m clusters $H_1, \ldots, H_m$, minimizing the total weighted squared error:

$$\text{Min} \quad S = \sum_{j=1}^{m} \sum_{i \in H_j} w_i \sum_{t=1}^{T} (c_{it} - c_t^j)^2,$$

where the hub price c_t^j is calculated as the weighted average of nodal prices in cluster H_j:

$$c_t^j = \frac{\sum_{i \in H_j} c_{it} w_i}{\sum_{i \in H_j} w_i}. \tag{3.26}$$

3.4.1 The H-Means Method

R. Howard [20] is considered to be the first one who outlined the clustering method H-means. Here we view this algorithm in adaptation to the Hubs Construction Problem with a given number of hubs. Starting with an initial set of points c_j, $j = 1, \ldots, m$ in T-dimensional Euclidean space, H-means algorithm iterates the following three steps:

1. For each node i, $i = 1, \ldots, n$, find the closest c_j, $j \in \{1, \ldots, m\}$ with respect to Euclidean distance and place the node i into the cluster H_j.
2. Let m' be the number of non-empty clusters and reorder the clusters so that $H_1, \ldots, H_{m'} \neq \emptyset$.
3. Recalculate c_j for all $j = 1, \ldots, m'$ according to (3.26) with the new partition $H_1, \ldots, H_{m'}$.

The algorithm terminates when the set of clusters does not change any more. If during the run of the algorithm the number of non-empty clusters m' falls below m, the assignment of nodes becomes degenerate.

In the case when all weights w_i are identical, it is well known [10] that Step 1 gives the optimal partition for the given centroids, and Step 3 gives the optimal centroids location for the given partition. In view of Proposition 1, it is easy to see that the same holds if w_i are not identical.

At each iteration, the value of objective function can not increase, so the algorithm will eventually reach some value of objective function it can not further improve. This will take only a finite number of iterations because there is only a finite number of partitions of a finite data set. The computational cost of each iteration is equal to $O(mnT)$.

The output of H-means depends on the initial set of centroids $c_1, \dots, c_m$ and it is not necessarily a global optimum of problem (3.13)–(3.17). The initial values $c_1, \dots, c_m$ may be provided by an expert in the form of cluster seeds (a set of clusters, each consisting of a single node) to direct the heuristic to some "reasonable" structure of hubs. Alternatively, one can run the H-means algorithm a number of times with different randomly chosen cluster seeds and choose the best output over all runs.

The computational study in [12] indicates an advantage of the following simple modification of H-means. The difference of the modification, named H-means+, from the original method consists in checking for degeneracy of the assignment found. The algorithm stops if the assignment is not changing any more and it is non-degenerate. If it is degenerate, the number of non-empty clusters is raised up to m as it was described in Sect. 3.3.1 and the iterations continue.

Minimization of Euclidean distance may be substituted by other criteria mentioned in Sect. 3.3, e.g. objective (3.20). The H-means algorithm is sufficiently flexible and it may be adjusted to use such criteria as well (on applicability of H-means see [10, 30] and references therein).

3.4.2 The *K*-Means Heuristic

R. Jancey [22] and J. MacQueen [24] proposed the K-means heuristic which is similar to the H-means but fits better into the standard local search scheme. Here we use the terminology of Hubs Construction Problem, presenting a slightly generalized version of H-means, which takes the nodal weights w_i into account.

The K-means starts from an initial partition $H_1, \dots, H_m$ of nonempty hubs and iteratively moves a node from one hub to another to decrease the value of objective function D. To choose a node to be moved, all possible reassignments are considered and the one with largest decrement of the objective function value is chosen. The iterations are performed until either no nodes can be moved, or the value of D decreases unsubstantially.

Without loss of generality, suppose that node k is moved from H_1 to H_2. Hubs $H_3, \dots, H_m$ are unaffected by the move, H_1 transforms into $\tilde{H}_1 = H_1\{k\}$, and H_2 becomes $\tilde{H}_2 = H_2 \cup \{k\}$. Denote $c_i = (c_{i1}, \dots, c_{iT})$, $c^j = (c_1^j, \dots, c_T^j)$ and

$$\|c_i - c^j\|^2 = \sum_{t=1}^{T} (c_{it} - c_t^j)^2$$

and find the difference between total weighted squared errors analogously to the computations in [32]:

$$\begin{aligned}\tilde{D} - D = &\sum_{i \in \tilde{H}_1} w_i \|c_i - \tilde{c}^1\|^2 + \sum_{i \in \tilde{H}_2} w_i \|c_i - \tilde{c}^2\|^2 \\ &- \sum_{i \in H_1} w_i \|c_i - c^1\|^2 - \sum_{i \in H_2} w_i \|c_i - c^2\|^2\end{aligned}$$

$$= \sum_{i \in H_1} w_i \|c_i - \tilde{c}^1\|^2 + \sum_{i \in H_2} w_i \|c_i - \tilde{c}^2\|^2$$
$$- \sum_{i \in H_1} w_i \|c_i - c^1\|^2 - \sum_{i \in H_2} w_i \|c_i - c^2\|^2$$
$$- w_k \|c_k - \tilde{c}^1\|^2 + w_k \|c_k - \tilde{c}^2\|^2$$
$$= \sum_{i \in H_1} w_i \left(\|c_i - \tilde{c}^1\|^2 - \|c_i - c^1\|^2\right)$$
$$+ \sum_{i \in H_2} w_i \left(\|c_i - \tilde{c}^2\|^2 - \|c_i - c^2\|^2\right)$$
$$- w_k \|c_k - \tilde{c}^1\|^2 + w_k \|c_k - \tilde{c}^2\|^2.$$

Due to the equality

$$\|c_i - b\|^2 - \|c_i - a\|^2 = \sum_{t=1}^{T} (a_t - b_t)(2c_{kt} - a_t - b_t)$$
$$= 2 \sum_{t=1}^{T} (a_t - b_t)(c_{kt} - a_t) + \|b - a\|^2,$$

one has

$$\sum_{i \in H_1} w_i \left(\|c_i - \tilde{c}^1\|^2 - \|c_i - c^1\|^2\right)$$
$$= \sum_{i \in H_1} w_i \|\tilde{c}^1 - c^1\|^2 + 2 \sum_{i \in H_1} w_i \sum_{t=1}^{T} (c_t^1 - \tilde{c}_t^1)(c_{it} - c_t^1)$$
$$= \|\tilde{c}^1 - c^1\|^2 \sum_{i \in H_1} w_i + 2 \sum_{t=1}^{T} (c_t^1 - \tilde{c}_t^1) \sum_{i \in H_1} w_i (c_{it} - c_t^1).$$

By definition of the hub price,

$$\sum_{i \in H_1} w_i (c_{it} - c_t^1) = 0.$$

Similar calculations are used for other summands to obtain,

$$\tilde{D} - D = \|\tilde{c}^1 - c^1\|^2 \sum_{i \in H_1} w_i + \|\tilde{c}^2 - c^2\|^2 \sum_{i \in H_2} w_i$$
$$- w_k \|c_k - \tilde{c}^1\|^2 + w_k \|c_k - \tilde{c}^2\|^2.$$

Therefore, one chooses the reassignment providing the minimal value of $\tilde{D} - D$ among all possible moves of each node.

The computational cost of one iteration is equal to $O(mnT)$, as well as in the H-means. This algorithm may be restarted a number of times from randomly chosen partitions.

The K-means and H-means clustering methods are probably the most widely used in practice. This is due to the simplicity and computational efficiency of these algorithms. The quality of their solutions, however, may be far from the optimal. This is demonstrated, e.g. in [12], where both of these algorithms were experimentally compared to the Variable Neighborhood Search. A significant improvement of output results of K-means and H-means is reported for a combined method, which firstly starts the H-means+, and the obtained solution is further optimized by the K-means. This finding is consistent with the fact that any solution, which is non-improvable for the K-means, is also non-improvable for the H-means, while the converse is not true [33].

We have considered only two well-known heuristics adapted to the Hubs Construction Problem. A number of other exact and heuristic approaches, such as the Branch and Bound algorithms, metaheuristics, hierarchical clustering heuristics and other methods (see, e.g. [11, 12] and references there) can be also applied to the Hubs Construction Problem and its modifications. However, the main limiting factor, which may hinder the usage of some of these methods, is a large dimensionality of typical instances of the Hubs Construction Problem. A generalization of the known methods to the case of time-dependent nodal weights w_{it} constitutes another challenge for algorithmic research.

3.4.3 Experimental Evaluation of the K-Means on PJM Data

Performance of the K-means method described in Sect. 3.4.2 is tested here on the real-time market hourly prices from PJM InterconnectionV, available at `http://www.pjm.com`. The input includes hourly data of 5 weeks, each week representing one of the months from January to May of year 2007, in total 840 records for each of $n = 7,599$ nodes. Analogously to the existing 11 PJM hubs, $m = 11$ is chosen. All nodal weights are set to 1.

The K-means is programmed in C++ and tested on Pentium-IV, 3 GHz machine. A series of 30 independent runs is made with random initial solutions, each run taking from 40 min to 1 h. The best outcome in terms of the objective function (3.13) consists of 11 hubs with sizes ranging from 152 to 1,898 nodes. It turns out that if the price of the new hubs were computed as the simple average of their nodal prices, then our largest hub would be the closest one to the existing Western hub. We denote this hub by LS-Hub1. The closest to LS-Hub1, among the existing hubs, is Dominion.

An 825-node hub, closest to PJM Eastern hub, we denote by LS-Hub2. It also tightly approximates the existing New Jersey hub. The hubs AEP Gen, AEP Dayton, Chicago Gen, Chicago, Northern Illinois and Ohio are approximated by other hubs found by the K-means. The Western Interface has no equivalent in the set of our hubs. At the same time the set of our hubs contains a high-price hub of 313 nodes with no equivalent among the existing hubs.

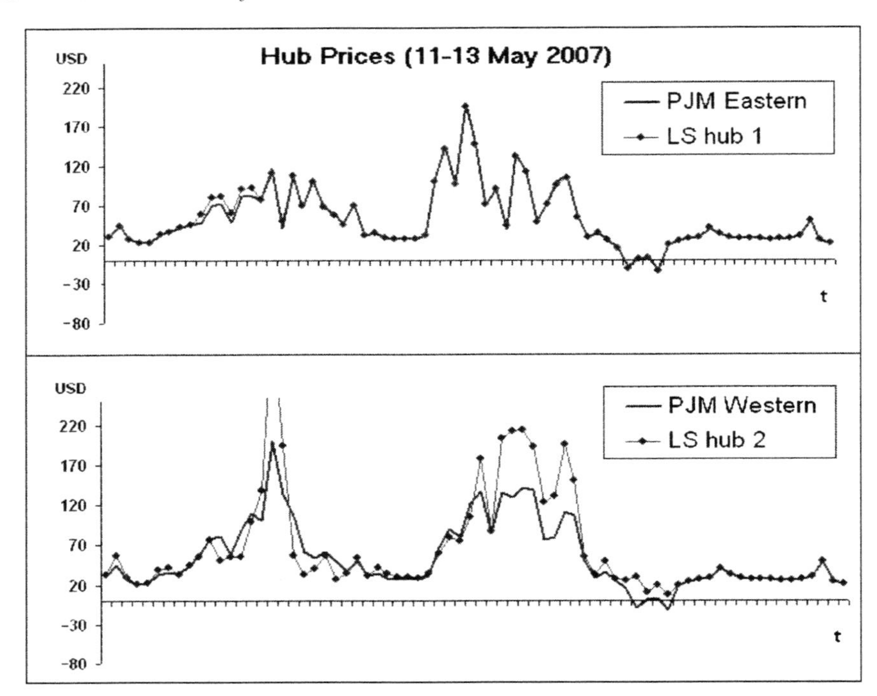

Fig. 3.2. Energy prices in PJM Western Hub, PJM Eastern Hub, LS-Hub1 and LS-Hub2 on May 11–13, 2007

Figure 3.2 demonstrates the behaviour of prices in Western, Eastern, LS-Hub1 and LS-Hub2 during 3 days from Friday, May 11 to Sunday, May 13, 2007. As it can be seen from the figure, the prices of PJM Eastern and LS-Hub2 are very close. However, PJM Western hub does not have the price peaks as high as the peaks of its counterpart LS-Hub1. The graph of Dominion exactly overlaps with LS-Hub1, so we not display it on the figure.

Absence of exact counterpart for the PJM Western hub in the output of the K-means heuristic may be due to the difference of clustering methods and their settings, different historical data and subsequent optimization and modifications of PJM Western hub, which followed the initial clustering stage. A higher volatility of our largest hub LS-Hub1, compared to PJM Western, may lead to lower liquidity of LS-Hub1 because the price of the latter may be more difficult to predict. At the same time, the companies located in BGE, Dominion and PEPCO, experiencing high on-peak prices could use the LS-Hub1 more actively for hedging their contracts, which is a positive factor for the hub liquidity.

This computational experiment illustrates that even a simple K-means clustering can produce a meaningful initial grouping of nodes. The running time of the K-means is not exceptionally high even for a system of about 7,000–8,000 nodes and the same approach could be applied to the data representing all 12 months of the year.

3.5 Solving the Single Hub Selection Problem

The nonlinear discrete optimization problem (3.21)–(3.25) can be transformed into a family of Boolean linear programming problems, each one with a different value of parameter L, $L = n_{\min}, \ldots, N$. Let us consider a term $(c_t - p_{rt})^2$ from (3.21) for any pair of r and t:

$$(c_t - p_{rt})^2 = \left(\frac{1}{L} \sum_{i=1}^{N} x_i c_{it} - p_{rt} \right)^2 =$$

$$\frac{1}{L^2} \sum_{i=1}^{N} (c_{it}^2 - 2Lc_{it}p_{rt})x_i + \frac{2}{L^2} \sum_{k=1}^{N} \sum_{l=1}^{k-1} c_{kt}c_{lt}x_k x_l + p_{rt}^2. \tag{3.27}$$

In view of this expression, one can remove the nonlinearity from the objective function (3.21) by introducing new variables y_{kl}, $k = 1, \ldots, N$, $l = 1, \ldots, k-1$ so that

$$y_{kl} = x_k x_l, \; k = 1, \ldots, N, \; l = 1, \ldots, k-1. \tag{3.28}$$

The set of equalities (3.28) may also be substituted by a system of linear constraints:

$$y_{kl} \leq x_k, \quad y_{kl} \leq x_l, \quad k = 1, \ldots, N, \; l = 1, \ldots, k-1, \tag{3.29}$$

$$y_{kl} \geq x_k + x_l - 1, \quad k = 1, \ldots, N, \; l = 1, \ldots, k-1. \tag{3.30}$$

Therefore, the Single Hub Selection Problem with given value L becomes a mixed-integer linear programming (MIP) problem. In view of (3.27), it is easy to notice that constraints (3.29) are always satisfied in the optimum, even if they were not included into problem formulation. Now we can conclude that problem (3.21)–(3.25) reduces to solving the following family of MIP problems:

$$\text{Min} \;\; C_0 + \sum_{i=1}^{N} C_i x_i + \sum_{k=1}^{N} \sum_{l=1}^{k-1} B_{kl} y_{kl} \tag{3.31}$$

s.t.

$$y_{kl} \geq x_k + x_l - 1, \; k = 1, \ldots, N, \; l = 1, \ldots, k-1, \tag{3.32}$$

$$\sum_{i=1}^{N} x_i = L, \tag{3.33}$$

$$x_i \in \{0, 1\}, \; i = 1, \ldots, N, \tag{3.34}$$

$$y_{k,l} \geq 0, \ k = 1, \ldots, N, \ l = 1, \ldots, k-1, \tag{3.35}$$

where $L \in \{n_{\min}, \ldots, N\}$ and the coefficients of objective function are:

$$C_0 = \sum_{r=1}^{R} \sum_{t=1}^{T} p_{rt}^2 W_{rt},$$

$$C_i = \frac{1}{L^2} \sum_{r=1}^{R} \sum_{t=1}^{T} (c_{it}^2 - 2L c_{it} p_{rt}) W_{rt},$$

$$B_{kl} = \frac{2}{L^2} \sum_{r=1}^{R} \sum_{t=1}^{T} c_{kt} c_{lt} W_{rt}.$$

Indeed, if one selects the value L which yields the optimum with minimal objective function among all problems (3.31)–(3.35) of the family, it will be the optimal solution to problem (3.21)–(3.25) as well.

An important property of this MIP formulation is that now the time dimension T and the total number of market participants R do not influence the dimensionality of the model because these parameters are excluded from consideration at the stage of computing the coefficients C_i and B_{kl}. This fact becomes important, e.g. when the historical data consist of the nodal prices of all hours of the previous year. Taking into account that when the number of variables is bounded, the MIP problems fall into the class of polynomially solvable problems [23], we conclude that the Single Hub Selection Problem is also polynomially solvable, if the number of nodes is bounded above by a constant.

3.5.1 Genetic Algorithm

Genetic algorithm (GA) originally proposed by Holland [18] is a random search method that models a process of evolving a population of individuals. Each individual corresponds to some solution of the problem (feasible or maybe infeasible) and it is characterized by the *fitness*, reflecting the goal function value and satisfaction of problem constraints. The higher is the fitness value, the more chances are given for the individual to be selected as a parent. New individuals are built by means of *crossover* and *mutation* procedures. The crossover procedure $Cross$ produces the offspring from two parent individuals by combining and exchanging their elements. The mutation procedure Mut adds small random changes to an individual. The size of population K is kept constant throughout the run of a GA. A detailed description of the GAs and their properties may be found, e.g. in [28].

For solving the Single Hub Selection Problem we use the *binary representation* of solutions in the GA, i.e. an individual in our case is a string g which coincides with the Boolean N-dimensional vector x. The fitness of individual is inversely proportional to the objective function value. Parent genotypes are selected by *s-tournament* selection operator: choose s individuals from the population at random and return the best of them (by default in this section "random" means random with uniform distribution). This selection operator is used to choose each of the two parents independently. New individuals are produced by the *2-point* crossover operator, which chooses randomly two breakpoints in parent genotypes and exchanges all bits in the middle part. The standard mutation inverses each bit independently with a fixed probability p_m. If after crossover and mutation the obtained genotype contains less than $n_{\min}$ ones then a repair procedure is applied. This procedure simply adds more ones to the child individuals at random positions. The overall scheme of the GA used here is as follows:

Genetic Algorithm

1. Generate K random genotypes and add into the initial population.
2. While the termination condition is not met, do
   2.1. Choose the parent genotypes g_u, g_v by s-tournament selection
   2.2. Produce g, h from g_u and g_v
        using 2-point crossover with probability p_c
        otherwise assign $g = g_u$, $h = g_v$
   2.3. Mutate each gene of g and h with probability p_m
   2.4. Apply repair procedure to g and h
   2.5. Choose two individuals of least fitness in the current population
        and substitute them by g and h, if they have greater fitness.
3. Return the best found solution as a result.

3.5.2 Experiments with the GA and CPLEX MIP-Solver

The genetic algorithm is tested here on the hourly electricity prices over 365 days from the day-ahead two-sector electricity market of the European zone of Russia collected in years 2004–2005. First of all, the H-means heuristic is applied to form a set of clusters (preliminary hubs), using identical weights $w_i = 1$, $i = 1, \ldots, n$. The GA is applied to form one hub in each cluster. The problem characteristics and the results are given in Table 3.1. Here the larger instances P3, P4, and P5 correspond to clusters located in Urals-Volga, Urals-Tyumen, and Center regions accordingly. The smaller ones, P1 and P2, are constructed as random subsets from the cluster of P3. The instances P6, P7, and P8 are based on the same clusters as P3, P4, and P5, but considered over half-year (4,343 h) time horizon. The required minimal number of nodes $n_{\min}$ for the GA is set to $[N/2]$ (here the brackets $[\cdot]$ denote rounding to the nearest integer). In our experiments, we set the following control parameters: $s = 20$, $p_c = 1/2$, $K = 200$ and $p_m = 1/N$. The actual number of nodes L in the computed hubs turns out to be equal or close to $n_{\min}$. This value of L is used in problem formulation (3.31)–(3.35), which is

Table 3.1. Comparison of the GA and CPLEX 11.0

Problem	N	R	T	CPLEX CPU time	CPLEX best sol	CPLEX lo. bound	GA CPU time	GA best sol
P1	15	14	500	15 sec	26.97	26.97	<1 s	26.97
P2	25	14	500	5 min	22.97	−8641	<1 s	22.95
P3	82	14	500	5 min	24.53	−69230	10 s	23.24
P4	118	15	500	5 min	23.8	−62137	30 s	23.53
P5	336	69	500	10 min	16.19	−72755	2 min	14.46
P6	82	14	4343	30 min	196.55	−519978	5 min	146.28
P7	118	15	4343	60 min	389.80	−545399	5 min	364.25
P8	336	69	4343	–	–	–	10 min	208.00

also solved by CPLEX 11.0. The amount of CPU time at Celeron 2.8 GHz is indicated in the table as well. After this time both algorithms are terminated.

The results show a clear advantage of the GA in terms of the running time and the solution quality. The lower bounds obtained by CPLEX in the given amount of time are negative in most of the cases and they cannot be of practical use. For problem P8 CPLEX fails because of memory limitation.

Evaluation and comparison of the obtained hubs.

In the clusters corresponding to P6, P7, and P8 the following different hubs are constructed and compared:

1. A hub constructed by the GA minimizing quadratic objective (3.21).
2. A hub constructed by the GA minimizing linear objective

$$\sum_{t=1}^{T}\sum_{r=1}^{R}|c_t - p_{rt}|W_{rt}. \tag{3.36}$$

3. A hub constructed by the GA *maximizing* linear objective (3.36). This hub gives a worst case in a linear model (for the sake of comparison only).
4. A hub containing all nodes of a cluster.
5. A hub containing a randomly chosen subset of nodes of a cluster (every node is included independently of the other nodes with probability 0.5).

The set of conditions (3.23)–(3.25) is never changed. The comparison is illustrated by Fig. 3.3. Here each hub is represented by a point on a plane where X and Y axes correspond to the values of linear and quadratic objective functions. The results show that the optimized hubs are far from the worst case hub in terms of both criteria and not so much distant from each other. In the cases of P7 and P8, the hub optimized with respect to linear objective (3.36) has even greater value of quadratic objective (3.21) than the hub consisting of all nodes of the cluster. This indicates that the choice of optimization criterion is important and it should be adjusted to the interests of participants.

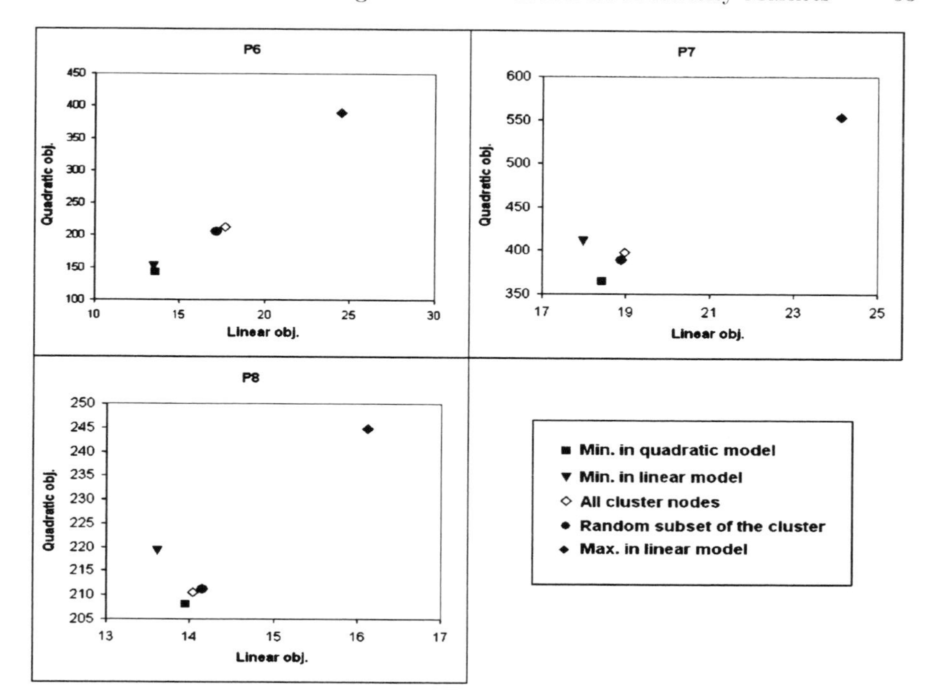

Fig. 3.3. Evaluation and comparison of different solutions to single hub selection problem

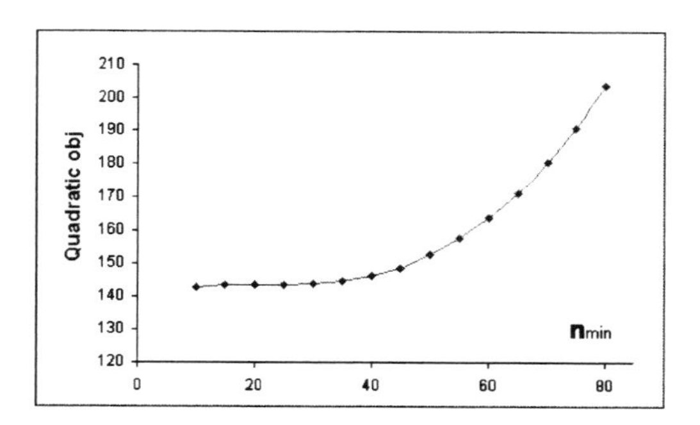

Fig. 3.4. Trade-off between the number of nodes in hub and the objective value

Figure 3.3 implies that some proper subsets of the cluster can constitute better hubs than the set of all nodes. The question about the trade-off between the number of nodes in hub and the attainable objective value is addressed in the next experiment, where we ran the GA minimizing quadratic objective separately for different values of $n_{\min}$. The results are shown on Fig. 3.4. One can see that a fast growth begins near the point $n_{\min} = 40$, so this setting is a plausible candidate to form a hub. In practice, a graph of such type can provide useful information for the decision-maker, who needs to choose a sufficiently large value of $n_{\min}$ without significant compromise in distance between the hub price and the locational prices of participants.

3.6 Conclusion

We have considered the hubs construction problem from the optimizational prospective. Our analysis of this problem indicated that even though its connections with clustering problems allow to apply the well-known clustering methods, some important constraints and criteria do not necessarily fit into the clustering framework. In view of large dimensionality of typical instances of the Hubs Construction Problem, further development of the specialized optimization algorithms is important to support decision making.

A set of heuristic algorithms, we have considered, allows to find practically useful solutions. Even if the clustering algorithm does not yield an acceptable solution for most of the market participants, each of the obtained hubs can be further refined by solving a Single Hub Selection Problem. A genetic heuristic is shown to be suitable for finding approximate solutions to this problem with different criteria.

An important direction for further research is evaluation of the Principal Component Analysis and its comparison to the optimization-based methods described in this chapter. In some situations the hubs need to be defined even before the new electricity market opens. The statistical data on LMP prices is absent in such cases and one has to construct the hubs using some kind of market simulation and/or the statistical data describing the transmission of the electric power in the grid. This presents another challenge for research. The issues of hubs construction in view of negative influence of the market power require a careful consideration as well.

Acknowledgement. The research is partially supported by Russian Foundation for Basic Research grant 07-01-00410 and Administrator of Trade System for United Energy System of Russia. The authors are grateful to A. Hartshorn, P. Hansen, W. Hogan, A. Komissarov, A. Ott and J. Popova for their comments, notes or helpful discussions. Also, we thank S. Alekseev and D. Tartynov for programming and technical support.

References

1. D. Aloise and P. Hansen. On the complexity of minimum sum-of-squares clustering. Les Cahiers du GERAD, G-2007-50, GERAD, Montréal, 2007
2. E. R. Braziel. Trading hubs: Where power is traded and why. PMA OnLine Magazine 12, 1998
3. X. Cai. *An evaluation of investment incentives for maintaining system adequacy in deregulated markets for electricity.* PhD thesis, Cornell University, 2005

4. J. F. Campbell, A. Ernst, and M. Krishnamoorthy. *Location analysis: Theory and applications*, chapter Hub location problems. Springer, Berlin, 2002
5. M. P. Davidson, U. V. Dogadushkina, E. M. Kreines, N. M. Novikova, U. A. Udaltsov, and L. V. Shirjaeva. Mathematical model of competitive wholesale electricity market in Russia. *Izvestija Akademii Nauk. Teorija i Sistemi Upravlenija*, 3:72–83, 2004. (in Russian)
6. Energy Information Administration. *Derivatives and risk management in energy industries*, chapter Derivatives and risk management in the petroleum, natural gas, and electricity industries. Report SR-SMG/2002-01, EIA, Washington, DC, 2002
7. P. Falbo and S. Stefani. Use and misuse of price indexes in the electricity markets. Working Papers Dipartamento Metodi Quantitativi Universitá di Milano Bicocca, Milan, 2004
8. FERC Docket No. RM01-12-000. *Summary standard market design notice of proposed rulemaking*, 2002
9. M. R. Garey and D. S. Johnson. *Computers and intractability. A guide to the theory of NP-completeness.* Freeman, San Francisco, 1979
10. A. Gersho and R. Gray. *Vector quantization and signal compression.* Kluwer, Norwell, MA, 1992
11. P. Hansen and B. Jaumard. Cluster analysis and mathematical programming. *Math Progr*, 79:191–215, 1997
12. P. Hansen and N. Mladenović. J-means: A new local search heuristic for minimum sum-of-squares clustering. *Pattern Recogn*, 34:405–413, 2001
13. A. Hartshorn and S. Chang. *MWISO hubs development.* LECG, LLC, Cambridge, MA, 2003
14. S. M. Harvey, A. P. Hartshorn, D. Bertagnolli, and R. Kowalski. Derivation of a trading hub in the New England Market. NEPOOL Joint CMS/MSS Group, 1999
15. W. W. Hogan. Contract networks for electric power transmission. *J Regul Econom*, 4(3):211–242, 1992
16. W. W. Hogan. *Private communication*, 1998
17. W. W. Hogan. Financial transmission right formulations. Working Paper, Belfer Center for Business and Government, Cambridge, MA, 2002
18. J. H. Holland. Adaptation in natural and artificial systems. University of Michigan Press, Ann Arbor, 1975
19. R.Horst, P. M. Pardalos, and N. V. Thoai. *Introduction to global optimization*, volume 3 of *Nonconvex optimization and its applications.* Kluwer, Dordrecht, 1995
20. R. Howard. *Operational research in the social sciences*, chapter Classifying a population into homogeneous groups. Tavistock, London, 1966
21. Hub Analysis Working Group NEPOOL Markets Committee. NEPOOL energy market hub white paper and proposal. October 2003
22. R. C. Jancey. Multidimensional group analysis. *Aust J Bot*, 14:127–130, 1966
23. H. W. Lenstra Jr. Integer programming with a fixed number of variables. *Math Oper Res*, 8:538–548, 1983
24. J. MacQueen. Some methods for classification and analysis of multivariate observations. In *Proc. of the 5-th Berkeley Symposium on Mathematical Statistics and Probability*, volume 2, pages 281–297, 1967
25. A. Ott. *Private communication*, 2006

26. A. Philpott and G. Pritchard. Financial transmission rights in convex pool markets. *Oper Res Lett*, 32:109–113, 2004
27. J. Popova. Spatial pattern in modeling electricity prices: evidence from the PJM market. In *Proc. of 24th USAEE and IAEE North American Conference, July 8–10*, Washington, DC, 2004
28. C. Reeves and J. Rowe. *Genetic algorithms – principles and perspectives. A guide to GA theory*. Kluwer, Boston, 2003
29. F. C. Schweppe, M. C. Caramanis, R. D. Tabors, and R. E. Bohn. *Spot pricing of electricity*. Kluwer, Norwell, MA, 1988
30. S. Z. Selim and M. A. Ismail. K-means type of algorithms: A generalized convergence theorem and characterization of local optimality. *IEEE Trans Pattern Anal Machine Intell*, 6(1):81–87, 1984
31. H. A. Shawky, A. Marathe, and C. L. Barrett. A First look at the empirical relation between spot and futures electricity prices in the United States. *J Futures Markets*, 23:931–955, 2003
32. H. Späth. *Cluster analysis algorithms for data reduction and classification of objects*. Ellis Horwood, Chichester, 1980
33. B. Zhang, G. Kleyner, and M. Hsu. A local search approach to K-clustering. Technical report, HP Labs Technical Report HPL-1999-119, 1999
34. W. R. Zwick and W. F. Velicer. Comparison of five rules for determining the number of components to retain. *Psychol Bull*, 99:432–442, 1986

Appendix

Proof of Proposition 2. The proof is by reduction from the following NP-complete version of Partition problem (see Appendix 3.2 in [9]): Given M integers $a_1, \dots, a_M$, recognize the existence of such subset $I \subseteq \{1, \dots, M\}$ of cardinality $M/2$ that $\sum_{i \in I} a_i = \frac{1}{2} \sum_{i=1}^{M} a_i$.

Given a set of integers $a_1, \dots, a_M$, we construct an instance of the Single Hub Selection Problem with $T = 2$, $N = M + 2$, and $n_{\min} = M/2 + 1$.

We assign the data of hour 1 in such a way that any optimal solution H^* to the Single Hub Selection Problem (1) will consist of $M/2+1$ nodes, (2) it will contain the node number $M+1$ and (3) it will not contain the node number $M+2$. To this end we put $W_{i,1} = 0$, $i = 1, \dots, M$; $W_{M+1,1} = 1$, $W_{M+2,1} = 0$; $c_{i,1} = K$, $i = 1, \dots, M$; $c_{M+1,1} = 0$, $c_{M+2,1} = 2K$, where the parameter K is sufficiently large (its value will be chosen later). Note that with these assumptions the price of a hub H at hour 1 will be

$$c_1 = \begin{cases} K, & \text{if } M+1 \notin H,\ M+2 \notin H; \\ K(1 - 1/|H|), & \text{if } M+1 \in H,\ M+2 \notin H; \\ K(1 + 1/|H|), & \text{if } M+1 \notin H,\ M+2 \in H; \\ K, & \text{if } M+1 \in H,\ M+2 \in H. \end{cases} \tag{3.37}$$

At the same time, the only non-zero term, associated with hour 1 in sum (3.21) equals $(c_1)^2$. Thus, if K is sufficiently large and the input data for hour 2 does not depend on K, then the optimal hub will always meet conditions (2), (3) and involve the minimal admissible number of nodes, i.e. condition (1) holds as well.

Now we proceed to the input data of hour 2, which will ensure equivalence of (3.21)–(3.25) to the given Partition problem, assuming that conditions (1)–(3)

are satisfied. Let $c_{i,2} = a_i(M/2+1)$, $i = 1, \ldots, M$; $c_{M+1,2} = 0$, $c_{M+2,2} = \frac{1}{2}\sum_{i=1}^{M} a_i$. Then in hour 2 the hub price for an optimal hub H^* will be $c_2 = \sum_{i \in H^*} a_i$. Finally, assign $W_{i,2} = 0$, $i = 1, \ldots, M+1$ and $W_{M+2,2} = 1$. It is clear that to ensure (1)–(3) it is sufficient to assign $K = M \sum_{i=1}^{M} a_i$.

Note that the only non-zero summand in (3.21) at hour $t = 2$ is $(c_2 - c_{M+2,2})^2$, which attains the minimum equal to 0 if and only if $\sum_{i \leq M,\ i \in H^*} a_i = \frac{1}{2}\sum_{i=1}^{M} a_i$. So, the optimal value of criterion (3.21) is equal to $K(1 - 2/(M+2))$ if and only if there exists the set I required in the Partition problem. Thus, we have reduced an NP-complete problem to the Single Hub Selection Problem, and the reduction can be computed in polynomial time. □

4

A Decision Support System to Analyze the Influence of Distributed Generation in Energy Distribution Networks

J.N. Fidalgo, Dalila B.M.M. Fontes, and Susana Silva

Summary. Recent changes in electric network infrastructure and government policies have created opportunities for the employment of distributed generation to achieve a variety of benefits. In this paper we propose a decisions support system to assess some of the technical benefits, namely: (1) voltage profile improvement; (2) power losses reduction; and (3) network capacity investment deferral, brought through branches congestion reduction. The simulation platform incorporates the classical Newton–Raphson algorithm to solve the power flow equations. Simulation results are given for a real Semiurban medium voltage network, considering different load scenarios (Summer, Winter, Valley, Peak and In Between Hours), different levels of microgeneration penetration, and different location distributions for the microgeneration units.

4.1 Introduction

Several benefits can be achieve by integrating Distributed Generation (DG) with utility networks. These benefits should be clearly understood, analyzed, and quantified in order to increase the potential and value of DG penetration. The benefits of DG have been evaluated and quantified in terms of capacity credit, energy value, and energy cost saving [17, 20]. DG is expected to play a major role in future power systems, since it is able to reduce transmission losses, improve power quality to end users, and smooth peaks in demand patterns.

Besides that, another main driver for DG penetration growth is the development of new renewable DG (wind turbines, photovoltaic, biomass, etc.) and some fossil DG that have combined heat and power capabilities. The European Union (EU) Commission has set a target of 12% by 2010 for microgeneration integration in Low Voltage (LV) and Medium Voltage (MV) networks. Proliferation of renewable energy sources is being encouraged in order to progress towards the Kyoto agreement. Bearing in mind this agreement and considering each country specificities, EU proposed in September 2001 the Directive

077/CE/2001 [19], concerning the electricity production based on renewables. This directive sets the goal of renewable energy production by 2010 to 22%. The same directive has set a goal of 39% for Portugal. These renewable energy targets are supposed to be partially fulfilled by DG.

Employment of DG in existing systems can cause several potential operating conflicts such as voltage flicker, misoperation of protection, and reverse power flow [2, 8, 13]. These affect network operation and planning practices with economic implications [14, 15]. However, many benefits, both economic and technical, can be achieved [3, 4]. Research has shown that DG has generally a beneficial impact in the power networks, contributing positively to losses decrease, enhanced voltage profiles, and retarding branches congestion [6, 7, 9, 16, 23]. Active power losses decrease significantly with the growth of DG penetration. The same applies to network congestion issues – load can be notably reduced in the branches with microgeneration[1]. Another advantage of DG is the capture of intermittent and peaking loads in residential and commercial cogeneration [18].

Recently, it has been shown that DG units location can have a significant impact on power systems performance [1,10]. In fact, the location of DG plays a vital role in improving the voltage profile and in reducing power losses [7]. Results have also shown a higher improvement in voltage profiles when DG is installed closer to the higher loads [6]. However, when DG is distributed, in LV networks the voltage and power flow constraints are more relieved than when DG is concentrated in a few number of buses [23].

The impact of DG on investment deferral has not yet been the subject of much research. For a discussion on the potential of DG to defer investments, see, e.g., [24]. In a recent work, Gil and Jöos [11] proposed an approach to quantify the value of the capability to defer planned or required investments in wires and transformers for the distribution network.

The present work was developed within the framework of a research project that intends to analyze the interaction dynamics of the DG actors, whose actions are motivated by their own profit but also subject to the regulatory directives. Diverse exploring situations have been simulated and the effects of the different parameters were evaluated (the DG location was one of them). Under this perspective, this work is better seen as a decision aid tool, i.e., a Decision Support System (DSS), to analyze some technical benefits of DG. The DSS is supported by a simulation platform that allows for the simulation of different exploration scenarios, different DG locations, and different levels of DG penetration. The impact on network performance is evaluated by assessing three major technical benefits, namely voltage profile improvement, power losses reduction, and branch congestion reduction. The investment deferral is evaluated through the branch congestion reduction.

[1] Here and hereafter microgeneration and DG are to be used interchangeably.

The paper is organized as follows: Sect. 4.2 presents the proposed methodology. In Sect. 4.3 the simulation parameters and the distribution system are given. Results and conclusions are presented in Sects. 4.4 and 4.5, respectively.

4.2 Methodology

Technical benefits of introducing DG can be divided in two broad categories:

- Improvement of a certain attribute such as voltage profile, reliability, power quality, etc.
- Reduction of an attribute such as power losses, emissions, congestion, etc.

In this work we are concerned with the benefits obtained from voltage profile improvements, reduction in power losses, and reduction in branches congestion. We also study the impact of DG locations on these benefits.

One of the justifications for introducing DG is to improve the voltage profile. This happens since DG can provide a portion of the real and reactive power to the load, thus helping to decrease current along a section of the distribution line, which in turn, will result in a boost in the voltage magnitude at the customer site [5]. In this study, the voltage profile analysis is performed through the calculus of the mean voltage in the network buses.

Other major potential benefit offered by DG is the reduction in power losses. The losses can be significant under heavy load conditions. The utility is forced to pass the cost of power losses to all customers in terms of higher energy costs. With the inclusion of DG, power flows are reduced leading to a power loss reduction in the distribution system. However, depending on the ratings and locations of DG units, it is possible to have an increase in power losses at very high (and unrealistic) penetration levels, also shown in our results. For the purpose of this study we consider the active power losses and also the total energy losses.

Another benefit is brought through the reduction in branches congestion since this, typically, allows to defer investment in network capacity. For the purpose of this study we analyze the DG impact on the most loaded branch (relatively to its capacity), in order to pinpoint the most sensitive section. The analysis is completed by estimating by how many years lines upgrade may be postponed due to the load reduction accomplished.

A Decision Support System was developed within the scope of a project called GENEDIS (acronym of distributed generation). This DSS was developed and implemented on an AMD Athlon^{TM} 208 GHz (1.6 GB Ram DDR 400 MHz) PC. The JAVA programming language (JDK 1.2) was used for creating a customized, user-friendly graphical interface.

The user interface has three main parts, corresponding to the three major functionalities of GENEDIS:

- **Network Data -** Through which we can load the network from a file. We can also edit the network and change some of its parameters. In particular, we can add, drop, or replace injectors, see Fig. 4.1.
- **Power Flow -** Where we can open a file with results previously obtained, get the solution of the power equations, and visualize the power flow results, as depicted in Fig. 4.2.

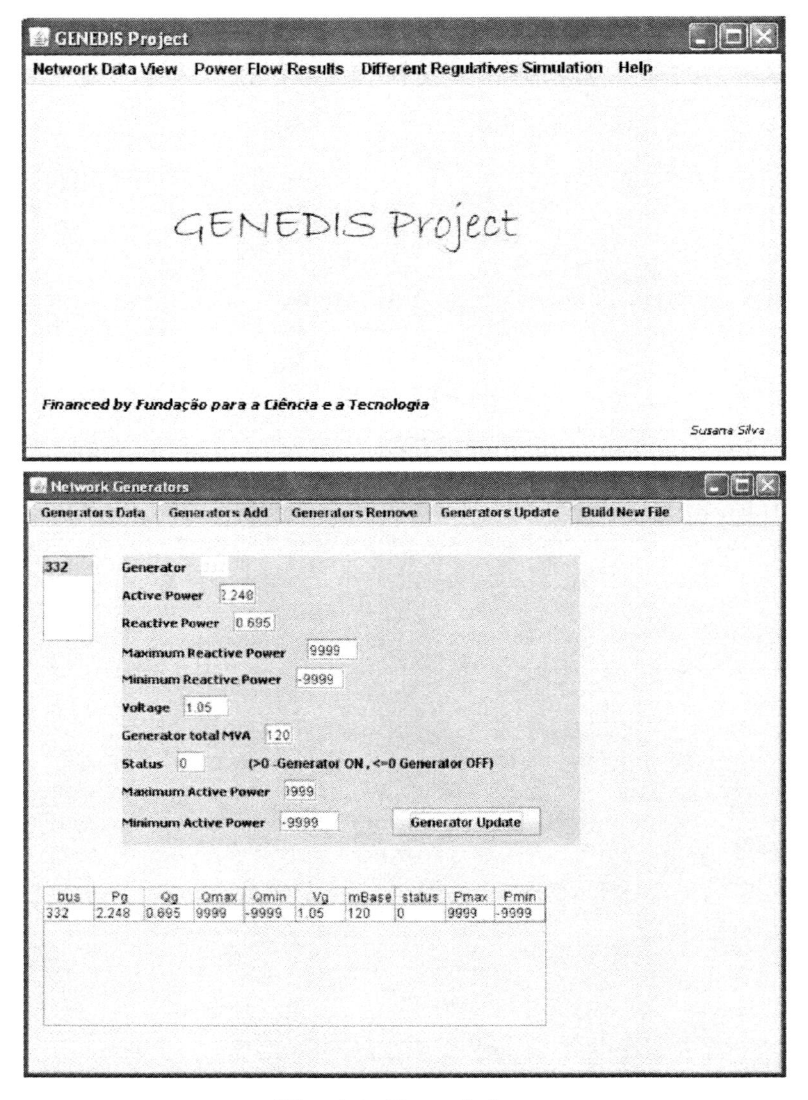

Fig. 4.1. Network data

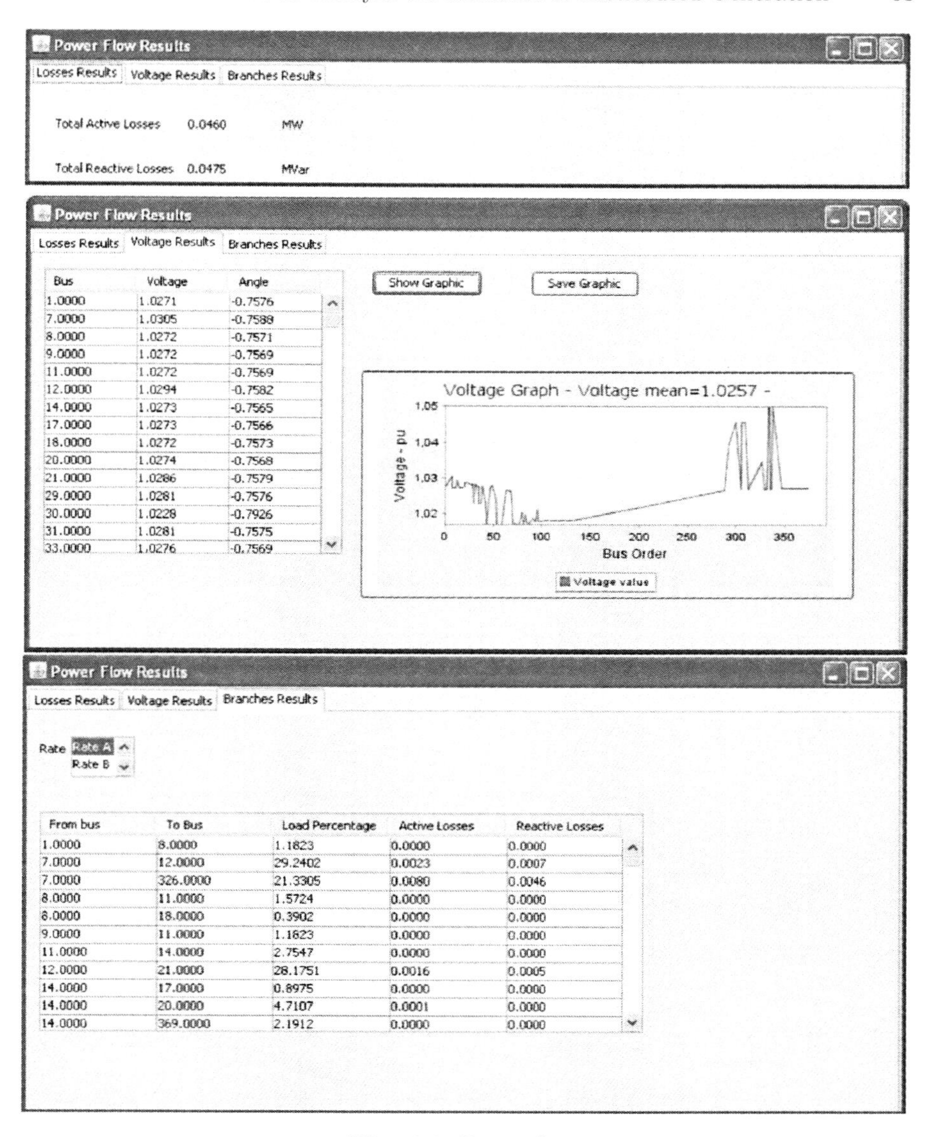

Fig. 4.2. Power flow

- **Simulation -** Where we choose the simulation parameters, that is load type, DG locations, and DG penetration levels, see Fig. 4.3.

The power flow module, that has been implemented in JAVA, uses a library that has been built by using MatLab. The Newton–Raphson algorithm [12, 21] is a classical and well known method, at least within the power system

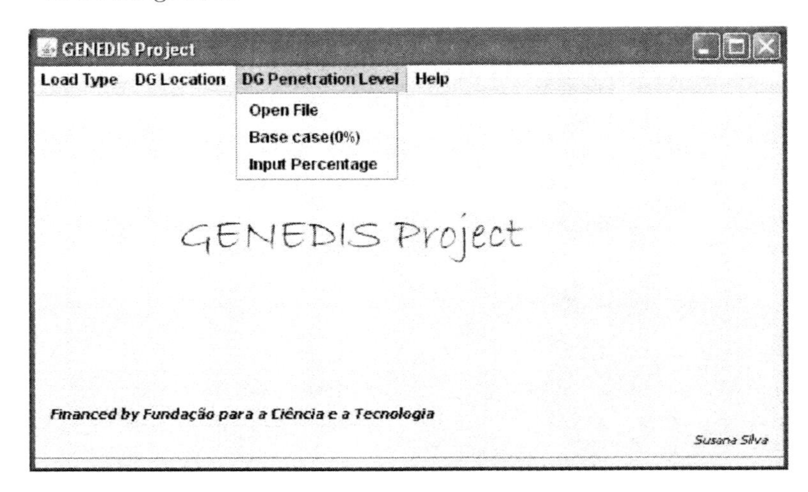

Fig. 4.3. Main simulation menu

community, for solving the power flow problem, a set of simultaneous nonlinear equations). This algorithm was implemented in MatLab and then by using Matlab Builder for Java *deploytool*, the library *jar* was built.

4.3 Simulation Study

The present study is based on the data of a real near-urban MV network. The simulation study considers several DG penetration levels. In order to achieved this we first select a set of buses to be the connection points for the DG machines. Then, in these buses the active power is decreased to several different values in order to simulate different penetration levels.

The simulation was performed for the following load scenarios:

1. Peak Hours of Winter (PHW)
2. Peak Hours of Summer (PHS)
3. In Between Values of Summer (BVS)
4. In Between Values of Winter (BVW)
5. Valley Hours of Summer (VHS)

As previously said we also study the impact of DG location, this is accomplished by considering four types of DG location distribution, as follows:

1. **Near -** DG machines are located close to the substation
2. **Far -** DG machines are located far away from the substation
3. **Random -** DG machine locations are randomly selected

Regarding the random location distribution we have generated two such scenarios since, as it can be seen in the results section, the power quality indexes may be rather different.

The network data is organized under a node-to-node scheme in order to allow a proper identification of every branch in the path connecting each pair of nodes. This data is, by default, read from a text file that contains all the information required for a full characterization of the entire network under simulation, such as network topology, lines and loads data, etc. Alternatively, the user may change manually any parameter using the DSS graphical interface.

For each exploration state, several settings for the power factor were considered ($tg = 0.0, 0.1, 0.6$), although the results reported in this paper refer only to the nominal case $tg = 0.4$. For full results the reader is referred to [22].

The DG production effect is simulated through the reduction of the value of the load (active power) in the corresponding bus bar. The value of the reduction of the active power, for the four DG location distributions, was chosen such that the percentage penetration of DG ranges from 0% up to 100%. For the lower limit no DG exists, while for the upper limit all loads are fed by DG machines and no power is supplied to by the substations. Although very high penetration levels are highly improbable, for the sake of completeness we also include them.

The percentage penetration value is computed as the ratio between the variation on the total load value, implied by the reduction in active power, and the value of the total load before the active power reduction.

4.3.1 Network General Characterization

The Semiurban MV network used in this study has 372 load nodes and three injectors of 15 kV that feed three (usually) independent sectors. The areas fed by different injectors were denominated by Area 1 (fed by node 332), Area 2 (fed by node 158) and Area 3 (fed by the node 267), see Fig. 4.4.

4.3.2 Algorithm

As referred to previously, the impact of DG is simulated by performing a reduction on the active power consumption in the load nodes where micro generators are connected. Network simulation and analysis is then performed throughout the following steps, which have been implemented by using PSS/E.

```
Set initial conditions:
(Base case: DG penetration is zero.)
For each typical scenario (PHS, PHW, BVS, BVW and VHS):
   For each type of DG location (Near, Far, Random1, Random2)
      Repeat until DG penetration ratio reaches 100%
         1. Run load flow (Newton--Raphson algorithm)
            obtaining network voltage profile
         2. Evaluate total losses and branches congestion
         3. Increase DG penetration ratio
```

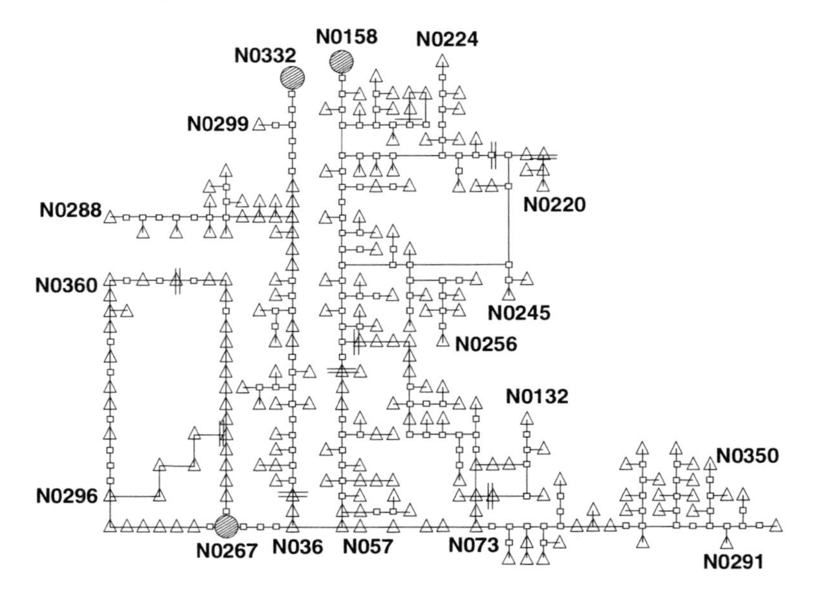

Fig. 4.4. Semiurban medium voltage network

This set of conditions provides an implicit characterization impact of DG in the network for the four different location distribution possibilities, parameterized in terms of DG penetration, for the exploration scenarios considered.

4.4 Computational Results

In this section we report on the results obtained in the tests performed. As referred to previously, the objective is to characterize the technical influence of DG in the distribution network in terms of power losses, voltage profiles, and branches congestion.

4.4.1 Active Power Losses

Figures 4.5 and 4.6 summarize the DG impact on the active power losses of the studied network. From these figures it can be seen that DG generally contributes to a decrease in reactive power losses. In fact, the power losses decrease with DG penetration for all the cases, at least for a penetration level of about 50%. For higher penetration level values, in the Near case, the losses start to increase, indicating power flow reverse in some branches.

Regarding the DG location distributions Random1 and Random2 the power losses always decrease. A possible interpretation for the difference in losses behavior may be that in the random cases, the nodes with DG are spread along the network while in the other cases the DG locations are constrained to be near to or far from the feeders. This way, the nodes with DG

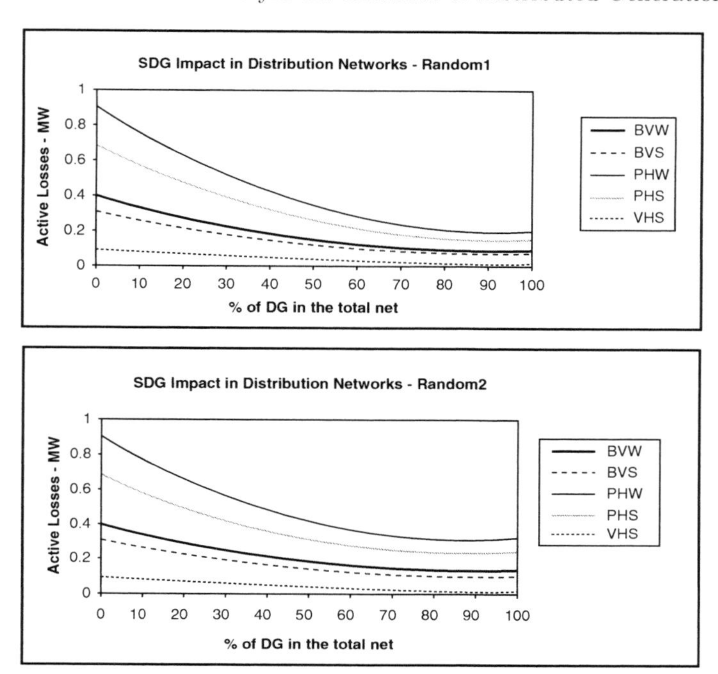

Fig. 4.5. Active power losses for cases Random1 and Random2, respectively

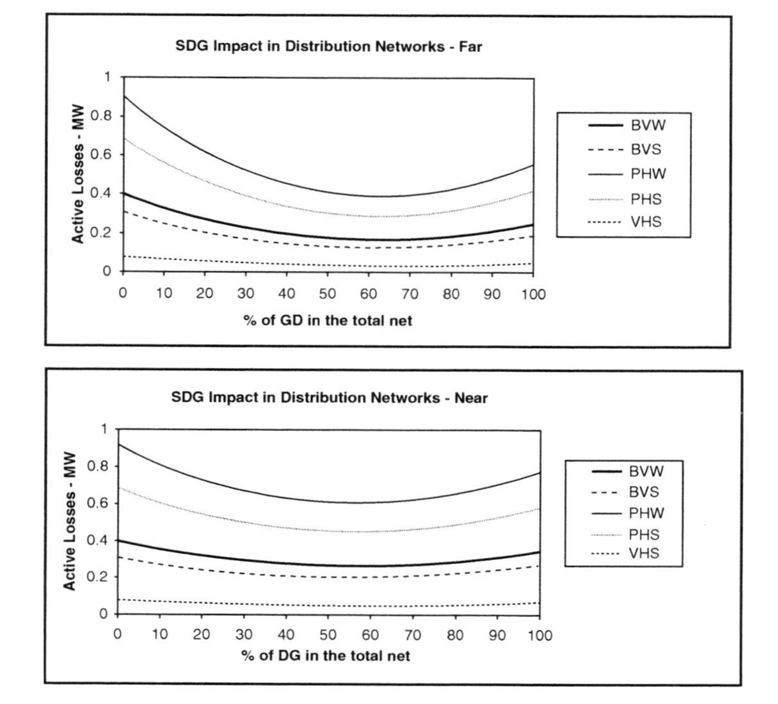

Fig. 4.6. Active power losses for the Far case and the Near case, respectively

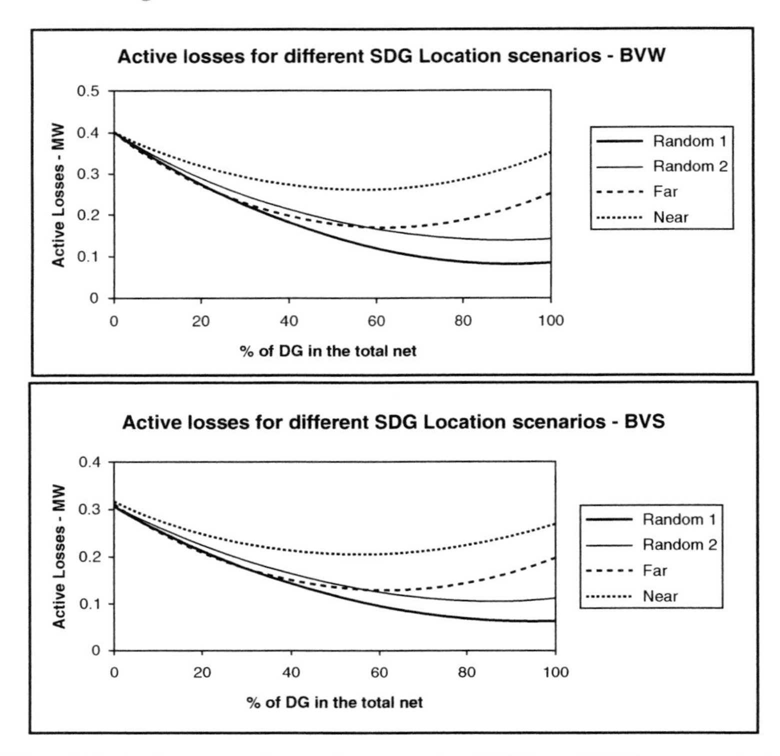

Fig. 4.7. Active power losses for scenarios BVW and BVS, respectively

are obviously close to each other, therefore concentrated in just a few network branches. So for high DG penetration levels, the microgeneration production will have a thorough impact in the branches close to the DG nodes.

Figures 4.7 and 4.8 show the active power losses for the different load scenarios considered. Although these figures contain the same information than previous ones, in this case the effect of location is much more obvious. The main conclusion to be drawn from these results, is that for penetration levels up to approximately 30%, the further away the micro generators are from the substations, the larger is the power loss reduction compared to the base case (DG penetration of 0%).

Table 4.1 provides a summary of the annual energetic losses, while Table 4.2 reports on the percentage gains[2] achieved on energy losses.

4.4.2 Voltage Profile

In this study, voltage values were computed for all nodes and for each load scenario and each DG location distribution. For the sake of simplicity and

[2] These values have been computed as $1 - \frac{loss(0\%GD)}{loss(10/20\%GD)}$.

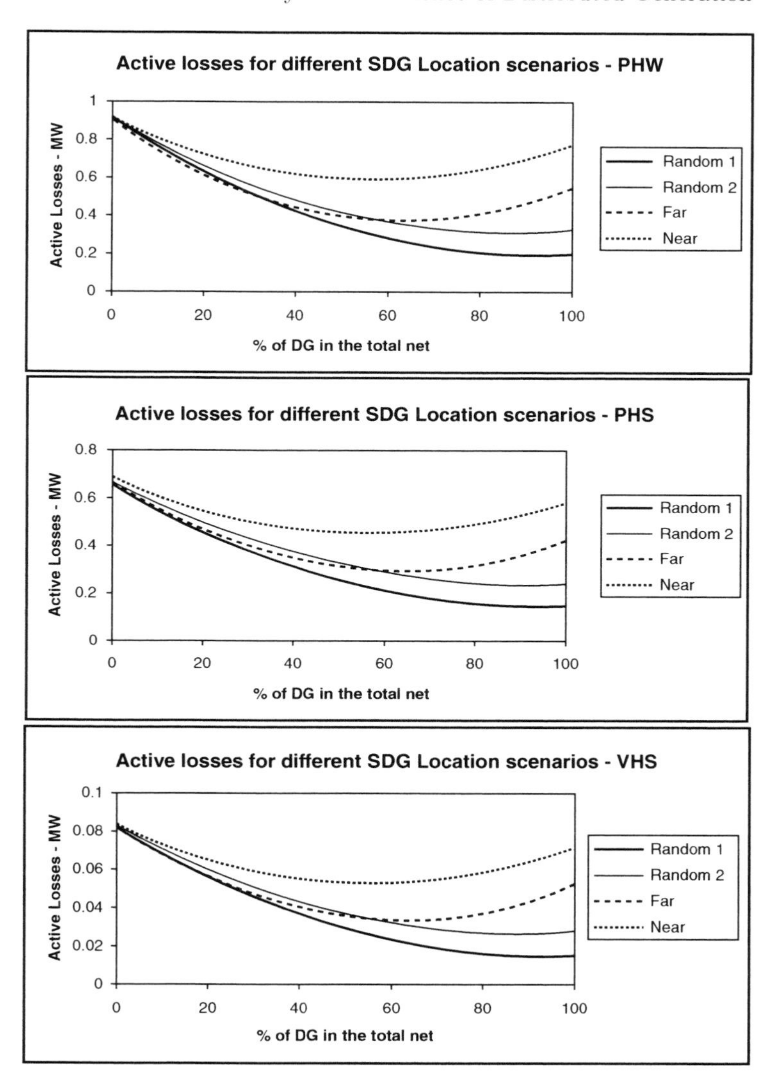

Fig. 4.8. Active power losses for scenarios PHW, PHS, and VHS respectively

Table 4.1. Annual energy losses, in MWh

DG (%)	Random1	Random2	Near	Far	Average
0	2,577.6	2,577.6	2,577.6	2,577.6	2,577.6
10	2,145.7	2,198.1	2,280.2	2,117.6	2,185.4
20	1,773.1	1,869.4	2,118.4	1,750.0	1,877.7

Table 4.2. Percentage annual energy loss gains

DG (%)	Random1	Random2	Near	Far	Average
10	16.8	14.7	11.5	17.9	15.2
20	31.2	27.5	17.8	32.1	27.2

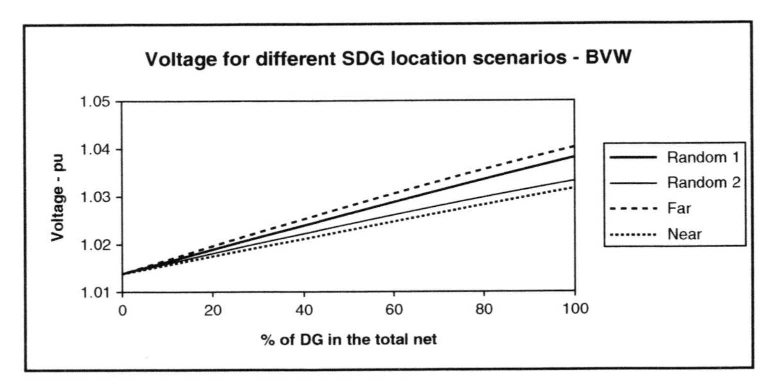

Fig. 4.9. Mean bus voltage vs DG penetration for scenario BVW

paper readability, we have decided to calculated and report only on the mean bus voltage. Figure 4.9 shows the variation of the mean bus voltage, for scenario BVW, as a function of DG penetration for each of the four DG location distributions considered. Figure 4.10 replicates this study for the other load scenarios.

The analysis of Figs. 4.9 and 4.10 shows that:

1. The mean voltage increases with DG penetration, as expected since DG production corresponds to a load decrease in the DG nodes
2. Mean voltage values are higher when the DG location distribution is far from the feeding nodes

Therefore, it may be conclude that, generally, DG producers located further away from the network substations have a shaper contribution towards voltage preservation. This is an important feature, especially for heavily loaded networks susceptible to the voltage collapse phenomenon.

4.4.3 Branches Congestion

Power distribution networks are usually planned according to a given forecasted load evolution scenario. However, evidence shows that the spatial growth of loads often mismatches the expected development. In such cases, some network branches may be operating close to capacity, resulting in high power losses or, in the worst case, in the actuation of protective overload relays and consequently in load shedding.

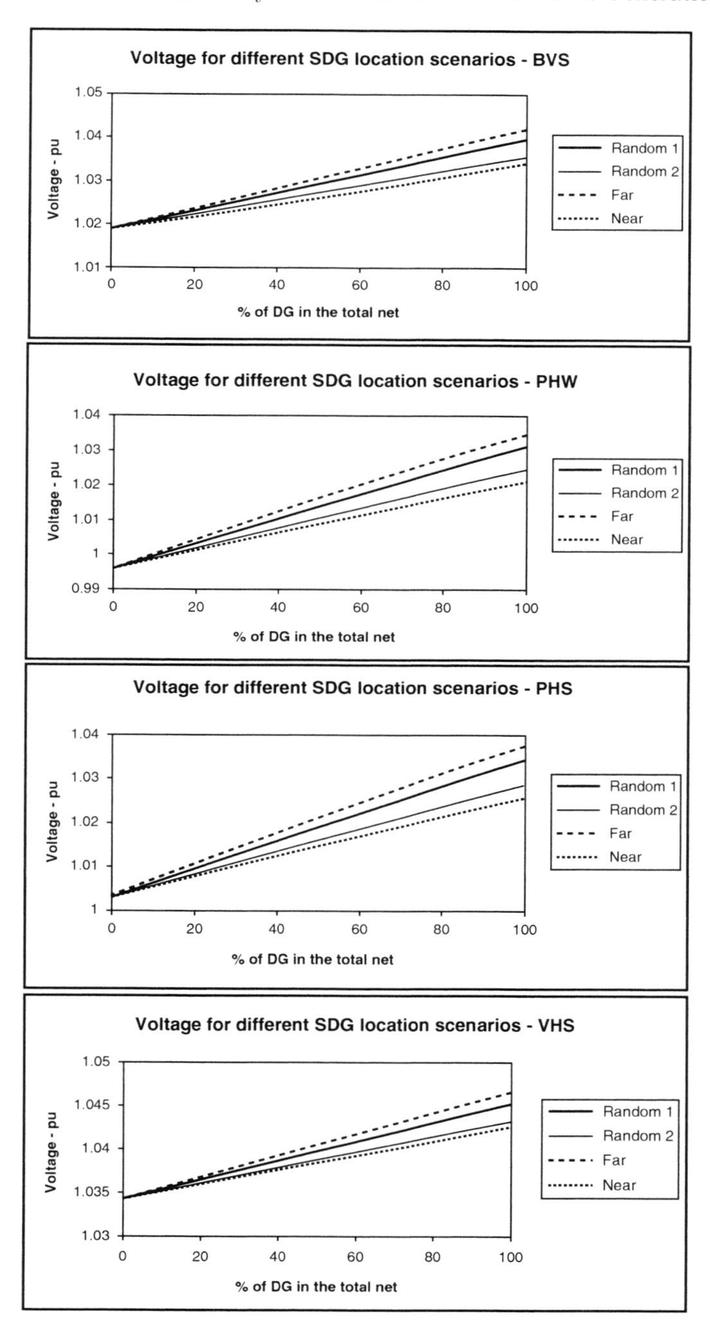

Fig. 4.10. Mean bus voltage vs DG penetration for scenarios BVS, PHW, PHS and VHS, respectively

In this work, we have started by detecting, for each load scenario, the heaviest loaded branch, see Table 4.3. DG may or may not contribute to relieve the most loaded branches depending on whether the flows of these branches are feeding the nodes where DG machines are connected to or not.

For the Near case, the most loaded branch is not "in the way" of any DG generator. Therefore, as it can be seen in Figs. 4.11 to 4.15, no effect is outputted for the Near case. Hence, in this case, the load remains constant regardless of the DG penetration. For the other DG locations, it can be conclude that higher levels of DG penetration lead to higher distress levels of the most loaded branch. It is not possible to draw general conclusions on the eventual benefits of DG generators location being close to or far from the

Table 4.3. The most loaded branch for each loading scenario

Scenario	Branch
BVW	36–57
PHW	36–57
BVS	15–338
PHS	15–338
VHS	15–338

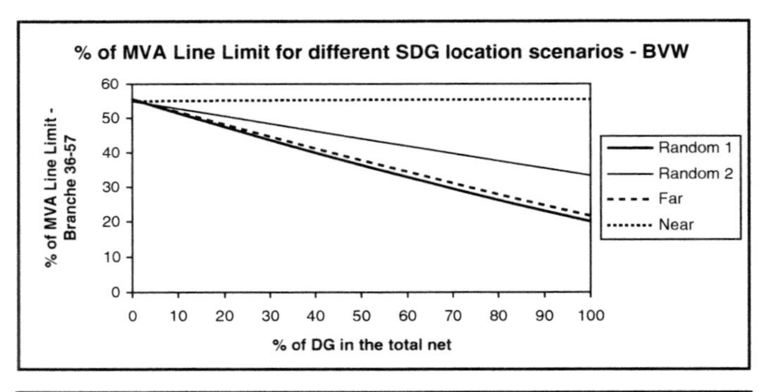

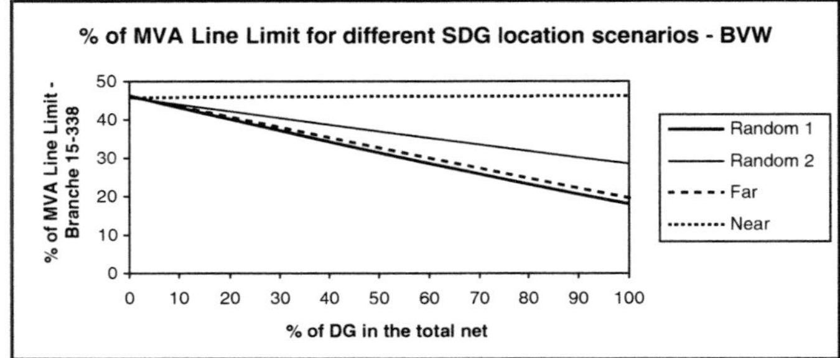

Fig. 4.11. Load charge of branches 36–57 and 15–338, respectively, for scenario BVW

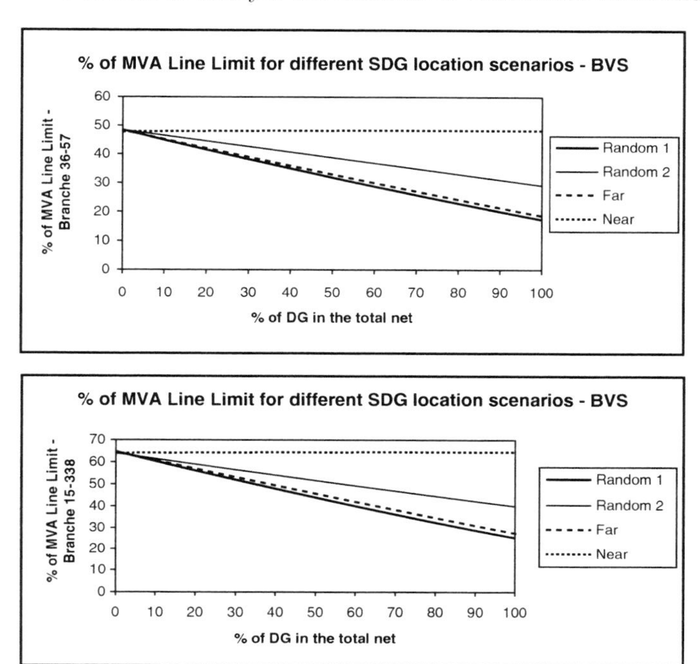

Fig. 4.12. Load charge of branches 36–57 and 15–338, respectively, for scenario BVS

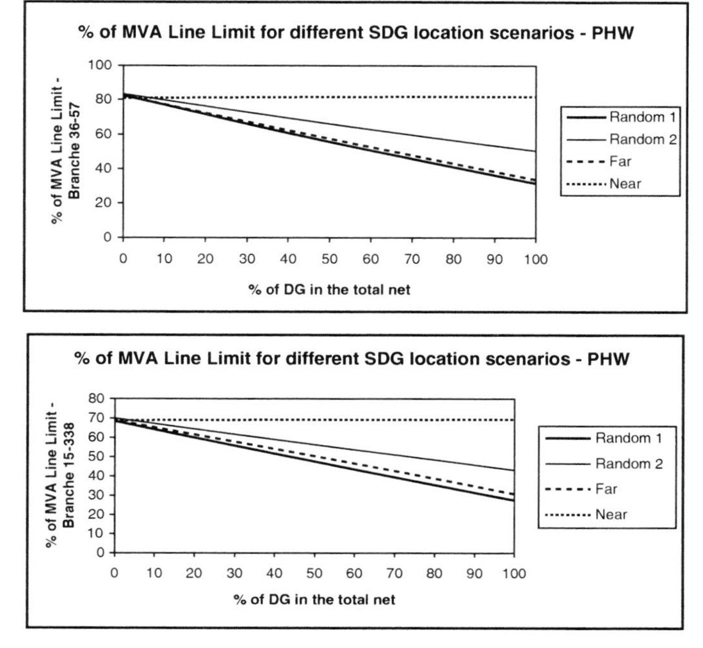

Fig. 4.13. Load charge of branches 36–57 and 15–338, respectively, for scenario PHW

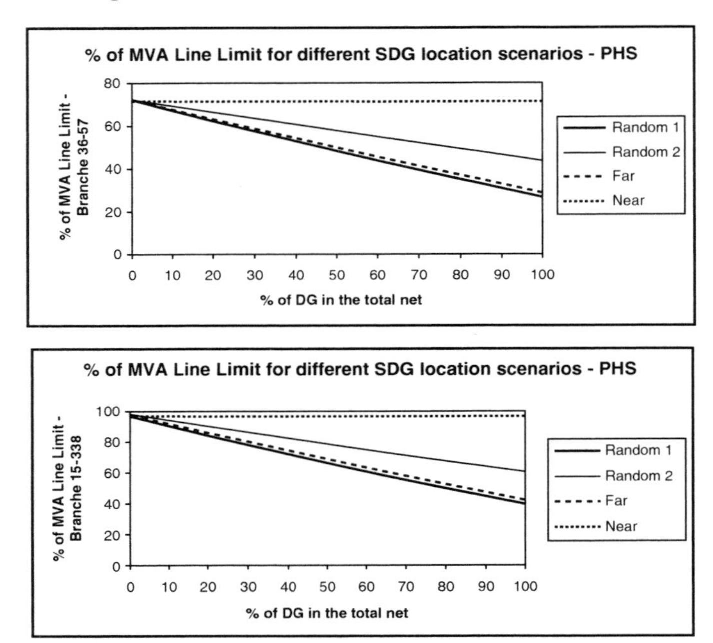

Fig. 4.14. Load charge of branches 36–57 and 15–338, respectively, for scenario PHS

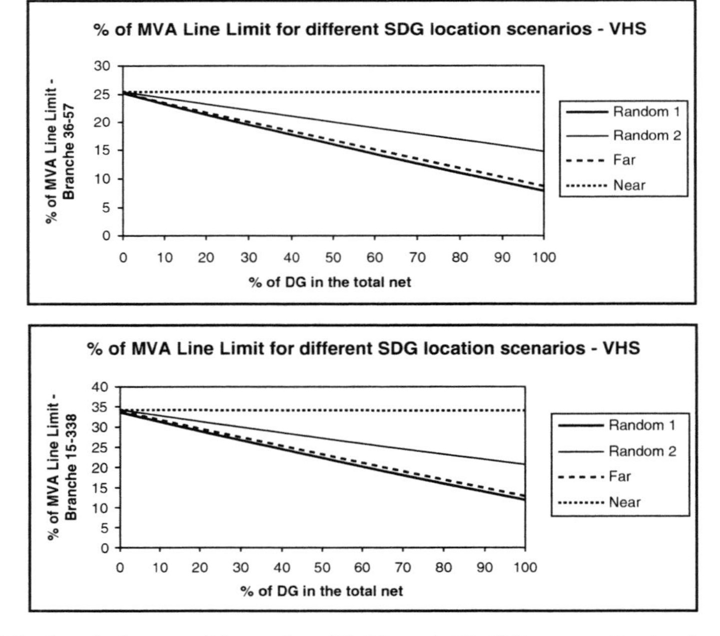

Fig. 4.15. Load charge of branches 36–57 and 15–338, respectively, for scenario VHS

Table 4.4. Years of branch reinforcement investment deferral

DG	Random1	Random2	Far	Near
10%	1	1	1	0
20%	3	2	3	0

feeding nodes. In fact, the changes on load flows in specific branches convey an absolute dependency on the electrical location of both items: DG machines and most loaded branch.

Table 4.4 shows by how many years the investment in branches reinforcement can be deferred. The years reported are for the worst case scenario (PHS) and assume a 5% annual load increase. With the exception of the Near case, it is obvious that DG is economically attractive, since it allows for the postponement of lines reinforcement investment.

4.5 Conclusions

This paper has proposed a decision support system to quantify some of the technical benefits of introducing DG. The application, which may be perceived as a decision aid tool, is supported by a simulation platform that uses the classical Newton–Raphson algorithm to solve the power flow equations.

The application proposed is very robust and simple to use as a decision aid toll. Besides that, it also has the advantage of providing the results very quickly. It should be noticed that it only takes about 5 s, in an AMD Athlon^{TM} 208 GHz PC, to output all the results required to plot each of the graphs shown in Fig. 4.5 through to Fig. 4.15.

Simulation results obtained using a real Semiurban MV network clearly show that DG can improve system voltage profile, reduce power losses, and reduce branches congestion. As expected, DG penetration level plays a vital role in determining the amount of voltage profile improvement, power losses, and branches congestion reduction. In general, voltage profile improvement, power losses reduction, and branches congestion reduction increase with DG penetration level. However, this trend is not always seen in the power loss reduction since for very high levels of penetration the power losses, typically start increasing. It should be notice, however, that this happens for penetration levels above 50%, which are very unlikely to occur.

The DG units' location distribution is an important factor in determining the benefits. The results clearly indicate that introducing DG units close to the substations (scenario Near) is the worst, since for any of the factors under analysis it is the least beneficial location. Regarding the random location distribution and the far from substation location distribution, the results indicate that Far is the location leading to the best improvements in the network.

Nevertheless, for very high DG penetration levels the voltage profiles are better for the random locations. The effects of the distance in branches congestion is not so clear as it depends not only on the specific location of the DG machines but also on the location of the most loaded branches.

Acknowledgement. Research partially supported by FCT/POCI 2010/FEDER through project POCTI/EGE/61823/2004.

References

1. A. R. Abdelaziz and W. M. Ali. Dispersed generation planning using a new evolutionary approach. In *IEEE Power Tech Conference Proceedings, 2003 IEEE Bologna*, volume 2, page 5, 2003
2. P. P. Barker. Determining the impact of distributed generation on power systems: Part i-radial distributed systems. In *IEEE Power Engineering Society Summer Meeting*, volume 3, pages 1645–1656, 2000
3. R. E. Brown and L. A. A. Freeman. Analyzing the reliability impact on distributed generation. In *IEEE Power Engineering Society Summer Meeting*, volume 2, pages 1013–1018, 2001
4. R. E. Brown, X. Feng, J. Pan, and K. Koutlev. Siting distributed generation to defer t&d expansion. In *Transmission and Distribution Conference and Expo*, volume 2, pages 622–627, 2001
5. P. Chiradeja and R. Ramakumar. A probabilistic approach to the analysis of voltage profile improvement with distributed wind electric generation. In *IEEE Frontiers of Power Conference*, pages XII 1–XII 10, 2001
6. P. Chiradeja and R. Ramakumar. Voltage profile improvement with distributed wind turbine generation – a case study. In *IEEE Power Engineering Society General Meeting*, volume 4, page 236, 2003
7. P. Chiradeja and R. Ramakumar. An approach to quantify the technical benefits of distributed generation. *IEEE Transactions On Energy Conversion*, 19(4):764–773, 2004
8. L. Dale. Distributed generation transmission. In *IEEE Power Engineering Society Winter Meeting*, volume 1, pages 132–134, 2002
9. J. Dolezal, P. Santarius, J. Tlusty, V. Valouch, and F. Vybiralik. The effect of dispersed generation on power quality in distribution system. In *Quality and Security of Electric Power Delivery Systems, 2003. CIGRE/PES 2003. CIGRE/IEEE PES International Symposium*, pages 204–207, 2003
10. M. Gandomkar, M. Vakilian, and M. Ehsan. Optimal distributed generation allocation in distribution network using hereford ranch algorithm. In *IEEE Electrical Machines and Systems, 2005. ICEMS 2005. Proceedings of the Eighth International Conference on*, volume 2, pages 916–918, 2005
11. H. A. Gil and G. Jöos. On the quantification of the network capacity deferral value of distributed generation. *IEEE Transactions on Power Systems*, 21:1592–1599, 2006
12. J. J. Grainger and W. D. Stevenson. *Power Systems Analysis.* McGraw-Hill, New York, 1994

13. N. Hadjsaid, J. F. Canard, and F. Dumas. Dispersed generation impact on distribution networks. *IEEE Computer Applications in Power*, 12:22–28, 1999
14. T. Hoff and D. S. Shugar. The value of grid-support photovoltaics in reducing distribution system losses. *IEEE Transactions on Energy Conversion*, 10:569–576, 1995
15. G. Jos, B. T. Ooi, D. McGillis, F. D. Galiana, and R. Marceau. The potential of distributed generation to provide ancillary services. In *IEEE Power Engineering Society Summer Meeting*, 2000
16. J. A. P. Lopes. Integration of dispersed generation on distribution networks-impact studies. In *IEEE Power Engineering Society Winter Meeting*, volume 1, pages 323–328, 2002
17. M. R. Milligan and M. S. Graham. An enumerated probabilistic simulation technique and case study: Integrating wind power into utility production cost models. In *National Renewable Energy Lab. for Wind Energy Program*, 1996
18. J. Oyarzabal, N. Hatziargyriou, J. Peas Lopes, A. Madureira, C. Moreira, and Aris Androutsos. Di3 – report on socio-economic evaluation of microgrids. Project Consortium European Commission, 2005
19. Parliament and Council of the European Union. Directive /77/ec of the european parliament and of the council of 27 september 2001 on the promotion of electricity produced from renewable energy sources in the internal electricity market. *Official Journal of the European Communities*, 44:33–40, 2001
20. S. Rahman. Fuel cell as a distributed generation technology. In *IEEE Power Engineering Society Summer Meeting*, volume 1, pages 551–552, 2001
21. H. Saadat. *Power Systems Analysis*. McGraw-Hill, New York, 2nd edition, 2002
22. S. Silva. Anlise do impacto da pequena gerao dispersa sob diferentes directivas de regulao. Dissertation, Faculdade de Economia da Universidade do Porto, 2007
23. T. Tran-Quoc, C. Andrieu, and N. Hadjsaid. Technical impacts of small distributed generation units on lv networks. In *IEEE Power Engineering Society General Meeting*, volume 4, page 2464, 2003
24. H. L. Willis and W. G. Scott. *Distributed power generation. Planning and evaluation.* Marcel Dekker, New York, 2000

5

New Effective Methods of Mathematical Programming and Their Applications to Energy Problems

Valerian P. Bulatov

Summary. Convex programming algorithms, which have polynomial-time complexity on the class of linear problems are considered. The paper addresses the Chebyshev points of bounded convex sets, algorithms of their search as well as their different applications in convex programming, for elementary approximations of attainability sets, optimal control, global optimization of additive functions on convex polyhedrons and in the integer programming.

New formulations of energy problems made possible by the following methods are discovered: minimal shutdown during power shortages in a power supply system, search for optimal states in thermodynamic systems, optimal allocation of water resources. The applicability of polynomial-time algorithms to such problems is demonstrated. Consideration is given to the problem of search for the Chebyshev points in multi-criteria models of electric power system expansion and operation.

5.1 Introduction

The paper represents a generalization of a number of results that were obtained by the author and some researchers from other laboratories in the past decades in the Department of Applied Mathematics at the Energy Systems Institute. The algorithms applied were first proposed at this Institute. The author confined himself only to the algorithms and their applications that have polynomial difficulty. Some methods use essentially the specific character of initial problem and are therefore presented with applications. Most of the presented applications are discussed in the special literature on energy and the author refers to this literature. The presented theorems and lemmas are proved by the author and his colleagues. The presented algorithms not only have the guaranteed polynomial convergence rate but are also efficient in solving applied problems.

In the 1970s of the last century the American mathematicians gave examples of the linear programming (LP) problems, in which the simplex method and all its modifications for calculation of the optimal value of a linear form ran through all the vertices of the admissible set. It means that for the simplest polyhedrons the optimal solution can be obtained by calculating a scalar product at no less than 2^n vertices (n – the problem dimension). This brought up the question that the solution of systems of linear inequalities was possibly more laborious than the solution of systems of linear algebraic equations. The problem was successfully solved by Khachiyan [25]. He was the first to show that the LP problem is solvable by the polynomial algorithm based on the space extension method [36]. The geometrical interpretation of this method, known as the ellipsoid method [11], was given by Yudin and Nemirovsky in the book [38]. In these methods the convergence rate of iteration processes depends only on the space dimension. Later Bulatov and Antsiferov suggested a similar approach, where simplexes played the role of ellipsoids [2, 14].

Subsequently a series of works were published that were devoted to the affine scaling methods. Karmakar gave the most effective estimate of the convergence rate of a similar type of methods for solving LP problems [23].

Similar works without proof of the polynomial convergence were published far in advance by Dikin [13,17] and also Evtushenko. Later these methods were developed by Zorkaltsev, Zhadan and Pardalos [33,39,40]. Here the centers of convex sets were represented by their centroids and the Chebyshev points.

The above methods have been intensively applied in Russia over decades to solve numerous applied problems in energy and physicochemical systems, in routing the pipeline systems, in optimization of farming industry, management of water resources, etc. Partially they are described below.

The paper logically considers polynomial algorithms for solving convex programming problems and presents the examples of their efficient application for solving important applied energy problems.

5.2 Polynomial-Time Algorithms in Convex Programming

5.2.1 Survey of Cutting-Circumscribing Methods

Cutting-circumscribing methods are based on two ideas. The first is cutting. This idea was used with great impact in the work of Levin [28], who proposed an algorithm to solve the problem

$$\min\{f_0(x) \mid x \in R \subset E^n\}, \tag{5.1}$$

$$R = \{x \in E^n : \ f_i(x) \leq 0, \ i = 1, ..., m\}, \tag{5.2}$$

where $f_0, f_1, ..., f_m$ are convex functions and R is a convex bounded set with a nonempty interior.

It follows from (5.2) that dimensionality of R coincides with dimensionality of E^n.

Let $k = 0, \ R^0 = R$.

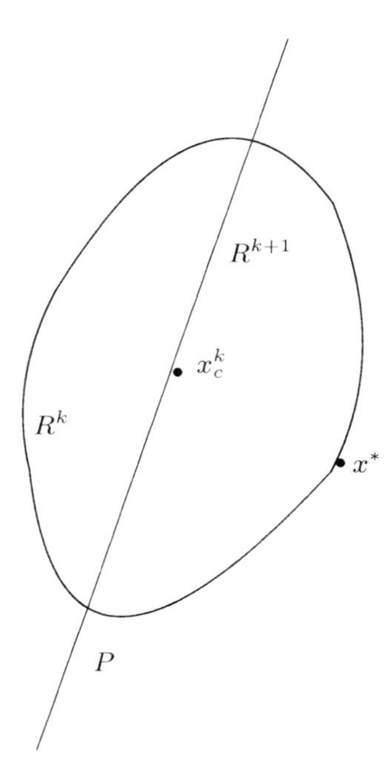

Fig. 5.1. The kth step of Levin's algorithm (geometrical sense of a method of the centres of gravity)

In step k, Levin's algorithm (Fig. 5.1) performs the following operations:

1. Find the point x_c^k – the center of mass of the set R^k.
2. Construct the cutting half-space

$$P = \{x \in E^n : \ a^T(x - x_c^k) \leq 0\}, \tag{5.3}$$

where $a \in E^n$ is the gradient of the objective function f_0 at the point x_c^k.

3. Construct the next set R^k containing the solution of problem (5.1), (5.2) by the superficially simple rule

$$R^{k+1} = R^k \cap P. \tag{5.4}$$

4. $k := k + 1$, go to 1.

By convexity of the function f_i, we have the obvious inclusions for all k:

$$R^{k+1} \subset R^k \tag{5.5}$$

$$x^* \in R^k \tag{5.6}$$

for any x^* from the set X^* of optimal solutions of the problem. We also know that the volumes V_k of the sets R^k satisfy the inequalities

$$[n/(n+1)]^n \leq V_{k+1}/V_k \leq 1 - [n/(n+1)]^n, \tag{5.7}$$

whence

$$1/e \leq V_{k+1}/V_k \leq 1 - 1/e, \tag{5.8}$$

where e is the base of natural logarithms.

This method successively localizes the optimal set X^* of the original convex problem inside contracting sets, whose volume tends monotonically to zero as the iteration index k increases. If we denote by q_n^k the volume contractor factor in iteration k, then by (5.8) this factor can be bounded from above by $\overline{q} = 1 - 1/e$. This bound is independent of the iteration index and of the dimension n of the space of independent variables.

For algorithms of this kind, we have the bound

$$f_c^k - f^* \leq \rho(q_n^k)^{1/n}, \ \forall k \in K' \subseteq K = \{0, 1, ..., \}, \tag{5.9}$$

where $f_c^k = f(x_c^k)$, f^* is the optimal value in problem (5.1), (5.2), K' is a subsequence of feasible centers: $x_c^k \in R$, $\forall k \in K'$, ρ is a constant independent of n and k.

From the bound (5.9) it follows that Levin's method converges to the optimal value f^* at a geometrical rate with common factor $\overline{q} \leq 0.632$ independently of the dimension of the space n. Despite this convergence property, Levin's method has only limited applicability in its pure form, because of the tremendous algorithmic difficulties associated with the implementation of step 1 of the algorithm. Indeed, even if the feasible set R in (5.2) is a polyhedron defined by a system of linear inequalities, the determination of its center of mass involves taking n-fold integrals, i.e. an exponential-time problem. This naturally raises the following question: can this operation be replaced by some other operation of lower time complexity which nevertheless preserves the convergence property of the method? Indeed, in some papers [2, 3, 29, 30] the operation of finding the center of mass of the set R^k is replaced with the operation of finding a different 'center' – the so-called Chebyshev center.

In the supporting cone algorithm (Fig. 5.2) [13, 14, 39, 40], the original convex-programming problem is reduced to canonical form

$$\min\{c^T x \mid x \in R\} \tag{5.10}$$

with a linear objective function. Let us describe this algorithm.

When the kth iteration starts ($k = 0, 1, ...$), the cone

$$C^k = \{x \in E^n \mid A^k(x - x_c^k) \leq 0\} \tag{5.11}$$

is assumed given. This cone contains the solution x^* of problem (5.1), (5.2) and has the property

$$\min\{c^T x \mid x \in C^k\} = c^T x_c^k. \tag{5.12}$$

Here A^k is a non-singular $n \times n$ matrix, $x_c^k \in E^n$ is the vertex of the cone C^k. Since the minimum of a linear objective function on C^k is attained

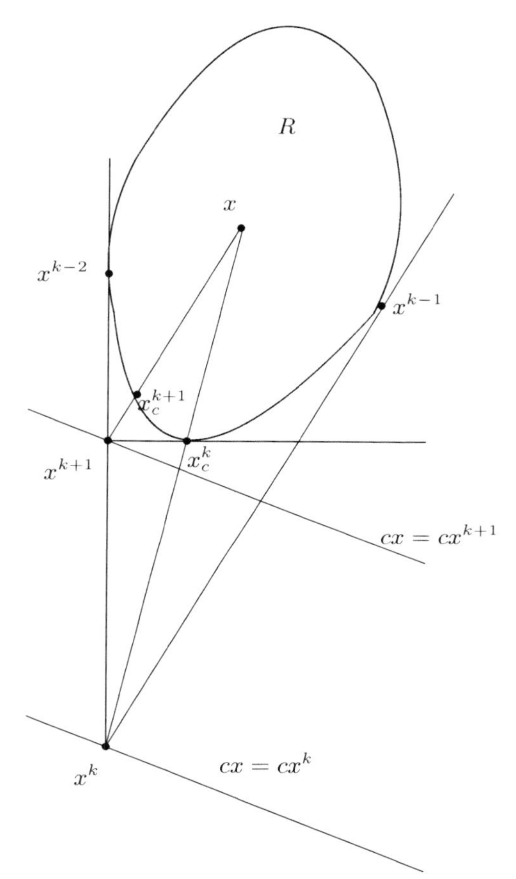

Fig. 5.2. The kth step of algorithm of the supporting cone method (geometrical sense)

at its vertex, the Kuhn–Tucker theorem implies the existence of non-positive multipliers $u^k \in E^n$ such that

$$c^T = (u^k)^T A^k. \tag{5.13}$$

Denoting $B_k = (A^k)^{-1}$, we rewrite (5.13) in the form

$$(u^k)^T = c^T B^k. \tag{5.14}$$

In the kth step of the supporting cone algorithm do:

1. Compute the 'maximum discrepancy' at the vertex of the cone x_c^k,

$$\Delta = \max_{1 \leq i \leq m} \{f_i(x_c^k)\} = f_s(x_c^k). \tag{5.15}$$

If $\Delta \leq \epsilon$, where ϵ is the accepted accuracy, then x_c^k is an almost feasible solution and the algorithm ends.

2. Construct the cutting inequality

$$(a^k)^T(x - x_c^k) \leq \nabla f_s(x_c^k)^T(x - x_c^k), \tag{5.16}$$

where ∇ is the subgradient of the function f_s.
3. Determine the expansion coefficients of the normal a^k in the basis A^k:

$$\mu^T = (a^k)^T B^k.$$

4. In the system of inequalities defining the cone C^k, identify the inequality l that can be replaced with the cutting inequality so that the linear form $c^T x$ attains its minimum at the vertex of the new cone formed by this replacement; formally l is determined from the condition

$$u_l^k/\mu_l = \max\{u_i^k/\mu_i \mid \mu_i > 0,\ i = 1, ..., n\}.$$

If the set of positive μ_i is empty, then the feasible set R in the original problem is also empty, and the algorithm stops.
5. Define the new cone vertex x_c^{k+1} by the formula

$$x_c^{k+1} = x_c^k - (\Delta/\mu_l)B_{.l}^k,$$

where B_l^k is the lth column of the matrix B^k.
6. Update the multiplier u^k and the inverse matrix B^k:

$$(u^{k+1})^T = (u^k)^T E^l,\ B^{k+1} = B^k E^l,$$

where the matrix factor E^l differs from the identify matrix only by its lth row, which is defined by the formula

$$E_{lj}^l = -\mu_j/\mu_l,\ j = 1, ..., n,\ j \neq l,\ E_{ll} = 1/\mu_l.$$

7. Set $k = K + 1$ and go to 1.

This scheme can be verbalized as follows: circumscribe the set containing the solution by a set from a given class **M**; cut away part of circumscribing set; again circumscribe the remaining part by a set of class **M**. This account for the name ‘cutting-circumscribing’ that we use for such methods.

In [4, 36–38], ellipsoids we are used as the class **M** of localizing sets. We will describe the method of [36].

Let x^* be contained in the ball $R^0 = \{x \mid\parallel x - x^0 \parallel\leq r\}$, $B^0 = I$ is the $n \times n$ identify matrix, $h_0 = r/(n+1)$.

Set $k = 0$.

In the kth step of the ellipsoid method (Fig. 5.3), do:

1. Construct the normal to the cutting plane a^k as in (5.3).
2.

$$\xi_k = (B^k)^T a^k / \parallel (B^k)^T a^k \parallel. \tag{5.17}$$

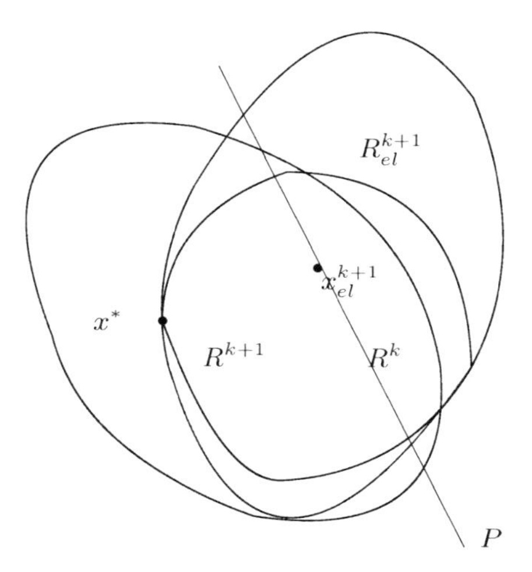

Fig. 5.3. The kth step of algorithm of the ellipsoid method (geometrical sense)

3.
$$x^{k+1} = x^k - h_k \xi^k. \tag{5.18}$$

4.
$$B^{k+1} = B^k R_\beta(\xi^k),\ \beta = [(n-1)/(n+1)]^{1/2}, \tag{5.19}$$

where $R_\beta(\xi^k)$ is the space stretching operator [36] in the direction ξ^k with stretching factor β.

5.
$$h_{k+1} = th_k,\ t = n/(n^2-1)^{1/2}. \tag{5.20}$$

Algorithm (5.17)–(5.20) has the following geometrical interpretation. Given the ball R^0, construct a half-ball which is the intersection of R^0 and the cutting half-space and circumscribe this half-ball by the minimum-volume ellipsoid. From geometrical considerations it is easy to show that the center of this ellipsoid is determined by formula (5.18). Now the ellipsoid localizing x^* consists of the points $x \in E^n$ that satisfy the inequality $(x - x^{k+1})^T(B^{k+1})(x - x^{k+1}) \leq 0$ and its volume is a factor of

$$q_c = [n/(n+1)]^{(n+1)/2}[n/(n-1)]^{(n-1)/2} \tag{5.21}$$

less than the volume of ellipsoid from the previous iteration. Note that contrary to Levin's method, the volume contraction factor in the ellipsoid method depends on the dimension of the space n. In the section dealing with the simplex circumscribing method, we present a table which describes the dependence $q_c(n)$. For large n, we have the asymptotic formula:

$$q_c \approx 1 - 1/(2n). \tag{5.22}$$

Since in (5.21) the right-hand side is strictly less than 1 for any n, the algorithm is strictly polynomial.

Formulas (5.21), (5.22) show that, first, the volume contraction factor rapidly tends to 1 as n increases (this is the price that we have to pay for using localizing sets of simple structure) and, second, it is constant in each iteration (Levin's method may produce both the left-hand inequality in (5.7) –the most favourable case, and the right-hand inequality-the worst case). In the following sections, we will consider computational cutting-circumscribing schemes (the simplex circumscribing method and a modification of the ellipsoid method) in which the volume contraction factor adapts to the specific situation in the course of the iterations.

Despite this shortcoming, the ellipsoid method is quite effective for medium-size problems ($n \leq 50$). Moreover, Khachiyan [24] demonstrated that the linear programming problem has polynomial-time complexity, because it can be solved by the ellipsoid algorithm in polynomial time. This result was highly unexpected: it showed that a slowly converging method, which however always converges, may be superior to the powerful and fast simplex method, which on some linear problems requires exhaustive enumeration of all vertices, i.e. runs in exponential time (such examples were constructed by Klee and Minty, 1972). Among later studies [22, 23, 25–27, 33] of polynomial-time algorithms, we should mention the work of Karmarkar [23], who proposed the so-called projective algorithm for solving the linear programming problem. It is remarkable that Karmarkar's formulas are very close to the formulas of Dikin's interior point method [17], proposed back in 1967.

5.2.2 Simplex Circumscribing Algorithm

We again turn to problem (5.1), (5.2) and construct for its solution cutting-circumscribing algorithm which uses the set of $n-$ dimensional simplexes in E^n as the class M of localizing sets.

Definitions

A simplex $S \subset E^n$ with a vertex at the point x_0 and edges $(x_1 - x_0, ..., x_n - x_0)$ forming a basis in E^n is the set

$$S(x_0, x_1, ..., x_n) = \left\{ x \in E^n \mid x = x_0 + \sum_{i=1}^{n} a_i (x_i - x_0),\ \sum_{i=1}^{n} a_i \leq 1,\ a_i \geq 0 \right\}. \tag{5.23}$$

Without loss of generality, we assume that $x_0 = 0$.
The volume of the simplex S is defined by the formula

$$V(S) = \mid det(X) \mid /n!,$$

where X is the $n \times n$ matrix whose columns are the vectors $x_1, ..., x_n$. The center of the simplex $S-$ the point x_c- is defined as

$$x_c = (x_1 + \cdots + x_n)/(n+1). \tag{5.24}$$

For any non-zero vector $a \in E^n$, we define, as in (5.3), the cutting plane

$$L = \{x \in E^n \mid a^T(x - x_c) = 0\},$$

the half-space

$$P = \{x \in E^n \mid a^T(x - x_c) \leq 0\},$$

and the truncated simplex S_p

$$S_p = S \cap P.$$

We say that the vertex $x_i (i = 0, ..., n+1)$ is not cut by the plane L (is an intact vertex) if

$$a_i = a^T(x_i - x_c) < 0,$$

otherwise, i.e. if $a_i \geq 0, x_i$ is called a cut vertex. The number of intact vertices is denoted by k. From this definition and from (5.24) we have

$$a_0 + a_1 + \cdots + a_n = 0$$

and by non-singularity of the matrix X and $a \neq 0$,

$$1 \leq k \leq n.$$

Without loss of generality, we assume that

$$a_i \leq 0, \ i = 0, ..., k-1, \tag{5.25}$$

$$a_l \geq 0, \ l = k, ..., n, \tag{5.26}$$

$$a_0 = \min_{0 \leq i \leq n} a_i. \tag{5.27}$$

The vertex x_0 is called the supporting vertex of the truncated simplex S_p.

Construction of Minimum-Volume Simplex Containing Given Truncated Simplex

We introduce a non-negative parameter vector $\tau = (\tau_1, ..., \tau_n)$ and define the simplex $S(\tau)$ with the vertices $\{0, (\tau_1 x_1), ..., (\tau_n x_n)\}$:

$$S(\tau) = \left\{ x \in E^n \mid x \in \sum_{i=1}^{n} \lambda_i(\tau_i x_i), \ \sum_{i=1}^{n} \lambda_i \leq 1, \ \lambda_i \geq 0, \ i = 1, ..., n \right\}. \tag{5.28}$$

Then the volumes $V(S(\tau))$ and $V(S)$ of the simplexes $S(\tau)$ and S are related by the equality

$$V(S(\tau)) = V(S) \prod_{i=1}^{n} \tau_i. \tag{5.29}$$

Let us now find the vector τ by solving the minimization problem

$$Q(\tau) = \prod_{i=1}^{n} \tau_i \to \min \tag{5.30}$$

subject to

$$S_p \subseteq S(\tau). \tag{5.31}$$

Let us rewrite condition (5.31) in terms of the sought parameters τ. To this end, we determine the points y_{il} where the plane L intersects the segments joining the cut vertices x_i with the intact vertices x_l:

$$y_{il} = x_i + t_{il}(x_l - x_i),\ t_{il} = a_i/(a_i - a_l), i = 0, ..., k-1,\ l = k, ..., n. \tag{5.32}$$

From (5.25)–(5.27) we obtain that $t_{il} \in [0.1]$.
Condition (5.31) is equivalent to the inclusions

$$x_i \in S(\tau),\ y_{il} \in S(\tau),\ i = 0, ..., k-1,\ l = k, ..., n.$$

On the other hand,

$$S(\tau) = conv\{0, \tau_i x_i \mid i = 1, ..., n\}, \tag{5.33}$$

where *conv* is the convex hull of the points. Thus the inclusions

$$y_{il} = (1 - t_{il})x_i + t_{il}x_l = [(1 - t_{il}/\tau_i](\tau_i x_i) +$$

$$+[t_{il}/\tau_l](\tau_l x_l) \in S(\tau),$$

$$y_{0l} = t_{0l}x_l = [t_{0l}/\tau_l](\tau_l x_l) \in S(\tau),$$

$$x_i = [1/\tau_i](\tau_i x_i) \in S(\tau)$$

lead to the following inequalities, which hold for all $1 \le i \le k-1,\ k \le l \le n$:

$$(1 - t_{il})/\tau_i + t_{il}\tau_l \le 1, \tag{5.34}$$

$$\tau_l \ge t_{0l}, \tag{5.35}$$

$$\tau_i \ge 1. \tag{5.36}$$

Lemma 1. *The function* $\phi(x) = (x_1 \cdot x_2 \cdot ... \cdot x_n)^{-1}$ *is convex on the set of positive* $x = (x_1, ..., x_n) \in E^n$.

Convexity of ϕ follows immediately from the mean value theorem:

$$\phi[\frac{1}{2}(x+y)] = \left[\prod_{i=1}^{n} \frac{1}{2}(x+y)\right]^{-1} \leq \left[\prod_{i=1}^{n} (x_i y_i)^{1/2}\right]^{-1} =$$
$$= [\phi(x)\phi(y)]^{1/2} \leq \frac{1}{2}[\phi(x)+\phi(y)].$$

In problem (5.30), (5.33)–(5.36) we make the change of variables

$$\tau_i = (1 + a_i \xi_i)^{-1},\ i = 1, ..., k-1,$$
$$\tau_l = (1 + a_l \eta_l)^{-1},\ l = k, ..., n.$$

The problem is rewritten in the variables (ξ, η) in the form

$$q = \prod_{i=1}^{k-1} (1 + a_i \xi_i)^{-1} \cdot \prod_{l=k}^{n} (1 + a_l \eta_l)^{-1} \to \min \tag{5.37}$$

subject to

$$\xi_i \geq \eta_l,\ 0 \leq \eta_l \leq -1/a_0,\ i = 1, ..., k-1,\ l = k, ..., n. \tag{5.38}$$

The function $q(\xi, \eta)$ in (5.37) is convex on the set (5.38) by Lemma 1. Now, from (5.25), (5.26) it follows that at the minimum point $(\xi^*,\ \eta^*)$ of the function q

$$\xi_i^* = \xi^*,\ i = 1, ..., k-1,\ \eta_l^* = \eta^*,\ l = k, ..., n.$$

Therefore, problem (5.37),(5.38) can be written in the form

$$q(\xi^*, \eta^*) = \prod_{i=1}^{k-1} (1 + a_i \xi^*)^{-1} \cdot \prod_{l=k}^{n} (1 + a_l \eta^*)^{-1} \to \min$$

subject to

$$\xi^* \geq \eta^*,\ 0 \leq \eta^* \leq -1/a_0.$$

From the last two formulas we see that the optimal values ξ^0 and η^0 in the last problem cannot satisfy the strict inequality $\xi^0 > \eta^0$ and therefore $\xi^0 = \eta^0 = t$. We have thus proved theorem.

Theorem 1. *The minimum-volume simplex $S(\tau^*)$ circumscribing the truncated simplex S_p is defined by the parameters*

$$\tau_i^* = (1 + a_i t^*)^{-1},\ i = 1, ..., n,$$

where the scalar τ^* is determined by solving the following one-dimensional convex minimization problem:

$$q_k^*(a) = \min \left\{ \prod_{i=1}^{n} (1 + a_i t)^{-1} \mid 0 \leq t \leq -1/a_0 \right\}. \tag{5.39}$$

Bound on Volume Contraction in the Simplex Circumscribing Method

Let

$$\beta_i = -a_i/a_0, \ i = 1, ..., n, \ q_k(t, \beta) = \prod_{i=1}^{n} (1 + \beta_i t)^{-1}.$$

Then (5.39) is equivalent to the equality

$$q^*(a) = \min\{q_k(t, \beta) \mid 0 \le t \le 1\} = \overline{q}_k(\beta). \tag{5.40}$$

The minimum value $\overline{q}_k(\beta)$ in (5.40) is a function of the natural parameter k and n real parameters $\beta = (\beta_1, ..., \beta_n)$, which may be arbitrary elements of the set

$$B_k = \{\beta \in E^n \mid -1 \le \beta_i < 0, \ i = 1, ..., k-1,$$
$$\beta_l \ge 0, \ l = k, ..., n, \ \sum_{i=1}^{n} \beta_i = 1\}.$$

Denote by $\overline{B}_k$ the closure of the set B_k. Then we have

Lemma 2. *For any* $t \in [0, 1]$ *and* $\beta \in \overline{B}_k$,

$$q_k(t, \beta) \le (1-t)^{-(k-1)}(1-kt)^{-1} = \Psi_k(t). \tag{5.41}$$

Proof. By convexity of the function $q_k(t, \beta)$ with respect to β for any $t \in [0, 1)$, its maximum in β for these t is attained at a vertex of the set $\overline{B}_k$. But at any vertex, $(n-1)$ values b_i should be 0 or -1. Let $k > 1$ and p be the number of -1 components of the vertex β. By definition of k, we should have $0 \le p \le k-1$. Denote

$$\psi(p, t) = (1-t)^{-p}[1 + (1+p)t]^{-1}.$$

Then for all $t \in [0, 1)$ and any $\beta \in \overline{B}_k$, we have the inequality

$$q_k(t, \beta) \le \max\{\psi(p, t) \mid 0 \le p \le k-1\}. \tag{5.42}$$

By the inequality

$$\psi(p+1, t) - \psi(p, t) = t^2(p+2)/(1-t)^{p+1} \ge 0$$

the maximum in the right-hand side of (5.42) is attained for $p = k-1$, which proves the lemma. □

Since the function $\Psi_k(t)$ is a majorant of $q_k(\beta, t)$ for $t \in [0, 1]$, $\overline{q}_k(\beta)$ in (5.40) does not exceed the minimum value Ψ_k^* of the majorant on [0,1]. Equating to zero the logarithmic derivative of $\Psi_k^*(t)$ with respect to t (for $k \ge 2$), we obtain the equation

$$-(k-1)/(1-t) + k/(1+kt) = 0,$$

which has a unique root $t_k = 1/k^2$.

For $k = 1$, the majorant attains its minimum value for $t_k = 1$. Substituting t_k in the expression for $\Psi_k(t)$, we obtain the following theorem.

Theorem 2. *Let $S \in E^n$ be an n-dimensional simplex, x_c its center, $S_p = \{x \in E^n \mid x \in S,\ a^T(x - x_c) \leq 0\}$ the truncated simplex corresponding to the non-zero normal a of the cutting plane and such that the precisely k vertices x_i of the simplex S satisfy the strict inequality $a_i = a^T(x_i - x_c) < 0$ with $a_0 = \min\{a_i \mid i = 0, 1, ..., n\}$.*

Then S_p can be circumscribed by the simplex S^* such that the ratio q_k^* of the volumes $V(S^*)$ and $V(S)$ of the simplexes S and S^* satisfies the inequality

$$q_k^* \leq \Psi_k^*, \tag{5.43}$$

where

$$\Psi_k^* = \begin{cases} 1/2 & \text{if } k = 1, \\ (k/(k+1))^k \cdot (k/(k-1))^{k-1} & \text{if } 2 \leq k \leq n. \end{cases}$$

The equality is attained if for $\beta_i = -a_i/a_0$ we have

$$\beta_i = -1,\ i = 1, ..., k-1,\ \beta_k = k, \beta_l = 0,\ l = k+1, ..., n.$$

Thus, Ψ_k^* is a minimal guaranteed upper bound for volume construction in the framework of our circumscribing contraction.

Algorithm of the Simplex Circumscribing Method

Set $l = 0,\ v_0 = 1,\ \phi_0 = +\infty,\ \triangle_0 = +\infty$.

Suppose that at the start of iteration $l (l = 0, 1, ...)$ the following are given:

X_l – a matrix with $n+1$ rows and n columns (the ith row of the matrix X_l, $i = 1, ..., n+1$, corresponds to the ith vertex of the simplex S_l which contains the solution of problem (5.1), (5.2);

v_l – the volume of the simplex S_l;

ϕ_l – the 'best value so far': $\phi_l = \min\{f_0(x_c^j) \mid x_c^j \in R,\ j = 0, ..., l\}$;

$\triangle_l$ – the length of the uncertainty interval: $0 \leq \phi_l - f_0^* \leq \triangle_l$, where f_0^* is the minimal value in problem (5.1), (5.2).

While $\triangle_l \geq \epsilon$ (ϵ is the specified accuracy) do:

1. Find x_c^l – the center of the simplex S_l:

$$x_{cj}^l = \left[\sum_{i=1}^{n+1} X_{ij}^l \right] \backslash (n+1),\ j = 1, ..., n.$$

2. Compute the 'maximum discrepancy' at the point x_c^l:

$$h_l = \max_{1 \leq i \leq m} f_i(x_c^l) = f_s(x_c^l).$$

Then the algorithm branches depending on the sign of the discrepancy h_l. If $h_l \leq 0$ ($x_c^l \in R$), then do 3–4(a).

3. Evaluate the minimand function and update the best value so far:

$$f_c^l = f_0(x_c^l), \ \phi_{l+1} = \min\{\phi_l, f_c^l\}.$$

4(a). Find the cutting plane normal a and the scalar h:

$$a = \nabla f_0(x_c^l), \ h = 0.$$

If $\| a \| < \epsilon$, then stop, x_c^l is the solution of the problem.
If $h_l > 0 \ (x_c^l \notin R)$, do 4(b).

4(b). $a := \nabla f_s(x_c^l), \ h := h_l$.

5. Find

$$a_i = \sum_{j=1}^{n} a_j (X_{ij}^l - x_{cj}^l) + h, \ a_p = \min_{l \le i \le n+1} a_i,$$

$$\beta_i = -a_i / a_p, \ i = 1, ..., n+1,$$

and the number k of negative a_i.

6. Find $\bar{t}$ either by solving the one-dimensional minimization problem (5.39) or by setting $\bar{t} = 1/k^2$.
7. Compute the stretching factor τ_i of the simplex edges that leave the supporting vertex p and the volume contraction factor q:

$$\tau_i = (1 + \bar{t}\beta_i)^{-1}, \ i = 1, ..., n+1, \ i \neq p, \ q = \prod_{i=1, i \neq p}^{n} \tau_i.$$

8. Update the matrix X by the formula

$$X_{ij}^{l+1} = X_{pj}^l + \tau_i (X_{ij}^l - X_{pj}^l).$$

9. $v_{l+1} = v_l q$.
10. If $x_c^l \in R$, then

$$\gamma := (v_{l+1})^{1/n}; \ \triangle_{l+1} = \gamma/(1-\gamma)[f_{0max} - \phi_{l+1}],$$

where f_{0max} is a number (a parameter of the algorithm) not less than the maximum value of the function $f_0(x)$ on R.

11. $l := l + 1$, go to 1.

Comparison with Ellipsoid Method

The guaranteed volume contraction bounds Ψ_k^* from (5.43) increase with the increase in the number of intact vertices k:

$$1/2 = \Psi_1^* \le \Psi_2^* \le ... \le \Psi_n^* =$$

$$= [n/(n+1)]^n \cdot [n/(n-1)]^{n-1}.$$

For large n, we have the asymptotic equality

$$\Psi_n^* \approx 1 - 1/(2n^2) + 1/(2n^3).$$

This is worse than the asymptotic bound of the ellipsoid method where the volume contraction coefficient depends only on the dimension of the space n and for large n is given by the formula

$$q_c = [n/(n+1)]^{(n+1)/2} \cdot [n/(n-1)]^{(n-1)/2} \approx 1 - 1/(2n).$$

The simplex circumscribing algorithm, however, has a definite advantage, in the sense that the volume contraction factor depends on the number k of intact vertices, i.e. on the specific situation arising in the process of solution. The worst bound Ψ_n^* is attained only when $n-1$ vertices remain intact and the cutting plane is parallel to the $(n-1)$-dimensional face of the simplex. Another, apparently equally likely case is when a single vertex remains intact. In this case, the volume contraction factor does not exceed $1/2$ and we obtain a process similar to dichotomy.

If we compare these methods on the class of convex programming problems, then for the simplex circumscribing method it is reasonable to take some mean value, e.g. the arithmetical mean of all Ψ_k^*:

$$\overline{q}_s = \left[\frac{1}{2} + \sum_{k=2}^{n} \Psi_k^*\right] / n.$$

For any $n, \overline{q}_s(n)$ satisfy the inequality

$$\overline{q}_s(n) \leq 1 - [\zeta(1)/2 - (\zeta(3) - 1)/6]/n + 1/(2n^2),$$

where $\zeta(x)$ is the Riemann zeta-function. Therefore for large n we have the asymptotic formula

$$\overline{q}_s(n) \approx 1 - C/n$$

which is of the same order as

$$\overline{q}_c(n) \approx 1 - 0.5/n$$

in the ellipsoid method, but with a better constant $C = 0.788791 > 0.5$.

Table 5.1 gives the volume contraction factors for various n: Ψ_n^* for the simplex circumscribing method, q_c for the ellipsoid method, and $\overline{q}_s$ the value the simplex circumscribing method averaged over the number of intact vertices. We see from Table 5.1 that $\overline{q}_s < q_c$ for all n.

5.2.3 Methods of the Chebyshev Points of Convex Sets

The Chebyshev Points of Convex Sets in E^n

Let us consider the system of linear inequalities:

Table 5.1. Comparison of ellipsoid method and simplex circumscribing method

n	Ψ_n^*	q_c	$\overline{q}_s$
2	0.88889	0.76980	0.69444
3	0.94922	0.84375	0.77937
4	0.97090	0.88132	0.82725
5	0.98115	0.90422	0.85803
6	0.98679	0.91968	0.87949
7	0.99023	0.93083	0.89531
8	0.99248	0.93925	0.90746
9	0.99404	0.94585	0.91708
10	0.99515	0.95114	0.92488
20	0.99877	0.97529	0.96131
50	0.99980	0.99005	0.98424
100	0.9995	0.99501	0.99207
200	0.99999	0.99750	0.99602
500	0.99999	0.99900	0.99840
1000	0.99999	0.99950	0.99920

$$\begin{aligned} &h_1(x) = a_{11}x_1 + a_{12}x_2 + ... + a_{1n}x_n \leq b_1, \\ &\dots\dots\dots\dots\dots \\ &h_m(x) = a_{m1}x_1 + a_{m2}x_2 + ... + a_{mn}x_n \leq b_m, \end{aligned} \tag{5.44}$$

which gives a bounded set in E^n and let

$$L = \min_x \max_{1 \leq i \leq m} h_i(x) = \max_{1 \leq i \leq m} h_i(x^*). \tag{5.45}$$

It is obvious that system (5.44) is solvable if and only if $L \leq 0$. In this case $|L|$ is a stability measure of the solution x^* of system (5.44) in the sense of (5.45). If $L > 0$, then system (5.44) has no solutions and $|L|$ is a measure of its incompatibility. Let us call the point $x^* = (x_1^*, ..., x_n^*)$ the Chebyshev point of system (5.44) if $\min_x \max_{1 \leq i \leq m} h_i(x) = \max_{1 \leq i \leq m} h_i(x^*) = L \leq 0$. In the case where $L > 0$ the point x^* is called the Chebyshev approximation of the system of linear inequalities (5.44).

Introduce a new variable x_{n+1} and pass to the extended space E^{n+1}. Compare problem (5.45) to the following linear programming problem:

Find

$$\min x_{n+1} \tag{5.46}$$

subject to

$$\sum_j a_{ij}x_j - b_i \leq x_{n+1} \ \forall i = \overline{1, m}. \tag{5.47}$$

It is clear that problem (5.46), (5.47) is equivalent to problem (5.44), (5.45). In case, if in (5.46), (5.47) the condition of Haar is satisfied [1, 41], the solution to the linear programming problem is unique and determines the unique Chebyshev point that corresponds to condition (5.45). Let us explain the geometrical sense of the Chebyshev points of the system of linear inequalities $Ax \leq b$, where $A - (m \times n)$ is a matrix, $m > n$.

Let the rows a^i of the matrix A in (5.44) be normalized and the pair $\{x^*, x^*_{n+1}\}$ be a solution to problem (5.46), (5.47). Then x^* is the center of a sphere of the maximum radius $r^* = |x^*_{n+1}|$, which is inscribed in the convex polyhedron $R = \{x : \; Ax \leq b\}$. In doing so if $||a^i|| = \sum_{j=1}^{n} |a_{ij}|$, then r^* is a radius of the maximum volume cube that belongs to R, if $||a_{ij}|| = \sqrt{\sum_j a_{ij}^2}$, then r^* is a radius of the maximum volume sphere that belongs to R, if $||a^i|| = \max_{1 \leq j \leq n} |a_{ij}|$, then r^* is a radius of the maximum volume rhombus that belongs to R.

Let $\| \; a^i \; \| \neq 1$ and the constraint $x^* \in D$, D be imposed on the Chebyshev center x^*, D is a convex compact.

Then the pair $\{x^*, x^*_{n+1}\}$ that solves the linear programming problem

$$\min\{x_{n+1} : \; a^i x - b_i \leq \| \; a^i \; \| \; x_{n+1}, \; x \in D, \; i = \overline{1, m}\}, \tag{5.48}$$

is the Chebyshev point of system (5.44) on the set D.

Similarly to (5.44), (5.45) it is possible to introduce the notion of the Chebyshev point of the system of inequalities with convex or quasiconvex functions in the left-hand sides.

Below we will consider the method of the Chebyshev points in the convex programming. The first constructions, related to the application of the Chebyshev points for construction of the iteration processes of solving the convex programming problem, as is known to the authors, were suggested in [12, 14].

These authors also suggested some ways of excluding inactive constraints at each step of iteration process and proved the theorem that any additional constraint is active for a finite number of times.

In [12] consideration is given to the problem:

$$\min\{\varphi(x) : x \in R\}, \tag{5.49}$$

$$R = \{x : \; Ax \leq b\}. \tag{5.50}$$

The step of the iteration process of solving problem (5.49), (5.50) that was described in [12] has the form

$$\min\{x_{n+1} : \nabla\varphi(x^j)^T(x - x^j) \leq x_{n+1}, \; x \in R, \; j = \overline{1, k}\}. \tag{5.51}$$

At each step in (5.51) the linear programming problem is solved and the dimension of the auxiliary problem increases from step to step: one more linear constraint is added. This disadvantage is typical practically of all cutting off methods. The geometrical sense of the linear programming problem in (5.51) consists in search for the Chebyshev point of the system of linear inequalities $\nabla\varphi(x^j)^T(x - x^j) \leq 0$, $j = \overline{1, k}$ under the additional constraint $x^* \in R$.

The authors of [12] prove the theorem for convergence of the sequence $\{x_{n+1}^k\}$ to zero and for existence of the sequence $\{x^{l_j}\} \subset \{x^j\}$ converging to the solution of problem (5.49), (5.50).

Further the Chebyshev points methods were unfolded in the works by the Kiev mathematicians S.I. Zukhovitsky, R.A. Polyak, M.E. Primak et al. [18, 31, 34, 35, 42].

In [3] the methods of type (5.51) are applied to search for the equilibrium points of n-person games that are reduced to the search of the normalized equilibrium points. The convergence of the method is substantiated and geometrical estimates of the convergence rate are presented.

Let us consider the solution of the convex programming problem

$$\min\{\varphi(x) : x \in R\}, \tag{5.52}$$

$$R = \{x : \ q_j(x) \leq 0, \ j = \overline{1, m}\}, \tag{5.53}$$

where $\varphi(x)$, $q_j(x)$ are convex functions, R satisfies the regularity conditions – this is one additional condition that is met by the set R, i.e. for each $j = 1, 2, ...m$ there exists the point $\overline{x} \in R$ such that $g_i(\overline{x}) < 0$.

Assume that R^0 is a convex polyhedron, $x^* \in R^0$ and the starting point $x^0 \in R^0$ is known. At each k-th step of the iteration process find

$$\alpha^k = \begin{cases} \nabla\varphi(x^k), \ \max\limits_j q_j(x^k) \leq 0, \\ \nabla q_{jk}(x^k), \max\limits_j q_j(x^k) = q_{jk}(x^k) > 0, \end{cases} \tag{5.54}$$

and the convex polyhedron

$$R^{k+1} = \{x, x_{n+1} : \ x, x_{n+1} \in R^k, \ (\alpha^k)^T(x - x^k) \leq \parallel \alpha^k \parallel x_{n+1}\}. \tag{5.55}$$

Then we will find the next approximation x^{k+1} as a solution to the linear programming problem:

$$x^{k+1} = arg\min\{x_{n+1} : x, x_{n+1} \in R^{k+1}\}. \tag{5.56}$$

The proof of convergence of the iteration process (5.11)–(5.13) can be found in [18]. There are different versions of the base algorithm. They belong to different authors.

Modification of the Chebyshev Point Method

Let us describe a new version of the Chebyshev point method for a general problem of convex programming, i.e. the problem of the form

$$\min\{\varphi(x) : x \in R\}, \tag{5.57}$$

$$R = \{x : \ g_j(x) \leq 0, \ j = \overline{(1, m)}\}, \tag{5.58}$$

where $\varphi(x)$, $g_j(x)$ are convex functions, R satisfies the regularity conditions.

Let the set

$$R^k = \{x:\ A^k x \le b^k\}, \tag{5.59}$$

that contains x^* be a simplex, x^k – the Chebyshev point R^k that satisfies the system of equations

$$\sum_{j=1}^{n} a_{ij}^k x_j - b_i^k = ||a^{ki}||\ x_{n+1},\ i = \overline{1, n+1}, \tag{5.60}$$

a^{ki} – the i-th row of the matrix A^k and $\alpha^{kT}(x - x^k) = 0$ – the equation of the cutting plane that passes through the center x^k, and

$$\alpha^k = \begin{cases} \nabla\varphi(x^k),\ \max\limits_j g_j(x^k) = g_{jk}(x^k) \le 0\ \forall_j = \overline{1, m}, \\ \nabla g_{jk}(x^k), \max\limits_j g_j(x^k) = g_{jk}(x^k) > 0. \end{cases} \tag{5.61}$$

Find

$$\overline{x}^k = argmin\{\alpha^{kT} x:\ x \in R^k\}. \tag{5.62}$$

The auxiliary problem (5.62) is a special linear programming problem on a simplex. At point $\overline{x}^k$, a vertex of the simplex, exactly n constraints are active, they form R^k. Adding to them the constraint $\alpha^{kT}(x - x^k) \le 0$ we obtain a new simplex R^{k+1}. Call the vertex $\overline{x}^k$ a basic vertex of the truncated simplex. From the system of linear equations of the form (5.60) find a new approximation x^{k+1}– the Chebyshev point R^{k+1}. It is easy to see that

$$R^{k+1} \ni x^* = argmin\{\varphi(x):\ x \in R\},\ \forall k = 1, 2, ...$$

Moreover the radii of the Chebyshev spheres (in case the Haar condition is met) monotonously decrease, $x_{n+1}^* > x_{n+1}^{k+1}$, and

$$\lim_{k\to\infty} x_{n+1}^k = 0 \ \text{ and } \ \lim_{k\to\infty} \varphi(x^k) = \varphi^*.$$

is true [15].

Now let us assess the computational complexity of the iteration. Two problems are solved at each step of the iteration process:

(1) The problem of simplex minimization of a linear function

(2) The solution of a system of linear algebraic equations of the form (5.60) (finding the Chebyshev point)

In doing so the sequence of matrices that determine simplexes and systems of linear algebraic equations differs from step to step only by one row, therefore the iteration consists of about n^2 arithmetic operations.

Let us study in more detail the algorithm of the method. Let R^k be a simplex of the k-th step, set by its vertices $\{x^{0k}, x^{1k}, ..., x^{nk}\}$, x^k – the Chebyshev point of R^k, $\{x:\ \alpha^{kT}(x - x^k) \le 0\}$ – a cutting halfspace. Then the base vertex $\overline{x}^k$ of simplex R^k is

$$\overline{x}^k = arg \min_{0 \le i \le n} \alpha^{k^T}(x^{ik} - x^k). \tag{5.63}$$

It is obvious from the geometrical considerations that $\overline{x}^k \in R^{k+1}$ – the simplex of the next step of the iteration process. Moreover, the next Chebyshev point x^{k+1} belongs to the angle bisector of simplex R^k at the vertex $\overline{x}^k$, i.e. to the ray $x = \overline{x}^k + \lambda(x^k - \overline{x}^k)$, $\lambda > 0$ and is equally distant both from the plane $\{x : \alpha^{k^T}(x - x^k) = 0\}$, and from any of the planes that form the simplex R^k and are connected to the vertex $\overline{x}^k$.

Assume that vector α^k and row a^{jk} of matrix A^k are normalized. Then x^k is the center of the sphere inscribed in R^k. The condition of equal distance from x^{k+1} to the faces of simplex R^{k+1} of dimension $n-1$ will be written in the form

$$a^{jk^T}(\overline{x}^k + \lambda(x^k - \overline{x}^k)) - b_{jk} = \alpha^{k^T}(\overline{x}^k + \lambda(x^k - \overline{x}^k)) - \beta_k, \tag{5.64}$$

where $\beta_k = \alpha^{k^T}x^k$, a^{jk^T} is an arbitrary row of matrix A^k, such that $a^{jk^T}\overline{x}^k - b_{jk} = 0$ (the number of such rows will be equal to n). Since $a^{jk^T}\overline{x}^k - b_{jk} = 0$ from (5.64), we obtain

$$\lambda^k = \frac{-\alpha^{k^T}\overline{x}^k + \beta_k}{|x^k_{n+1}| + (\alpha^k)^T(x - \overline{x}^k)}. \tag{5.65}$$

Then the Chebyshev point x^{k+1} of the simplex R^{k+1} that satisfies the system of algebraic equations (5.60) is determined by the formula

$$x^{k+1} = \overline{x}^k + \frac{-\alpha^{k^T}\overline{x}^k + \beta_k}{|x^k_{n+1}| + (\alpha^k)^T(x^k - \overline{x}^k)}(x^k - \overline{x}^k). \tag{5.66}$$

Hence, the algorithm of solving problem (5.57), (5.58) consists in the following.

Let R^k be a simplex that contains the solution to problem (5.57), (5.58), x^k is its Chebyshev point, $|x^k_{n+1}|$ is a radius of its inscribed 'maximal' sphere.

Using (5.61) determine the cutting halfspace. Then, according to (5.63) we find the basic vertex $\overline{x}^k$ of the simplex R^k, which is a solution to system (5.60); n linear constraints are active at point $\overline{x}^k$ as well as an additional constraint, that corresponds to the cutting halfspace. Then we form a new simplex R^{k+1} and thus determine its Chebyshev point (5.66).

According to the construction $|x^k_{n+1}| \ge |x^{k+1}_{n+1}|$. Assume that all simplexes R^k are in the sphere of diameter $|D|$.

Then it is obvious that $\max_{x \in R^k}\{\alpha^{k^T}x - \beta_k\} \le |D|$.

Besides, assume that $r^k = |x^k_{n+1}| \ge r > 0$, $\forall_k$.

Theorem 3. *The sequence r^k meets the condition*

$$\frac{r^k}{r^{k-1}} \le \frac{1}{1 + \dfrac{|r|}{|D|}}.$$

Proof. Using the technique for searching of the common point of the system of linear inequalities we obtain

$$|x_{n+1}^{k+1}| = -a^{jk^T}(\overline{x}^k + \frac{-\alpha^{k^T}\overline{x}^k + \beta_k}{|x_{n+1}^k| + \alpha^{k^T}(x^k - \overline{x}^k)}(x^k - \overline{x}^k) + b_{jk}$$

or by virtue of the equality $-a^{jkT}\overline{x}^k + b_{jk} = 0$, obtain:

$$|x_{n+1}^{k+1}| = \frac{-\alpha^{k^T}\overline{x}^k + \beta_k}{|x_{n+1}^k| + \alpha^{k^T}(x^k - \overline{x}^k)}|x_{n+1}^k| = |x_{n+1}^k| \cdot \frac{-\alpha^{k^T}\overline{x}^k + \beta_k}{|x_{n+1}^k| - \alpha^{k^T}\overline{x}^k + \beta_k} =$$

$$= |x_{n+1}^k| \cdot \frac{1}{1 + \frac{|x_{n+1}^k|}{(-\alpha^{k^T}\overline{x}^k + \beta_k)}},$$

or

$$\frac{|x_{n+1}^{k+1}|}{|x_{n+1}^k|} \leq \frac{1}{1 + \frac{|\tau|}{|D|}}, \tag{5.67}$$

which completes the proof. □

Estimate (5.67) depends on the structure of the problem and if $|D| \gg |\tau|$, the rate of the iteration process convergence can be made arbitrarily slow.

At each step of the iteration process the linear transformation is employed to consider new variables $\{y_1, ..., y_n\}$, in which the current simplex will be correct. The linear transformation maintains the relation between the volumes and then the relation between the radius of the sphere, circumscribed around the current simplex, and the radius of the maximum sphere inscribed in this simplex is $\frac{1}{n}$. Hence estimate (5.67) will have the form:

$$\frac{|y_{n+1}^{k+1}|}{|y_{n+1}^k|} = 1 - \frac{1}{1 + \frac{1}{2n}}. \tag{5.68}$$

and will not depend on the problem structure.

Applications of the Chebyshev Point Methods

On Elementary Approximations of the Attainability Sets of Linear Systems with Phase Constraints

Let the evolution of some object be described by linear difference equations

$$x^{i+1} = x^i + A_i x_i + B_i u^i, \; x(0) = x^0, \; i \in \overline{0, T}; \tag{5.69}$$

$$x^i \in R^i_x = \{x^i \in E^n;\ C^i x^i \le c^i\} \subset E^n,$$
$$c^i \in E^{r_i},\ u^i \in R^i_u = \{u^i \in E^m:\ A^i u^i \le b^i\} \subset E^m, \tag{5.70}$$

where $C^i - (r_i \times n)$ is a matrix, $A^i - (l_i \times m)$ is a matrix, $b^i \in E^{m_i}$. Let X^i denote the attainability set of system (5.69), (5.70) at the moment i. The problem consists in inscribing the n-dimensional sphere (cube) $\overline{B}^i$ of maximum volume in $X^i \cap R^i_x$ and circumscribe the parallelepiped $\overline{B}^i$ of minimum volume around $X^i \cap R^i_x$. We know more accurate approximations of the attainability sets, for example, using ellipsoids [16, 32] and convex polyhedrons, however it is precisely the approximations using spheres, cubes and parallelepipeds, despite their roughness, most often suit the practitioners best.

Recurrently expressing the right hand side of system (5.69) through the initial conditions, obtain

$$x^i = D^i \xi^i + d^i,$$

where $D^i - [n \times (m \times i)]$ is a matrix, $d^i \in E^n$ is a set vector determined by the initial conditions $x(0) = x^0$, $\xi^i = (u^0, ..., u^{i-1}) \in E^{m \times i}$. Hence the initial problem is equivalent to the approximation by the inscribed sphere (cube) $\underline{B}^i$ of the maximum volume of the set

$$R^i = \{x^i \in E^n:\ x^i = D^i \xi^i + d^i,\ \overline{A}^i \xi^i \le \overline{b}^i,\ x^i \in R^i_x,\ A^i u^i \le b^i\}, \tag{5.71}$$

or after substitution of $x^i = D^i \xi^i + d^i$ into (5.70) we will have

$$R^i = \{\xi^i \in E^{m \times i}:\ C^i (D^i \xi^i + d^i) \le c^i;\ A^i u^i \le b^i\},\ i \in [0, T].$$

Assume that R^i is bounded and not empty. Then $\underline{B}^i$ is bounded and determined from the solution to the linear programming problem of the form (5.48).

Construction of the n-dimensional parallelepiped $\overline{B}^i$ of the minimum volume that contains the set R^i is, obviously, reduced to the solution of $2n$ linear programming problems: find

$$\underline{x}_i = arg \min\{x_i:\ x, \xi \in R^i\},\ i = \overline{1, n},$$

$$\overline{x}_i = arg \max\{x_i:\ x, \xi \in R^i\},\ i = \overline{1, n}.$$

Then $\overline{B}^i = \{x:\ \underline{x}_i \le x_i \le \overline{x}_i,\ \forall_i = \overline{1, n}\} \supset R^i \supset \underline{B}^i$.

The elementary approximations of attainability sets of linear dynamic systems are used to evaluate the consequences of disturbances in different models of energy system expansion under uncertainty.

The Chebyshev Points in the Study of System Survivability Problems

The basic notions of system survivability that are discussed here follow from Chap. 4 (written by Ashchepkov) "The study of operations and survivability of controlled systems" in the book [10]. The term survivability was introduced for the scientific use by the well-known Russian scientist and naval commander Makarov in the 1870s as applied to shipbuilding. In shipbuilding it corresponded to the ability of a ship to survive damage and remain afloat. In a wider interpretation survivability is the ability of a system to resist disturbances. When designing different technical systems and objects great attention is paid to the functionality of the object, which is called a designated purpose in [10]. The designated purpose is normally determined by the vector x of the object state. When considering survivability of systems it is useful to consider also the variables of control u and variables of disturbance v. Let us assume that the elements x, u and v belong to some sets X, U and V. Under the designated purpose of a system we will understand the fact that some functions and operators of variables of state, control and disturbance took the desired values, or had some previously set properties. This implies that the operator F is set, it is determined on the set $R = X \times U \times V$, and Q is the set of its desired values. Then the designated purpose of the system consists in observing the inclusion of $F(x, u, v) \in Q$ at $x, u, v \in R$.

Obviously, there can exist $v^* \in V$ such that $\forall x \in X,\ u \in U\ F(x, u, v^*) \notin Q$, i.e. at some disturbances the system losses its designated purpose. The idea consists in the fact that the appropriate control and the state x corresponding to this control are chosen for the system to maintain its designated purpose for the 'largest', in some sense, subset from the set of disturbances.

Now let us consider formal constructions. For fixed $u \in U$ form the set $V(u) \in V$ such that for any $v \in V(u)$ there exists $x \in X$ such that $F(x, u, v) \in Q$.

By definition $V(u)$ is a set of non-dangerous disturbances, i.e. the disturbances for which there exists control $u \in U$ and state $x \in X$ that provide the designated purpose of the system.

It is obvious that the larger the set of safe controls the higher survivability of the system.

Let some measure μ be introduced on V. Then the control $u \in U$ corresponds to $\mu(V(u))$-a measure of the set of non-dangerous disturbances. Let us introduce real function $J(u) = \dfrac{\mu(V(u))}{\mu(V)}$ in order to compare $V(u)$ and V. If $J(u) = 1$, then the sets $V(u)$ and V coincide in measure, i.e. the control neutralizes almost all disturbances. Vice versa, if J(U) $= 0$, then the set of non-dangerous disturbances has the measure 0 and for this control almost all disturbances violate the designated purpose of the system.

The best control that provides the maximum system survivability is the solution to the extreme problem max $\max\{J(u) : u \in U\}$. Implementation of the approach calls for accurate description of $V(u)$- a set of non-dangerous disturbances, which normally causes considerable computational difficulty. The approach implies the approximation of a set of non-dangerous disturbances $V(u)$ using the simpler set $\tilde{V}(w) \subset V(u)$ with an easily computable measure. In applications the sets of disturbances have, as a rule, a simple structure (cubes, spheres, ellipsoids). Therefore, it is expedient to construct the parametric family $\tilde{V}(w)$, $w \in W \subset E^m$ from the same objects.

Since

$$\tilde{J}(w) = \frac{\mu(\tilde{V}(w))}{\mu(V)} \leq \frac{\mu(V(u))}{\mu(V)} = J(u),$$

$\tilde{J}(w)$ serves as a lower bound of survivability.

Let us give the example. The model of system expansion will be considered in the form

$$d(x) + C(x)v \leq b, \; x \in R_x, \; v \in R_v. \tag{5.72}$$

Here $d(x)$, $C(x)$ – m are an m vector and $(m \times s)$ is a matrix with continuous elements $d_i(x)$, $c_{ij}(x)$ respectively, $R_v = \{v: \; 0 \leq v_i \leq 1, \; i = \overline{1,s}\}$. We will search for $\tilde{V}(w)$ in the form of the cube $\tilde{V}(w) = \{v: \; w \leq v \leq 1 - w\}$, which is determined by the Chebyshev point of the system of inequalities (5.72). Then the problem of search for the maximum cube $\tilde{V}(w)$, contained in the set of non-dangerous disturbances $V(x) = \{v \in R_v; \; d(x) + C(x)v \leq b\}$, has the form:

$$\max\{w: \; d(x) + C(x)v + N(x)w \leq b; \;\; w \leq v \leq 1 - w, \; x \in R_x\},$$

where

$$N_i(x) = \sum_{j=1}^{s} |c_{ij}(x)|, \; i = \overline{1,m}, \; N(x) = \{N_1(x), ..., N_m(x)\},$$

or in the standard form:

$$\min\{v_{s+1}: \; d_i(x) + \sum_{j=1}^{s} c_{ij}(x)v_j - b_i \leq \sum_{j=1}^{s} |c_{ij}(x)|v_{s+1};$$

$$v_j - 1 \leq v_{s+1}, \; -v_j \leq v_{s+1}, \; x \in R_x, \; i = \overline{1,m}, \; j = \overline{1,s}\}.$$

Assume that $d_i(x)$ and $c_{ij}(x)$ do not depend on x.

Then the linear model of the system expansion is considered in the form:

$$Ax + Cv \leq b, \; v \in V = \{v: \; 0 \leq v \leq 1\}, \tag{5.73}$$

where A, C are matrices of sizes $(m \times n)$ and $(m \times s)$, respectively. Compare the linear model (5.73) and the following auxiliary linear programming problem of search for the Chebyshev point:

$$\min\{v_{s+1}: \; \sum_{j=1}^{n} a_{ji}x_j + \sum_{j=1}^{s} c_{ij}(x)v_j - b_i \leq \sum_{j=1}^{s} |c_{ij}(x)|v_{s+1}, \; \forall i = \overline{1,m},$$

$$v_j - 1 \leq v_{s+1}, \; -v_j \leq v_{s+1}, \; j = \overline{1,s}\}. \tag{5.74}$$

The following theorem [10] is true.

Let $\{x^*, v^*, v^*_{s+1}\}$ be the solution to the auxiliary problem (5.74). Then one of the following three statements are true.

1^0. If $v^*_{s+1} < 0$, then the system of inequalities $Ax^* + Cv \leq b$ is compatible for all v from the cube $\tilde{V}(v)$, inscribed in the polyhedron $V(x) = \{v \in V; Ax + Cv \leq b, x \in E^n\}$ of non-dangerous disturbances; v^* is the cube center, $2|v^*_{s+1}|$ – its side.

2^0. If $v^*_{s+1} = 0$, then there is no pair x, v that meets the strict inequalities $Ax + Cv - b < 0,\ 0 < v < 1$.

3^0. If $v^*_{s+1} > 0$, then the system is incompatible. When replacing vector b in (5.73) by vector $b + 2v^*_{s+1}N$ the inequalities $Ax + Cv < b + 2v^*_{s+1}n$ are satisfied for $x = x^*$ and any $v \in \tilde{V}(v^*), |v^*_{s+1}|)$.

Application of the Chebyshev Points in Global Optimization

Let us consider the minimization problem of scalar additive Lipschitz function $\varphi(x)$ on the convex polyhedron R:

Find

$$\varphi^* = \min\{\sum_i \varphi_i(x_i) : x \in R\}, \tag{5.75}$$

Where $R = \{A : \leq b\}$. Applying the previously described technique inscribe in R the maximum volume cube with sides parallel to the coordinate axes:

$$\min x_{n+1} \tag{5.76}$$

subject to

$$\sum_{j=1}^{n} a_{ij}x_j - b_i \leq \sum_{j=1}^{n} |a_{ij}|x_{n+1}, \quad \forall_i = \overline{1,m},\ j = \overline{1,n}. \tag{5.77}$$

Let $\{x^*, x^*_{n+1}\}$ be a solution to problem (5.76), (5.77). Then x^* is a center of the maximum cube $\underline{R}^0$, inscribed in R^i, $|2x^*_{n+1}|$ is the length of its sides. Determine the set $R^0 = \{x : \alpha_i^0 \leq x_i \leq \beta_i^0\}$, where $\alpha_i^0 = x_i^* - x^*_{n+1}, \beta_i^0 = x_i^* + x^*_{n+1}$, and find

$$\varphi(x^0) = \min\{\sum_{i=1}^{n} \varphi_i(x_i) : x \in R^0\}. \tag{5.78}$$

To solve this problem it is obviously necessary to solve n one-dimensional problems: min $\min\{\varphi_i(x_i) :\ \alpha_i^0 \leq x_i \leq \beta_i^0\}$.

Since $R^0 \subset R$, then $\varphi^* \leq \varphi(x^0) = \overline{\varphi}^0$.

Now let us find a lower bound for φ^*. For this, as previously done, embed R in the minimum volume parallelepiped.

It is obviously necessary to solve $2n$ linear programming problems: find

$$\underline{x}^{0i} = arg\min\{x_i :\ x \in R\},\ i = \overline{1,n}, \tag{5.79}$$

$$\overline{x}^{0i} = arg \max\{x_i : x \in R\},\ i = \overline{1,n}, \tag{5.80}$$

The set $\overline{R}^0 = \{x : \underline{x}^0 \leq x \leq \overline{x}^0\} \supset R$, therefore $\overline{\varphi}_0 = \min \sum_{i=1}^{n} \varphi_i(x_i) : x \in \overline{R}^0\} \leq \varphi^*$, and since $\overline{R}^0 \supset R \supset \underline{R}^0$ we obtain a two-sided estimate $\underline{\varphi}_0 \leq \varphi^* \leq \overline{\varphi}_0$.

By virtue of the fact that $\underline{x}^{01}, ..., \underline{x}^{0n}, \overline{x}^{01}, ..., \overline{x}^{0n}$ are extreme points of R we can make more precise the upper bound φ^0 for φ^*, by calculating the value $\varphi(x) = \sum_i \varphi_i(x_i)$ at these points and choosing the minimal one. This is particularly important if the function $\varphi(x)$ is concave, since it attains both local and global minima at the extreme points of R. If in doing so $|\underline{\varphi}^0 - \overline{\varphi}^0| \leq \epsilon$, then any of the points being either upper or lower bound is taken as an approximate solution. Otherwise, we partition the set R as follows:

Let us form $2n$ convex polyhedrons

$$\overline{R}^{0i} = \{x : x \in R,\ x_i \geq \beta_i^0\},\ \ \forall_i = \overline{1,n},\ \text{ and}$$

$$\underline{R}^{0i} = \{x : x \in R,\ x_i \leq \alpha_i^0\},\ \ \forall_i = \overline{1,n},$$

It is obvious that $R = \cup_i \underline{R}^{0i} \cup_i \overline{R}^{0i} \cup \underline{R}^0$.

In each of the sets $\underline{R}^{0i}, \overline{R}^{0i}$, $i = \overline{1,n}$, we inscribe a parallelepiped or cube $\underline{R}^{1i}$, $i = \overline{1,2n}$, of the maximum volume similarly to (5.76), (5.77).

We solve $2n$ problems of the form:

$$\min\{\sum_i \varphi_i(x_i) : x \in R^{1i}\},\ i = \overline{2,n}.$$

Then we obtain the lower bound $\varphi^* : \underline{\varphi}^1 = \min\{\underline{\varphi}^0, \underline{\varphi}^{1,1}, ..., \underline{\varphi}^{1,2n}\}$. It is easy to see that $\underline{R}^0 \subset \underline{R}^1 = \cup_i \underline{R}^{1i} \cup_i \overline{R}^{1i} \cup \underline{R}^0 \subset R$ and, hence, $\varphi^0 \geq \varphi^1 = \min\{\sum_{i=1}^{n} \varphi_i(x_i) : x \in \underline{R}^1\} \geq \varphi^*$. Now we make more precise the upper bound of φ^*. For this purpose each of the sets $\overline{R}^{0i}$, $\underline{R}^{0i}$, $i = \overline{2,n}$, is embedded in the minimum volume cube $\overline{R}^{1i}$, solving $2n$ linear programming problems of the form (5.79), (5.80).

Let

$$\overline{\varphi}^{1i} = \min\{\sum_{i=1}^{n} \varphi_i(x_i) : x \in \overline{R}^{1i}\},\ i = \overline{2,n},\ \ \overline{\varphi}^1 = \min\{\overline{\varphi}^{1,1}, ..., \overline{\varphi}^{1,2n}\}.$$

It is obvious that $\overline{\varphi}^1 = \min\ \{\sum_{i=1}^{n} \varphi_i(x_i) :\ \ x \in \cup_{i=1}^{2n} \overline{R}^{1i} \cap \overline{R}^0\}$, and since $\overline{R}^0 \supset \overline{R}^1 = \cup_{i=1}^{2n} \overline{R}^{1i} \cap \overline{R}^0 \supset R \supset \underline{R}^1 \supset \underline{R}^0$, then $\underline{\varphi}^0 \geq \underline{\varphi}^1 \geq \varphi^* \geq \overline{\varphi}^1 \geq \overline{\varphi}^0$. Continuing the iteration process we obtain the sequence of embedded sets $\overline{R}^0 \supset \overline{R}^1 \supset ... \supset \overline{R}^k \supset ... \supset R \supset ... \supset \underline{R}^k \supset ... \supset \underline{R}^1 \supset \underline{R}^0$, each of which represents a union of parallelepipeds on which the additive function is minimized quite simply.

Moreover, the volumes $|\overline{R}^k|$ of sets $\overline{R}^k$ satisfy the strict inequality:

$$|\overline{R}^0| > |\overline{R}^1| > ... > |\overline{R}^k| > |R| > ...|\underline{R}^k|... > |\underline{R}^1| > |\underline{R}^0|,$$

and therefore the method has a two-sided estimate of the error in the approximated solution $\underline{\varphi}^k \geq \varphi^* \geq \overline{\varphi}^k$, i.e. a constructive rule of the process completion.

Now let us consider the concave separable programming. Since the concave function attains its minimum at the limit points of the admissible set, the minimization of the function $\sum_{i=1}^{n} \varphi_i(x_i)$ on parallelepipeds becomes essentially simpler and requires only two calculations of each of the functions $\varphi_i(x_i)$.

In order to improve the values of the upper bound we can involve simplex partitions which do not require that the problems of the form

$$\underline{x}^{0i} = arg\min\{x_i : \ x \in R\}, \ i = \overline{1,n}$$

$$\overline{x}^{0i} = arg\max\{x_i : \ x \in R\}, \ i = \overline{1,n},$$

where $R = \{x : \ Ax \leq b\}$ be solved in each iteration to construct the sequence of minimum volume parallelepipeds.

The Chebyshev Point Methods in the Control Theory

1. *Statement of the problem.* Suppose that expansion of an object controlled over the time interval [0,T] is described by linear differential equations

$$\dot{x} = Ax + Bu, \ x(0) = x^0, \tag{5.81}$$

where A and B are the data of matrices of dimension $n \times n$ and $n \times m$, respectively, with the elements $\{a_{ij}(t)\}$ and $\{b_{ij}(t)\}$, $u(t) \in E^m$ is a vector of control from the class of piecewise continuous functions that satisfy the condition $u(t) \in R_u$, $x(t) \in E^n$ is a phase vector such that $x(t_l) \in R_x(t_l) = \{x(t_l);\ g(x(t_l)) \leq 0\}$ in time instant $t_l \in [0,T]$, $g(x)$ is a convex smooth function.

The problem consists in finding the control $u^* \in R_u$ such that the corresponding trajectory $x^*(t)$ belongs to the convex set $R_x(t_i)$ and, besides, the convex function $\varphi(x(T))$ attains the minimum in $x^*(T) \in R_x(T) \cap \overline{X}$, where $\overline{X}$ is the attainability set (5.81).

2. *Variation of functional and variation of control.* Suppose that the current control $u^k(t) \in R_u$ is known. Let us find a corresponding trajectory $x^k(t)$ from the Cauchy problem $\dot{x} = Ax + Bu^k$, $x(0) = x^0$.

Determine the gradient $\nabla\varphi(x^k(T))$.

Now let us consider the system of differential equations

$$\dot{x} = Ax + Bu \tag{5.82}$$

and the conjugate system

$$\dot{p} = -A^T p \tag{5.83}$$

with the right-hand boundary

$$p(T) = \nabla\varphi(x^k(T)). \tag{5.84}$$

Performing scalar multiplication of equations (5.82), (5.83) $p(t)$ and $x(t)$, respectively and summing up, we obtain in the k-th iteration

$$\dot{x}^{kT} p^k + \dot{p}^{kT} x^k = \frac{d(x^{kT} p^k)}{dt} = p^{kT} B u^k.$$

Using (5.84) we have

$$\nabla\varphi(x^k(T))^T x^k(T) - \nabla\varphi(x(0))^T x(0) = \int_0^T p^{kT}(t)B(t)u^k(t)dt. \tag{5.85}$$

Repeating the same transformations for the arbitrary control $u(t)$, corresponding trajectory $x(t)$ and solution of conjugate system (5.83) with constraints $p^k(T) = \nabla\varphi(x^k(T))$ we obtain

$$\nabla\varphi(x^k(T))^T x(T) - \nabla\varphi(x(0))^T x(0) = \int_0^T p^k(t)^T B(t)u(t)dt. \tag{5.86}$$

Subtracting (5.85) from (5.86) we finally have

$$\delta\varphi^k = \nabla\varphi(x^k(T))^T (x(T) - x^k(T)) = \int_0^T p^k(t)^T B(t)\delta u^k(t)dt. \tag{5.87}$$

Formula (5.87) relating the variation $\delta\varphi^k$ of the function $\varphi(x(T))$ with variables of control $u(t)$ ($\delta u^k = u(t) - u^k(t)$), can be considered as a basis for the Chebyshev point algorithm.

Suppose that the sequence of controls $\{u^1(t), ..., u^k(t)\} \in R_u$ has been already obtained. Find the trajectory $x^k(t)$, that corresponds to the control $u^k(t)$ as a solution to the Cauchy problem $\dot{x} = Ax + Bu^k$, $x(0) = x^0$.

Calculate $\beta_{tk} = \max_{0 \le l \le T} g(x^k(t_l))$. If $\beta_{tk} > 0$, write the variation δg^k of the function $g(x^k(t_{j_k}))$ similarly to (5.87):

$$\delta g^k = \nabla g(x^k(t_{jk}))^T (x - x^k(t_{jk})) = \int_0^{t_{jk}} p^k(t)^T B(t)\delta u^k(t)dt, \tag{5.88}$$

where $p^k(t)$ is a solution to the conjugate system of differential equations $\dot{p} = -A^T p$ on the section $[0, t_{jk}]$ with the boundary condition $p(t_{jk}) = \nabla g(x^k(t_{jk}))$.

Now assume that $\beta_{jk} \leq 0$. In this case write the variation $\delta\varphi^k$ of the function $\varphi(x(t))$:

$$\delta\varphi^k = \nabla\varphi(x^k(T))^T(x - x^k(T)) = \int_0^T p^k(t)^T B(t)\delta u^k(t)dt, \tag{5.89}$$

where $p^k(t)$ is a solution to the conjugate system of differential equations $\dot{p} = -A^T p$ on the section $[0, T]$ with the boundary condition $p(T) = \nabla\varphi(x^k(T))$.

Now let us write the analogue of the method (5.59)–(5.62) for solution of problem (5.57), (5.58) for the above written optimal control problem of linear system.

Let $R_u^k = \{u(t) : A^k(t)u(t) \leq b(t)$ be a simplex, $u^k \in R_u^k$. Find its Chebyshev point $u^k(t)$ from the system of equations $A^k(t)u(t) - b(t) = ||a^{kj}(t)||u_{m+1}(t)$ at each time instant. Then find the base vertex of the simplex $\mathrm{R}R^k(t)\overline{u}^k(t)$, in which at each time instant there are exactly m active constraints from $m + 1$ constraints (5.62) that form the simplex R_u^k. For this purpose we solve the Cauchy problem $\dot{x} = Ax + Bu^k$, $x(0) = x^0$ and the conjugate system, find $x^k(t)p^k(t)$ and write the functional variation (5.88) or (5.89). Then linearize it on R_u^k at each time instant. Construct a new simplex R_u^{k+1}, by adding to active constraints at point $\overline{u}^k(t)$ of the simplex R_u^k the inequality

$$\int_0^{t_{jk}} p^k(t)^T B\delta u^k(t)dt \leq \parallel \alpha^k \parallel u_{m+1},$$

where $||\alpha^k|| = ||\nabla g(x^k(t_{jk}))||$, $u_{m+1} \in E^1$ and $\beta_{tk} > 0$, or if $\beta_{tk} \leq 0$, the inequality

$$\delta\varphi^k = \int_0^T p^k(t)^T B(t)\delta u^k(t)dt \leq \parallel \alpha^k \parallel u_{m+1},$$

where $\parallel \alpha^k \parallel = ||\nabla\varphi(x^k(T))||$, $u_{m+1} \in E^1$.

Now find $u^{k+1}(t)$ as the Chebyshev point of the simplex R_u^{k+1}, passing to the step $k + 1$.

Convergence of the iteration process follows from the convergence of the finite-dimensional analogue [14]. The author of [14] also presents similar cutting schemes to solve the problems of optimal control of transient processes in electric power systems as well as similar methods of solving the Goursat-Darboux problems.

Integer programming

Let it be necessary to find

$$\varphi^* = \min\{\varphi(x) = \sum_{i=1}^{n} \varphi_i(x_i) : x \in R\}, \quad x_i \text{ is integer,}$$

$$R = \{x : Ax \le b\}. \tag{5.90}$$

The main operations of the method consist in solving the linear programming problems of the form

$$\underline{x}^{0i} = arg\min\{x_i : x \in R\}, \; i = \overline{1,n}$$

$$\overline{x}^{0i} = arg\max\{x_i : x \in R\}, \; i = \overline{1,n}.$$

In doing so, to obtain the upper bound it is necessary to solve the problem: $\overline{\varphi}^0 = \min\{\sum_{i=1}^{n} \varphi_i(x_i) : x \in \overline{R}^0\}$, where $\overline{R}^0 = \{x : \underline{x}^0 \le x \le \overline{x}^0\}$. Since $\overline{R}^0 \supset R$, then $\overline{\varphi}^0 \le \varphi^*$.

To obtain the lower bound $\underline{\varphi}^0$ it is necessary to inscribe the maximum volume cube in R, i.e. to solve the problem: $\min\{x_{n+1} : \sum_{j=1}^{n} a_{ij}x_j \le ||a^i||x_{n+1}$, $i = \overline{1,m}\}$.

Assume that the pair $\{x^0_{n+1}, x^0\}$ is its solution. Then $\underline{R}^0 = \{x : x^0 + x^0_{n+1} \le x \le x^0 - x^0_{n+1}\} \subset R$, where $x^0_{n+1} < 0$, $2|x^0_{n+1}|$ is the side length of the cube $\underline{R}^0$, then find

$$\underline{\varphi}^0 = \min\{\sum_{i=1}^{n} \varphi_i(x_i) : x \in \underline{R}^0\}.$$

It is obvious that $\overline{\varphi}^0 \le \varphi^* \le \underline{\varphi}^0$. Call $\overline{R}^0$, $\underline{R}^0$ an initial external and internal approximation of the set R.

Further to make precise the external and internal approximation of the set R by uniting parallelepipeds $\overline{R}^k$ and cubes $\underline{R}^k$ we apply the technique presented in (5.71) of the work.

Here, the minimization problem of $\varphi_i(x_i)$ on the section $\underline{x}_i^k \le x_i \le \overline{x}_i^k$, that corresponds to the external approximation has an integer solution always, if the initial problem has a solution. It is obvious that minimization problem of $\varphi_i(x_i)$ on the internal approximation does not always have a solution since the union of parallelepipeds contained in R, may not have integer points. However, as the approximation is made more precise from below the internal problems also become solvable. The technique of solving problem (5.90) repeats absolutely the one presented in (5.71) for the search of global minimum (maximum) of additive functions on a convex polyhedron.

5.3 Solution of Energy Problems by Polynomial-Time Algorithms

5.3.1 Minimal Shutdown During Power Shortage

Consider an electric supply network consisting of nodes $0, 1, 2, ..., m$. Each node, with the exception of node 0, is identified with one of the users of electric power; node 0 is called the base node and it corresponds to the source of power. To each node $i (i = 0, 1, ..., m)$ is associated a non-negative number u_i, the voltage at node i. The voltage at node 0, u_0, is assumed fixed.

The users receive electric power through the network, which consists of n branches. Each branch $k (k = 1, ..., n)$ is defined by a tuple of the following values:

i_k – the index of the initial node in the branch;
j_k – the index of the final node in the branch;
$r_{i_k j_k}$ – the electrical resistance of the branch;
$\overline{P}_{i_k j_k}$ – the maximum active power that can be transmitted through the branch (i_k, j_k) from node i_k to node j_k or from node j_k to node i_k.

The resistances r_{ij} are replaced with so-called mutual conductances $y_{ij} = 1/r_{ij}$. The self-conductances y_{ii} of each non-base node i is defined as

$$y_{ii} = P_i/u_i^2, \tag{5.91}$$

where P_i is the power consumed in node i.

Using this notation, we can write the current-balance equations for each non-base node:

$$0 = -y_{ii}(u_i - 0) + \sum_{j \in I_i} y_{ij}(u_i - u_j), \; i = 1, ..., m, \tag{5.92}$$

where $I_i \subseteq \{0, 1, ..., m\}$ is the set of nodes to which node i is connected by branches (i, j).

Substituting (5.91) in (5.92) and expressing the consumed power in terms of node voltages, we obtain

$$P_i = u_i \left[\sum_{j \in I_i} y_{ij}(u_i - u_j) \right], \; i = 1, ..., m. \tag{5.93}$$

We denote $(i, j = 1, ..., m)$

$$a_{ii} = \sum_{j \in I_i} y_{ij},$$

$$a_{ij} = \begin{cases} -y_{ij} & \text{if } j \in I_i, \\ 0 & \text{otherwise,} \end{cases}$$

$$b_i = \begin{cases} y_{i0}u_0 & \text{if } 0 \in I_i, \\ 0 & \text{otherwise.} \end{cases}$$

In this notation, (5.93) is rewritten as

$$P_i = P_i(u) = u_i\left(b_i - \sum_{j=1}^{m} a_{ij}u_j\right), \; i = 1, ..., m. \tag{5.94}$$

It is easy to see that a_{ij} and b_i satisfy the following conditions:

$$b_i \geq 0, \; a_{ii} > 0, \; a_{ij} = a_{ji} \leq 0, \; i \neq j, \quad \sum_{j=1, j\neq i}^{m} a_{ij} \leq a_{ii}. \tag{5.95}$$

Assuming that the node voltage vector $u \in E^m$ is used as the control, we will determine the system of admissible controls U. First, the controls u_i should be within the allowed bounds u_i^-, u_i^+:

$$u_i \in U_i = \{u_i | u_i^- \leq u_i \leq u_i^+\}, \; i = 1, ..., m. \tag{5.96}$$

Second, the power transmitted by the branch (i, j) is bounded from above. This upper bound is expressed by the following conditions:

$$(u_i, u_j) \in R_{ij} = \{(u_i, u_j) | y_{ij}u_i|u_i - u_j| \leq \bar{P}_{ij}\}, \; i = 1, ..., m, \; j \in I_i. \tag{5.97}$$

From physical considerations, all variables u_i, P_i in the model are non-negative. Therefore the conditions (5.96), (5.97) defining U can be augmented with the following conditions:

$$u \in L_i = \left\{u \in E^n | b_i - \sum_{j=1}^{m} a_{ij}u_i \geq 0\right\}, \; i = 1, ..., m. \tag{5.98}$$

Now the feasible set U can be represented as the Cartesian product

$$U = \prod_{i=1}^{m} U_i \times \prod_{k=1}^{n} R_{i_k j_k} \times \prod_{i=1}^{m} L_i. \tag{5.99}$$

Let us now describe the control goals. Ideally, each user should receive the demanded power P_i^0 and the corresponding node voltage vector u^0 should be in the feasible set U. In this case, we say that the pair (u^0, P^0) defines 'normal operation'.

Let us now consider a so-called emergency (abnormal operation). It arises when the elements of the mutual conductance matrix change so that the system of conditions

$$P_i(u) = P_i^0, \; i = 1, ..., m, \tag{5.100}$$

$$u \in U \tag{5.101}$$

becomes inconsistent. We will only consider the case when the system (5.101) is consistent and the inconsistency of (5.100), (5.101) is attributable to the fact that for all $u \in U$ there is a node i where $P_i(u)$ is less than the demanded power P_i^0:

$$\Delta(u) = \min_{1 \le i \le m} (P_i(u) - P_i^0) < 0 \; for \; all \; u \in U. \tag{5.102}$$

Here $-\Delta(u)$ is called the maximum power shortage corresponding to the control u. It is natural to choose the control u so that the maximum power shortage is minimized.

We thus obtain the maximization problem

$$\max\{\Delta(u) | u \in U\}. \tag{5.103}$$

Problem (5.103) can be modified by introducing for each node i its priority or weight $w_i > 0$ and maximizing in (5.103) the shortage $\Delta_w(u)$ instead of $\Delta(u)$:

$$\Delta_w(u) = \min_{1 \le i \le m} (P_i(u) - wP_i^0).$$

Let us analyze problem (5.103).

Lemma 3. *Given is a pair of points $u,\ \bar{u} \in U$. Then for all $i = 1, ..., m$, we have the implication*

$$P_i(u) \ge P_i(\bar{u}) \Rightarrow \nabla^T P_i(\bar{u})(u - \bar{u}) \ge 0, \tag{5.104}$$

where $\nabla P_i(\bar{u})$ is the gradient of the function P_i evaluated at the point $\bar{u}$:

$$\nabla^T P_i(\bar{u}) = (\partial P_i / \partial u_1, ..., \partial P_i / \partial u_m)_{u = \bar{u}},$$

$$\partial P_i / \partial u_i = b_i - \sum_{j=1}^{m} a_{ij} u_j - a_{ii} u_i,$$

$$\partial P_i / \partial u_j = -a_{ij} u_i, \; i \ne j.$$

Proof. For simplicity, let

$$P_i(u) = P(x, y), \; x = u_i, \; b = b_i, \; a = a_{ii},$$

$$y^T = (y_1, ..., y_{m-1})^T = (u_1, ..., u_{i-1}, u_{i+1}, ..., u_m),$$

$$s(y) = - \sum_{j=I, j \ne i}^{m} a_{ij} u_j.$$

In this notation, (5.104) takes the form

$$x[b - ax + s(y)] \ge \bar{x}[b - a\bar{x} + s(\bar{y})] \Rightarrow$$

$$R = [b - a\bar{x} + s(\bar{y}) - a\bar{x}](x - \bar{x}) + \bar{x}[s(y) - s(\bar{y})] \ge 0.$$

Thus, we have

$$x[b - ax + s(y)] \geq \bar{x}[b - a\bar{x} + s(\bar{y})] \tag{5.105}$$

Since $u \in U$ and $\bar{u} \in U$, we have

$$x \geq 0,\ b - ax + s(y) \geq 0, \tag{5.106}$$

$$\bar{x} \geq 0,\ b - a\bar{x} + s(\bar{y}) \geq 0. \tag{5.107}$$

From (5.105) and $x \geq 0$, we obtain

$$s(y) \geq (\bar{x}/x)[b - a\bar{x} + s(\bar{y})] - [b - ax]. \tag{5.108}$$

From (5.108) and $\bar{x} \geq 0$ we obtain a lower bound for R:

$$R \geq [b - a\bar{x} + s(\bar{y}) - a\bar{x}](x - \bar{x})$$

$$+\bar{x}\{\bar{x}[b - a\bar{x} + s(\bar{y})]/x - [b - ax] - s(\bar{y})\}.$$

Combining the terms with x in the last inequality, we obtain

$$R \geq [b - a\bar{x} + s(\bar{y})][x - \bar{x}^2/x - 2\bar{x}] = \bar{R}(x).$$

By inequality (5.107), $\bar{R}(x)$ is a convex function. Let us find its minimum by equating to zero the derivative $\bar{R}'(x)$:

$$1 - (\bar{x}/x)^2 = 0 \Rightarrow x = \bar{x} \Rightarrow R \geq \bar{R}(\bar{x}) = 0.$$

□

The Lemma has the following geometrical interpretation: the gradient of the function $P_i(u)$ at the point $\bar{u} \in U$ is used to identify the half-space which contains all the points $u \in U$ with values $P_i(u)$ not less than $P_i(\bar{u})$. In other words, the half-space

$$\nabla^T P_i(\bar{u})(u - \bar{u}) \geq 0$$

is the supporting half-space to the set

$$\bar{R}_i = \{u \in U | P_i(u) \geq P_i(\bar{u})\}.$$

At the same time, it is impossible to construct supporting planes to the sets R_{ij} from (5.99) because they are non-convex. To avoid this difficulty, we approximate R_{ij} by the polyhedron $\bar{R}_{ij}$ so that

1. $\bar{R}_{ij}$ includes R_{ij}.
2. Any point $(u_i, u_j) \in \bar{R}_{ij}$ is 'sufficiently close' to R_{ij}.
   We define the approximation accuracy by

$$\nu = \max_{(u_i,u_j) \in \bar{R}_{ij}} \left\{ \max\left\{ \frac{u_i(u_i - u_j) - q_{ij}}{q_{ij}}, \frac{u_j(u_j - u_i) - q_{ij}}{q_{ij}} \right\}\right\}, \tag{5.109}$$

where $q_{ij} = \bar{P}_{ij}/y_{ij}$.

Let us describe one of the possible constructions of the approximating polyhedron $\bar{R}_{ij}$, assuming for simplicity that

$$u_1^- = ... = u_m^- = a,\ u_1^+ = ... = u_m^+ = b. \qquad (5.110)$$

Below, u_i, u_j, and q_{ij} are denoted simply by u, v, q. Find the intersection points of the curve:

$$u(u - v) = q \qquad (5.111)$$

with the boundary of the parallelepiped

$$\Pi = \{(u, v)|a \leq u \leq b,\ a \leq v \leq b\}.$$

This intersection points are $M_1(u_0, a)$ and $M_2(b, v_0)$, where

$$u_0 = [a + (a \cdot a + 4q)^{1/2}]/2,\ v_0 = b - q/b. \qquad (5.112)$$

We naturally assume that $v_0 - a > 0$, because otherwise the constraint on power flow through the branch (i, j) is superfluous.

Let us write the equation of the line through the points M_1 and M_2:

$$v = \alpha u + \beta,\ \alpha = (v_0 - a)/(b - u_0),\ \beta = (a - \alpha)u_0. \qquad (5.113)$$

It is easily seen that $\alpha > 1$ and $\beta < 0$. The maximum of the function $f(u, v) = u(u - v)$ on the line (5.113) is attained at the point

$$M[\beta/2(1 - \alpha),\ \beta(2 - \alpha)/(1 - \alpha)]$$

and its value is $\bar{f} = \beta^2/4(1 - \alpha)$.

The approximation accuracy is then determined by the expression $\nu = \bar{f}/q$ and can be viewed as satisfactory if $\nu \leq 1 + \epsilon$, where ϵ is a given small number.

The set R_q is thus replaced with the polyhedron

$$\bar{R}_q = \{(u_i, u_j)|u_j \geq \alpha u_i + \beta,\ u_i \geq \alpha u_j + \beta\},$$

where α and β are given by (5.113). Practical examples show that the degree of approximation ν is usually between 1.001 and 1.05.

Having replaced R_{ij} with $\bar{R}_{ij}$, we are ready to apply the simplex circumscribing method of Sect. 5.2.

Remark. *The sets R_{i0} corresponding to power flow through the branch joining node i with the base node 0 obviously do not have to be approximated, because R_{i0} are defined by linear inequalities.*

In conclusion of this section note that the simplex circumscribing method has been applied by us to solve problems of the form (5.103) with up to 34 nodes and up to 35 branches. The calculation results show that in 50–100 iterations the method finds a simplex whose vertex coordinates differ in the third decimal place, i.e. it confidently converges to the optimal solution.

5.3.2 Finding Optimal Non-Equilibrium States in Closed Thermodynamic System

A closed thermodynamic system is defined in [8, 9] as follows. Given is a universal list consisting of the indices of n chemical substances $J = \{1, ..., n\}$, and to every index $j \in J$ is associated a non- negative value x_j – the quantity of substance j in moles. Also given is the list of initial substances $J_\nu \subseteq J$ which contains the indices j of the substances charged into the reactor in quantities y_j. We assume that s substances are charged and $J_\nu = \{1, ..., s\}$, $s \leq n$. The quantities $y_1, y_2, ..., y_s$ define the initial state vector of the thermodynamic system x^0:

$$x^0 = (y_1, y_2, ..., y_s, 0, ..., 0) = x^0(y). \tag{5.114}$$

In addition to the list of substances J, we also have the list of chemical elements $I = \{1, ..., m\}$ so that every substance j consists only of atoms from the list I.

Let a_{ij} be the number of atoms of species i in one mole of substance j. Then we can determine $b_i = b_i(y)$, equal to the number of atoms of species i in the initial state of the system:

$$b_i = b_i(y) = \sum_{j=1}^{s} a_{ij} y_j, \; i = 1, ..., m \tag{5.115}$$

The system is closed in the sense that it does not exchange matter with its environment. Therefore, any vector of quantities

$$x = (x_1, ...x_n), \; x_j \geq 0, \; j = 1, ..., n,$$

which in what follows is called the state vector or simply the state of the system, satisfies the mass balance equation

$$\sum_{j=1}^{n} a_{ij} x_j = b_i, \; i = 1, ..., m. \tag{5.116}$$

Let A be an $m \times n$ matrix with elements a_{ij}, $b(y)$ an m -dimensional column vector with components b_i. Then the system of conditions imposed on any state x may be written in the form

$$Ax = b(y), \tag{5.117}$$

$$x \geq 0. \tag{5.118}$$

The set of states $x \in E^n$ satisfying (5.117), (5.118) will be called the balance polyhedron and denoted by $D(y)$. This notation emphasizes that the set of all possible states depends on the initial state y. If there is no danger of confusion, we will omit the symbol y and simply write b for $b(y)$ and D for $D(y)$.

On the state set D we define a scalar function $G(x)$, called the Gibbs free energy:

$$G(x) = \sum_{j=1}^{n} [G_j^0 + RTln(Px_j/\sigma(x))]x_j,\ \sigma(x) = \sum_{i=1}^{n} x_i. \tag{5.119}$$

Here G_j^0 is the given molar energy of substance j, R is the universal gas constant, T is the constant temperature, P is the constant pressure.

The chemical reaction occurring in this thermodynamic system will be interpreted as a thermodynamically allowed transition of the system from state $x' \in D$ to state $x'' \in D$. This implies the existence of a continuous vector function $x(\tau)$ defined for $\tau \in [0, 1]$ such that

$$Ax(\tau) = b,\ x(0) = x',\ x(1) = x'',\ x(\tau) \geq 0, \tag{5.120}$$

$$G(x(\tau')) \geq G(x(\tau''))\ for\ all\ \tau' \leq \tau'' \in [0, 1]. \tag{5.121}$$

Condition (5.121) indicates that the Gibbs free energy is non-increasing for any chemical reactions in a closed thermodynamic system.

We state without proof the following important property of the Gibbs function: it is convex on the set of all non-negative $x(E_+^n)$ and is strictly convex on any segment in E_+^n which does not lie on the straight line through the origin.

From physical considerations it follows that $a_{ij} \geq 0,\ b_i > 0$. Therefore, the balance polyhedron does not contain a segment which lies on a line through the origin, as otherwise we would have

$$Ax' = b,\ Ax'' = b,\ x' \geq 0,\ x'' \geq 0,\ x'' \neq 0,\ x'' - x' = kx'',\ k \neq 0,$$

whence $0 = A(x' - x') = kAx'' = b,\ i.e.\ b = 0$. Thus, $G(x)$ is strictly convex on D.

The gradient $\nabla G(x)$ of the function $G(x)$ is defined at all points $x \in D$ such that $x_j > 0,\ j = 1, ..., n$ (i.e. in the interior points of D). The components of the gradient vector are given by

$$\partial G(x)/\partial x_j = G_j^0 + RTln(Px_j/\sigma(x)),\ j = 1, ..., n. \tag{5.122}$$

Assume that D has an interior point $\bar{x}$. The state x^e where $G(x)$ attains its minimum will be called an equilibrium state, or simply an equilibrium. By strict convexity of $G(x)$ on $D,\ x^e$ is an interior point of D. The latter is easily proved by contradiction.

Let $x_j^e = 0\ for\ j \in J_0 \subset J$. Consider the interval

$$I = \{x \in D | x = x(t) = x^e + t(\bar{x} - x^e),\ 0 \leq t < 1\}.$$

All the coordinates of the point $\bar{x}$ are strictly positive, and therefore the same is also true for all points $x(t)$ from I and we can compute

$$\frac{d}{dt}G(x(t)) = \sum_{j\in J/J_0} \frac{\partial G(x(t))}{\partial X_j}(\bar{x}_j - x_j^e) + \sum_{j\in J_0} \frac{\partial G(x(t))}{\partial x_j}\bar{x}_j.$$

The first term in the right-hand side is bounded for all $t \in [0, 1]$; the second term becomes an artbitrarily large negative value by (5.122), which contradicts the continuity of G and the definition of the point x^e.

States other than x^e will be called non-equilibrium states in what follows.

We now define the subset $J^* \subset J$ and the utility function of the state x:

$$L_p(x) = \sum_{j\in J}^{*} c_j x_j \tag{5.123}$$

This definition identifies the subset of useful substances from the universal list J. Here $c_j > 0$ is the utility associated with substance j.

It is easy to find the vertex of the balance polyhedron $D(y)$ where the function $L_p(x)$ attains its maximum. To this end, we have to solve the linear programming problem

$$\max\{L_p(x)|x \in D(y)\}. \tag{5.124}$$

Since $x^0(y) \in D(y)$ and $D(y)$ is a bounded set, the function

$$f_p^M(y) = \max_{x\in D(y)} L_p(x) \tag{5.125}$$

is defined. The value $f_p^M(y)$ provides an upper bound for the yield of useful substances corresponding to the initial state $x^0(y)$. In other words, introducing the substances $j (j \in J_y)$ in quantities y_j into the system, we cannot obtain more than $f_p^M(y)$ of the useful product.

The reachability set $D'(y)$ from the state $x^0(y)$ is the set of all states x which can be reached by a continuous transition from $x^0(y)$ satisfying conditions (5.120), (5.121). Denote by $G^0(y)$ the energy in state $x^0(y)$ and define the set

$$D^0(y) = \{x \in D(y)|G(x) \leq G^0(y)\}. \tag{5.126}$$

We have the obvious inclusions

$$D'(y) \subseteq D^0(y) \subseteq D(y). \tag{5.127}$$

We can now state the main problem of this section:

$$\max\{L_p(x)|x \in D'(y)\}. \tag{5.128}$$

The maximum value in (5.128) is denoted by $f_p'(y)$ and the point where this maximum is attained is denoted by $x'(y)$. The state $x'(y)$ is called optimal non-equilibrium state. By inclusion (5.127)

$$f_p'(y) \leq f_p^M(y). \tag{5.129}$$

Let us proceed with the analysis of problem (5.128).

Lemma 4. *The reachability set $D'(y)$ is convex.*

Proof. Let $x' \in D'(y)$ and $x'' \in D'(y)$. This means that there exist thermodynamically allowed transitions from $x^0(y)$ to x' and from $x^0(y)$ to x''. But then there also exists a transition from $x^0(y)$ to the state $z = (x' + x'')/2$, because one of the composite transitions $x^0(y) \to x' \to z$ or $x^0(y) \to x'' \to z$ is thermodynamically allowed by convexity of the function G(x). □

Thus, (5.128) is formally a convex programming problem. The main difficulty here is that, unlike the set $D^0(y)$, the feasible set $D'(y)$ is defined implicitly.

Let $x^*(y)$ be one of the solutions of problem (5.124) and $G^*(y) = G(x^*(y))$. Assume that $G^*(y) \leq G^0(y)$. Then $x^*(y) \in D^0(y)$. It would appear that the point $x^*(y)$ could be accepted as a solution of problem (5.124). This would be a mistake, however, because there is no transition to the vertex of the balanced polyhedron from any point other than the vertex itself.

To prove this assertion, we represent the system of conditions defining D in the form of inequalities, expressing the variables that correspond to the basis components of the vector $x^*(y)$ in terms of the variables corresponding to the non-basis components:

$$\left.\begin{array}{ll} x_i = \bar{b}_i - \sum_{j=m+1}^{n} \bar{a}_{ij} x_{ij} \geq 0, \; i = 1, ..., m, & \\ x_j \geq 0, & j = m+1, ..., n. \end{array}\right\} \tag{5.130}$$

We re-indexed the variables x_j so that the first m components correspond to the basis components of $x^*(y)$. Then

$$\left.\begin{array}{l} x_i^*(y) = \bar{b}_i, \; i = 1, ..., m, \\ x_i^*(y) = 0, \;\; i = m+1, ..., n. \end{array}\right\} \tag{5.131}$$

Rewrite inequalities (5.130) in matrix form, setting $z \in E^{n-m}$, $z = (x_{m+1}, ..., x_n)$:

$$\bar{A}z \leq \bar{b}, \; z \geq 0, \tag{5.132}$$

where $\bar{A}$ is an $m \times (n-m)$ matrix, $\bar{b} \in E^m$.

Denote the convex polyhedron (5.132) by $\bar{D}(\bar{D}(y))$. The point $z = 0$ is the vertex of $\bar{D}$. Assume that the first k components of the vector $\bar{b}$ are 0 and the remaining $m - k$ components of $\bar{b}$ are strictly positive. At the vertex $z = 0$ define the cone of feasible directions S:

$$S = \{x \in E^{n-m} | \sum_{j=1}^{n-m} \bar{a}_{ij} s_j \leq 0, \; i = 1, ..., k, \; s_j \geq 0, \; j = 1, ..., n-m\}. \tag{5.133}$$

Also define the functions

$$\bar{G}(z) = G\left(\bar{b}_1 - \sum_{j=1}^{n-m} \bar{a}_{ij} z_j, ..., \bar{b}_m - \sum_{j=1}^{n-m} \bar{a}_{mj} z_j, \; z_1, ..., z_{n-m}\right), H(s) = \min_{i \geq 0} \bar{G}(ts). \tag{5.134}$$

The function $\bar{G}$ obtained from G by replacing the basis variables with their linear expressions in terms of non-basis variables is obviously convex. The function $H(s)$, in general, does not have this property, but the following lemma is true.

Lemma 5. *$H(s)$ is quasiconvex on S.*

We have to show that for any $s_1, s_2 \in S$,

$$H((s_1 + s_2)/2) \leq \max\{H(s_1), H(s_2)\}. \tag{5.135}$$

Proof. Take arbitrary $s_1, s_2 \in S$. By convexity of S, $\bar{s} = (s_1 + s_2)/2 \in S$. Let $H(s_1) = \bar{G}(t_1 s_2)$, $H(s_2) = \bar{G}(t_2 s_2)$. If at least one of the multipliers t_1, t_2 is zero, then (5.135) is obviously true. We therefore assume that $t_1 > 0$, $t_2 > 0$ and we denote $\lambda = t_2/(t_1 + t_2)$. Then $0 < \lambda < 1$ and $1 - \lambda = t_1/(t_1 + t_2)$,

$$H\left(\frac{s_1 + s_2}{2}\right) = \min_{t \geq 0} \bar{G}\left(t\frac{s_1 + s_2}{2}\right) \leq \bar{G}\left(\frac{2t_1 t_2}{t_1 + t_2} \cdot \frac{s_1 + s_2}{2}\right)$$

$$= \bar{G}\left(\frac{t_2}{t_1 + t_2}(t_1 s_1) + \frac{t_1}{t_1 + t_2}(t_2 s_2)\right)$$

$$= \bar{G}((1 - \lambda)(t_2 s_2) + \lambda(t_1 s_1)).$$

Now by convexity of $\bar{G}$ we conclude that

$$H((s_1 + s_2)/2) \leq \lambda \bar{G}(t_1 s_1) + (1 - \lambda)\bar{G}(t_2 s_2)$$

$$= \lambda H(s_1) + (1 - \lambda)H(s_2) \leq (1 - \lambda + \lambda)\max\{H(s_1), H(s_2)\}$$

$$= \max\{H(s_1), H(s_2)\}.$$

□

The unreachability of the vertex is proved by Lemma 2.

Lemma 6. *The vertex $z = 0$ of the cone S is a local maximum point of the function $\bar{G}(z)$ on S.*

Proof. To prove the lemma, find the total partial derivatives of $G(z)$:

$$\frac{\partial \bar{G}}{\partial z_j} = \sum_{i=1}^{m} \frac{\partial G(x)}{\partial X_i} \cdot (-\bar{a}_{ij}) + \frac{\partial G(x)}{\partial X_{j+m}}, \; j = 1, ..., n - m, \tag{5.136}$$

where $x \in E^n$ is obtained by substituting $z = 0$ in (5.130). From (5.136) and formulas (5.122) for the derivatives of G we obtain that in a sufficiently small neighborhood of the point $z = 0$ the derivatives $\partial \bar{G}/\partial z_j$ become negative. □

Thus, for any direction $s \in S$, there exists a number $\bar{t}(s) > 0$ such that $\bar{G}(ts) > \bar{G}(\bar{t}(s)s)$ for all $t \in [0, \bar{t}(s)]$.

Let us find the value $\bar{G}(y)$, equal to the maximum value in the problem

$$\hat{G}(y) = \max_{s \in S}\{\min_{i \geq 0, ts \in D} \bar{G}(ts)\}. \tag{5.137}$$

By Lemma 5, the inner minimum in (5.137) is a quasiconvex function of s. Therefore, to find $\hat{G}(y)$, it suffices to enumerate all the extreme directions (edges) of the cone S.

Let $G^e = G(x^e)$ be the energy at the equilibrium point and G_{max} the maximum value of $G(x)$ on D. For any $g \in [G^e, G_{max}]$, define the set

$$\bar{D}(y, g) = \{x \in D(y) | G(x) \leq g\} \tag{5.138}$$

and the function

$$\hat{f}_p(y, g) = \max\{L_p(x) | x \in \bar{D}(y, g)\}. \tag{5.139}$$

In particular, $D^0(y) = \bar{D}(y, G^0(y))$.

Since $L_p(x)$ is a linear function, its maximum is achieved on the boundary of the set $\bar{D}(y, g)$. We thus have the implication

$$g'' > g'' \Rightarrow \bar{D}(y, g'') \supseteq \bar{D}(y, g') \Rightarrow \bar{f}_p(y, g'') \geq \bar{f}_p(y, g'), \tag{5.140}$$

which shows that $\bar{f}_p(y, g)$ is a non-decreasing function of the parameter g. Construct the segment $v(t)$ joining the points $x^0(y)$ and $x^*(y)$,

$$v(t) = (1 - t)x^0(y) + tx^*(y) \tag{5.141}$$

and find the intersection point of this segment with the surface

$$G(x) = \hat{G}(y). \tag{5.142}$$

Two cases are possible.

Case A. There are no such intersection points or the point $x^0(y)$ is on the surface (5.142).

In this case, the solution of problem (5.128) is also attained on this surface: $f_p'(y) = \bar{f}_p(y, G^0(y))$. Indeed, from the definition of the function $\hat{G}(y)$ it follows that on any ray originating from the vertex $x^*(y)$ the function $G(x)$ is monotone decreasing to its intersection with the surface (5.142) and thus certainly to its intersection with the surface $G^0(y)$, because in this case $G^0(y) \geq \hat{G}(y)$. But by the same argument the function $L_p(x)$ is non-increasing on this ray. Therefore

$$f_p'(y) \leq \bar{f}_p(y, G^0(y)). \tag{5.143}$$

Let $\hat{x}(y)$ be a point on the surface (5.142) where the utility function attains its maximum, i.e.

$$\hat{f}_p(y, G^0(y)) = L_p(\bar{x}(y)).$$

Construct a thermodynamically allowed transition from $x^0(y)$ to $\hat{x}(y)$. To this end, construct the segments $AB = [x^0(y), x^*(y)]$, $CB = [\hat{x}(y), x^*(y)]$, $AC = [x^0(y), \hat{x}(y)]$. By convexity of $G(x)$ it follows that $G(x)$ is less than $G^0(y)$ at all points of the segment AC except the ends. At the same time, $G(x^*(y)) > G^0(y)$ in the case being considered. If we continuously move the point x of the segment AC in the direction from $x^0(y)$ to $\bar{x}(y)$ and for each point x find the intersection points $\bar{x}$ of the segment $[x, x^*(y)]$ with the surface (5.142), then these points $\bar{x}$ describe a continuous are which is entirely contained in (5.142) and defines the sought transition.

Since we have constructed a thermodynamic transition, we may write

$$f'_p(y) \geq \hat{f}_p(y, G^0(y)). \tag{5.144}$$

Combining (5.133) and (5.144), we conclude that

$$f'_p(y) = \hat{f}_p(y, G^0(y)). \tag{5.145}$$

Thus in case A, the problem is equivalent to the explicit convex programming problem

$$\max\{L_p(x) | x \in \bar{D}(y, G^0(y))\}. \tag{5.146}$$

Case B. In this case, as we move along the segment AC, the function $G(x)$ first decreases from $G^0(y)$ to some value $\bar{G}$ and then starts increasing. In this case, we can only obtain a two-sided bound on the optimal value $f'_p(y)$:

$$\hat{f}_p(y, \bar{G}) \leq f'_p(y) \leq \hat{f}_p(y, \bar{G}(y)). \tag{5.147}$$

In both cases, we have to solve convex programming problems of a special kind (g-problems):

$$\max\{L_p(x) | x \in D(y),\ G(x) \leq g\}. \tag{5.148}$$

In case A, problem (5.148) is solved only for one value of g, $g = G^0(y)$, while in case B it is solved for two values: $g = \bar{G}$ and $g = \hat{G}(y)$.

Given the information obtained by solving the linear programming problem (5.124) (determination of the maximum yield of useful products ignoring the thermodynamic constraints) by the supporting cone algorithm of Sect. 5.2, we can continue to apply this algorithm to the g-problem augmenting the linear problem constraints with the constraint $G(x) \leq g$.

5.3.3 Optimal Allocation of Water Resources

Many optimization problems with an economic performance criterion are representable as linear programming models of the form

$$\min\{c^T x | Ax \leq b,\ x \geq 0\}, \tag{5.149}$$

This model is perfectly appropriate for problems with deterministic parameters, but it becomes undefined when the input information is stochastic. In some problems studied at Energy Systems Institute SB RAS, the parameters A, b and c of the model (5.149) are random variables, e.g. the yield of crop i in soil j, failure of unit i at time j, price of product i in period j, river flow, head water level in a reservoir, used load, etc. If averaged parameters are substituted, the model may become totally inadequate for the phenomenon being observed.

The transition to crude game-theoretical criteria is justified only if violation of the constraints incurs a severe penalty that wipes out the gain achieved by optimization of the objective function. The simplest techniques of allowing for the stochastic nature of the input information are therefore not always rigorously justified. In stochastic programming, it is essential to define clearly what is meant by feasible and optimal solutions.

For the problems considered in this section, the constraints $Ax \leq b$ are naturally replaced with stochastic constraints of the form

$$A: \qquad P\left\{\sum_{j=1}^{n} a_{ij}x_j \leq b_i\right\} \geq p_i, \ i = 1, ..., m.$$

This format indicates that a feasible stochastic program is a vector $x \geq 0$ which ensures that the first constraint is satisfied with probability P not less than a given p_i.

Such models are called models with row probability constraints. For the problem of optimal allocation of water resources considered in this section, feasible solutions are defined by the probability constraints

$$B: \qquad P\{Ax \leq b\} \geq p, \ 0 < p \leq 1, \ x \geq 0.$$

This format allows correlations between all random variables, but ignores the relative value or importance of satisfying separate inequalities. Format B appears the most appropriate when the researcher is mainly interested in the operating reliability of the system as a whole.

Linear problems with stochastic constraints are not new, and they have been studied in detail. Deterministic equivalents have been obtained in some cases. We will demonstrate this for the problem with row probability constraints.

Suppose that the matrix A is fixed and b is a random vector with multivariable density function $\Phi(b_1, ..., b_n)$. Then the density function of the component b_i is given by

$$\Phi_i(b_i) = \int_{-\infty}^{+\infty} ... \int_{-\infty}^{+\infty} \Phi(b_1, ..., b_m) \prod_{j \neq i} db_j.$$

We define $\overline{b}_i$ from the equation

$$f(b_i) = \int_{b_i}^{+\infty} \Phi_i(t)dt = p_i.$$

The condition

$$P\left\{\sum_{j=1}^{n} a_{ij}x_j \le b_i\right\} \ge p_i$$

is obviously equivalent to the inequality

$$\sum_{j=1}^{n} a_{ij}x_j \le \overline{b}_i.$$

This equivalence enables us to reduce the linear stochastic programming problem

$$\min M(c^T x)$$

subject to

$$P\left\{\sum_{j=1}^{n} a_{ij}x_j \le b_i\right\} \ge p_i,\ i = 1, ..., m,\ x \ge 0$$

with random b and c and a deterministic matrix A to a linear programming problem

$$\min\{\overline{c}^T x | Ax \le \overline{b},\ x \ge 0\},$$

where $\overline{c}$ is the expectation of c.

Now assume that the components b_i of the vector b are independent random variables and the constraints are defined in the form

$$P\{Ax \le b\} \ge p,\ 0 < p \le 1,\ x \ge 0.$$

Then the equivalent deterministic problem takes the form

$$\min \overline{c}^T x \tag{5.150}$$

subject to

$$Ax \le b,\ x \ge 0, \tag{5.151}$$

$$\prod_{i=1}^{m} [1 - F_{b_i}(\overline{b}_i)] \ge p, \tag{5.152}$$

where $F_{b_i}(\overline{b}_i)$ is the distribution function of the component b_i of the vector b. The unknowns in problem (5.150)–(5.152) are the vectors x and b.

If the components of the vector b follow Weibull, normal, uniform, exponential or gamma distributions, then the problem (5.150)–(5.152) is a convex programming problem. Although the left-hand side of inequality (5.152) is not a convex function, its replacement with the equivalent inequality:

$$\sum_{i=1}^{m} ln[1 - F_{b_i}(\bar{b}_i)] \geq lnp$$

makes it possible to pass for all these distributions to deterministic problems in which the left-hand sides of the inequality constraints ($\leq$) are linear or convex functions.

Similar considerations suggest reduction of other, essentially more complex, stochastic programming problems to their deterministic equivalents.

A detailed model formulated in the form (5.149) contains 5,700 variables, more than 1,000 constraints, and around 27,000 non-zero elements in the matrix A. The model includes constraints that represent the demand for main agricultural products; the relationship of livestock production with the feed base; labor, land, and water resources; capital investments; and product distribution. The constraints on distribution and capital investments are systemwide, while the other constraints have block structure, corresponding to the decomposition of the system into 29 economic and 25 water-resource regions. The objective function is discounted cost minimization.

An aggregated model is being developed. In this model, the number of regions is reduced to 10 and the various agricultural products are partly consolidated. The aggregated model is expected not to exceed 1,000 variables, 300–400 linear constraints, and up to 5,000 non-zero elements in the matrix A. Two versions of the model are proposed – deterministic and stochastic.

The stochastic problem of optimal allocation of water resources in the context of planning the location of agricultural production is formulated in the form

$$\min \bar{c}^T x \tag{5.153}$$

subject to

$$P\{Ax \leq b\} \geq p, \ 0 < p \leq 1, \ x \geq 0, \tag{5.154}$$

$$x \in R, \tag{5.155}$$

where R is a convex polyhedron. Some elements a_{ij} of the matrix A and the components b_i of the vector b are normal or gamma-distributed random variables.

A software system was developed for automatic generation and solution of problem (5.153)–(5.155). The system includes the following programs:

1. The program MODEL, which automatically generates the matrix A and the set R using given technical-economic tables and characteristics and also constructs a deterministic analog of problem (5.153)–(5.155) for given distribution functions.
2. The program CONE, which solves the convex programming problem by the supporting cone algorithm and produces a two-sided bound on the error of the approximate solution.

The sought variables in this problem are the irrigated and dry crop areas, livestock production volumes by climatic regions, river flows, and water transfer between reservoirs.

The feasible set in problem (5.153)–(5.155) is defined by constraints on production volumes, on land, water, and labor resources, on crop rotation, and on conversion of plant material into feed. Flows are considered as gamma-distributed random variables; crop yields and labor resources are assumed normally distributed.

Two models are considered. The first model minimizes the cost of production, land amelioration, and water transfer for a given gross product level. The second model maximizes the gross product subject to cost constraints.

The aggregated version of the first model contains the same groups of variables as the detailed model. In future, in will include constraints on production assets, fertilizer usage, energy resources, conjunctive water use, and location of energy facilities in agricultural regions.

5.3.4 The Chebyshev Points in Trade-Off Control of Electric Power Systems

With the electric power industry entering the market economy the need arises to revise the principles of control of bulk electric power systems [19, 20]. The authors of [19–21] show the trade-off control models and methods appropriate for the new conditions that can be used to find a set of trade-off solutions for two partners in the market (Pareto-optimal set) and suggest a choice of a trade-off on this set under different criteria of partners. In [21] these approaches were generalized for n partners.

One of the authors of this work suggested trying to state this problem as a search for the Chebyshev point on the pareto set of solutions and use the method of simplex embeddings for its solution [7].

Let us assume that the considered electric power system consists of n nodes and m ties among them. Everywhere further agree to denote the numbers of nodes by the letter i and the numbers of ties by the letter j. The node can contain either power plant, i.e. a power generator or a power consumer, or both.

Description of the model can be divided into two stages: physical and economic.

Let us start with the description of the physical part of the model.

Variables and Constants of the Physical Model

Variables and constants of the model are divided into two groups: variables and constants associated with nodes and variables and constants associated with ties. Each node i is characterized by generation or production of electric power in the volume of x_i. If node i has a power plant then its maximal capacity is set by the constant $\overline{x}_i$. If node i has only a consumer, then suppose that $\overline{x}_i = 0$. Thus, x_i is the variable that satisfies the inequalities $0 \leq x_i \leq \overline{x}_i$.

If node i has a consumer, its capacity is set by the constant P_i. If node i has only a power plant, then $P_i = 0$.

Thus, each node is characterized by the variable x_i, constant $\overline{x}_i$ and constant P_i.

Now let us consider the variables and constants describing ties. Each tie j is characterized by the power flow y_j along this tie and constant $\overline{y}_j$ indicative of the maximal transfer capability of the tie j, i.e. y_j is the variable that satisfies the inequality

$$y_j \leq \overline{y}_j. \tag{5.156}$$

Thus, each tie is described by the variable y_j, and constant $\overline{y}_j$.

Relationships Relating Variables and Constants of the Physical Model

The main and the only relationship here is the first law of Kirchhoff which can be interpreted as follows: the node i, should use or transmit to other nodes and use to cover power P_i. as much of electric power as was produced or received by the node i. In order to write this law formally it is necessary to know the direction of flows in the entire system which is unknown in advance. Let us set some a priori direction of the flows. If it is necessary to change the flow to the inverse one, allow negative values of the variables y_j. Here inequality (5.156) will have the form:

$$-\overline{y}_j \leq y_j \leq \overline{y}_j.$$

When constructing the model we proceed from the fact that the information on ties among nodes is set in the form of matrix A

$$a_{ij} = \begin{cases} 1, & \text{if there is tie } i-k \\ 0, & \text{otherwise.} \end{cases}$$

In the notions assumed the first law of Kirchhoff has the following form:

$$x_i = P_i + \sum_{j=1}^{m} a_{ij} y_j, \quad i = \overline{1,n}$$

or in the matrix form $x = P + Ay$, where $x = (x_1, ..., x_n)$, $P = (P_1, ..., P_n)$, m is the number of ties.

Thus, the physical operation of system is described by the following system of equations and inequalities

$$x_i = P_i + \sum_{j=1}^{m} a_{ij} y_j, \quad i = \overline{1,n},$$

$$0 \le x_i \le \overline{x}_i, \; i = \overline{1,n}$$

$$-\overline{y}_j \le y_j \le \overline{y}_j, \; j = \overline{1,m}.$$

Note that in this physical model we neglect power losses in the ties.

Description of Economic Part of the Model, Variables of the Economic Model

The variables introduced at this stage are associated only with ties. The first variable z_j corresponds to the transfer capability reserves of each tie j for the case of some contingency (emergency situation). Since the reserves of transfer capability can not be more than the transfer capability itself, then

$$-\overline{y}_j \le z_j \le \overline{y}_j, \; j = \overline{1,m}.$$

We believe that z_j can take negative values based on the considerations we applied to y_j.

Let us assume that the maximal price of the flow along the tie line j is constant $\overline{c}_j$. Then the price itself c_j is the variable that satisfies the inequalities

$$0 \le c_j \le \overline{c}_j.$$

Thus, two more groups of variables z_j and c_j appeared in the economic part of the model.

The minimal reserve capacity of each node is set by the constant $\underline{z}_j$. The system of constraints to be introduced has the following form:

$$\overline{x}_i - x_i + \sum_{j=1}^{m} a_{ij} z_j \ge \underline{z}_i, \; i = \overline{1,n}.$$

Each tie line j is assigned to both flow y_j and reserve z_j. Hence, there exists the inequality

$$-\overline{y}_j \le y_j + z_j \le \overline{y}_j, \; j = \overline{1,m}.$$

Each node that generates electric power has certain costs of this generation. In this model it is considered that the relationship between the costs $B_i(x_i)$ in node i and the generated power x_i has the following form:

$$B_i(x_i) = \overline{B}_i + p_i x_i + d_i x_i^2, \; i = \overline{1,n} \tag{5.157}$$

where $\overline{B}_i, p_i, d_i$ are some constants. And, in the end, the flow y_j, that comes to or goes from node i at price c_j leads to a loss or profit of node i in the amount of $a_{ij} y_j c_j$. Hence, the total profit or loss of node i related to flows y_j along all tie lines is $\sum_{j=1}^{m} a_{ij} y_j c_j$, and costs $L_i(x_i, c, y)$ of node i depend on generation x_i, a set of prices $c = (c_1, ..., c_m)$, a set of flows $y = (y_1, ..., y_m)$ and have the form

$$L_i(x_i, c, y) = B_i(x_i) + \sum_{j=1}^{m} a_{ij} y_j c_j,$$

where $B_i(x_i)$ is determined in (5.157).

The problem consists in the search for the Chebyshev point of the system of inequalities

$$L_i(x_i, c, y) \leq 0$$

subject to

$$x_i = P_i + \sum_{j=1}^{m} a_{ij} y_j, \ i = \overline{1, n},$$

$$\overline{x}_i - x_i + \sum_{j=1}^{m} a_{ij} z_j \geq \underline{z}_i \ i = \overline{1, n},$$

$$-\overline{y}_j \leq y_j + z_j \leq \overline{y}_j, \ j = \overline{1, m},$$

$$0 \leq x_i \leq \overline{x}_i \ i = \overline{1, n},$$

$$-\overline{y}_j \leq y_j \leq \overline{y}_j, \ j = \overline{1, m},$$

$$-\overline{y}_j \leq z_j \leq \overline{y}_j, \ j = \overline{1, m},$$

$$0 \leq c_j \leq \overline{c}_j \ j = \overline{1, m}.$$

The authors of [21] present the relevant calculations on this model using the method of simplex embeddings. The calculations are performed on the example of a three-node EPS consisting of three thermal power plants.

5.4 Conclusion

In 1980 Khachian showed a polynomial solution to linear programming problems. One would think it was a revolution in the applied mathematics, but everything was not so simple. The algorithm suggested by Khachiyan was efficient only for the problems of medium dimensionality and could not be applied to high-dimensional problems that were solved by modified simplex methods. The matter is that the guaranteed estimate of convergence rate of the Khachian method coincided with the average estimate.

At the same time our Institute performed analogous work, in which the sets containing initial data coincided with simplexes or cones. For these methods, as is shown in the present work, the estimates on the average were essentially better than the guaranteed estimates. However, in real situations it was difficult to expect the problems in which the methods could obtain guaranteed estimates in each iteration. This is confirmed by numerous examples, in particular, by those presented in this paper. Further Karmarkar and others [23, 39, 40] also obtained the improved method of interior points which made it possible to solve the problems of higher dimensionality than those that could be solved by the simplex method and its modifications.

References

1. *The mathematical encyclopedia.* Editor-in-chief I.M.Vinogradov. Moscow, 1985. V. 5, p. 744.
2. I. A. Alcksundrov, E. G. Antsiferov, and V. P. Bulalov. Centered cutting methods. In *Conference on Mathematical Progress, abstracts of papers*, pages 162–163, 1981. Sverdlovsk: IMM UNTs AN SSSR.
3. I. A. Aleksandrov, E. G. Antsiferov, and V. P. Bulatov. *Centered cutting methods in convex programming.* Preprint. Irkutsk: SEI SO AN SSSR, 1983.
4. E. G. Antsiferov. The ellipsoid method in quadratic programming. In *Conference on Mathematical Progress, abstracts of papers*, pages 9–10. Sverdlovsk: IMM UNTs AN SSSR, 1987.
5. E. G. Antsiferov. *Numerical methods of analysis and their applications (Chislennye Metody Analiza i ikh Prilozheniya)*, chapter The ellipsoid method in convex programming, pages 5–29. Irkutsk: SEI SO AN SSSR, 1987.
6. E. G. Antsiferov. *Models and methods of operations research (Modeli i Melody Issledovaniya Operatsii)*, chapter The ellipsoid method in convex programming, pages 4–22. Novosibirsk: Nauka, 1988.
7. E. G. Antsiferov and V. P. Bulatov. An algorithm of simplex embeddings in convex programming. *J. comput. Math. Math. Phys.*, 27(3):377–385, 1987.
8. E. G. Antsiferov, B. M. Kaganovich, P. T. Semenei, and M. K. Takaishvili. *Numerical methods of analysis and their applications (Chislennye Metody Analiza i ikh Prilozheniya)*, chapter Search for intermediate thermodynamic states in physico-chemical systems, pages 150–170. Irkutsk: SEI SO AN SSSR, 1987.
9. E. G. Antsiferov, V. M. Kaganovich, and G. S. Yablonskii. Thermodynamic limitations in searching for regions of optimal performance of complex chemical reactions. *React. Kinet. Lett.*, 31(1), 1988.
10. E. G. Antsiferov, L. T. Ashchepkov, and V. P. Bulatov. *Optimization methods and their applications*, chapter Part I. Mathematical programming, page 157. Novosibirsk, Nauka, 1990.
11. R. G. Bland, D. Goldfarb, and M. J. Todd. The ellipsoid method: A survey. *Oper. Res.*, 29(6):1039–1091, 1981.
12. V. P. Bulatov. *Methods of approximation for solution of some extreme problems.* Dissertation, Tomsk University, Russia, 1967.
13. V. P. Bulatov and I. I. Dikin. Methods of optimization. In *Collected papers "Applied mathematics"*, pages 1–171. Preprint, Irkutsk: SEI SO AN SSSR, P., 1974.
14. V. P. Bulatov and Shepotko. *Methods of orthogonal simplexes in convex programming.* Collected papers "Applied mathematics." Preprint, Irkutsk: SEI SO AN SSSR, 1982.
15. V. Bulatov, T. Belykh, and A. Burdukovskaja. Methods of the chebyshev points for some problems of operations research. In *Proceedings of the 13-th*

Baikal International School-seminar. Optimization methods and their applications, volume 1, Irkutsk, 2005.
16. F. L. Chernousko. Optimal guaranteed bounds on uncertainty using ellipsoids. *Izv. Akad. Nauk SSSR, Tekh. Kibern.*, 4:3–11, 1980.
17. I. I. Dikin. Iterative solution of linear and quadratic programming problems. *Dokl. Akad., Nauk SSSR*, 174(4):747–748, 1967.
18. J. Elzinda and T. G. Moore. A central cutting plane algorithm for the convex programming problems. *Math. Program.*, 8:134–145, 1975.
19. A. Z. Gamm. Optimization of power interconnection operation under new economic conditions. *Elektrichestvo*, 11:1–8, 1993.
20. A. Z. Gamm. A trade-off control of economically independent electric power systems. *Izv. RAS. Energetika*, 11:46–57, 1993.
21. A. Z. Gamm, E. V. Tairova, and O. V. Khamisov. *Pareto-optimal and Chebyshev points in the trade-off control of electric power systems.* ESI SB RAS, Irkutsk, 1998.
22. D. Goldfarb and S. Liu. *An o(n'l) primal interior point algorithm for convex quadratic programming.* Technical report, Dept. IEOR, Columbia University, New York, NY, 1968.
23. N. Karmarkar. A new polynomial-time algorithm for linear programming. *Combinatorica*, 4:373–395, 1984.
24. L. G. Khachiyan. Polynomial-time algorithm in linear programming. *Dokl. Akad. Nauk SSSR*, 244(5):1093–1096, 1979.
25. L. G. Khachiyan. Polynomial-time algorithm in linear programming. *Zh. Vychisl. Matem. i Mat. Fiz.*, 20(1):51–58, 1980.
26. M. Kojima, S. Mizuno, and A. Yoshise. *An o $(n^{1/2}l)$ iteration potential reduction algorithm for linear complementary problems.* Technical report, Res. Rep. Dept. Inform. Sci., Tokyo Inst. Technol., Tokyo, Japan, 1988.
27. M. K. Kozlov, S. P. Tarasov, and L. G. Khachiyan. Polynomial-time solvability of convex quadratic programming. *Dokl. Akad. Nauk SSSR*, 20(1):51–58, 1979.
28. V. Yu. Levin. An algorithm for minimization of convex functions. *Dokl. Akad. Nauk SSSR*, 160(6):1244–1247, 1965.
29. E. I. Nenakhov and M. E. Primak. *Multipoint nonsmooth optimization problems.* Preprint. Kiev: IK AN UkrSSR, 1983.
30. E. I. Nenakhov and M. E. Primak. *Chebyshev center method in a model to find an economic equilibrium.* Preprint. Kiev: IK AN UkrSSR, 1985.
31. E I. Nenakhov and M. E. Primak. On convergence of the chebyshev center method and some of its applications. *Kibernetika*, 2:60–65, 1986.
32. A. I. Ovsevich and Yu. N. Reshetnyak. *Approximation of intersection of ellipsoids in the guaranteed estimation problems.* Number 4. Tekhnicheskaya kibernetika, 1988.
33. P. M. Pardalos, Y. Ye, and G. G. Han. *Algorithms for the solution of quadratic knapsack problem.* Technical report, CS-89-10, Computer Science Department, Pennsylvania State University, Pennsylvania, USA, April, 1989.
34. M. E. Primak. On convergence of the modified method of the chebyshev centers for solving the concave programming problems. *Kibernetika*, 5:100–102, 1977.
35. M. E. Primak. On convergence of the cutting method with refinement at each stage. *Kibernetika*, 1:119–121, 1980.
36. N. A. Shor. Cutting methods with space stretching for solving convex programming problems. *Kibernetika*, 1:94–95, 1977.

37. D. B. Yudin and A. S. Nemirovskii. Information complexity and effective methods of solution of convex extremal problems. *Ekon. i Matem. Metody*, 2:357–369, 1976.
38. D. B. Yudin and A. S. Nemirovskii. *Information Complexity and Effective Methods in Convex Programming (Informatsionnaya Slozhnost' i Effektivnve Melody Vypuklogo Programmirovaniya)*. Moscow: Nauka, 1977.
39. V. Zhadan. Interior point method with steepest descent for linear complementarity problem. In *Proceedings of the 13-th Baikal International School-seminar. Optimization methods and their applications*, volume 1, Irkutsk, 2005.
40. V. Zorkaltsev. The combined algorithms of interior points. In *Proceedings of the 13-th Baikal International School-seminar. Optimization methods and their applications*, volume 1, pages 37–51, Irkutsk, 2005.
41. S. I. Zukhvitsky and L. I. Avdeyeva. *Linear and convex programming*. Moscow: Nauka, 1967.
42. S. I. Zukhovitsky, R. A. Polyak, and M. E. Primak. Two methods of joining the equilibrium points of concave n-person games. *Rep. Acad. Sci.*, 185(1):24–27, 1969.

6

Improving Combustion Performance by Online Learning

Andrew Kusiak and Zhe Song

Summary. In this chapter, combustion process is improved by computing control settings with clustering algorithms. The framework involves learning from a high-dimensional data stream generated by the combustion process. Thus the system's dynamics is captured. The concepts of virtual age of the boiler and the control settings are introduced. The confidence of applying a control setting to improve boiler performance is quantified. The framework is easy to implement and it handles a large number of process variables. The ideas introduced in this paper have been implemented at a 20 MW boiler controlled with a standard control system. That system makes run-time recommendations to the standard control system.

6.1 Introduction

Modern power plants are equipped with sensors generating large volumes of process data reflecting the status of combustion and energy generation processes. Analyzing and extracting knowledge from the data stored in a data historian offers a potential to improve the efficiency of power plants. Data mining algorithms are designed to discover knowledge and models from large data sets. An optimization framework based on data mining algorithms closes the gap between limited knowledge and a rich data environment.

In this paper, the variables of the boiler combustion process are represented by a triplet $(\mathbf{x}, \mathbf{u}, \mathbf{v})$, where $\mathbf{u} \in R^l$ is a vector of l controllable variables (i.e., an operator can adjust these variables from the control panel, e.g., the fuel input, air input), $\mathbf{v} \in R^n$ is a vector of n noncontrollable variables (e.g., outside air temperature, river water temperature, coal quality), $\mathbf{x} \in R^k$ is a vector of k system response variables (e.g., efficiency, temperature, megawatt load, unit heat rate, and turbine heat rate). Most response variables are measured, but some are computed. The response variables change with controllable and noncontrollable variables. Highly correlated response variables offering redundant information should be excluded from modeling. All real-time variables are stored in a data historian. Each variable can be regarded as a data stream.

The underlying process is represented as $\mathbf{x} = f(\mathbf{u}, \mathbf{v}) + \varepsilon$, where $f(.)$ is a function capturing the process in the steady state, and it may change in time. For example, refurbishing and maintaining a boiler usually increases combustion performance. ε is some Gaussian white noise with a zero means. The online clustering algorithm captures knowledge about the function $f(.)$ from the process data. The knowledge will be used to improve the process. It is also assumed that $f(.)$ can be written as k multi-input-single-output functions, where $x_1 = f_1(\mathbf{u}, \mathbf{v}) + \varepsilon_1, ..., = f_k(\mathbf{u}, \mathbf{v}) + \varepsilon_k$, where $\varepsilon_1, ..., \varepsilon_k$ are zero-mean Gaussian white noises for the corresponding response variables.

The performance index is regarded as a response variable. Assume that x_1 is a valid performance index of a boiler that is accurate enough for the purpose of optimization. Some performance indexes (e.g., boiler efficiency) may be computed with error. Process optimization based on indexes barring errors is not likely to lead to consistent improvements. For example, optimization could aim at minimizing fuel consumption, minimizing the environmental impact, extending the equipment lifespan of a boiler, and so on.

To improve combustion performance, one has to find optimal $\mathbf{u}$ based on $\mathbf{v}$ while satisfying the constraints of other response variables. For example, the megawatt load can not exceed the demand imposed by a dispatcher.

Data mining is an emerging science oriented towards finding useful patterns in large volumes of data [26]. Data mining algorithms have been successfully applied to different industries to improve performance or extract useful knowledge for decision-making [15]. Applications of data mining in the power industry are discussed in [16] and [18]. By mining combustion data the interactions among $\mathbf{u}$, $\mathbf{v}$ and x can be revealed to some extent. In other words, knowledge about the underlying function $f(.)$ can be extracted. Note that the function $f(.)$ is not constructed from the process data, rather patterns generated from the function are stored in the database. This makes the proposed approach different from the model-based optimization approach. In the model-based optimization approach usually the process function needs to be identified first.

Of the many data mining algorithms, in this research the K-means clustering algorithm [19] has been selected. It is an unsupervised learning algorithm, and can be easily modified to meet the online processing requirements. Previous research has shown that the clustering algorithms can improve boiler efficiency [17] in an offline mode. The research reported in this paper focuses on optimizing the temporal combustion process by online learning from the data streams. The approach captures the time-shifting combustion process. The proposed framework for online clustering high dimensional data streams can be easily implemented in any power plant.

6.1.1 Related Research

There are two major research components in this paper, the real-time optimization and online stream clustering.

Data stream clustering algorithms fall into two categories, based on the way the data streams are handled. Most data stream clustering algorithms compute similarity among streams and group similar streams within a fixed time window [2, 5]. Other algorithms [1] treat data streams as multidimensional data points with the dimensionality equal to the number of streams. In this paper, the data streams coming from various process variables are considered as high-dimensional data points, and similar points are grouped together. In addition, the concept of virtual age is presented to quantify the difference between the previous extracted knowledge of the combustion process and the current boiler's status.

From the process control and real-time optimization perspective, there are two major approaches to improve the process: direct search methods and model-based methods [9]. For model-based methods an assumption is made that the underlying combustion process model is known or identifiable. The recent research and applications of model-based methods are included in [7,10, 12,22,23,29,30]. Besides analytical approaches, a model can be constructed by neural networks and fuzzy logic [13]. Updating process models using current process data is necessary in model-based methods.

When the process model is difficult to construct, direct search methods can be used [28]. The main idea of this method in using limited measurements of the process to estimate the possible direction of minimizing some cost function.

Other methods to improve the combustion process are: trial-and-error method and the ASME performance test code. [24] discussed how to change control settings of a boiler to identify ones leading to pollution reduction. [8] used the performance test code to generate fuel savings in a power plant.

Compared with the model-based methods, the online clustering approach presented in this paper does not need to identify the model. The knowledge base (control settings) is automatically updated by the new process data. The online clustering approach to some degree resembles the direct-search methods. Compared with the direct-search methods, the online clustering approach may find better solutions by the virtue of using more information. All the solutions generated from the online clustering approach must have happened in the past, while direct-search methods use a simplex-based approach to find the optimal settings that may not exist in the history [27]. In addition, direct-search methods attempt to use as few measurements as possible, while online clustering may use as much data as possible. An assumption made in the online clustering approach is that the historical process data contains useful patterns to improve the process performance. If current process performance would be better than the one reflected in the history, online clustering approach would simply not make any recommendation.

The main components of the online boiler combustion optimization are shown in Fig. 6.1. However, other necessary components of the real-time process optimization should be implemented in actual applications, such as steady-state detection, gross-error detection, data reconciliation, and results

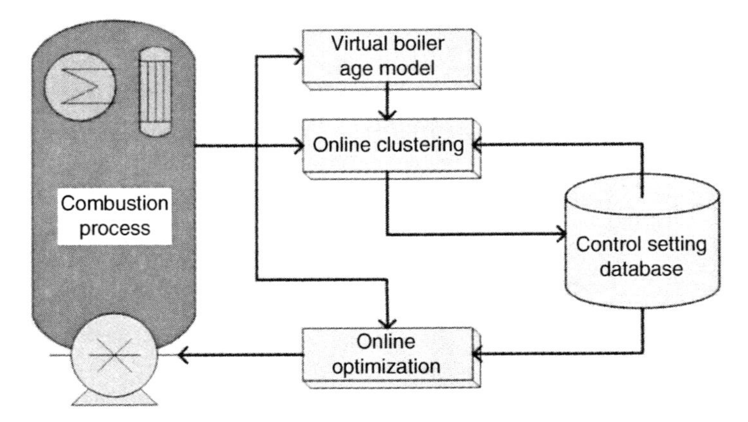

Fig. 6.1. Information flow of the boiler combustion optimization

analysis [9, 20, 25, 29]. The virtual age model computes virtual age of the boiler and the control settings and they are stored in the database. The online clustering module learns and generates new knowledge to update the control setting database (see Fig. 6.1). The online optimization module receives combustion process data and performs a search of the control setting database to find an optimal control setting to control the process. Thus, the boiler combustion performance is improved. The virtual age is used to compute the confidence index of the recommended control settings.

6.2 High Dimensional Combustion Data Streams

The combustion data stream is regarded as a set of multidimensional points (records) $P_0, P_1, P_2, ..., P_t, ...$ with the corresponding time stamp $0, 1, 2, ..., t, ...$ Here, it is assumed that the initial time is 0 and the time interval is 1 min. The combustion data stream could be passed through a steady-state filter and only steady-state data points could be considered for clustering. Each data point is further represented as

$$P_t = [x_1(t), ..., x_k(t), u_1(t), ..., u_l(t), v_1(t), ..., v_n(t)]^T.$$

Each data point includes $(k + l + n)$ entries, k response variables, l directly controlled variables, and n noncontrollable variables. In particular, $x_1(t)$ is the first response variable's value at time t, $u_1(t)$ and $v_1(t)$ are the first controllable and noncontrollable variables' values at time t respectively. The response variables change according to the controllable and noncontrollable variables (inputs).

Since the data points are generated from the combustion process, they should satisfy the equations $x_1(t) = f_1(u_1(t), ..., u_l(t), v_1(t), ..., v_n(t)) + \varepsilon_1(t), ...,$

$x_k(t) = f_k(u_1(t), ..., u_l(t), v_1(t), ..., v_n(t)) + \varepsilon_k(t)$. Here $\varepsilon_1(t), ..., \varepsilon_k(t)$ are zero-mean Gaussian white noise at time t for corresponding response variables.

Each data point has a time stamp. Thus one can assign an age for each point; for example the age of P_t is t. Thus the age for each control setting (i.e., cluster centroids with some statistics) is computed through methods discussed later. Also the age of a control setting is updated by the incoming data points. If new points show up, and they are within the limiting radius of a cluster, the age of the control setting (i.e., the cluster centroid) is updated by using the age of each new point.

6.3 Virtual Age of a Boiler

The published research does not formally model the time-decay effects of the boiler system that impacts combustion process. A boiler ages, like a human, and the combustion process changes with time. For example, boiler efficiency normally decreases the longer a unit operates. The heat transfer surfaces become fouled and less heat is transferred to the steam for the same fuel input. During outages the units are cleaned of slag and ash with the use of explosives, water, or sandblasting. This restores the boiler to a "clean" condition where boiler efficiency approaches that of the newly designed one. Air in-leakage is also a problem because cracks in the boiler casing (recall most boilers operate in a vacuum) allow additional air into the furnace, which is heated (wasted energy), as well as change of the desired stoichiometric relationships of the combustion materials.

Assume the age of a brand-new boiler is zero. The time unit set here is one minute, to reflect the de facto standard sampling frequency. It is also assumed that during cleaning, repair, or maintenance of the boiler, the impact of the accumulated virtual age is diminished. Figure 6.2 illustrates that as the general trend of a boiler's age increases, each repair (maintenance) task decreases the total accumulated age.

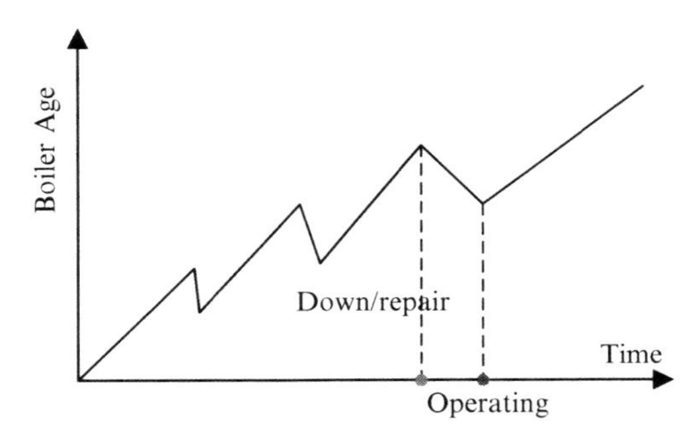

Fig. 6.2. Virtual age of a boiler

Kijima's virtual age model [3] can be used here, i.e.:

$$BoilerAge_N = BoilerAge_{N-1} + \alpha X_N \tag{6.1}$$

where α is some constant such that $0 \leq \alpha \leq 1$, X_N is the duration of the period between the $(N-1)^{th}$ boiler repair (or maintenance) completion (with time stamp $Comp_{N-1}$), and the Nth boiler down time (with time stamp $Down_N$). It can be easily seen that $X_N = Down_N - Comp_{N-1}$. $BoilerAge_N$ denotes the virtual age of the boiler at the time the Nth repair (maintenance) has been completed. After each down time, repair (maintenance) diminishes some of the aging effect. $BoilerAge_t$ denotes the boiler age at time t. If $Down_{N+1} > t > Comp_N$, $BoilerAge_t$ can be computed as

$$BoilerAge_t = BoilerAge_N + (t - Comp_N) \tag{6.2}$$

Note that model (6.1) is only an approximation of the real boiler age but provides insight into the underlying time-shifting combustion process.

6.4 Stream Clustering

Since the combustion data streams are treated as multidimensional points, clustering such data involves two major steps. One is offline clustering, which uses historical data points to build an initial number of clusters and centroids. The other one is online clustering, which assigns new points to the existing clusters or creates new clusters based on the new points.

6.4.1 Offline Clustering

The clustering algorithm used for offline clustering is mainly based on the K-means algorithm [19]. A basic K-means algorithm is as follows [26]:

1. Select K points as initial centroids.
2. Repeat
3. Form K clusters by assigning each point to its closest centroid.
4. Recompute the centroid of each cluster.
5. Until the centroids do not change.

Table 6.1 lists the basic notation used by the K-means clustering algorithm. The Euclidean distance is used to measure the similarity (closeness) between two multidimensional data points. For any two combustion data points P_t and P_r, the Euclidean distance is computed as $\| P_t - P_r \|$.

If needed, a weight is assigned to each variable to compute the weighted Euclidean distance. Besides other efficient clustering algorithms [4, 11] can be used to improve the computational performance.

The centroid of cluster C_i is computed from $c_i = \frac{1}{m_i} \sum_{P_i \in C_i} P_t$, which can also be written as $c_i = [\overline{x}_{1,i}, ..., \overline{x}_{k,i}, \overline{u}_{1,i}, ..., \overline{u}_{l,i}, \overline{v}_{1,i}, ..., \overline{v}_{n,i}]^T$.

Table 6.1. Notation used by the K-means clustering algorithm

Symbol	*Description*
C_j	The jth cluster
c_j	The centroid of cluster C_j
K	The number of clusters
m_j	The number of points in C_j

6.4.2 Virtual Age of a Centroid

Centroids computed from historical process data impact to a different degree the future behavior of a combustion process. For example, if a centroid was computed from the data that was collected 3 years ago, the confidence in the control setting represented by such a centroid is rather low, as the process conditions have likely changed between the time the data was collected and the control action was taken. As each point in a cluster is time stamped, based on this time stamp, the virtual age of the boiler can be computed.

The virtual age $CAge_i$ (Centroid Age) of centroid c_i is computed as follows:

$$CAge_i = \frac{1}{m_i} \sum_{P_i \in C_i} BoilerAge_t \tag{6.3}$$

The centroid age $CAge_i$ indicates the average virtual boiler age based on the time the data points making the corresponding cluster have been generated. In other words, as the control settings represented by a centroid reflect the boiler status, its virtual age should be similar to the centroid's virtual age.

6.4.3 Limiting Radius and Boundary of a Cluster

A cluster can be regarded as a high-dimensional sphere. The points within this sphere are homogenous (similar), and they are represented by the cluster's centroid with a certain variance. Thus, a metric is needed to quantify the appropriateness of assigning a data point to a cluster.

The *limiting radius* is the maximum allowable radius for a cluster. The limiting boundary is determined by the limiting radius (Fig. 6.3). If the distance between a point and a centroid is smaller than the limiting radius, it can be assumed that the point is similar to the centroid, and can be assigned to that cluster. If a maximum allowable variance is set for each variable, the limiting radius is easily obtained.

The radius of a cluster C_i is computed by

$$r_i = \sqrt{\frac{1}{m_i} \sum_{P_i \in C_i} (P_t - C_i)^2}.$$

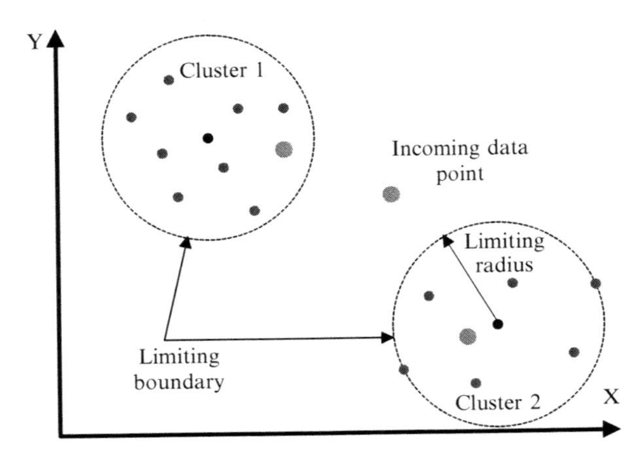

Fig. 6.3. Limiting radius and boundary of clusters

It is easy to see that the cluster radius is the square root of the variance. Given the maximum variance for each variable, the limiting radius of a cluster is computed as

$$\sqrt{\underbrace{r_{x_1}^2 + \ldots + r_{x_k}^2}_{x} + \underbrace{r_{u_1}^2 + \ldots + r_{u_l}^2}_{u} + \underbrace{r_{v_1}^2 + \ldots + r_{v_n}^2}_{v}} \tag{6.4}$$

where $r_{x_1}^2, \ldots, r_{x_k}^2, r_{u_1}^2, \ldots, r_{u_l}^2, r_{v_1}^2, \ldots, r_{v_n}^2$ are the maximum allowable variances for the corresponding variables. The radius in two-dimensional space defined in Fig. 6.3 can be used in both offline and online clustering. During the offline clustering, the limiting radius checks whether all K clusters' radiuses are smaller than the limiting radius. Any cluster with a radius greater than the limiting radius is split into two smaller clusters. This process continues until all clusters have radiuses smaller or equal to the limiting radius.

6.4.4 Online Clustering

The main goal of online clustering is to determine whether to assign a new steady-state point to an existing cluster or to create a new cluster for this new point. The basic steps of the online clustering algorithm are:

1. Given all the centroids stored in the database and a new data point, find the nearest centroid for the new point using Euclidean distance.
2. If the new point is within the limiting boundary of the cluster, add this point to this cluster and recompute the centroid and related statistics.
3. If the new point does not lie within the limiting boundary, this new point will become a new centroid and its corresponding cluster will be created in the database.

Figure 6.3 illustrates a case where new points (large dots) may or may not fall into the clusters' boundaries.

6.4.5 Updating or Creating a New Centroid

If a new point is added to a cluster, its centroid needs to be updated. Consider cluster C_j where centroid c_j has m_j points in it. A new point P_t is added to the cluster; the new centroid is computed from $c_j = \frac{c_j \times m_j + P_t}{m_j + 1}$.

The new virtual age of the centroid is updated as follows: $CAge_j = \frac{CAge_j \times m_j + BoilerAge_t}{m_j + 1}$.

If a new point does not fall into the limiting boundary of any centroid (control setting) stored in the control setting database, it becomes a new centroid of a cluster. New points may be incorporated into this new cluster from future data streams.

6.4.6 Centroids Database Maintenance

The centroids database needs to be periodically maintained. One task is deleting some bad centroids, for example, centroids generated from bad process data and centroids that have become too old. The other task is to merge centroids with distances that are smaller than the limiting radius.

6.5 Determining the Best Centroid

The ultimate goal is to find a centroid from the database of control settings that improves the boiler combustion performance. Recall that a multidimensional data point is represented as

$$P_t = [x_1(t), ..., x_k(t), u_1(t), ..., u_l(t), ..., v_n(t)]^T,$$

where $x_1(t)$ is assumed to be the performance index. The control setting search algorithm is as follows:

1. Given an incoming data point P_t and the control setting database.
2. Select from the control setting database all centroids with $\overline{x}_{1,i}$ greater than $x_1(t)$.
3. Among the selected centroids, select a centroid which satisfies all constraints and maximizes some predefined performance index (objective function).

The constraints are determined by a particular application. Typically the constraints are related to response variables x and noncontrollable variables v. For example, the amount of SO_2 emissions (response variable) should not exceed a certain threshold while the limestone consumption should be as small as possible. The distance between the noncontrollable variables $\sum_{i=1}^{n}(v_i(t) - \overline{v}_i)^2$ should be as small as possible, and not exceed the preset threshold value.

Suppose an optimal centroid $c_i = [\overline{x}_{1,i}, ..., \overline{x}_{k,i}, \overline{u}_{1,i}, ..., \overline{u}_{l,i}, \overline{v}_{1,i}, ..., \overline{v}_{n,i}]^T$ is found for P_t, and $\sum_{i=1}^{n}(v_i(t) - \overline{v}_i)^2$, by changing $\{u_1(t), ..., u_l(t)\}$ to the control setting $\{\overline{u}_{1,i}, ..., \overline{u}_{l,i}\}$, it is expected that $\{x_1(t), ..., x_k(t)\}$ go to

$\{\overline{x}_{1,i}, ..., \overline{x}_{k,i}\}$ within bounded errors. Once an optimal centroid is retrieved from the database, it is desirable to know the confidence in this centroid in terms of its ability to improve boiler performance. Based on the virtual age of the boiler and the centroid, a confidence index is computed. The confidence index is in the interval $[0, 1]$, and it decreases as the difference between the centroid's virtual age and the current boiler's virtual age increases.

The constant failure rate reliability model [6] is borrowed here to compute the confidence index. The constant failure reliability model is:

$R(t) = e^{-\lambda t}$, $t \geq 0$, where $R(t)$ is the reliability of the system (component) at time t, λ is the constant failure rate. The mean time to failure ($MTTF$) is the expected time to fail, and the relationship $MTTF = \frac{1}{\lambda}$ holds for the constant failure rate reliability model with $R(MTTF) = e^{-1} = 0.368$.

The absolute difference Δt between $BoilerAge_t$ and the searched optimal centroid's virtual age $CAge$ is defined as $\Delta t = |BoilerAge_t - CAge|$. Thus, the confidence index of applying a centroid with a virtual age of $CAge$ to a boiler with a virtual age of $BoilerAge_t$ is $R(t) = e^{-\lambda \Delta t}$. As Δt increases, the confidence of applying this centroid to the current boiler decreases. To determine the failure rate λ, the $MTTF$ (e.g., 7 days) can heuristically be assumed. In other words, if the Δt is greater than 7 days, it would expect that the control setting would not have much effect on the boiler.

6.6 Industrial Case Study

6.6.1 Background

A computational tool based on the ideas introduced in this paper has been implemented at The University of Iowa Power Plant (UI PP). The fluidized-bed boiler at the UI PP burns coal and oat hull (OH). The tool recommends optimal control settings (e.g., primary air flow, secondary air flow) for the boiler operators. The operators can adjust the bias of the controllers based on the recommended values displayed through a webpage.

Domain experts suggested starting with four controllable variables and three noncontrollable variables to optimize the boiler's efficiency (i.e., reducing fuel consumption while keeping steam load stable) and the limestone consumption. Limestone is expensive material used to react with SO_2, thus keeping SO_2 in the stack at an allowable level. The process variables considered in this case study are listed in Table 6.2.

The variable u_1 is defined as $\frac{\text{primaryair}}{\text{coal}}$. The variable u_2 is defined as $\frac{\text{secondery air}}{\text{coal}}$ and v_1 is defined as $\frac{\text{coal}}{\text{oat hull}}$. The variable v_1 is assumed to be noncontrollable because how much coal is burned is determined by the availability of oat hull and the steam load demand. If an optimal centroid could be found in the setting database, the optimal air flow could be computed from the $u_1 - u_2$ based on the current boiler's coal flows.

Table 6.2. Process variables

$\boldsymbol{u}$	u_1	Primary air-to-coal ratio
	u_2	Secondary air-to-coal ratio
	u_3	Boiler bed pressure
	u_4	ID fan inlet damper
$\boldsymbol{v}$	v_1	Coal-to-oat hull ratio (COHR)
	v_2	Coal quality index, BTU/lb
	v_3	Oat hull quality index, BTU/lb
$\boldsymbol{x}$	x_1	Boiler efficiency
	x_2	Limestone-to-coal ratio

The values v_2 and v_3 of the quality indices of the coal and oat hull are obtained through daily lab tests of the coal and oat hull. They are assumed to be noncontrollable variables. The boiler efficiency x_1 is a response variable. Here, the boiler efficiency is computed by a simple input–output method with economic performance considered.

$$x_1 = \frac{SteamLoad}{Coal \times CoalQuality \times 3 + OH \times OHQuality \times 1.05},$$

where the steam load is measured in $klbs/hr$; the coal and OH are measured by $klbs/hr$; the coal quality and OH quality are measured by BTU/lb; "3" is the price for coal in \$, "1.05" is the price for oat hull in \$.

For example coal is priced at \$3/$MBTU$, oat hull is priced \$1.05/$MBTU$. "$MBTU$" refers to mega BTU. The final unit for boiler efficiency is $klbs$/\$. Although the actual boiler efficiency is not calculated in this paper, however x_1 provides information important to boiler operations.

The variable x_2 is the limestone-to-coal ratio defined as $\frac{\text{lime stone}}{\text{coal}}$. The limestone is priced at \$24.63/ton which equals \$12.32/$klbs$ (here 1 ton = 2,000 lbs). According to the UI PP data, ash disposal fee is another cost for limestone consumption. Current ash disposal cost for one ton of limestone is \$11.34. Thus the limestone cost including the disposal fee is priced at \$17.99/$klbs$.

6.6.2 Offline Clustering

About 21-day historical data points (sampled each minute) were used to do the offline clustering and generate a certain number of initial clusters. The software tool receives real-time data streams from the boiler, performs online clustering, and recommends optimal control settings.

6.6.3 Performance Index for Centroids

The search of an optimal centroid can be formulated as a multiple-objective optimization model where the objective function is a weighted summation of several objective functions [14]. In order to facilitate the search process, a performance index is defined as a weighted combination of important metrics and the "hard" weighting scheme is used. Other weighting schemes (e.g., fuzzy weighting) can certainly be formulated. But considering the computational time cost, a "hard" weighting scheme is preferred in this case study. The search algorithm maximizes the performance index by finding a centroid i in the database. $Performance\_Index(i)$ is calculated from (6.5).

$$w_1 \times \overline{x}_{1,i} + w_2 \times \frac{1}{\overline{x}_{2,i}} + w_3 \times \min(Support_i, 50) + w_4 \times \frac{1}{(|\overline{v}_{1,i} - v_1(t)|, 0.01)} \quad (6.5)$$

where $\overline{x}_{1,i}$ is the boiler efficiency of the ith centroid, $\overline{x}_{2,i}$ is the limestone-to-coal ratio of the ith centroid, and $Support_i$ is the number of points assigned to that centroid. Large support indicates the control settings represented by the centroid are not a casual event or noise. It is a strong pattern shown through the data. The value $\min(Support_i, 50)$ makes sure that the support does not overwhelm the performance index. The absolute *COHR* difference between the current boiler status and the centroid is $|\overline{v}_{1,i} - v_1(t)|$.

Ideally one would like to find a centroid that exactly matches the current *COHR* of the boiler, which is not always possible. Thus some tolerance is allowed (see model (6.6)). The threshold $\max(|\overline{v}_{1,i} - v_1(t)|, 0.01)$ is to ensure the *COHR* difference does not dominate the performance index. The weights $w_1 - w_4$ indicate the importance of the four metrics. In this case, boiler efficiency is much more important than the *support* and $|\overline{v}_{1,i} - v_1(t)|$. The values "50" and "0.01" are heuristically determined constants ensuring that both efficiency and limestone consumption are improved, rather than the support or the *COHR* difference. Model (6.6) is used to find an optimal centroid, which will be recommended to the boiler operators.

$$arg \max_i \quad Performance\_Index(i) \quad (6.6)$$

Subject to:

$$\overline{x}_{1,i} > x_1(t); \overline{x}_{2,i} - v_t(t) \leq x_2(t)$$
$$|\overline{v}_{1,i} - v_1(t)| \leq Cnst_1; |\overline{v}_{2,i} - v_2(t)| \leq Cnst_2; |\overline{v}_{3,i} - v_3(t)| \leq Cnst_3$$

The following notation is used in model (6.6):

- i is the centroid index stored in the centroids database.
- $Cnst_1$, $Cnst_2$, $Cnst_3$ are constants to make sure that those three noncontrollable variables (i.e., coal and *OH* mixture ratio, the coal and oat hull qualities) do not change too much between the optimal centroid and the current boiler status.

Model (6.6) can be solved at steady states to generate the optimal control settings. If no feasible solution is available, the system does not recommend any control settings.

6.6.4 Computation of Primary Air (PA) and Secondary Air (SA)

Let the optimal centroid based on model (6.6) is

$$c^* = [\overline{x}_1, \overline{x}_2, \overline{u}_1, ..., \overline{u}_4, \overline{v}_1, \overline{v}_2, \overline{v}_3]^T.$$

The current boiler's coal flow is $Coal_t$. The optimal PA and SA computed for the coal are $PA^*_{coal} = \overline{u}_1 \times Coal_t$ and $SA^*_{coal} = \overline{u}_2 \times Coal_t$. If the computed values of PA^* and SA^* exceed the limits of the boiler's current air system capacity, the recommended values have to be set at the current limits. For example, if PA^* is 114 $klbs\, h^{-1}$ and the PA fan of the boiler has a maximum limit of 112 $klbs\, h^{-1}$, the recommended optimal airflow PA^* is reduced to 112 $klbs\, h^{-1}$. Knowing that the optimal air flow exceeds the boiler's current air system's capacity helps the power plant identify its limitations, thus achieving better power plant performance by adding more air capacities in the future.

6.6.5 Industrial Experiment with the System

The algorithms proposed in this paper were implemented as a web-based software system and operators of a boiler were asked to follow the recommendations by changing the bias of PA, SA, the boiler bed pressure, and the ID fan inlet damper. In one of the experiments, the control room began to follow the recommendations on 21 September 2006, at 3:00 P.M.

The data points used in Fig. 6.4 are based on the time period from 9/21/2006, 3:00 A.M. to 9/22/2006, 10:45 A.M. The boiler efficiency trend is represented by the plot in the center of Fig. 6.4. The steam load ($klbs\, h^{-1}$) is the upper plot and the lime/coal ratio is the bottom one. The three variables plotted in Fig. 6.4 have been scaled so that they fit one graph. The two vertical lines in Fig. 6.4 indicate the start time and the end time of the experiment

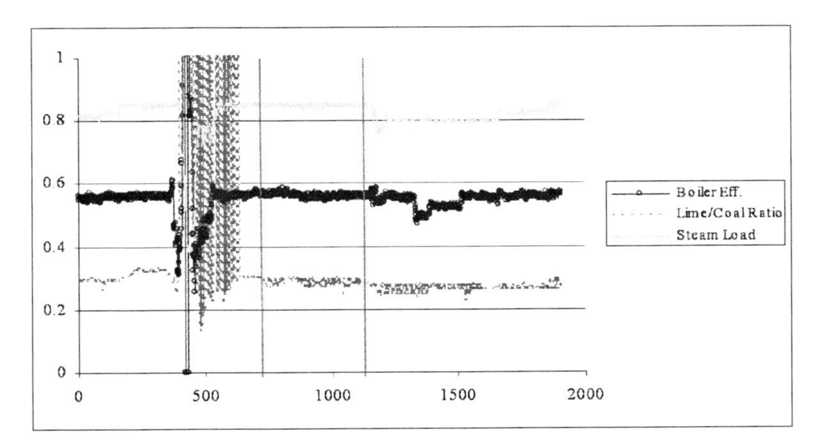

Fig. 6.4. Trends analysis of boiler efficiency, steam load and limestone-to-coal ratio

under constant steam load (for the fairness of comparison). Between the two vertical lines in Fig. 6.4, the boiler was operated at 170 $klbs$ steam load, 6.15 $klbs\,h^{-1}$ coal input and 13.34 $klbs\,h^{-1}$ OH input. The coal and OH quality were 11,070 $BTU\,lb^{-1}$ and 7,022 $BTU\,lb^{-1}$, respectively.

The two vertical lines in Fig. 6.4 correspond to 420 data points (7 h). A reference data set collected under similar operating conditions was extracted to perform a comparative analysis. In this reference data set, the values of steam load, coal and OH were about 170, 6.15, and 13.34 $klbs\,h^{-1}$, respectively. The coal and OH quality indices of the reference data set were 11,070 and 7,022 $BTU\,lb^{-1}$. The reference data set included 4,372 data points extracted from the time period between 8/17/2006, 11:01 P.M. and 9/29/2006 10:09 A.M. The boiler efficiency and lime/coal ratio of these 4,372 data points were analyzed and compared with the experimental sample of 420 data points.

To test whether the mean efficiency of Fig. 6.6 is statistically significant compared with the mean of Fig. 6.5, the control chart approach is used [21] to test whether the sample mean is significantly different from the reference mean.

Consider

$$\begin{aligned} UCL &= \mu_{reference} + 3\frac{\sigma_{reference}}{\sqrt{M}} \\ CenterLine &= \mu_{reference} \\ LCL &= \mu_{reference} - 3\frac{\sigma_{reference}}{\sqrt{M}} \end{aligned} \tag{6.7}$$

where $\mu_{reference}$ is the mean value of the reference data set, $\sigma_{reference}$ is the standard deviation of the reference data set; M is the sample size (in this case $M = 420$). The mean efficiency of the data represented in Fig. 6.6 is 0.5616,

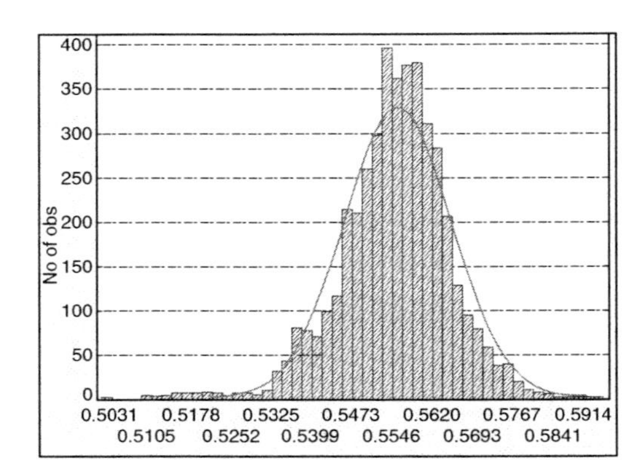

Fig. 6.5. Boiler efficiency distribution of the reference data set with $mean = 0.557$ and $standard\ deviation = 0.0098$

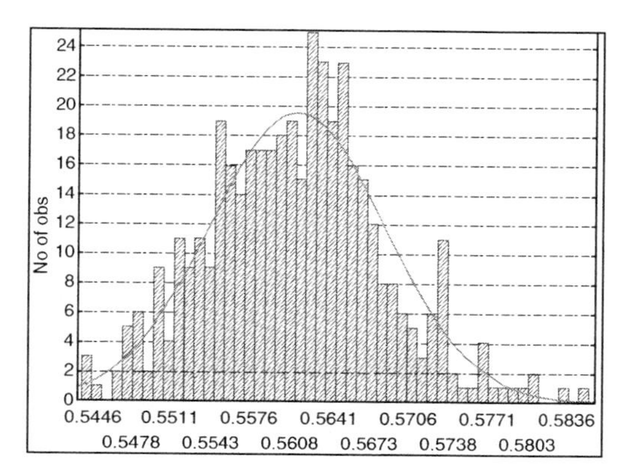

Fig. 6.6. Boiler efficiency distribution of the experimental data set with *mean* = 0.5616 and *standard deviation* = 0.007

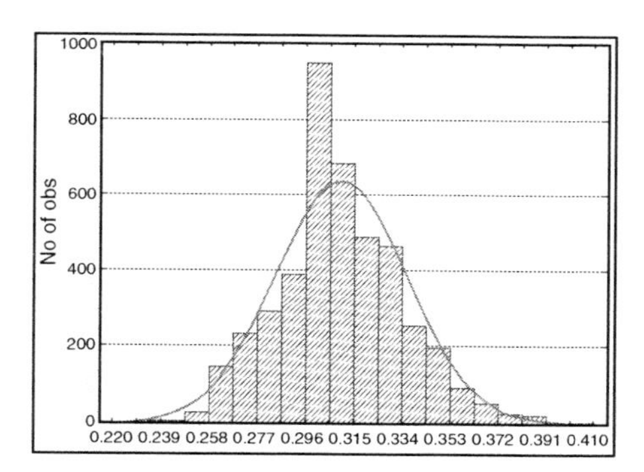

Fig. 6.7. Lime/coal distribution of the reference data set with *mean* = 0.3093 and *standard deviation* = 0.0258

which is greater than the upper limit $UCL = 0.557 + 3\frac{0.0098}{\sqrt{420}}$ (Fig. 6.5). Therefore the efficiency improvement is statistically significant.

Analysis of the charts in Figs. 6.5 and 6.6 indicates that controlling the boiler according to the computed recommendations has increased the average boiler efficiency from 0.557 to 0.5616.

The cost savings from the improved boiler efficiency improvement can be computed from $SteamLoad \times (\frac{1}{OriginalEfficiency} - \frac{1}{ImprovedEfficiency})$.

Thus, based on Figs. 6.5 and 6.6, the potential savings are \$2.5 per hour for the 170 $klbs\, h^{-1}$ steam load.

In addition to the fuel savings, more savings are generated from decreased limestone consumption. The data depicted in Figs. 6.7 and 6.8 illustrate that the average lime/coal ratio has been reduced from 0.3093 to 0.2891. Based on 6.7, the value $LCL = 0.3093 - 3\frac{0.0258}{\sqrt{420}} = 0.3055$ is higher than the mean lime/coal ratio 0.2891 in Fig. 6.8. This indicates that the mean lime/coal reduction is statistically significant.

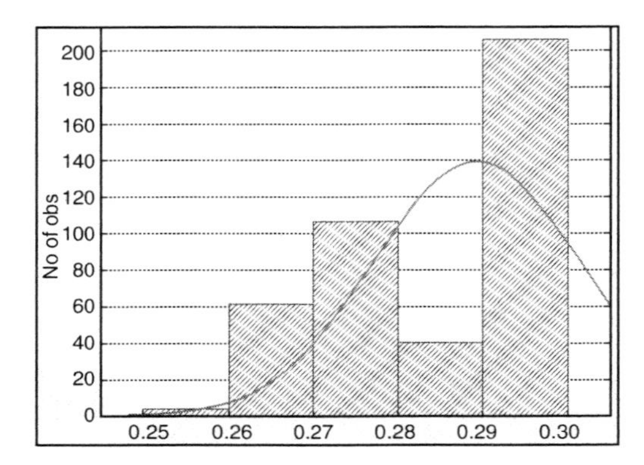

Fig. 6.8. Lime/coal ratio distribution of the data points when following the recommendations with $mean = 0.2891$ and $standard\ deviation = 0.012$

The cost savings resulting from the lime/coal ratio reduction can be computed from $CoalInput \times (OriginalLime/CoalRatio - NewLime/CoalRatio) \times LimeStonePrice$. Thus, based on Figs. 6.7 and 6.8, the potential savings is \$1.53 per hour if the boiler is running with 6.15 $klbs\,h^{-1}$ coal input. If the limestone cost, which incorporates the ash disposal fee, is used to calculate the savings, the potential savings is \$2.25 per hour.

6.7 Conclusion

An online framework for boiler performance optimization based on an online clustering technique was presented. The recommendation system, DACOMO was successfully applied to generate optimal control settings for an industrial boiler. Significant savings were observed from a live demonstration of the system. The confidence index, which was developed based on the concept of a virtual age, was used to help the operators determine when to follow the system recommendations. The boiler used in the experiment was relatively small (20 MW). The savings from the proposed methodology will multiply when the system is deployed for large commercial boilers.

Acknowledgement. The research published in this paper has been partially supported by funds from the Iowa Energy Center (IEC Grant No. 04-06). Many thanks to Hui Chen for formatting the chapter.

References

1. C.C. Aggarwal, J.W. Han, J.Y. Wang, and P.S. Yu. On high dimensional projected clustering of data streams. *Data Mining and Knowledge Discovery*, 10(3):251–273, 2005.
2. J. Beringer and E. Hűllermeier. Online clustering of parallel data streams. *Data Mining and Knowledge Engineering*, 58(2):180–204, 2006.

3. C.R. Cassady, I.M. Iyoob, K. Schneider, and E.A. Pohl. A generic model of equipment availability under imperfect maintenance. *IEEE Transactions on Reliability*, 54(4):564–571, 2005.
4. K. L. Chung and J.S. Lin. Faster and more robust point symmetry-based k-means algorithm. *Pattern Recognition*, 40(2):410–422, 2007.
5. B.R. Dai, J.W. Huang, M.Y. Yeh, and M.S. Chen. Adaptive clustering for multiple evolving streams. *IEEE Transactions on Knowledge and Data Engineering*, 18(9):1166–1180, 2006.
6. C.E. Ebeling. *An Introduction to Reliability and Maintainability Engineering.* Waveland, Long Grove, IL, 2005.
7. S. Engell. Feedback control for optimal process operation. *Journal of Process Control*, 17(3):203–219, 2007.
8. S. Farhad, M. Younessi-Sinaki, and M. Saffar-Avval. Energy saving in operating steam power plants based on asme performance test code. *ASME Power Conference 2005*, 2005.
9. S. Ferrer-Nadal, I. Yelamos-Ruiz, M. Graells, and L. Puigjaner. An integrated framework for on-line supervise d optimization. *Computers and Chemical Engineering*, 31(1):32–40, 2006.
10. M. Guay and J.F. Forbes. Real-time dynamic optimization of controllable linear systems. *Journal of Guidance, Control, and Dynamics*, 29(4):929–935, 2006.
11. P. Hansen and N. Mladenovic. J-means: a new local search heuristic for minimum sum of squares clustering. *Pattern Recognition*, 34(2):405–413, 2001.
12. V. Havlena and J. Findejs. Application of model predictive control to advanced combustion control. *Control Engineering Practice*, 13(6):671–680, 2005.
13. S.A. Kalogirou. Artificial intelligence for the modeling and control of combustion processes: A review. *Progress in Energy and Combustion Science*, 29(6):515–566, 2003.
14. A. Kusiak. *Engineering Design: Products, Processes, and Systems.* Academic, San Diego, CA, 1999.
15. A. Kusiak. Data mining: Manufacturing and service applications. *International Journal of Production Research*, 44(18/19):4175–4191, 2006.
16. A. Kusiak and S. Shah. A data-mining-based system for prediction of water chemistry faults. *IEEE Transactions on Industrial Electronics*, 53(2):593–603, 2006.
17. A. Kusiak and Z. Song. Combustion efficiency optimization and virtual testing: A data-mining approach. *IEEE Transactions on Industrial Informatics*, 2(3):176–184, 2006.
18. A. Kusiak, A. Burns, and F. Milster. Optimizing combustion efficiency of a circulating fluidized boiler: A data mining approach. *International Journal of Knowledge-Based and Intelligent Engineering Systems*, 9(4):263–274, 2005.
19. J.B. MacQueen. *Some Methods for Classification and Analysis of Multivariate Observations*, volume 1. University of California Press, Berkeley, CA, 1967. Proceedings of 5th Berkeley Symposium on Mathematical Statistics and Probability, Vol. 1, pp. 281–297.
20. I.P. Miletic and T.E. Marlin. On-line statistical results analysis in real-time operations optimization. *Industrial and Engineering Chemistry Research*, 37(9):3670–3684, 1998.
21. D. Montgomery. *Introduction to Statistical Quality Control*, 5th Edition. Wiley, New York, NY, 2005.

22. M. Guay, N.Peters, and D. DeHaan. Real-time dynamic optimization of batch systems. *Journal of Process Control*, 17, 2007. Vol. 29, No. 4, pp. 929–935.
23. G. Poncia and S. Bittanti. Multivariable model predictive control of a thermal power plant with built-in classical regulation. *International Journal of Control*, 74(11):1118–1130, 2001.
24. C.E. Romero, Y. Li, H. Bilirgen, N. Sarunac, and E.K. Levy. Modification of boiler operating conditions in coal-fired utility boilers. *Fuel*, 85(2):204–212, 2006.
25. S.E. Sequeira, M. Herrera, M. Graells, and L. Puigjaner. On-line process optimization: Parameter tuning for the real time evolution (rte) approach. *Computers and Chemical Engineering*, 28(5):661–672, 2004.
26. P.N. Tan, M. Steinbach, and V. Kumar. *Introduction to Data Mining*. Pearson, Readings, MA, 2006.
27. M.H. Wright. *Direct search methods: once scorned, now respectable*, in D.F. Grioths and G.A. Watson (Eds.), Numerical Analysis 1995 (Proceedings of the 1995 Dundee Biennial Conference in Numerical Analysis). Addison-Wesley, UK, 1995.
28. Q. Xiong and A. Jutan. Continuous optimization using a dynamic simplex method. *Chemical Engineering Science*, 58(16):3817–3828, 2003.
29. W.S. Yip and T.E. Marlin. Multiple data sets for model updating in real-time operations optimization. *Computers and Chemical Engineering*, 26(10):1345–1362, 2002.
30. H. Zhao, J. Guiver, R. Neelakantan, and L.T. Biegler. A non-linear industrial model predictive controller using integrated pls and neural net state-space model. *Control Engineering Practice*, 9(2):125–133, 2001.

7

Critical States of Nuclear Power Plant Reactors and Bilinear Modeling

Vitaliy A. Yatsenko, Panos M. Pardalos, and Steffen Rebennack

Summary. We present a new system methodology for modeling of nonlinear processes in nuclear power plant cores. This methodology makes use of a variety of different approaches from different mathematical fields. The problem of modeling critical states is reduced to a bilinear subproblem. A scheme which provides stable parameter identification and adaptive control for the nuclear nuclear power plant described by the bilinear differential equation is presented. Abnormal events are found via a system-theoretical approach. Transitions to critical states can be detected by bilinear analysis of observed characteristics and by optimization of sensory measurements. Latent conditions and critical parameters in the reactor core are estimated trough a bilinear modeling.

7.1 Introduction

The use of nuclear power plants is very controversially discussed in our society. However, nuclear power is a sustainable energy source. It emits almost no greenhouse gases and, according to the 2006 annual report of the International Energy Agency [16], when it replaces coal-fired plants, the CO_2 emission can be reduced by about 6–7 million tonnes per year per 1 GW. After coal, the uranium fuels are the second most abundant sources of electric energy in the world, and it is, in contrast to oil, distributed among many countries. However, terrible accidents in the past, like in the Chernobyl power plant on April 26 in 1986, together with the enormous capital cost involved in nuclear power plants caused an end of the nuclear power boom in the 1970 and 1980. As the politicians get more sensitive to climate change issues and the fears for the security of the supply of fossil fuels grows, the nuclear power gets a new boost nowadays. An increase of the carbon prices to between \$10 and \$25/tCO_2 makes nuclear power economically competitive against coal or natural gas-based power generation; estimated by the International Energy Agency [16]. The International Atomic Energy Agency (IAEA) reported at the end of 2006 that 29 nuclear power plant reactors were under construction [15]. Just recently, The New York Times reported that the company "NRG Energy" applied to build two new nuclear power plant reactors in Texas, the first time in the United States after the "Three Mile Island" accident in 1979 [25]. This trend also includes other nations like Japan or England.

In 2006, about 16% of the global electricity was produced by nuclear power plants and at the end of 2006, 435 nuclear power plants were in operation in the world [15]. There are two mayor challenge for nuclear power plants: First, the waste management and second the safety of the reactors. Greenpeace provides a list of over 100 serious incidents in nuclear power plants since December 1952, reflecting only a small subset of all accidents [12]. All this together with the "revival" of the nuclear power, increases the demand for the save control of the nuclear power reactors.

One of the difficulties in safety engineering for nuclear power plant reactors is the problem of modeling and optimization [2, 5–7, 17, 20, 24, 29]. Very often, the mathematical models include highly nonlinear differential equations, for which design techniques are a complicated problem. Accurate models are known for physical processes, that can be accurately simulated together. However, the equations of motion consist of partial and ordinary differential equations coupled via their boundary conditions, a model that offers little to the control designer [10, 21, 23]. It is therefore a problem of considerable interest in developing explicit low-order models; once a design has been constructed using such a low-order model, it can be tested by comparing with a full high-order simulation [13, 21, 26].

An example is given by the nuclear power plant reactor. The channels of nuclear reactor cores, boilers and other chemical processes very often present problems embodying thermo-hydraulic systems. The dynamics of a channel system characterized by coupling between fuel pin and coolant flow in a reactor core may be represented by either linear or bilinear differential equations, depending on what is chosen as control variables. The instrument chosen to characterize the process control system is for instance the valves, which are often used in their quality of a simple tool for furnishing disturbance to plants. These valves often provide a simple means of modulating the input signals, given in the form of maximum-sequence or other binary signals.

Bilinear models (BM) can approximate a wide class of nonlinear systems. They are used to model nonlinear processes in signal and image processing and communication systems modeling. In particular, they arise in areas such as channel equalization, echo cancellation, nonlinear tracking, multiplicative disturbance tracking, and many other areas of engineering, socioeconomics, and biology. BM represent a mathematically tractable structure over Volterra models for a nonlinear system. Also, a bilinear model can obviously represent the dynamics of a nonlinear system more accurately than a linear model. Hence, modeling and control of nonlinear systems in a bilinear framework are fundamental problems in engineering.

This chapter proposes new methodologies for analysis and modeling of nuclear power plant reactors as controlled systems using algebraic and geometric methods. These can be subdivided into methods that attempt to treat the system as a bilinear system in a limited range of operation and use bilinear design methods for each region. The most important aspect of these methodologies is transformation of a nonlinear control system into a bilinear system.

The controllability, observability, and invertibility of nonlinear control systems using Lie algebras of vector fields are considered. The study of this type of systems was initiated by Brockett [4]. Brockett's observability results are generalized and necessary and sufficient conditions for observability are presented. Effective algorithms are proposed to verify such conditions.

Local and global bilinear realizations of nonlinear control systems were studied in the literature. For a controlled nonlinear system with control appearing linearly, there exist necessary and sufficient conditions for the existence of a dynamically equivalent bilinear system. It was also shown that every nonlinear realization can be approximated by a bilinear realization [18, 19].

This chapter is organized as follows. In Sect. 7.2, we discuss the principles of nuclear reactor dynamics. This enables us to formulate a bilinear model describing the nuclear power plant reactor in Sect. 7.3. Critical states of the nuclear power plant core are represented via versal models, Sect. 7.4. Thermal-hydraulic systems in the reactor core are modeled via bilinear models, Sect. 7.5. The coefficients of this bilinear model are obtained via an identification algorithm, discussed in Sect. 7.5.1. In Sect. 7.6, we discuss the simulation of nuclear power plant reactor core accidents.

7.2 System-Theoretical Description of Nuclear Reactor Dynamics

Bilinear systems are one of the simplest nonlinear systems and therefore particularly applicable to analysis of much more complicated nonlinear systems. They can be used to represent a wide range of physical, chemical, biological, and social systems, as well as manufacturing processes that cannot be effectively modeled under the assumption of linearity [26, 28, 29].

We emphasize the role of three disciplines that modified our outlook on bilinear system theory. The first one is modern differential geometry. The second discipline is the modern theory of control dynamical systems. The third discipline is optimization theory. Bilinear systems can approximate a wide class of nonlinear control systems. They can be represented as state space models or as systems of input–output).

The wide spectrum of the above-mentioned problems can be represented by the following theoretical schemes.

1. Construction of a set of states accessible from a given initial state
2. Identification of the set of controls steering the system from a given initial state to a desired accessible state with the greatest or specified probability
3. Stability analysis for adaptive bilinear systems

4. Identification of a control that is optimal with respect to a given criterion, for example, the response time or the minimum of switches (in bang-bang control)
5. Control and optimization of nonlinear systems
6. Construction of a system of a feedback providing for the possibility of control with accumulation of data

A global change in coordinates for transforming the system are used for finding a lower-order nonlinear subsystem. A constructive system analysis of such systems on the base of geometric and algebraic methods is conducted. The specific examples of nonlinear systems reduction to bilinear systems (BS) and dynamical systems (DS) with known physical properties are given. It is also shown that every nonlinear realization can be locally approximated by a bilinear realization, with an error that grows as a function of time t.

Necessary and sufficient conditions for the invertibility of a class of nonlinear systems, which includes matrix bilinear systems, were also obtained. Lie algebraic invertibility criteria are obtained for bilinear systems in $\mathbb{R}^n$, which generalize standard tests for single input linear systems. These results are used to construct nonlinear systems that act as left-inverses for bilinear systems.

Due to the widespread use of bilinear models, there is strong motivation to develop identification algorithms for such systems given noisy observations Fnaiech, Ljung and Fliess's paper [8] presents methods for parameter identification of bilinear systems. These methods are directly transferred from linear system identification methods, such as least squares and recursive prediction error methods. A conjugate gradient method for identification of bilinear systems has been developed by Bose and Chen [3]. Most studies of the identification problem of bilinear systems have assumed an input–output formulation. Standard methods such as recursive least squares, extended least squares, recursive auxiliary variable, and recursive prediction error algorithms, have been applied to identifying bilinear systems.

In this chapter we describe new principles of monitoring control, and optimization of a large class of nonlinear objects including nuclear reactor cores [5–7, 9, 11, 14, 22, 29]. Nonlinear physics and bilinear control are two rich and well-developed theories. Their efficient unification requires joint efforts of specialists in both fields. When two such abundant theories are joined, the effect is multiplicative rather than additive because they amplify each other's potential in proportion to their range of development.

7.3 Bilinear Logic-Dynamical Models

Suppose that a nonlinear process in the nuclear power plant reactor can be described by equation

$$\dot{y}(t) = b_0(y) + \sum_{i=1}^{h} u_i(t) b_i(y),$$
$$z(t) = f(y(t)), \quad y(0) = y^0, \quad u(t) \in \Omega, \quad y \in Y, \tag{7.1}$$

where $y = (y_1, \dots, y_n)$ is a state vector; $z = (z_1, \dots, z_n)$ is a vector of sensor outputs; $b_0(y), \dots, b_n(y)$ are analytical vector fields; f is an infinite differentiable $\mathbb{R}^1$ vector-function; Y is a compact manifold, and $u(t) \in \Omega = \{u : |u_i| \le 1, i = 1, \dots, h\}$.

By using coordinate transformations we want to construct a logic-dynamical system, i.e., a system describing the processes evolving according to continuous dynamics, discrete dynamics, and logic rules.

Consider the system

$$\dot{x}(t) = \sum_{j=1}^{r} L_j \left[A_{0j} + \sum_{i=1}^{h} u_1(t) A_{ij} \right] x(t),$$
$$\omega(t) = \sum_{j=1}^{r} L_j C_j x(t), \quad x(0) = x, \quad u(t) \in \Omega \tag{7.2}$$

and consider the matrix equation

$$\dot{X}(t) = \left(A_0 + \sum_{i=1}^{h} u_i(t) A_i \right) X(t),$$
$$W(t) = CX(t), \quad X(0) = I, \quad u(t) \in \Omega, \tag{7.3}$$

where $X(t)$ is a matrix, which evolves in $Gl(m, \mathbb{R})$, of invertible $(m \times m)$ matrices. Each column of this equation is a system in the form (7.1).

The Lie algebra of the group $Gl(m, \mathbb{R})$ is finite-dimensional over the real field $\mathbb{R}$. There is a closed Lie subgroup G of $Gl(m, \mathbb{R})$ which corresponds to the subalgebra g of the algebra $gl(m, \mathbb{R})$. This algebra is defined by the Lie bracket and the matrices $\{A_0, \dots, A_h\}$ are characterized by the solution of the equation

$$\dot{X}(t) = \left(\sum_{i=1}^{h} u_i(t) A_i \right) X(t),$$
$$(X(0) = I, \quad |u_i| \le 1, \quad i = 0, \dots, h).$$

The group G contains the set of all accessible matrices of (7.3). The set of accessible matrices of the system is a subset of G with nonempty interior in the relative topology of G, hence G is the smallest subgroup of $Gl(m, \mathbb{R})$ containing all accessible matrices of (7.3).

Let S_j be some neighborhood of the point y_j^0; then $W_j(S_j)$ is a minimal subalgebra of the Lie algebra C^∞ of all vector fields on S_j over $\mathbb{R}$ containing $\{b_0, \dots, b_h\}$, and a submanifold Y_j containing y_j^0, is an integral manifold $\widetilde{W}_j(S_j)$, whereas the dimension of Y_j is equal to the rank $\widetilde{W}_j(S_j)$ at the y_j^0. Then, according to Chow's theorem, the set of all points tY_j is accessible by the system (7.1) from y_j^0.

Because Y is a compact manifold, there exist submanifolds Y'_j, such that $Y = \cup_{j=1}^{r} Y'_j$. If the subalgebra $\widetilde{W}_j(Y'_j)$ is finite-dimensional, then there exists a Lie subalgebra g_j of the algebra $gl_j(m_j, \mathbb{R})$ for some m_j, and according to the Ado's theorem (Ado 1947), an isomorphism of Lie algebras $\varphi_j \colon \widetilde{W}_j(Y'_j) \mapsto g_j$. We define the matrix bilinear system (7.3) by the map $A_{ij} = \varphi_j(b_i)$. Let l_j be the map

$$l_j : W_j(Y'_j) \mapsto \widetilde{W}_j(y_j^0),$$

such that $l_j(c) = c(y_j^0)$ for $c \in \widetilde{W}_j(Y'_j)$. Then the linear map $l'_j = l_j \circ \varphi_j^{-1}$ satisfies the condition

$$l'_j = \left(\left[A_{i_1 j} \ldots \left[A_{i_{\nu-1} j}, A_{i_\nu j}\right] \ldots\right]\right) = \left[b_{i_1 j} \ldots \left[b_{i_{\nu-1} j}, b_{i_\nu j}\right] \ldots\right](y_j^0)$$

for any ν_i, $0 \leq i_1, \ldots, i_\nu \leq h$. By Krener's theorem [19], there exists a neighborhood M of I and maps $\lambda_j \colon M_j \mapsto Y'_j$, that preserve the solutions.

By Brockett's theorem [4], we can find the following result. If (7.1) satisfies the above stated conditions and the map $f \circ \lambda_j \colon X \mapsto Z$ is polynomial, then there exists a logic-dynamical realization (7.2) of $u(t) \mapsto \omega(t)$ and a constant $T \geq 0$, such that for any input $u(t)$, the corresponding outputs satisfy $\omega(t) = z(t)$ for $t \in [0, T]$.

Remark 7.1 The dimension of a state space of LDS is the maximal dimension of Euclidean space, corresponding to some submanifold M_j.

We define a logic variable L_j for each integral submanifold Y'_j of the compact space state Y by the following;

$$L_j = \begin{cases} 0, \text{ if } y \in Y'_j, & j = 1, \ldots, r, \\ 1, \text{ if } y \notin Y'_j & \text{otherwise}. \end{cases} \tag{7.4}$$

We suppose that the logic function L_j can be realized by a finite automaton. For each value $z_i \in Z$, $i = 1, \ldots, r$ we can find a submanifold Y_t by the map $\gamma_t \colon T \times Y \mapsto Z$. This map satisfies the condition

$$\gamma_t(Y'_j) = z_j, \quad Y'_i \cap Y'_j = \phi, \quad i \neq j.$$

If the system (7.1) satisfies the above hypothesis, then there exists a logic-dynamical system (7.2), such that for any input $u(t)$, the corresponding outputs satisfy $z(t) = \omega(t)$, $t \in [0, T]$.

7.4 Versal Models of Critical States

Mathematical model of critical states in nuclear plant core can be described by versal or universal models. The concept of a versal or universal mapping was introduced in Arnold [1], however, the methods for calculation of the parameters of a versal or universal model using an initially given model of a time-varying system are important for engineering applications. In other words, the case in point is the construction of analytical dependence of parameters of a universal model as a function of parameters of an a priori given model, e.g., of its controlling part. This problem can also be interpreted as the problem of robust decomposition of sets of dynamical systems. It should be pointed out that each subsystem forming a part of the universal model contains a minimum admissible number of parameters from the point of view of completeness of consideration of possible variants of subsystem interaction in the initial model and admits an independent investigation. In this case, interaction between the subsystems in the initial model is reduced to parametric interaction (self-operation) in these subsystems. Interactions between the initial subsystems that cannot be removed in this way appear only in the cases where there are singularities in the initial subsystems (symmetry, close eigenfrequencies, singularity of the matrix of higher derivatives of differential equations in the initial model, and possibly some others). In addition to the circumstances mentioned above, selection of dimension of universal subsystems is determined by computing resources used for calculation of parameters of universal models from preset interaction coefficients and for investigation of the models themselves. Once such dependencies are obtained, investigation of a universal model becomes practically manageable and can be easily performed analytically.

Let us point out also that the construction of a universal model admits its extension by connecting new subsystems. In this case, algorithms for calculation of universal model parameters are arranged so that they allow us to refine the parameters of the initial universal model with regard to the presence of new subsystems and, at the same time, to determine parameters of the universal model of the connected subsystem as a function of the initial varied parameters of the whole system.

Methods for calculation of versal model parameters based on the Campbell–Hausdorff decomposition are well known.

Let $A = A_0 + B$, where A_0 is the constant principal matrix of the object, B is the matrix of the interaction constant or is analytically depending on the parameters. We apply to the matrix A, the homothetic transformation e^S parameterized by means of a matrix exponential curve and obtain

$$\widehat{A} = e^{-S} A e^{S} = e^{-S}(A_0 + B)e^{S} = A_0 + X.$$

The matrices S and X should be determined from the known matrix B.

Let us consider a formal expansion of S and X in terms of degrees of the matrix B:

$$S = S^1 + S^2 + \cdots, \quad X = X^1 + X^2 = \cdots,$$

where the superscript is the exponent of the expansion with respect to B. To obtain the component of this expansion, we expand the matrix A into the Campbell–Hausdorff series:

$$\widehat{A} = A_0 + X = e^s A e^s = A + [A, S] + \frac{1}{2!}[[AS]S] + \frac{1}{3!}[[[AS]S]S] + \cdots ,$$

where $[A, S] = AS - SA$ is a Lie bracket. We substitute the expansions of the matrices S and X into this expansion and obtain an infinite system of relations by comparing the terms with equal indices of homogeneity:

$$[A_0 S^1 + B^1] = X^1, \quad B^1 \equiv B,$$
$$[A_0 S^2] + [B^1 S^1] + \frac{1}{2}[[A_0 S^{-1}]S^1] = X^2,$$
$$[A_0 S^3] + [B^1 S^2] + \frac{1}{2}[[A_0 S^2]S^1] + \frac{1}{2}[[B^1 S^1]S^1] + \frac{1}{2}[[A_0 S^1]S^2] = X^3.$$

The formal algorithm for the solution of these equations with respect to the homogeneous components S^i and X^i can be described as follows:

1. Select the first-degree component S^1 in such a way that a maximum number of terms of nonzero elements of the matrix B^1 are annihilated and then determine the first-degree component X^1; the known component $[B^1 S^1] + \frac{1}{2}[[A_0 S^1]S^1]$ appears in this case in the second-degree equations,
2. Select the component S^2 of the transformation so as to annihilate a maximum number of elements in the appeared component and then determine the second-degree component X^2.

The same method should be applied to the third-degree components by selecting S^3, and so on. The algorithm of the transformation e^s is reduced to compensation of as many as possible degrees of perturbation of B, and thus, to decrease its influence in the transformed matrix A. As a whole, this process turns out to be infinite. If we terminate it in N steps, then the terms of degree $N + 1$ and higher with respect to B will remain in the transformed matrix, which symbolically can be written as

$$e^{-s}(A + B)e^s = A_0 + X \pmod{B^{N+1}}.$$

A practical implementation of this algorithm is difficult, inasmuch as it is not clear how to perform its first step.

Based on the versal model theory, an alternate, more constructive algorithm can be proposed for calculation of the transformation e^s and the component X that is not annihilated in principle by this transformation.

Essentially it can be reduced to the solution of equations obtained from the Campbell–Hausdorff expansion, simultaneously for the matrices S and X, using the structure of these matrices known from the versal model theory. In other words, we search for the matrices S in the form of expansion in terms of the base $\{S\}$ from matrices transversal to the centralizer of the matrix A_0:

$$S = \sum_{i=1}^{m} \omega_i S_i \equiv S^1 + S^2 + \cdots + S^m.$$

The basic matrices S, for different types of the matrices A_0 can be constructed in an explicit form. We search for the matrices X in the form of $\{x_k\}$-base expansion of the normal to the orbit:

$$X = \sum_{k=1}^{p} \lambda_k X_k \equiv X^1 + X^2 + \cdots + X^p, \quad p = n^2 - m.$$

Let us point out that each matrix of the infinite sequences of the matrices $S^1, S^2, \ldots$ (or $X^1, X^2, \ldots$) can be decomposed in terms of a finite base $\{S_i\}$ or $\{X_i\}$, respectively.

If the matrix B is given numerically, then we have the following system of equations for determination of the homogeneous components S^i and X^i from the Campbell–Hausdorff expansion,

$$X^1 - [A_0 S^1] = B^1 \equiv B,$$

$$X^2 - [A_0 S^2] = B^2 = [B^1 S^1] + \frac{1}{2}[[A_0 S^{-1}] S^1],$$

$$X^3 - [A_0 S^3] = B^3 = [B^1 S^2] + \frac{1}{2}[[A_0 S^2] S^1]$$

$$+ \frac{1}{2}[[B^1 S^1] S^1] + \frac{1}{2}[[A_0 S^1] S^2],$$

which can be solved recurrently. With a given structure of the matrices S^i and X^i, each equation of this system is of the same type and they differ only by their right-hand sides. A solution of each equation can be obtained by parts using a block representation of the matrices A, S^i, and X^i. The required result is obtained through summation of a finite number of the matrices S^i and X^i with the selected degree N of homogeneity.

Let us consider the algorithm of construction of the solution in the form of an explicit dependence on varied parameters. Let the matrix B of dimension $(n \times n)$ be a linear function of parameters

$$B(\mu) = \sum_{i=1}^{S} \mu_i B_i, \quad S \leq n^2,$$

where B_i are constant matrices.

We present homogeneous components of the matrices X and S in the form

$$X^1 = \sum_{j=1}^{s} \mu_j Y_j, \quad X^2 = \sum_{j,k=1}^{s} \mu_j \mu_k Y_{jk},$$

$$X^3 = \sum_{j,k,l=1}^{s} \mu_j \mu_k \mu_i Y_{j,k,i}, \ldots,$$

$$S^1 = \sum_{j=1}^{s} \mu_j Q_j, \quad S^2 = \sum_{j,k=1}^{s} \mu_j \mu_k Q_{jk},$$

$$S^3 = \sum_{j,k,l=1}^{s} \mu_j \mu_k \mu_i Q_{j,k,i}, \ldots, \tag{7.5}$$

where $Y_j, Y_{jk}, \ldots,$ $Q_j, Q_{jk}, \ldots$ are two infinite sequences of matrices from finite-dimensional spaces $X = \{X_1 \ldots X_p\}$ and $S = \{S_1 \ldots S_m\}$.

Having substituted these expansions into the Campbell–Hausdorff expansion, we obtain the equations for determining the matrices Y_j, Q_j, Y_{jk}, and Q_{jk}:

$$Y_j - [A_0 Q_j] = B_j,$$

$$Y_{jk} - [A_0 Q_{jk}] = B_{jk} = [B_j Q_k] + \frac{1}{2}[[A_0 Q_j] Q_k],$$

$$Y_{jki} - [A_0 Q_{jkl}] = B_{jkl} = [B_j Q_{kl}] + \frac{1}{2}[[A_0 Q_{kl}] Q_j]$$

$$+ \frac{1}{2}[[B_j Q_k] Q_l] + \frac{1}{2}[[A_0 Q_l] Q_{kl}],$$

$$j, k, l = 1, \ldots, S.$$

Because the spaces of the matrices X and S are of finite dimension, each of the two infinite sequences of the matrices $\{Y_j, Y_{jk}, \ldots\}$ and $\{Q_j, Q_{jk}, \ldots\}$ is a finite-dimensional linear combination of the basic sequences:

$$Y_j = \sum_{q=1}^{p} a_{jq} X_q, \quad Y_{jk} = \sum_{q=1}^{p} a_{jkq} X_q, \quad Y_{jkl} = \sum_{q=1}^{p} a_{jklq} X_q,$$

$$Q_j = \sum_{r=1}^{m} b_{jr} S_r, \quad Q_{jk} = \sum_{r=1}^{m} b_{jkr} S_r, \quad Q_{jklr} = \sum_{r=1}^{m} b_{jklr} S_r, \tag{7.6}$$

where $\{a_{iq}, a_{jkq}, \ldots\}$, $\{b_{ir}, b_{jkr}, \ldots\}$ are constant coefficients that can be calculated from the systems of linear algebraic equations of the type

$$\sum_{q=1}^{p} a_q Sp X_q X_{q'}^* = Y, \quad q' = 1, \ldots, p,$$

$$\sum_{r=1}^{m} b_r Sp S_r S_{r'}^* = Q, \quad r' = 1, \ldots, m,$$

after substitution of the matrices $\{Y_j, Y_{jk}, \ldots\}$ for the coefficients $\{a_j, a_{jq}, \ldots\}$ and matrices $\{Q_j, Q_{jk}, \ldots\}$ for the coefficients $\{b_j, b_{jk}, \ldots\}$ into their right-hand sides.

Having substituted expansions (7.6) into expansion (7.5), we obtain expressions for the parameters of the universal model in the form of power series in parameters of the initial strain:

$$\omega_r(\mu) = \sum_{j=1}^{s} b_{jr}\mu_j + \sum_{j,k=1}^{s} b_{jkr}\mu_j\mu_k + \sum_{j,k,l=1}^{s} b_{jklr}\mu_j\mu_k\mu_l + \cdots,$$

$$\lambda_q(\mu) = \sum_{j=1}^{s} a_{jq}\mu_j + \sum_{j,k=1}^{s} a_{jkq}\mu_j\mu_k + \sum_{j,k,l=1}^{s} f_{jklq}\mu_j\mu_k\mu_l + \cdots.$$

If we restrict ourselves to the terms of the Nth degree in these series, then we can speak about universal models of the orders $1, 2, \ldots, M$.

7.5 Bilinear Model of the Thermal-Hydraulic Systems

The dynamics of a reactor core channel of nuclear plants may be represented by a set of ordinary differential equations. For system analysis of critical states associated with the channel, three cases can be considered in respect of the variable to be modulated:

(a) The inlet coolant temperature
(b) The reactivity and
(c) The coolant flow rate

In case (a), the dynamics become linear if we choose the inlet coolant temperature as identification input with the other variables considered fixed. The same applies to case (b), but the neutron dynamics require to be known when the control rods are displaced to change the reactivity. In case (c), the dynamics become bilinear, if besides the above the coolant flow rate is chosen as a reference input to identify the parameters. Now, for practical considerations, the reference input designated for case (a) is not very convenient in actual implementation. Case (b) has been widely utilized in the past for reactor identificationand particularly for examining reactor core dynamics. In the last-mentioned case, however, it is not desirable to apply reactivity changes of large amplitude, which are liable to impair the neutron flux balance.

A mathematical model of a channel in a reactor core can be represented by the ordinary differential equations

$$\dot{x}_1 = -2h(x_1 - x_2)/\rho ca + p/\rho c,$$
$$\dot{x}_2 = -2ah(x_1 - x_2)/(b_2 - a_2)de - v(x_{2|_{y=L}} - x_{2|_{y=0}}/L), \qquad (7.7)$$

where x_1 is the average temperature of fuel pin; x_2 is the average temperature of coolant; p is the average power; ρ is the density of fuel pin; a is the radius of el pin; d is the density of coolant; e is the specific heat of coolant; a is the radius of fuel pin; L is the core height; c is the specific heat of fuel pin; h is the heat transfer coefficient; b is the radius of coolant flow channel tube; v is the coolant flow velocity.

In deriving the above equation the following assumption have been used:

(a) No boiling
(b) The heat conduction through coolant flow along the fuel pin neglected.

Now, letting

$$x_2 = (x_{2|_{y=L}} + x_{2|_{y=0}})/2, \tag{7.8}$$

and with the outlet coolant temperature $x_{2|_{y=L}}$ as output and the coolant flow velocity v as input, we obtain the bilinear equation

$$\begin{aligned}
\dot{x}_1 &= -2h(x_1 - x_2)/\rho c a + p/\rho c, \\
\dot{y}_2 &= -\lambda_2 y_2 + y_1, \\
\dot{y}_3 &= -\lambda_2 y_3 + v y_1, \\
\dot{y}_4 &= -\lambda_2 y_4 + u_2, \\
\dot{y}_5 &= -\lambda_2 y_5 + u_1 v, \qquad (7.9) \\
z_1 &= y_1 - \lambda_2 y_2, \\
z_2 &= v(y_2 - u_1), \\
z_3 &= y_3 - y_5, \\
z_4 &= -z_4 - y_4, \\
z_5 &= 2y_1 - u_1. \qquad (7.10)
\end{aligned}$$

We can show that the bilinear equation (7.10) is equivalent to (7.7) with initial condition

$$y_1(0) = y_{10}, \quad y_2(0) = y_{20}, \quad y_3(0) = y_4(0) = y_4(0) = 0. \tag{7.11}$$

Next section describes an algorithm which provides stable parameter identification and adaptive control for the thermo-hydraulic system described by the bilinear differential equation (7.10).

7.5.1 Identification Algorithm

In this section we describe an identification method based on the expansion of signal processes over an orthogonal basis. Using this methodology we can obtain a system of linear algebraic equations, which is used to determine the coefficients of the bilinear model. By means of the least squares method we obtain estimates of the unknown parameters of the model. It is based on a discrete approximation of the input–output map of a nonlinear object [27].

Consider the bilinear model

$$\dot{x}(t) = Ax(t) + Lu(t) + \sum_{j=1}^{n} B_j x(t) u_j(t), \tag{7.12}$$

where A, L, and B_j are unknown parameters to be estimated; u is a control. By the generalized product of orthogonal series we mean

$$u_j(t) = \sum_{t=0}^{m-1} u_{jl} t^l,$$

$$x(t)u_j(t) = \sum_{l=0}^{m-1} u_{jl} X t^l \Pi(t) = \sum_{l=0}^{m-1} u_{jl} X R_l \Pi(t).$$

The integration of (7.8) gives

$$\begin{aligned} x(t) - x(0) &= A \int_0^t x(t')dt' + L \int_0^t u(t')dt' \\ &+ \sum_{j=1}^{n} L_j \int_0^t x(t') u_j(t') u_j(t') dt'. \end{aligned} \tag{7.13}$$

Using this result, we obtain

$$X\Pi - X(0)\Pi = AXE\Pi LUE\Pi + \sum_{j=1}^{n} B_j X \left[\sum_{t=0}^{m-1} u_{jl} R_l \right] E\Pi. \tag{7.14}$$

Substituting the expression for Θ into (18.20) gives

$$\begin{aligned} XG\Pi - \sum_{j=1}^{n} X(0)G\Pi &= AXEG\Pi + LUEG\Pi \\ &+ \sum_{j=1}^{n} B_j X [\sum_{j=1}^{m-1} u_j R_j] EG\Pi(t) \end{aligned}$$

or

$$\begin{aligned} XG - X(0)G = AXEG + LUEG + \sum_{j=1}^{n} B_j X \left[\sum_{l=0}^{m-2} u_{jl} R_l \right] EG, \\ ZS = (X - X(0))G, \end{aligned} \tag{7.15}$$

where Z is the parameter vector; that is,

$$Z = [ALB_1 B_2 \ldots B_n]. \tag{7.16}$$

7.6 Bilinear Simulation of Reactor Core Accidents

The power increase in the reactor core during a control system accident can be effectively described by bilinear model in terms of reactivity released, the Doppler reactivity feedback coefficient, the delayed neutron fraction and the lifetime of prompt neutrons [6, 23]. Given the speed of the transient, two assumptions have to be made:

1. Energy cannot be transferred from fuel to water and
2. There is no time for delayed neutrons to be emitted

These assumptions match the adiabatic point model and apply when reactivity is very high. When feasible, these assumptions provide good illustrations to safety problems demonstration. The neutron bilinear balance equation (7.17) is expressed simply as

$$\dot{x} = \frac{\rho(t) - \beta}{\lambda} x(t), \tag{7.17}$$

with λ representing prompt lifetime, and β being the delayed neutrons fraction [2,21].

The increase in reactivity

$$\rho(t) = \rho_0 - \alpha \left(T(t) - T(0)\right) \tag{7.18}$$

is expressed in terms of the total reactivity ρ_0 of the control rods assembly, while the Doppler reactivity feedback effect of the fuel is expressed through the coefficient α and the mean temperature increase in the core T.

The mean temperature increase in the core T depends directly on the energy produced

$$M \cdot C \left(T(t) - T(0)\right) = \int_0^1 x(\tau) d\tau, \tag{7.19}$$

where M represents the mass and C is the specific heat of the fuel.

It is also demonstrated that the produced energy E and the maximum power $x_{\max}$ reached during the transition satisfy the expressions [14]

$$E = 2M \cdot C \frac{\rho_0 - \beta}{\alpha} \tag{7.20}$$

and

$$x_{\max} = M \cdot C \frac{(\rho_0 - \beta)^2}{\alpha\gamma}. \tag{7.21}$$

The results depend on the estimation of the constants used in the mathematical applications. It can also be directly demonstrated that the reactivity at maximum power equals β. In this way, the key aspects of the power increase, see Fig. 7.1, during the first moment of the transition can be obtained. For example, important variables for studying the thermo-mechanical behavior of the fuel, such as the rate of power increase and the width of the curve $x(t)$ are obtained explicitly.

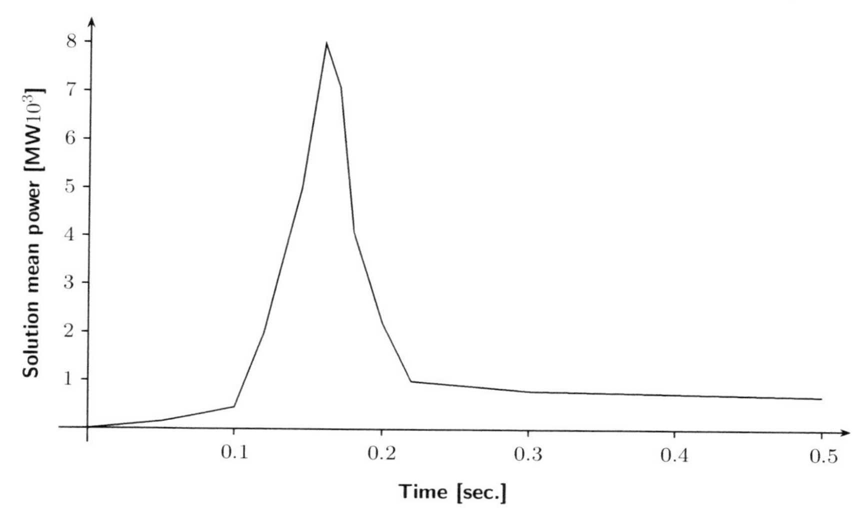

Fig. 7.1. Core power increase. The dependence of the average power of the reactor containment on time

7.7 Conclusions

In this chapter, we consider the problem of determining critical states of nuclear plant reactors using bilinear modeling. Mathematical bilinear modeling and numerical analysis of initial critical events are proposed. We considered the ways in which the disposition of the phase curves of a vector field of the dynamical model can alter in a neighborhood of a singularity as the parameters on which the vector field depends vary. A technical convenience in the study of such changes are certain deformations having a special universality property – the so-called versal families. Our results are presented mainly in the form of explicit formulae for versal families and an analysis of the corresponding bifurcation diagrams.

A bilinear model of thermo-hydraulic system is proposed. We described an identification method based on the expansion of signal processes over an orthogonal basis. Using this methodology we can obtain a system of linear algebraic equations, which is used to determine the coefficients of the bilinear model. By means of the least squares method we obtain estimates of the unknown parameters of the model. The computational algorithm obtained has quite good accuracy. An algorithm for identification of the bilinear discrete models is obtained. It is based on a discrete approximation of the input–output map of a nonlinear object.

Acknowledgement. Vitaliy A. Yatsenko are partially supported by STCU grant G033. Panos M. Pardalos and Steffen Rebennack are partially supported by Air-Force and CRDF grants. The support is greatly appreciated.

References

1. V. Arnold. Singularities of smooth mappings. *Uspekhi Mat. Nauk.*, 23(1):3–44, 1968.
2. D. Bell and S. Glesston. *Theory of nuclear reactors.* Moscow, Atomizdat, 1974.
3. T. Bose and M. Chen. Conjugate gradient method in adaptive bilinear filtering. *IEEE Trans. Signal Process.*, 43:349–355, 1995.
4. R. Brockett. System theory of group manifolds and coset spaces. *SIAM J. Contr.*, 10:265–284, 1972.
5. K. Chitkara and J. Weisman. Equilibrium approach to optimal in-core fuel management for pressurized water reactors. *Nucl. Technol.*, 24(1):33–49, 1974.
6. F. D'Auria, B. Gabaraev, S. Soloviev, O. Novoselsky, A. Moskalev, E. Uspuras, G. M. Galassi, C. Parisi, A. Petrov, V. Radkevich, L. Parafilo, and D. Kryuchkov. Deterministic accident next term analysis for RBMK. *Nucl. Eng. Des.*, 238(4):975–1001, 2008.
7. E. De Klerk, C. Roos, T. Terlaky, H. T. Illés, I. A. J. De Jong, J. Valkó, and J. E. Hoogenboom. Optimization of nuclear reactor reloading patterns. *Ann. Oper. Res.*, 69(0):65–84, 1997.
8. F. Fnaiech, L. Ljung, and Fliess M. Hoogenboom. Recursive identifcation of bilinear systems. *Int. J. Control*, 45(2):453–470, 1987.
9. R. R. Fullwood and R. E. Hall. *Probabilistic risk assessment in the nuclear power industry: fundamentals and applications.* Butterworth-Heinemann, New York, 1988.
10. V. Goldin, G. Pestriakova, Y. Troishchev, and E. Aristova. Neutron and nuclear regime with self-organisation in reactor with the hard spectrum and carbide fuel. *Math. Model.*, 14(1):27–39, 2002.
11. C. S. Gordelier. Nuclear energy risks and benefits in perspective. *NEA News*, 25(2):4–8, 2007.
12. Greenpeace. *Subject: Calender of Nuclear Accidents and Events (Updated 21st March)*, 2007. http://archieve.greenpeace.org/comms/nukes/chernob/rep02.html.
13. L. Hunt, R. Su, and G. Meyer. Global transformations of nonlinear systems. *IEEE Trans. Autom. Contr.*, 25(2):4–8, 2007.
14. International Atomic Energy Agency. *Accident analysis for RBMKs.* Safety Reports Series No. 43, IAEA, Vienna, 2005.
15. International Atomic Energy Agency. *Annual Report 2006.* IEA, Vienna, 2006.
16. International Energy Agency. *IEA energy technology essentials: Nuclear power.* IEA, Vienna, March 2007.
17. R. Kozma, S. Sato, M. Sakuma, M. Kitamura, and T. Sugiyama. Generalization of knowledge acquired by a reactor core monitoring system based on a neuro-fuzzy algorithm. *Prog. Nucl. Energy*, 29:203–214, 1995.
18. J. Lo. Global bilinearizastion of systems with control appearing linearly. *SIAM J. Control*, 13:879–884, 1975.
19. A. Krener. Bilinear and nonlinear realizations of input-output maps. *SIAM J. Control*, 13(4):827–834, 1975.
20. Zhian Li, P. M. Pardalos, and S. H. Levine. *Space-covering approach and modified Frank-Wolfe algorithm for optimal nuclear reactor reload design.* Recent advances in global optimization. Princeton University Press, New Jersey, 1992.
21. G. Marchuk. *Methods of nuclear reactors calculations.* Samizdat, Moscow, 1961.

22. N. J. McCormick. *Reliability and risk analysis: methods and nuclear power applications*. Academic, New York, 1981.
23. M. F. Robbe, M. Lepareux, E. Treille, and Y. Cariouc. Numerical simulation of a hypothetical core disruptive accident in a small-scale model of a nuclear reactor. *Nucl. Eng. Des.*, 223(2):159–196, 2003.
24. A. Veinberg and E. Vigner. *Physical theory of nuclear reactors [Russian translation]*. IL, Moscow, 1961.
25. M. L. Wald. Approval is sought for reactors. *The New York Times*, pages C1–C11, September 25, 2007.
26. V. Yatsenko. An engineering design method for automatic control of transverse magnetic field in tokamaks. *Proceedings of Conference on The 2nd All-Union Conference on the Engineering Problems of Thermonuclear Reactors*, pages 272–273, 1981.
27. V. Yatsenko. Dynamic equivalent systems in the solution of some optimal control problems. *Avtomatika*, 4:59–65, 1984.
28. V. Yatsenko. Methods of risk analysis for energy objects. *Proceedings of Conference on International Energy Conference*, July 23–28, Las Vegas, Nevada, USA pages 272–273, 2000.
29. V. Yatsenko. Reliability forecasting of nuclear reactor in fuzzy environment. *Proceedings of Conference on Problems of Decision Making Under Uncertainties*, pages 54–57, 2003.

8

Mixed-Integer Optimization for Polygeneration Energy Systems Design

Pei Liu and Efstratios N. Pistikopoulos

Summary. In this chapter we introduce polygeneration energy systems in the context of future energy systems, and modeling and optimization issues involved in planning and configuration design of polygeneration processes. A mixed-integer nonlinear programming (MINLP) model is developed for the design optimization of polygeneration energy systems. A suitable superstructure is introduced, based on partitioning a general polygeneration energy system into four major blocks, for each of which alternative available technologies and types of equipment are considered. A detailed case study, involving a coal-based polygeneration plant producing electricity and methanol, is presented to demonstrate the key features and applicability of the proposed approach.

Key words: Polygeneration, Mixed-integer nonlinear programming, Design optimization

8.1 An Overview of Polygeneration Energy Systems

Global energy consumption has been rising since 1970s, and according to the projection of Department of Energy (DOE) of the U.S., it will keep on rising for quite a long period in the future [25], see in Fig. 8.1. However, the global greenhouse gas (GHG) emissions have to be restricted to a certain level since most countries around the world (excluding the U.S.) had signed the Kyoto Protocol by 2005, which requires its participating countries to reduce their GHG emissions to below emission levels in 1990 by 2012 [25], see in Fig. 8.2.

A severe and lasting global energy problem is the shortage of liquid fuels. Worldwide proved oil reserves amount to 1,293 billion barrels by 2006, and the daily consumption in 2003 was 80 million barrels [25]. Even if this consumption rate were not to increase, all global oil reserves would be depleted in about 44 years. Moreover, 57% of the oil reserves are found in the Middle East, the most politically unstable region around the world. Thus, countries that

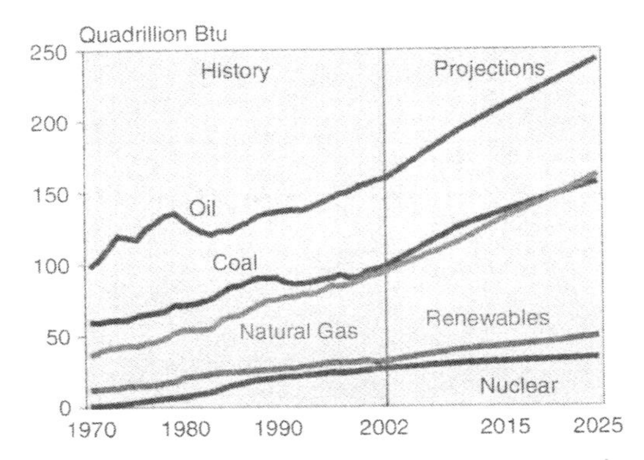

Fig. 8.1. World marketed energy use by energy type, 1970–2025

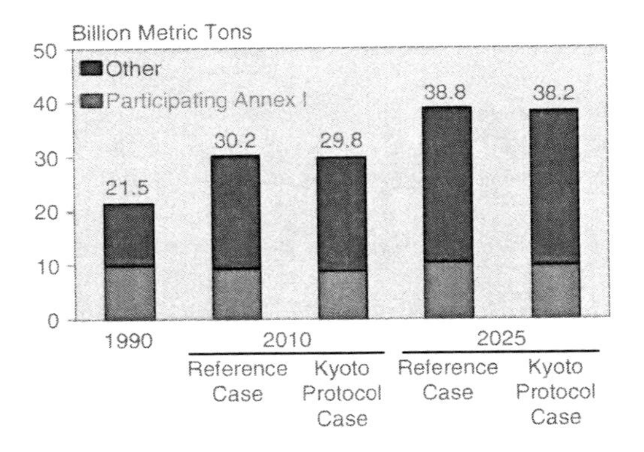

Fig. 8.2. World carbon dioxide emissions

depend heavily on oil importation need to seek diversification of liquid fuel suppliers to enhance national energy security.

One possible solution to the current acute energy and environmental problems is to utilize more advanced, innovative, and efficient primary energy technologies. Power generation is the largest primary energy consumer, accounting for 40% of the primary energy and using all energy resources, like coal, natural gas and oil. Consequently, it is a colossal source of GHG emissions, being the cause for the release of more than 7.7 billion tons of carbon dioxide annually; thus, power generation accounts for 37.5% of the total annual carbon dioxide emissions [24]. Innovations and improvements of power generation technologies for higher efficiency and lower emissions have never ceased to emerge and be licensed over the decades. The Integrated Gasification Combined Cycle (IGCC), which combines a gasifier with a gas turbine cycle and a steam turbine cycle, is one of the most promising alternatives.

Fortunately, oil is not the only energy source for the production of liquid fuels. They can also be synthesized from other fossil fuels, like coal, natural gas, and petroleum coke, as well as renewable energy sources such as biomass. Synthetic liquid fuels have the potential to substitute conventional, oil based liquid fuels, for example, methanol and dimethyl ether (DME) can be used as gasoline and diesel oil, respectively.

The concept of polygeneration comes from the similarities between liquid fuel synthesis processes and combined cycle power generation processes. Both processes require synthesis gas (syngas), mainly consisting of carbon dioxide and hydrogen, as an intermediate product. It is shown in Fig. 8.3 a conventional process for methanol synthesis from natural gas [9]. Natural gas is first fed into a reactor together with adequate amount of oxygen and steam to produce syngas. The syngas is then cooled down to remove extra amount of steam and impurities like sulfureted hydrogen. The cooled syngas is compressed to certain pressure, fed into a synthesis reactor, and catalyzed to produce crude methanol. The effluent is condensed and distillated to produce the final product, either fuel-class or chemical-class methanol. Unconverted gas is recycled to the synthesis reactor. Figure 8.4 shows a typical conceptional structure of an IGCC power plant [8]. It usually uses coal as main fuel, but can also use

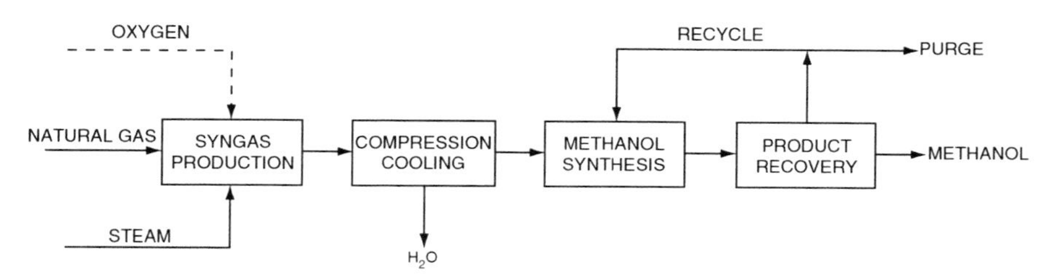

Fig. 8.3. Methanol synthesis from natural gas

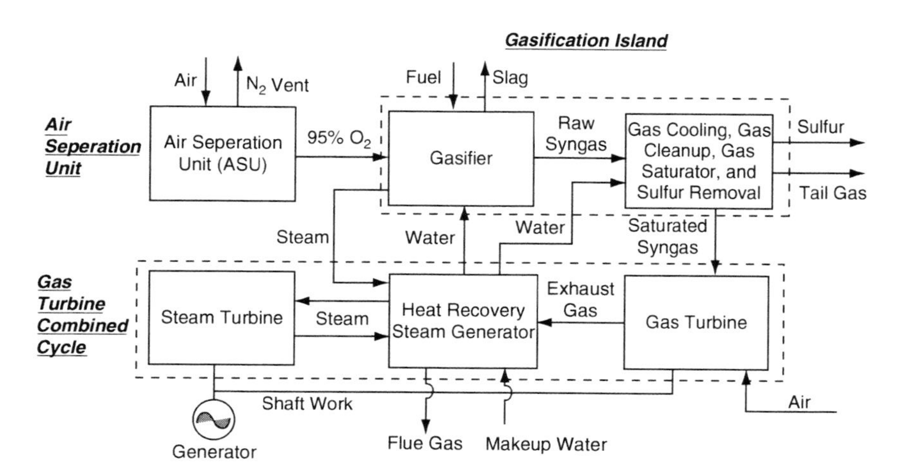

Fig. 8.4. Conceptional process illustration of an IGCC power plant

petroleum coke or biomass. Fuel is fed into a gasifier first, where it is gasified to produce syngas. The syngas goes through a series of cleaning modules to remove solid particles and acid gases. The clean syngas is then fed into the combustion chamber of a gas turbine, where it combusts with a large amount of air, producing hot flue gas. This hot gas expands in the gas turbine first, and then goes through a heat recovery steam generator (HRSG), producing high, medium and low pressure steam to drive corresponding pressured steam turbines. The similarities existing between liquid fuel synthesis processes and IGCC processes indicates a possibility to coproduce electricity, synthetic fuels, heat and other chemicals in one process, with higher conversion efficiency that will result in lower polluting emission levels.

A polygeneration energy system can improve profit margins and market penetration, decrease the overall capital investment cost, reduce GHG emissions, increase feedstock flexibility and alleviate the current grave dependence on crude oil and all refinery fuels. An exemplary polygeneration energy system for integrated production of methanol and electricity is shown in Fig. 8.5, in which coal or other carbon-based fuels are fed to a gasifier, where they react with oxygen to produce syngas; part of it is fed to a chemical synthesis plant to produce methanol, which can be either sold, stored in the plant for peak-time power generation, or transported to other power plants for peak-time power generation. The flue gas from the chemical synthesis plant, together with the other part of the fresh syngas flow, undergoes combustion in a power generation plant to generate electricity [18].

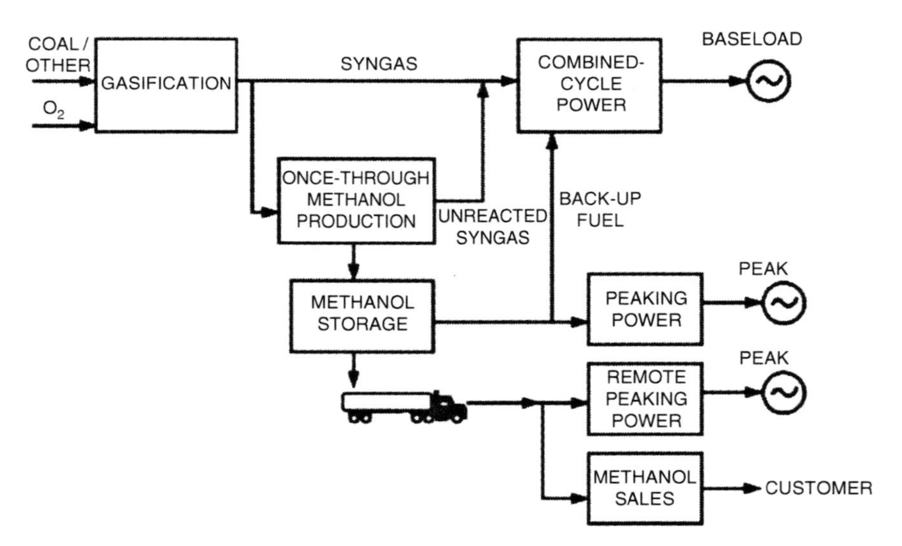

Fig. 8.5. A polygeneration plant coproducing methanol and electricity

Polygeneration energy systems have many advantages over conventional stand-alone power or chemical plants. The production cost for methanol can be reduced by 40% in a polygeneration plant coproducing methanol, heat and electricity. For a quad-generation plant coproducing syngas, methanol, heat and power, the reduction over conventional plants is 46% for syngas production cost, 38% for capital investment, 31% for operating cost per energy unit, and 22.6% for GHG emission [19]. In a polygeneration plant coproducing DME and electricity, the DME production cost will be $6–6.5/GJ, thus comparable with conventional fuel prices [5].

However, polygeneration energy systems also have some disadvantages to be overcome before large-scale installation. First, its investment cost is higher than a conventional stand-alone power plant or a chemical plant, although it may be lower than the combination of two stand-alone plants with the same production rates. Secondly, the high degree of integration between the power generation and chemical synthesis parts may reduce the availability of the process as a whole, thus more advanced control scheme may be required.

Overall, the advantages of polygeneration energy systems over conventional stand-alone power generation or chemical synthesis technologies, like higher energy conversion efficiency and lower emissions, are superior to the few disadvantages discussed above. All these advantages make polygeneration a very competitive technology. Its advantages lie in three main aspects:

- Energy efficiency – due to the tight integration of the power generation and the chemical synthesis sections, the overall energy utilization of a polygeneration plant is expected to be higher than the overall efficiency of stand-alone plants, producing the same products.
- Alternative fuels and energy carriers chemical products produced by a typical polygeneration plant can be used as substitutions for traditional liquid fuels; for example, methanol for gasoline, DME for diesel oil. Hydrogen can also be a product.
- Cost effective emissions reduction the large-scale of polygeneration energy systems is expected to result in cost-effective solutions for the implementation of CO_2 capture and sequestration (CCS) units.

8.2 Studies and Existing Problems

Due to the high degree of integration and coupling between the power generation and chemical synthesis parts, determining the optimal configuration and design of a polygeneration energy system is quite a challenging task. Different process designs have been reported in literature. Ma et al. [15, 16] proposed a group of sequential and parallel process designs for a coal-based polygeneration plant producing electricity and methanol. By comparing energy efficiency and economic characteristics, they concluded that the sequential design with a once-through methanol synthesis unit exhibits optimal overall performance.

P. Liu et al. [14] tested the dynamic behavior of the processes designed by Ma et al. under varying power loads, and concluded that a parallel process design will have better performance under certain operating conditions. G.J. Liu et al. [12] developed a novel process design producing electricity and DME from natural gas, in the context of determining a better way to transport natural gas from West China to East China. Chen et al. [4] compared the energy and exergy efficiencies between polygeneration plants producing electricity and DME and stand-alone DME plants, and concluded that the energy saving ratio in a polygeneration plant could be as high as 16.6%. Besides general processes producing electricity and chemical fuels, there are also other forms of polygeneration process designs for specific purposes, such as exploring the potential of coal-gas generated in coke ovens in iron and steel industry, and combining an ammonia process with a coal-based power generation process for higher energy utilization rates [26, 27].

While the reported works above have significantly advanced our understanding of polygeneration from a design perspective, they share a common limitation – they either focus on specific technologies, or mostly focus on specific requirements/conditions.

In this context, it is important to provide a general systematic methodology for the design of polygeneration energy systems, which could be applicable for different technology, design and operational requirements.

In this work, building on our earlier work for the strategic planning of polygeneration energy system [13], we present the building blocks of such a general methodology, featuring a superstructure representation and a comprehensive mixed integer optimization model formulation.

The paper is structured as following. The superstructure representation is described. The mathematical model is presented then, followed by detailed study of a polygeneration plant for the production of methanol and electricity.

8.3 Superstructure Representation

A general superstructure representation of a polygeneration plant is shown in Fig. 8.6, consisting of four blocks: gasification, chemical synthesis, gas turbine, and heat recovery steam generator (HRSG) and steam turbine. The superstructure acts as the overriding model, capturing all the possible alternatives and intersections between process components. For each block, several alternative technologies and types of equipment are available for selection. All combinations of these technologies and types of equipment form the design space of the plant. The optimal process design will then correspond to the best combination of these components, obtained by eliminating existence of units and links between them.

To further illustrate the model superstructure and its utilization in modeling, a four-block superstructure of a coal-based polygeneration process producing electricity and methanol is discussed in some detail below.

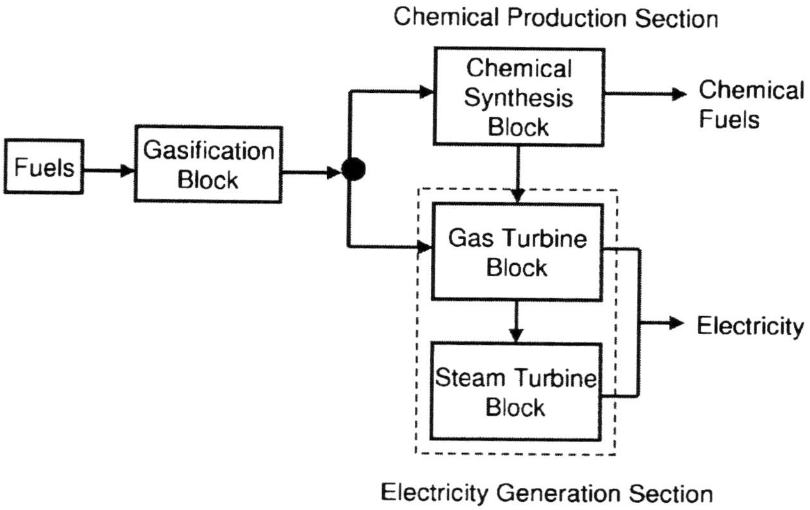

Fig. 8.6. General polygeneration energy system superstructure

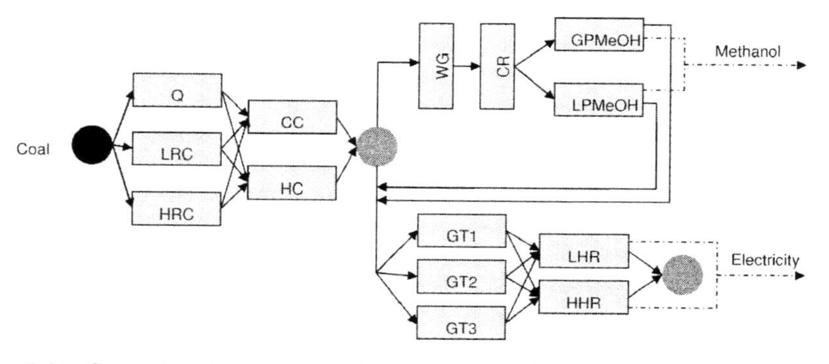

Fig. 8.7. Superstructure representation of a polygeneration plant producing electricity and methanol

Figure 8.7 shows the superstructure and all alternative technologies and types of equipment for each block.

The function of the gasification block is to prepare clean synthesis gas (syngas) for downstream utilization by gasifying feedstocks, usually coal, in a high temperature, high pressure, and reductive atmosphere. The crude syngas consists mainly of hydrogen, carbon monoxide, carbon dioxide, hydrogen sulphide (H_2S), carbonyl sulphide (COS), unconverted carbon, and ash. The hot crude syngas can either be quenched by cold water or cooled through a series of radiative and convective heat exchangers where heat can be recovered and used for power generation. Once it is cooled down, slag is removed and fine solid particles of unburned carbon are separated and recycled. After that, the syngas goes through a cleanup process to remove acid components which are extremely hazardous to downstream units and catalysts. Depending on the temperature of the syngas entering the cleanup process, two types of cleanup technologies are available, cold gas cleanup (CGCU) and hot gas cleanup (HGCU). In the model superstructure, the gasification block is further divided into two subblocks, representing the cooling part and the cleanup part. Technologies and types of equipment for the gasification block are denoted by:

- Q: Quench
- LRC: Low temperature radiative and convective cooling
- HRC: High temperature radiative and convective cooling
- CC: Cold syngas cleanup
- HC: Hot syngas cleanup

Syngas leaving the gasification block enters the methanol synthesis block. There are two kinds of commercially matured methanol synthesis technologies. According to the phase of synthesis reaction, they are known as gas phase methanol synthesis (GPMeOH) and liquid phase methanol synthesis (LPMeOH). In a GPMeOH reactor, reactants are in gas phase and react with each other on the surface of solid catalysts. In an LPMeOH reactor, gaseous reactants resolve in inert oil with solid catalyst particles being suspended in.

The methanol synthesis progress typically consists of mainly three reactions, where only two of them are independent, as follows:

$$CO + 2H_2 \longrightarrow CH_3OH \tag{8.1}$$

$$CO_2 + 3H_2 \longrightarrow CH_3OH + H_2O \tag{8.2}$$

$$CO + H_2O \longrightarrow CO_2 + H_2 \tag{8.3}$$

Besides the main reactor, some auxiliary units are needed to ensure an optimal performance for the reactor. First of all, since the synthesis reactions are highly exothermic, heat released in the synthesis reaction should be either recovered for power generation or absorbed by cooling water to obtain an isothermal operation. For the ease of controlling the reaction heat, GPMeOH has an upper limit for the carbon monoxide content in reactants and needs a water gas shift reactor before it to adjust the composition of the feeding syngas. However, LPMeOH reactors do not have such a constraint. Therefore, a water gas shift reactor always exists before a GPMeOH reactor. Secondly, both GPMeOH and LPMeOH reactors can achieve maximum conversion rate at approximately five percent for the carbon dioxide volume fraction, making the catalyst staying at the most active level. With this requirement, a carbon dioxide removal unit is usually needed before the reactor. Typical technologies and types of equipment for the methanol synthesis block are denoted by:

- WG: Water gas shift reactor
- CR: Carbon dioxide removal unit
- GPMeOH: Gas phase methanol synthesis

- LPMeOH: Liquid phase methanol synthesis

Fluegas leaving the methanol synthesis block enters the gas turbine block for power generation. This block consists of a combustion chamber where fuel burns with pressurized air to produce pressurized hot gas, an air compressor that compresses air into the combustion chamber, and a turbine that transforms the thermal energy of the hot gas to mechanical work. Typical technologies and types of equipment for this block are given by:

- GT1: Gas turbine with first-stage inlet temperature at 1,703 K
- GT2: Gas turbine with first-stage inlet temperature at 1,589 K
- GT3: Gas turbine with first-stage inlet temperature at 1,473 K

The exhausted gas leaving the gas turbine block enters HRSG where its heat is recovered to generate steam for the steam turbine, where the thermal energy in the steam is transformed into mechanical work. Technologies and types of equipment for this block are given by:

- LHR: Low heat recovery technology with exhaust gas temperature of 450 K
- HHR: High heat recovery technology with exhaust gas temperature of 400 K

Based on such a superstructure representation, a mathematical model can be developed, as discussed next.

8.4 Mathematical Model

The mathematical model comprises the physical representation of each one of the four blocks in the superstructure representation discussed in the previous section, along with an appropriate objective function. Mixed-integer logical conditions are also employed, associated, for example, with selection of technologies, types of equipment and connectivity restrictions. Nomenclature notation is listed in Appendix A.

8.4.1 Gasification Block

Mass composition, temperature and pressure of the fuel stream fed to the gasification block are given by

$$z(ie) = \sum_{ft} z_0(ft, ie) * y_f(ft) \tag{8.4}$$

$$T_f = \sum_{ft} T_{f,0}(ft) * y_f(ft) \tag{8.5}$$

$$P_f = \sum_{ft} P_{f,0}(ft) * y_f(ft) \quad (8.6)$$

$$\sum_{ft} y_f(ft) = 1 \quad (8.7)$$

Physical properties of other feeding streams, like water or steam, and oxygen or air, can be expresses in a similar way.

Key operational parameters of the gasifier are given as follows.

$$R_{waterfuel} = \sum_{gft} R_{waterfuel,0} * y_{gas}(gft) \quad (8.8)$$

$$R_{O_2 fuel} = \sum_{gft} R_{O_2 fuel,0} * y_{gas}(gft) \quad (8.9)$$

$$C_{gas}(rs) = \sum_{gft} C_{gas,0}(rs, gft) * y_{gas}(gft) \quad (8.10)$$

$$T_{gas} = \sum_{gft} T_{gas,0} * y_{gas}(gft) \quad (8.11)$$

$$P_{gas} = \sum_{gft} P_{gas,0} * y_{gas}(gft) \quad (8.12)$$

$$\sum_{gft} y_{gas}(gft) = 1 \quad (8.13)$$

Using these parameters, mass relations between the feeding steams of the gasifier can be set up, as follows:

$$ma_{gfwater} = ma_f * R_{waterfuel} \quad (8.14)$$

$$ma_{gfO_2} = ma_f * R_{O_2 fuel} \quad (8.15)$$

Mass balances connecting the feedstocks of the gasifier and the crude syngas are then built on an elementary basis:

$$f(z(ie), ma_f, ma_{gfwater}, ma_{gfO_2}, ma_{rawsg}) = 0 \quad (8.16)$$

The mole flowrate and the mass flowrate of the crude syngas can be connected to each other through its mole composition and the molecular weight of its components:

$$f(ma_{rawsg}, mo_{rawsg}, \mathbf{x}_{rawsg}) = 0 \quad (8.17)$$

Equation (8.16) is on an elementary basis, while (8.17) is on a component basis. Considering the fact that more types of components exist in the crude syngas than elements in the feedstocks, ratios of mole fractions between certain components in the crude syngas, which are associated with a particular type of gasification technology, are added.

$$f(\mathbf{x}_{rawsg}, C_{gas}) = 0 \tag{8.18}$$

Specific enthalpy and enthalpy can be expressed as a function of mole composition, temperature and pressure, as follows:

$$h_{rawsg} = h(T_{gas}, p_{gas}, \mathbf{x}_{rawsg}) \tag{8.19}$$

$$H_{rawsg} = mo_{rawsg} * h_{rawsg} \tag{8.20}$$

A capacity constraint is added to size the gasifier, as follows:

$$ma_f - F_{gas} \leq 0 \tag{8.21}$$

Selection of technologies for the gasifier cooler, sizing of the cooler and physical properties of the syngas leaving the cooler were also modelled in a similar way as for the gasifier.

Heat recovered in the gasifier cooler is given by

$$\Delta H_{cooler} = H_{coolsg} - H_{rawsg} \tag{8.22}$$

The recovered heat is used for power generation. The amount of power generation depends on the temperature and pressure of the working fluid carrying it and the working process in the HRSG and steam turbine block. Instead of going into extensive technical details of heat transfer and fluid engineering, which is not the focus of this model, all the influential factors involved in generating power from the recovered heat are incorporated in a single parameter , defined as the ratio of the power generated by the recovered heat to the total amount of the recovered heat. Thus the power generated indirectly from the gasifier cooler is given by

$$W_{cooler} = \Delta H_{cooler} * \eta_{cooler} \tag{8.23}$$

Note that in case various technologies are not compatible, mixed integer logical constraints are added. For example, a hot gas cleanup unit can never be used after a quench cooler, instead, it requires a cooler using high-temperature radiative and convective technology, which can be represented as follows:

$$y_{cleanup}(HC) - y_{cooler}(HRC) \leq 0 \tag{8.24}$$

Calculations of physical properties, mole composition, mass flowrate, and enthalpy for streams in the other blocks are derived in a similar way, hence omitted in the following. Only mathematical expressions with unique characteristics to the particular case are depicted below.

8.4.2 Chemical Synthesis Block

Leaving the cleanup unit, the clean syngas is split into two streams. One goes through an optional water gas shift reactor, and the other is bypassed, both mixing together again after the water gas reactor. Mole composition of the stream going through the reactor is changed through the water gas shift reaction

$$CO + H_2O \longrightarrow CO_2 + H_2 \tag{8.25}$$

while the bypassed stream keeping unchanged. This is a means of adjusting the mole composition of the syngas according to the requirements of the methanol synthesis reactor. The degree of adjustment depends on the design parameter of split ratio R_{split} given by the following equation:

$$f(R_{split}, ma_{clsg}, \mathbf{x}_{clsg}, \mathbf{x}_{wgsg}) = 0 \tag{8.26}$$

The syngas then goes through a carbon dioxide removal unit, where the fraction of carbon dioxide in the syngas is adjusted to an appropriate level for the best performance of the catalysts in the methanol synthesis reaction, given by

$$x_{sg}(CO_2) = x_{sg,0}(CO_2) \tag{8.27}$$

After the carbon dioxide removal, the syngas goes to the methanol synthesis reactor to produce methanol. Gas phase synthesis technology has a strict upper limit on the mole fraction of carbon monoxide in the syngas, given by reactions (8.1) to (8.3), as follows:

$$x_{sg}(H_2) - (2x_{sg}(CO) + 3x_{sg}(CO_2)) \geq (y_{meoh}(GPMEOH) - 1) * U \tag{8.28}$$

Parameters of the reactor, such as the conversion rate of reactants, depend on the selection of synthesis technologies, given by

$$R_{meoh}(pmeoh) = \sum_{meoh} R_{meoh,0}(pmeoh, meoh) * y_{meoh} \tag{8.29}$$

Using these parameters, mass balance between the incoming syngas and product gas is given by

$$f(mo_{sg}, \mathbf{x}_{sg}, mo_{pg}, \mathbf{x}_{pg}, R_{meoh}) = 0 \tag{8.30}$$

One realization of (8.30), based on mass balance, is shown below as an example:

$$mo_{sg} * (x_{sg}(H_2) - 2R_{meoh}(CO)x_{sg}(CO) - 3R_{meoh}(CO_2)x_{sg}(CO_2)) = \\ mo_{pg} * x_{pg}(H_2) \tag{8.31}$$

Crude methanol produced in the synthesis reactor goes through a series of distillation columns to produce methanol as a final product, either of fuel degree or chemical degree. Mathematically, this process is formulated as splitting the crude product into two streams. One stream contains mainly methanol and a minor content of water, depending on the product degree, whilst the other stream includes all the other components in the crude methanol. The mass flowrate of the final product methanol, or its production rate, must meet its market demand, given by (8.47).

8.4.3 Gas Turbine Block

The exhausted gas leaving the synthesis block, also known as fuel gas, goes to the gas turbine block, where it combusts in the combustion chamber with a large amount of compressed air to produce gas with sufficient high temperature and pressure. Mathematically, the combustion procedure is expressed as an oxidation reaction with excessive oxygen. Assuming complete combustion takes place in the combustion chamber, all carbon monoxide and hydrogen in the fuel gas is converted to carbon dioxide and steam.

The selection of gasification technologies determines the temperature and pressure of the gas entering and leaving the gas turbine, denoted by T_1 and p_1, and T_4 and p_4, respectively. Through energy balance, the mass flowrate of the air flowing into the compressor of the gas turbine is a function with respect to T1, the mass flowrate of the fuel gas, and its mole composition, given by

$$f(ma_{air}, ma_{fg}, \mathbf{x}_{fg}, T_1) = 0 \tag{8.32}$$

Now that the flowrate of the air flowing through the compressor and its physical properties at the inlet and outlet point of compressor are known, the compression work consumed by the compressor can be expressed as a function of them:

$$W_{gc} = f(ma_{air}, T_1, p_1, p_2) \tag{8.33}$$

The realization of (8.33) is shown below [3]:

$$W_{gc} = \frac{1}{\eta_{isen}} ma_{air} C_p T_1 \left(\left(\frac{p_2}{p_1} \right)^{\frac{\gamma-1}{\gamma}} - 1 \right) \tag{8.34}$$

The mechanical work generated by the gas turbine is a function of the mass flowrate of the gas flowing through the gas turbine, its composition, and its physical properties at the point before and after the turbine, denoted by

$$W_{gt} = f(ma_{gas1}, \mathbf{x}_{gas1}, T_1, p_1, T_4, p_4) \tag{8.35}$$

8.4.4 HRSG and Steam Turbine Block

Gas leaving the gas turbine enters the HRSG and steam turbine block, where its heat is recovered in the HRSG and transformed to mechanical work in the steam turbine. An overall efficiency, denoted by η_{st}, is used to represent different technologies for HRSG and the steam turbine, given by

$$\eta_{st} = \sum_{st} \eta_{st,0}(hst) * y_{st}(hst) \tag{8.36}$$

Work generated by the steam turbine is thus given by

$$W_{st} = H_{gas4} * \eta_{st} \tag{8.37}$$

So far, all streams of the mechanical work consumed and generated in the process have been presented, based on which the net work generated by the process is given by

$$W = W_{gt} + W_{st} + W_{cooler} + W_{meoh} - W_{ASU} - W_{gc} \tag{8.38}$$

where W_{ASU} is the work consumption in the air separation unit (if there is one) which provides oxygen for the gasifier. It is a function of the mass flowrate of the oxygen steam to the gasifier [17].

The mechanical work is transformed to electricity through a generator, and the electricity generation is given by

$$E = W * \eta_G \tag{8.39}$$

The electricity generation should meet its market demand, given by (8.47).

8.4.5 Objective Function

The objective function of the model is the annual profit of the polygeneration plant over lifetime, given by

$$Profit = Income - CostEquip - CostFuel \tag{8.40}$$

Income from the sale of products is given by:

$$Income = \sum_{p} PriceP(p) * ProRate(p) * OpTime \tag{8.41}$$

Total costs of equipment include annual depreciated investment cost, fixed O&M cost, and variable O&M cost, as follows:

$$CostEquip = \sum_{e} Inv(e) + OMFix(e) + OMVar(e) \tag{8.42}$$

Assuming there are ei kinds of technologies or types of equipment e available for a certain block or unit, the investment costs are expressed as

$$Inv(e) = \sum_{ei} UInv_0(e, ei) * F(e, ei) \tag{8.43}$$

$$0 \leq F(e, ei) \leq y(e, ei) * UL \tag{8.44}$$

$$\sum_{ei} y(e, ei) = 1 \tag{8.45}$$

Equation (8.44) ensures that if a technology or type of equipment is not selected, its corresponding capacity is zero, whilst if it is selected, the operation capacity can take any value between zero and the upper limit. Equation (8.45) makes sure that one and only one kind of technology or type is selected for a piece of equipment. Equations for calculating the fixed and variable O&M costs are similar, omitted here for conciseness.

Expense on purchase of fuels is expressed as below:

$$CostFuel = \sum_{ft} PriceF(ft) * FuelRate(ft) * OpTime \tag{8.46}$$

Assuming there is an upper bound and lower bound for market demands, the production rate should meet the following constraints:

$$LDemand(p) \leq ProRate(p) \leq UDemand(p) \tag{8.47}$$

8.4.6 Overall Model

By gathering all the terms together, we obtain the following mathematical model, shown in (8.48).

$$\begin{aligned}
max \quad & Profit \\
s.t. \quad & z(ie) = \sum_{ft} z_0(ft, ie) * y_f(ft) \\
& T_f = \sum_{ft} T_{f,0}(ft) * y_f(ft) \\
& P_f = \sum_{ft} P_{f,0}(ft) * y_f(ft) \\
& \sum_{ft} y_f(ft) = 1 \\
& R_{waterfuel} = \sum_{gft} R_{waterfuel,0} * y_{gas}(gft) \\
& R_{O_2fuel} = \sum_{gft} R_{O_2fuel,0} * y_{gas}(gft) \\
& C_{gas}(rs) = \sum_{gft} C_{gas,0}(rs, gft) * y_{gas}(gft) \\
& T_{gas} = \sum_{gft} T_{gas,0} * y_{gas}(gft) \\
& P_{gas} = \sum_{gft} P_{gas,0} * y_{gas}(gft)
\end{aligned}$$

$$\sum_{gft} y_{gas}(gft) = 1$$
$$\dots$$
$$y_{cleanup}(HC) - y_{cooler}(HRC) \leq 0$$
$$x_{sg}(H_2) - (2x_{sg}(CO) + 3x_{sg}(CO_2)) \geq (y_{meoh}(GPMEOH) - 1) * U$$
$$R_{meoh}(pmeoh) = \sum_{meoh} R_{meoh,0}(pmeoh, meoh) * y_{meoh}$$
$$\eta_{st} = \sum_{st} \eta_{st,0}(hst) * y_{st}(hst)$$
$$Inv(e) = \sum_{ei} UInv_0(e, ei) * F(e, ei)$$
$$0 \leq F(e, ei) \leq y(e, ei) * UL$$
$$\sum_{ei} y(e, ei) = 1 \tag{8.48}$$

Note that some approximations and simplifications have been made in the model, based on which we can remove redundant technical details and focus on the most important points, listed below:

- We assume that reactions (8.1)–(8.3) are all the chemical reactions taking place in a methanol synthesis reactor, and there are no side reactions. This assumption is valid for most cases as only trace amounts of reactants take place in some side reactions besides the three main reactions.
- Some minor quantities are neglected as they are too small compared with others and have little impact on the mass and energy balance of the process, even though technically they may be crucial. For example, steam streams extracted from steam turbines to cool gas turbine blades are essential for the operation of a gas turbine, but they are small in quantity and have little influence on energy efficiency, thus these cooling streams are not taken into account in the node.

Equation (8.48) is a mixed-integer nonlinear programming (MINLP) formulation. It is also nonconvex due to the presence of the bilinear terms in mass balance calculations. Therefore, global optimization techniques for MINLP problems are needed here to obtain a global optimum. These techniques can either be applied directed to the model, or indirectly through commercial solvers such as BARON [1, 2, 7, 22].

Equation (8.48) can be primarily used in the context of scenario analysis of various options for every polygeneration systems. Different objectives, other than profit, can also be included, such as environmental indicators (see for example Hugo et al. [10]), thereby transforming (8.48) into a multiobjective optimization model. The effect of uncertainty in the model parameters can also be studies C here developments in the area of optimization under uncertainty and uncertainty analysis can be explored, such as multiparametric programming [20], or global sensitivity analysis [11, 21]. These topics contribute our current research focus in this area.

In the following, a detailed case study will be presented for a polygeneration plant producing electricity and methanol.

8.5 A Polygeneration Plant for Electricity and Methanol – A Case Study

We consider a polygeneration plant as shown in Fig. 8.5, to produce electricity and methanol. The market demand for methanol is assumed to vary between 400 and 700 tons per day, and the electricity demand is between 100 and 300 MW. The following specifications are considered:

- All four blocks of technologies and types of equipment as outlined in the previous sections are considered for selection.
- Eleven chemical compounds are involved, namely O_2, N_2, H_2, CO, CO_2, H_2O, CH_4, H_2S, SO_2, COS, and CH_3OH.
- In the gasification block, Texaco gasification technology is applied to the gasifier, which uses dry pulverized coal, pure oxygen, and steam from power generation sector as main feedstocks. The gasification temperature and pressure is 1,371°C and 42 bar, respectively. Parameters of the gasification and power generation units are from NETLs report of Texaco IGCC case study [23].
- Technical parameters used in the model are listed in Table 8.1–8.3. Table 8.1 depicts the characteristic of the coal considered. Table 8.2 shows the conversion ratios of gas phase and liquid phase methanol synthesis technologies, whereas Table 8.3 outlines the corresponding operating conditions.
- Economic parameters for prices and unit costs are listed in Table 8.4. A time horizon of 30 years is considered for depreciation, with an annual operating time of 6,500 h.

Table 8.1. Ultimate analysis of Illinois #6 coal (wt. %, dry)

C	H	O	N	S	Ash
71.72	5.06	7.75	1.41	2.82	11.24

Table 8.2. Conversion rates of methanol synthesis

Technology	CO to methanol	CO_2 to methanol
Gas phase	0.446	0.199
Liquid phase	0.128	0.0075

Table 8.3. Temperature and pressure loss for each technology

Unit	Technology	Temp (K)	Pressure/ pressure loss (bar)
Syngas cooler	Quench	491	−1
	Low temperature radiative and convective	477	−3.3
	High temperature radiative and convective	813	−3.3
Syngas cleanup unit	Cold cleanup	320	−4.6
	Hot cleanup	840	−5.6
Water gas shift reactor	Water gas shift	473	−1
Methanol synthesis	Gas phase	523	−5.5
	Liquid phase	523	−5.5
Gas turbine	Gas turbine technology 1	1,703	19
	Gas turbine technology 2	1,589	18
	Gas turbine technology 3	1,473	17
HRSG and steam turbines	High heat recovery	400	1.05
	Low heat recovery	450	1.05

The overall model is implemented in GAMS [6]. See http://polygeneration.spaces.live.com for the details. The model involves 15 binary variables, 299 continuous variables, 293 equations (107 nonlinear) and 20 inequality constraints. It was solved using DICOPT. The solving procedure comprises a presolving step and a main one. In the presolving step, a relaxed MINLP problem is solved to provide for the main step a feasible initial point, in which all integer variables have continuous values between 0 and 1. In the main step, it usually takes more iterations than in the presolving step to get the optimal integer solution. In this case, for instance, the presolving step requires 153 iterations, whilst the main step need 364.

The total computation time of the whole solving procedure is in the order of seconds. It makes the model suitable to solve practical problems of different sizes. Moreover, this general model can also be further developed in different case studies by adding more constraints according to specific situations, showing the value of a general model in its wide range of application.

In this case study, model results indicate the following:

- The plant uses low temperature radiative and convective technology for the cooling of the crude syngas, followed by a low temperature cleanup unit. The methanol synthesis part uses gas phase synthesis technology with a water gas shift reactor before it. The power generation unit uses gas turbine technology one which has the highest first stage temperature and pressure, together with the technology of high heat recovery for the HRSG and the steam turbine.

Table 8.4. Economic parameters

Parameter	Value
Coal price ($/ton)	35
Methanol ($/ton)	340
Electricity price ($/kWh)	0.06
Investment cost for the gasifier ($/((kg/s coal)*y))	28,500
Investment cost for the cooler, quench ($/((kg/s syngas)*y))	3,000
Investment cost for the cooler, low temperature radiative and connective ($/((kg/s syngas)*y))	45,000
Investment cost for the cooler, high temperature radiative and connective ($/((kg/s syngas)*y))	30,000
Investment cost for the cleanup unit, low temperature ($/((kg/s syngas)*y))	20,000
Investment cost for the cleanup unit, high temperature ($/((kg/s syngas)*y))	40,000
Investment cost for the water gas shift reactor ($/((kg/s syngas)*y))	5,000
Investment cost for the CO_2 removal unit ($/((kg/s syngas)*y))	5,000
Investment cost for the methanol synthesis unit, gas phase ($/((kg/s syngas)*y))	15,000
Investment cost for the methanol synthesis unit, liquid phase ($/((kg/s syngas)*y))	20,000
Investment cost for the gas turbine compressor ($/((kg/s air)*y))	2,000
Investment cost for the gas turbine, technology 1 ($/((kg/s gas)*y))	3,000
Investment cost for the gas turbine, technology 2 ($/((kg/s gas)*y))	2,500
Investment cost for the gas turbine, technology 3 ($/((kg/s gas)*y))	2,000
Investment cost for the HRSG and steam turbines, technology 1 ($/((kg/s gas)*y))	3,000
Investment cost for the HRSG and steam turbines, technology 2 ($/((kg/s gas)*y))	2,500

- 2,991 tons of coal per day are consumed for the production of 300 MW electricity and 700 tons per day of methanol, with an annual profit of $140.6 m (electricity – $117.0 m, methanol – $64.5 m, fuel expense $28.4 m, equipment $12.6 m).
- Table 8.5 summarizes the results of the analysis for different combinations of technologies employed for comparison purpose. The results indicate that the combination of gas phase methanol synthesis and high efficient power generation on technologies are preferable. In most configurations, the use of liquid phase methanol synthesis results in methanol production below its maximum market demand value of 700 tons per day. On the other hand, liquid phase methanol synthesis options are in general more cost-effective due to its low operating pressure, leading to less consumption of compression work in the air separation unit.

Table 8.5. Model results for technology combinations

Technology combination	Power (MW)	Methanol (ton/d)	Coal (ton/d)	Profit (million dollar)
LRC-CC-WG-GPMeOH-GT1-HHR	300	700	2,991	140.6
LRC-CC-WG-GPMeOH-GT1-LHR	300	700	3,050	139.9
LRC-CC-WG-GPMeOH-GT2-HHR	300	700	3,173	137.7
LRC-CC-WG-GPMeOH-GT3-HHR	300	700	3,460	133.2
Q-CC-WG-GPMeOH-GT1-HHR	300	700	3,618	136.9
LRC-CC-LPMeOH-GT1-HHR	300	474	2,567	125.0
Q-CC-LPMeOH-GT1-HHR	300	588	3,182	131.4
Q-CC-LPMeOH-GT2-LHR	300	673	3,643	133.5
Q-CC-WG-LPMeOH-GT3-LHR	300	700	4,113	129.6

- Table 8.6 summarizes the results of a simple sensitivity analysis that was carried out on the effect of a change of a key parameter on the profitability and coal consumption of the best technology combination observed. We considered that parameters follow normal distribution and have nominal values with known standard deviations, as given in Table 8.6. Note that amid operating time, electricity price and demand greatly influence the profitability of the plant, whereas the price of methanol and its market demand play a less dominant role. Note also that the price of coal has the most dominant effect.
- An interesting observation can be made in relation to the conversion ratio of the methanol synthesis. Note that its increase does not lead to a decrease of the coal consumption but rather to an increase. This can be explained as follows. Since the conversion ratio is already sufficiently high, its further increase will only result in making the fluegas stream exiling the synthesis reactor to have a lower value of its flowrate and heating value. As a result, more coal consumption is required to generate more syngas for power generation. This extra amount of coal consumption cannot be compensated by the enhanced efficiency of the methanol synthesis block, hence the overall coal consumption increases.

Interestingly, increasing the conversion ratio of the methanol synthesis does not lead to less coal consumption but the other way round. This is because the conversion ratio is already sufficiently high, further increase will only make the fluegas leaving the synthesis reactor to have lower flowrate and heating

Table 8.6. Sensitivity analysis

Parameters	Nominal Value	Standard Deviation	Coal Consumption	Profit
Annual operating time (hour)	6,500	650	0	0.656
Market demand for electricity, upper bound (MW)	300	30	0.628	0.363
Market demand for electricity, lower bound (MW)	100	10	0	0
Market demand for methanol, upper bound (t/d)	700	70	0.190	0.239
Market demand for electricity, lower bound	400	40	0	0
Coal price ($/t)	35	3.5	0	−0.121
Electricity price ($/kWh)	0.06	0.006	0	0.501
Methanol price ($/t)	340	34	0	0.276
Investment cost (k$/(kg/s)/y)	227	22.7	0	−0.045
O&M costs (k$/(kg/s)/y)	22.7	2.27	0	−0.009
Temperature at first stage of gas turbine (K)	1,703	170.3	−0.738	0.190
Energy efficiency of HRSG and steam turbines	0.454	0.0454	−0.157	0.035
Conversion ratio of methanol synthesis	0.645	0.0645	0.031	−0.005

value. More coal consumption is therefore required to generate more syngas to keep the power generation. This extra amount of coal consumption can not be compensated by the enhanced efficiency of methanol synthesis, thus the overall coal consumption increases.

8.6 Conclusions

In this chapter we discussed the application of mixed-integer programming methodology into designing of polygeneration energy systems. It shows that MIP/MINLP algorithms are quite capable of handling the many alternatives and integrations that exist in a polygeneration complex. Discrete desisions, such as the selection of technologies and types of equipment, are denoted by binary variables. Together with continuous formulations such as mass and balances in a process scale, calculation of investment and O&M costs, a general model for designing a polygeneration complex is presented. Its application in a case study illustrates that the model is applicable to address complexities that arise in real world applications.

Acknowledgement. The authors would like to gratefully acknowledge the financial support from BP and its contribution in the inception, progress, and completion of this research study. Pei Liu would also like to thank Kwoks' Foundation for providing scholarship.

References

1. C. S. Adjiman, I. P. Androulakis, and C. A. Floudas. Global optimization of MINLP problems in process synthesis and design. *Computers and Chemical Engineering*, 21:S445–S450, 1997.
2. C. S. Adjiman, I. P. Androulakis, and C. A. Floudas. Global optimization of mixed-integer nonlinear problems. *AIChE Journal*, 46(9):1769–1797, 2000.
3. Q. Z. Al-Hamdan and M. S. Y. Ebaid. Modeling and simulation of a gas turbine engine for power generation. *Journal of Engineering for Gas Turbines and Power-Transactions of the Asme*, 128(2):302–311, 2006.
4. B. Chen, H. G. Jin, and L. Gao. Study of DME/power individual generation and polygeneration. *Kung Cheng Je Wu Li Hsueh Pao/Journal of Engineering Thermophysics*, 27(5):721–724, 2006.
5. D. Cocco, A. Pettinau, and G. Cau. Energy and economic assessment of IGCC power plants integrated with DME synthesis processes. *Proceedings of the Institution of Mechanical Engineers, Part A: Journal of Power and Energy*, 220(2):95–102, 2006.
6. GAMS Development Corporation. GAMS – A user's guide. GAMS, Washington DC, USA
7. C. A. Floudas, A. Aggarwal, and A. R. Ciric. Global optimum search for nonconvex NLP and MINLP problems. *Computers and Chemical Engineering*, 13(10):1117–1132, 1989.
8. H. C. Frey and Y. H. Zhu. Improved system integration for integrated gasification combined cycle (IGCC) systems. *Environmental Science and Technology*, 40(5):1693–1699, 2006.
9. Michael J. Gradassi and N. Wayne Green. Economics of natural gas conversion processes. *Fuel Processing Technology Trends in Natural Gas Utilisation*, 42(2-3):65–83, 1995.
10. Andre Hugo, Paul Rutter, Stratos Pistikopoulos, Angelo Amorelli, and Giorgio Zoia. Hydrogen infrastructure strategic planning using multi-objective optimization. *International Journal of Hydrogen Energy*, 30(15):1523–1534, 2005.
11. C. Kontoravdi, S. P. Asprey, E. N. Pistikopoulos, and A. Mantalaris. Application of global sensitivity analysis to determine goals for design of experiments: An example study on antibody-producing cell cultures. *Biotechnology Progress*, 21(4):1128–1135, 2005.
12. Guang Jian Liu, Yu Liu, Zheng Li, Wei Dou Ni, and Heng Yong Xu. Design and analysis of a new dimethyl ether-electric power polygeneration system. *Dongli Gongcheng/Power Engineering*, 26(2):295–299, 2006.
13. P. Liu, D. I. Gerogiorgis, and E. N. Pistikopoulos. Modeling and optimization of polygeneration energy systems. *Catalysis Today*, 127(1–4):347–359, 2007.
14. Pei Liu, Jian Gao, and Zheng Li. Performance alteration of methanol/electricity polygeneration systems for various modes of operation. *Dongli Gongcheng/Power Engineering*, 26(4):587–591, 2006.

15. Lin Wei Ma, Weidou Ni, Zheng Li, and Ting Jin Ren. Analysis of the polygeneration system of methanol and electricity based on coal gasification (1). *Power Engineering*, 24(3):451–456, 2004.
16. Lin Wei Ma, Weidou Ni, Zheng Li, and Ting Jin Ren. Analysis of the polygeneration system of methanol and electricity based on coal gasification (2). *Power Engineering*, 24(4):603–608, 2004.
17. J. M. Martinez-Frias, S. M. Aceves, J. R. Smith, and H. Brandt. Thermodynamic analysis of zero-atmospheric emissions power plant. *Journal of Engineering for Gas Turbines and Power-Transactions of the Asme*, 126(1):2–8, 2004.
18. National Energy Technology Laboratory. Commercial-scale demonstration of the liquid phase methanol (LPMEOH) process (a DOE assessment). Technical Report DOE/NETL-2004/1199, October 27 2003.
19. W. Ni, Z. Li, and X. Yuan. National energy futures analysis and energy security perspectives in china –strategic thinking on the energy issue in the tenth 5-year plan (FYP). In *Workshop on East Asia Energy Futures*, Beijing, 2000.
20. E. N. Pistikopoulos, M. C. Georgiadis, and V. Dua. *Multi-parametric programming: theory, algorithms, and applications, vol. 1.* Wiley, Weinheim, 2007.
21. M. Rodriguez-Fernandez, S. Kucherenko, C. Pantelides, and N. Shah. Optimal experimental design based on global sensitivity analysis. *Computer-Aided Chemical Engineering*, 24, 2007.
22. N. V. Sahinidis and M. Tawarmalani. BARON 7.2.5: Global optimization of mixed-integer nonlinear programs, user's manual, 2005.
23. W. Shelton and J. Lyons. Texaco gasifier igcc base cases. Technical Report PED-IGCC-98-001, US Department of Energy, July 1998.
24. Ralph E. H. Sims, H. -H. Rogner, and K. Gregory. Carbon emission and mitigation cost comparisons between fossil fuel, nuclear and renewable energy resources for electricity generation. *Energy Policy*, 31(13):1315–1326, 2003.
25. U.S. Department of Energy, International energy outlook 2006. Technical report, U.S. Department of Energy, 2006.
26. X. Zhang, L. Gao, H. Jin, and R. Cai. Design and analysis of coal based ammonia-power polygeneration. *Dongli Gongcheng/Power Engineering*, 26(2): 289–294, 2006.
27. Y. Zhang, W. Ni, and Z. Li. Research on superclean polygeneration energy system of iron and steel industry. In *Proceedings of the Seventh IASTED International Conference on Power and Energy Systems, Nov 28-Dec 1 2004*, Series on Energy and Power Systems, pages 152–157, Clearwater Beach, FL, United States, 2004. Acta Press, Anaheim, CA, USA.

Appendix A – Nomenclature

Sets	
e	Equipment
ei	Technology for a piece of equipment
ft	Available fuel feedstocks
gft	Gasification technologies
hst	Available technologies for HRSG and steam turbines

ie	Elements in a fuel feedstock
p	Product
rs	Key chemical compounds in the syngas

Variables

Binary Variables

y

Continuous Variables

η	Energy efficiency
C	Key mole ratios in the crude syngas
$CostEquip$	Investment cost and O&M cost on equipment
$CostFuel$	Cost on fuel
E	Electricity generation rate
F	Equipment capacity
$FuelRate$	Fuel consumption rate
H	Enthalpy
$Income$	Sale of products
Inv	Investment cost
$OMFix$	Fixed O&M cost
$OMVar$	Variable O&M cost
P	Pressure
$Profit$	Annual profit of a polygeneration plant
$ProRate$	Production rate
R	Ratio
T	Temperature
W	Mechanical work
ma	Mass flowrate
mo	Mole flowrate
h	Specific enthalpy
x	Mole composition
z	Mass fraction for an element in the fuel Feedstock of the gasifier

Parameters

η_0	Energy efficiency
γ	Adiabatic coefficient
C_0	Key mole ratios in crude syngas
C_p	Specific heat capacity at a Constant pressure
$LDemand$	Lower bound for market demand
$OpTime$	Operation time per year
P_0	Pressure
$PriceF$	Fuel price
$PriceP$	Product price
R_0	Ratio
T_0	Temperature
U	A large positive number
$UDemand$	Upper bound for market demand
$UInv_0$	Unit investment cost
UL	Upper limit for process capacity
z_0	Mass fraction for an element in a fuel feedstock

Subscripts

ASU	Air separation unit
G	Generator
O_2fuel	Oxygen and fuel feeding streams to a gasifer
air	Air entering a gas turbine
$cleanup$	Syngas cleanup unit
$clsg$	clean syngas
$cooler$	Syngas cooler
$coolsg$	Cooled syngas
f	Feeding fuel stream to a gasifier
fg	Fuel gas entering a gas turbine
gas	Gasifier
$gas1$	Gas at the inlet of a gas turbine
$gas4$	Gas at the outlet of a gas turbine
gc	Gas turbine compressor
gfO_2	Feeding oxygen stream to a gasifier
$gfwater$	Feeding water stream to a gasifier
gt	Gas turbine
$isen$	Isentropic procedure
$meoh$	Methanol synthesis
pg	Product gas after methanol synthesis
$pmeoh$	Parameters for methanol synthesis
$rawsg$	Raw syngas
sg	Syngas for the chemical synthesis reaction
$split$	Split ratio for the water gas shift reaction
st	Steam turbine
$waterfuel$	Water and fuel feeding streams to a gasifier
$wgsg$	Syngas after water gas shift reaction
1	Inlet point of a gas turbine
4	Outlet point of a gas turbine

9

Optimization of the Design and Partial-Load Operation of Power Plants Using Mixed-Integer Nonlinear Programming

Marc Jüdes, Stefan Vigerske, and George Tsatsaronis

Summary. This paper focuses on the optimization of the design and operation of combined heat and power plants (cogeneration plants). Due to the complexity of such an optimization task, conventional optimization methods consider only one operation point that is usually the full-load case. However, the frequent changes in demand lead to operation in several partial-load conditions. To guarantee a technically feasible and economically sound operation, we present a mathematical programming formulation of a model that considers the partial-load operation already in the design phase of the plant. This leads to a nonconvex mixed-integer nonlinear program (MINLP) due to discrete decisions in the design phase and discrete variables and nonlinear equations describing the thermodynamic status and behavior of the plant. The model is solved using an extended Branch and Cut algorithm that is implemented in the solver LaGO. We describe conventional optimization approaches and show that without consideration of different operation points, a flexible operation of the plant may be impossible. Further, we address the problem associated with the uncertain cost functions for plant components.

9.1 Introduction

In deregulated energy markets the optimization of the design and operation of energy conversion plants becomes increasingly important. To reduce the product cost during the entire operation time of a plant, both selection of an optimal plant structure and selection of optimal operating parameters in different load situations are necessary. Several design optimization methods were developed and applied to energy conversion systems in the past, e.g., exergoeconomic methods [8, 16, 36, 47, 54–57], evolutionary algorithms [3, 7, 12, 13, 49], and mathematical programming methods [3–5, 14, 50].

All these approaches are based on deterministic models. Thus, the effect of data uncertainties is not considered. In this case, the optimization could lead to a solution that is not feasible when some variations in the data apply. Often, heuristics are used to adapt the solution to a new situation. However,

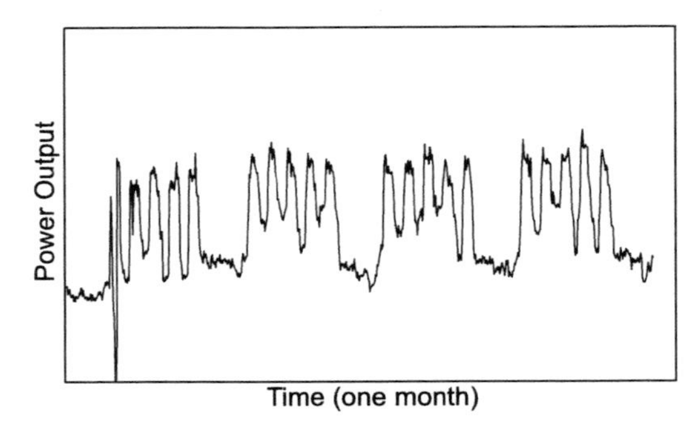

Fig. 9.1. Typical load diagram of a cogeneration plant

guarantees on the quality of the so obtained solution with respect to optimality are in general not available.

We consider two sources of uncertainty in this work: The first one is caused by frequently changing operating conditions while the second one is associated with the cost model. Due to the volatility of demand (see Fig. 9.1 for a typical load curve of a power plant), the plant operators are forced to operate a plant at operation points away from the usual design point, the full-load case. This effect is further reinforced by discontinuous and unsteady energy supplies from renewable energy sources such as wind energy (e.g., [17, 19, 34]).

An important observation is, that the consideration of partial-load operation points already in the design phase is not only meaningful for economical reasons, but also necessary to actually ensure a feasible plant operation under different partial-load conditions. This becomes even more important for cogeneration power plants, since here different amounts of each product can be requested at each time. So far only few approaches exist that could consider the partial-load operation within the design optimization, e.g., [23,36,37]. These approaches require in practical applications strong simplifications, such as a high linearity of the resulting problem. To handle the discrete decisions that are necessary to model different structures of a plant, often heuristic approaches such as genetic algorithms are applied.

Other approaches deal with the application of MINLP-techniques for the optimization of small-scale combined heat and power plants with fixed pressures and thus simplified working fluid properties within the cycle considering again only one operation point [50]. For the optimization of the operation of an *existing* plant with strong simplifications in the turbine models and the performance of heat exchangers also MINLP-techniques were applied [46]. First steps towards the design optimization of power plants with consideration of their partial-load performance are discussed in [30].

The cost model is another important cause of uncertainty. A detailed knowledge of the required investment costs and of the development of interest rates and fuel prices in the future is necessary to calculate the objective function of the optimization problem, here the levelized total revenue requirement

(TRR_{lev}) [8]. However, none of them is known with the required precision at the time when the optimization must be conducted. For estimating investment costs, several cost models can be found in the literature (e.g., [3, 26, 58, 60]), each of them leading after optimization to a different structure and different operation parameters.

The formulation of a model for plant-design optimization leads to a mixed-integer nonlinear program due to (a) required discrete decisions (existence, connection, and operation states of plant components) in the design part, and (b) nonlinear equations that describe, e.g., the thermodynamic properties of the working fluids and the off-design performance of components. While the number of discrete variables is still moderate, main challenges are posed by the nonconvexity of some equations. Both, discrete variables and nonconvex equations can lead to a feasible region that is disconnected and possesses many local minimal points. Thus, standard local search methods or an "easy" transformation into a mixed-integer linear model by linearization is prohibited and efficient mathematical algorithms are needed that can deal with the inherent nonconvexity of the search region to find global or good local optimal points.

Next to the already mentioned stochastic methods [7, 10], several approaches exist for the deterministic global optimization of a (nonconvex) MINLP problem [22, 27, 40, 42]. In successive outer-approximation algorithms [11, 18, 21, 62], an initial relaxation of the MINLP problem is iteratively solved and improved until a feasible point of the MINLP problem is found. If the problem is convex, a linear relaxation can be generated by linearizing nonlinear equations. However, working with linearizations of nonconvex equations can easily cut off global optimal points or lead to an infeasible relaxation. For such problems much effort is spend on finding good convex underestimators of nonconvex functions [2, 53] since they allow us to generate a convex relaxation of the problem that can be solved efficiently. To further achieve convergence to a global optimum, convex relaxation-based methods are often embedded in a Branch and Bound framework [1]. Such methods subdivide the feasible set into smaller subregions (branching) to allow for tighter convex underestimators on the corresponding subproblems. Comparing lower bounds given by evaluating the relaxation of a subregion with upper bounds calculated from feasible points of the original problem then allows coordinating the search for a global optimum [52]. The open source software package LaGO (**L**agrangian **G**lobal **O**ptimizer) [42–45] is an implementation of such a method and is used for the plant design optimization discussed in this paper (cf. Sect. 9.3).

9.2 Model of a Cogeneration Power Plant

We consider a simplified gas-fired combined cycle plant with steam extraction for a subsequent desalination unit. Different publications (e.g., [28]) discuss the relatively low importance of obtaining high electric efficiencies at these plants due to some specific local conditions, e.g., low gas prices. Therefore the

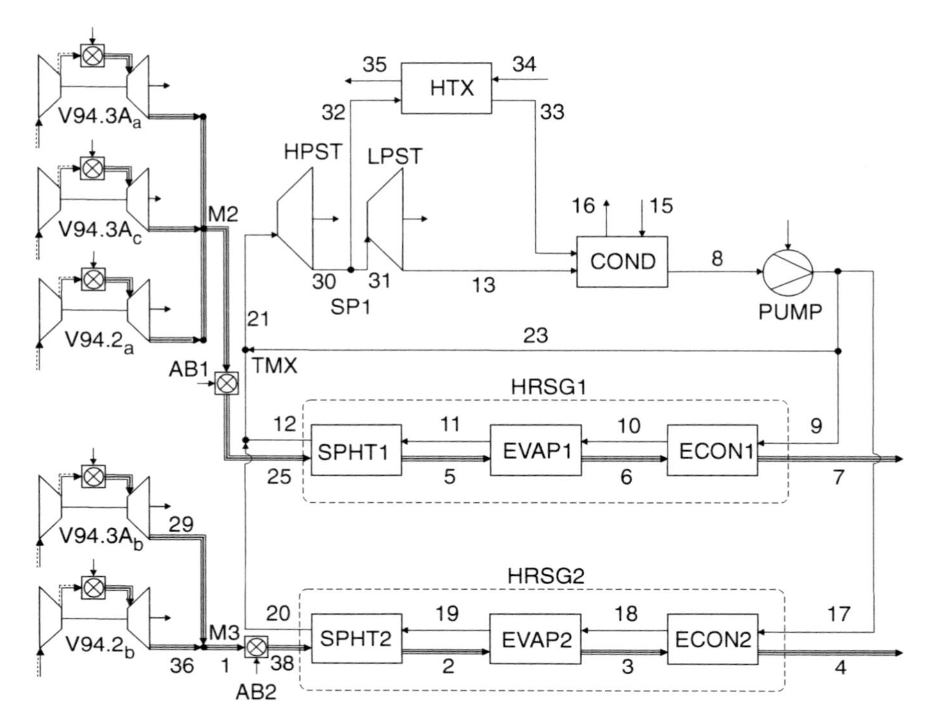

Fig. 9.2. Superstructure of the cogeneration power plant

optimization method is applied to a simplified single-pressure combined cycle plant with a supplementary firing for each heat recovery steam generator (HRSG). Figure 9.2 shows the superstructure of such a plant. The superstructure is based on realized power plant designs (e.g., [28, 48]).

In accordance with [25] and industrial information, the model considers two different types of gas turbines: Three Siemens V94.3A gas turbines and two Siemens V94.2 gas turbines. The V94.3A$_a$, V94.3A$_c$, and V94.2$_a$ gas turbines can feed the first heat-recovery steam generator HRSG1. Due to earlier studies, e.g. [31], the second heat-recovery steam generator HRSG2 is fed by only two gas turbines: The V94.3A$_b$ and the V94.2$_b$ turbines. Each heat-recovery steam generator is operated independently and can be fed by a free combination of these gas turbines. The optional additional burners AB1 and AB2 can increase the exhaust gas temperature.

The two heat-recovery steam generators consist of an economizer ECON, an evaporator EVAP, and a superheater SPHT. To simplify the model of the plant, only one subsequent water injector TMX is optionally used for both heat-recovery steam generators to regulate the steam temperature. The steam is supplied to the high, the intermediate (both indicated with HPST), and the low-pressure (LPST) sections of the steam turbine. After the intermediate-pressure section of the steam turbine, the steam for the

Table 9.1. Operation points of the cogeneration power plant shown in Fig. 9.2

Name	Operating hours h/a	Electric output MW	$\dot{m}_{32}$ t/h
OP1	1972	750	133.1
OP2	1972	600	86.7
OP3	1972	500	78.7
OP4	1972	400	86.7

subsequent desalination unit is extracted at SP1. The condensate returning from the desalination plant is mixed with the outlet stream of the low-pressure steam turbine in the condenser COND. The feedwater pump compresses the feedwater to the required sliding pressure. Only steady-state operation points are considered.

Next to the operation at full load (operation point OP1), we considered three more characteristic load conditions, each with a different demand for electricity and extracted steam, cf. Table 9.1. This, in reality uncertain, load information can be obtained from statistical methods, expert-knowledge, or existing data sets by using data mining techniques. We achieved good results using so-called self organizing maps (SOM, see also [33]), a special kind of artificial neural networks for the classification of data sets [20, 29], e.g., the load profile shown in Fig. 9.1.

In the following sections we give some insight into the thermodynamic and economic parts of the model and its formulation as a mathematical program.

9.2.1 Thermodynamic Model

The thermodynamic part of the model describes the physical behavior of the plant. We start with a discussion of the design phase of the plant, the calculation of the thermodynamic properties of the working fluids and the component sizing. After completing the plant design we can calculate the investment costs (cf. Sect. 9.2.2). In the subsequent phase the off-design performance of single components and of the overall plant can be computed. Considering partial-load operation points in the design optimization allows us to evaluate the operation costs also for these operation points. In this way, the optimization model becomes more realistic, but it requires an integration into the model of the off-design behavior for every considered operation point.

A possible result of the design optimization is, for example, the suggestion to install two gas turbines for each heat-recovery steam generator. In this case and when the power plant is operated under partial-load conditions, one of these gas turbines could be switched off while the other one would be running under full-load conditions. If so, the overall efficiency increases (lower fuel cost), but the investment costs increase too.

Exemplary, we illustrate the design and off-design model of a heat exchanger, see Fig. 9.3.

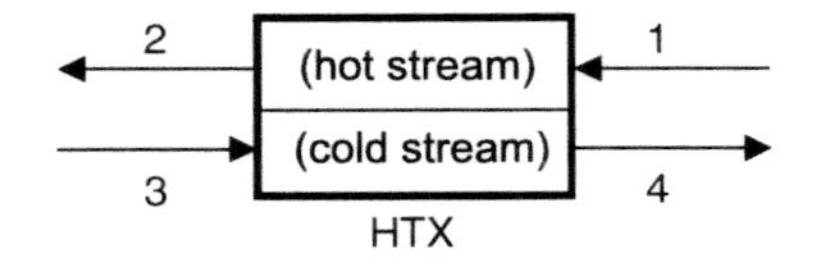

Fig. 9.3. Schematic illustration of a heat exchanger (HTX)

Design Modeling using a Heat Exchanger

As an example for the design modeling of components, an adiabatic heat exchanger is discussed. To build a model of a heat exchanger, its independent variables have to be known. Equations (9.1)–(9.6) are necessary for specifying the heat exchanger performance. The subscripts indicate the stream numbers shown in Fig. 9.3:

$$\dot{Q} = \dot{m}_1\,(h_1 - h_2) \tag{9.1}$$

$$\dot{Q} = \dot{m}_3\,(h_4 - h_3) \tag{9.2}$$

$$\dot{Q} = k\,A\,\Delta T_{\text{log}} \tag{9.3}$$

$$\Delta T_{\text{log}} = \frac{(T_2 - T_3) - (T_1 - T_4)}{\ln(T_2 - T_3) - \ln(T_1 - T_4)} \tag{9.4}$$

$$h_i = f(T_i, p_i \text{ or } x_i), \qquad i = 1, \ldots, 4 \tag{9.5}$$

$$p_{\text{exit},j} = f(p_{\text{inlet},j}), \qquad j = \text{cold, hot} \tag{9.6}$$

where $\dot{Q}$ denotes the rate or heat transfer heat rate, k is the overall heat transfer coefficient, A represents the heat exchanger surface area, h denotes the enthalpy, T is the temperature, p is the pressure, and x represents the steam quality.

Eight of these 18 ($\dot{m}_1$, $\dot{m}_3$, h_1, h_2, h_3, h_4, T_1, T_2, T_3, T_4, p_1, p_2, p_3, p_4, $\dot{Q}$, ΔT_{log}, k, A) variables can be selected more or less independently. In the design optimization, the values of these independent decision variables have to be determined to maximize the overall plant efficiency or to minimize the overall product cost using the respective objective function. Here, the sizing and costing of the components refers to the design case where the maximal values of pressure, temperature, and mass flow rate are used. The information obtained from the design case is used to calculate the off-design performance, where additional equations have to be considered.

Off-Design Modeling

The description of the off-design performance of a heat exchanger requires additional equations, some of which are taken from the commercial software EBSILONProfessional 7.00 [51].

$$\dot{Q} = k\,A\,\Delta T_{\text{log}} \tag{9.7}$$

$$\frac{1}{k} = \frac{1}{\alpha_{\text{c,N}}\,FK_1} + \frac{1}{\alpha_{\text{h,N}}\,FK_2} \tag{9.8}$$

$$FK_1 = \left(\frac{\dot{m}_{\text{c}}}{\dot{m}_{\text{c,N}}}\right)^{\text{Exp}_1} \tag{9.9}$$

$$FK_2 = \left[1 - 0.0005\left(\frac{(T_{\text{h,i,N}} + T_{\text{h,e,N}}) - (T_{\text{h,i}} + T_{\text{h,e}})}{2}\right)\right]\left(\frac{\dot{m}_{\text{h}}}{\dot{m}_{\text{h,N}}}\right)^{\text{Exp}_2} \tag{9.10}$$

Here, α represents the heat transfer coefficient, c and h stand for cold and hot, and Exp_1 and Exp_2 denote some component specific exponents.

For a steam turbine operating at off-design conditions, the characteristic curve describing its partial-load performance is modelled in accordance with Equation (9.11) which describes the so-called Stodola law. It correlates the inlet and outlet pressures p_{i} and p_{e}, the inlet temperature T_{i}, and the mass flow rate $\dot{m}$ at the actual (partial load) and nominal (design point) conditions:

$$\frac{\dot{m}}{\dot{m}_{\text{N}}} = \frac{p_{\text{e}}}{p_{\text{e,N}}}\sqrt{\frac{T_{\text{i,N}}}{T_{\text{i}}}}\sqrt{\frac{1-(p_{\text{i}}/p_{\text{e}})^2}{1-(p_{\text{i,N}}/p_{\text{e,N}})^2}} \tag{9.11}$$

The isentropic efficiency η_{s} of the turbine, that compares the real and ideal expansion in the turbine, is a function of the mass flow rate:

$$\begin{aligned}\frac{\eta_{\text{s}}}{\eta_{\text{s,N}}} =& -1.0176\left(\frac{\dot{m}}{\dot{m}_{\text{N}}}\right)^4 + 2.4443\left(\frac{\dot{m}}{\dot{m}_{\text{N}}}\right)^3 - 2.1812\left(\frac{\dot{m}}{\dot{m}_{\text{N}}}\right)^2 \\ &+ 1.0535\left(\frac{\dot{m}}{\dot{m}_{\text{N}}}\right) + 0.701.\end{aligned} \tag{9.12}$$

At partial-load operation, the efficiency η_s must be adjusted with respect to changes in the outlet steam quality Δx_{e}. When the exiting steam quality x_{e} is lower than 1, this adjustment is carried out using the following approximation

$$\eta_{\text{s,corr}} = \eta_{\text{s}} - \frac{1}{2}\,\Delta x_{\text{e}}, \tag{9.13}$$

where η_{s} denotes the isentropic efficiency in accordance with (9.12) and $\eta_{\text{s,corr}}$ denotes the resulting isentropic efficiency after the correction for steam quality.

Working Fluid Properties

The thermodynamic properties of the exhaust gases are calculated using equations from Knacke, Kubaschewski, and Hesselmann [32]. For example, the molar enthalpy of a pure ideal gas stream i is calculated with the following equation:

$$h_i = 10^3\left(h_i^{\text{ref}} + a_i y + \frac{b_i}{2}y^2 - c_i y^{-1} + \frac{d_i}{3}y^3\right), \tag{9.14}$$

where $y = T/1000$, h^{ref} is associated with the reference value calculating the enthalpy and a_i, b_i, c_i, and d_i are constants depending on the substance being considered. All gas streams in the process are treated as mixtures of ideal gases. Therefore the molar enthalpy h of these gas streams is calculated with the aid of the respective mole fractions x_j of the j components:

$$h = \sum_j x_j \, \dot{n} \, h_j \tag{9.15}$$

Here is $\dot{n}$ the molar flow rate that can be calculated using the mass flow rate $\dot{m}$ and the respective molar mass M

$$\dot{n} = \frac{\dot{m}}{M} \,. \tag{9.16}$$

To enable the software to find a good solution, the high degree polynomials from the original water steam properties (IAPWS IF97 [61]) are simplified here to polynomials of a degree at most four. A detailed discussion of these polynomials is presented in [3].

9.2.2 Economic Model and Uncertainty in Investment Costs

In addition to the thermodynamic model of the power plant, an economic analysis is needed to calculate its objective function, the annual levelized **T**otal **R**evenue **R**equirement (TRR) (TRR_{lev} [8]). TRR includes the fuel costs, the operating and maintenance expenses as well as the carrying charges (which consider the capital recovery, interest, dividends, taxes, and insurances). The levelized TRR_{lev} is a function of the annual values TRR_{n},

$$TRR_{\text{lev}} = i_{\text{eff}} \frac{(1 + i_{\text{eff}})^n}{(1 + i_{\text{eff}})^n - 1} \sum_n \frac{TRR_n}{(1 + i_{\text{eff}})^n} \,, \tag{9.17}$$

where i_{eff} denotes the effective interest rate and n the number of years considered in the analysis.

Purchased **E**quipment **C**ost (PEC in the following), fuel cost ($\dot{C}_{\text{f}}$ in the following), and operating hours have a strong influence on a cost-effective design of the plant. PEC are calculated for the full-load case, since here the largest pressures, temperatures, and mass flow rates occur (see also Sect. 9.2.1), whereas $\dot{C}_{\text{f}}$ and other operating cost are calculated at every operation point.

For calculating the PEC, cost functions for each component are used. These functions depend on the characteristic variables of a component, e.g., the surface area of a heat exchanger A or the power output of a turbine $\dot{W}$. However, in most cases the "real" cost function, if there is one, is not known. Instead, different cost models are discussed in the literature (e.g., [3,26,58,60]). Two very different cost models for a steam turbine [3,26] are shown in Fig. 9.4.

Although the differences in the PEC shown in Fig. 9.4 seem to be too large from the engineer's point of view, we consider these two very different models

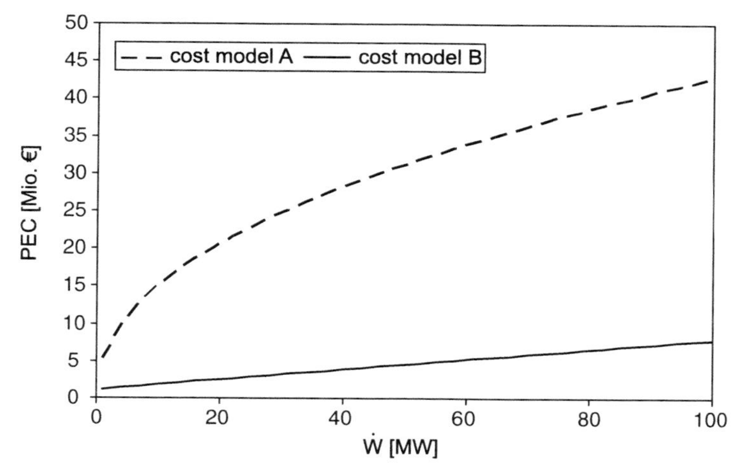

Fig. 9.4. *PEC* calculated with two extremely different cost models for a steam turbine. Here, *PEC* is a function of the power output $\dot{W}$.

as an academic example for the uncertainties associated with cost functions. It must be emphasized that such differences in a "real world" engineering problem are unrealistic. For considering the effects of the different cost approaches (Sect. 9.4.2) we employ in the objective function the assumed reliability of each function as a weighting factor.

9.2.3 Formulation as Mathematical Program

Two different types of variables are used to formulate the optimization problem as a MINLP: Binary and continuous variables. Binary variables are needed to decide, (a) which of two different types of gas turbines are used for each heat-recovery steam generator, (b) which of two possible heat-recovery steam generators are used, and (c) whether there is a need for the additional burners AB1 and AB2. Additionally we use binary variables to determine the actual state of some streams. Here, these variables indicate whether the working fluid is superheated or its thermodynamic state is within the liquid–vapor region and thus the steam quality lies strictly between 0 and 1.

Isentropic efficiencies, heat exchanger surfaces, and the thermodynamic properties of each working fluid represent continuous variables. The latter are calculated by nonlinear equations such as (9.14) and (9.15). Table 9.2 shows the decision variables for the design and the off-design models.

The entire model with its 41 independent continuous and 43 independent binary variables is formulated in GAMS [24] as one system of equations in form of a large mixed-integer nonlinear program. Here, the following equations are used: Mass, energy, and impulse balances, equations for calculating the working fluid properties, equations for calculating the components full- and

Table 9.2. Binary (y_i) and continuous decision variables in the model with full-load operation conditions OP1 and partial-load operation conditions OP2–OP4 (see Table 9.1) indicated by superscripts. GT1-3 indicate the electric power output of the gas turbines V94.3A$_a$–V94.3A$_c$ and GT4-5 indicate the electric power output of the gas turbines V94.2$_a$ and V94.2$_b$, respectively.

Component	Variable	Design conditions (OP1)	Off design conditions (OP2-OP4)	Cost part
GT1-5	operation	$y^1_{GT1} - y^1_{GT5}$	$y^{2,3,4}_{GT1} - y^{2,3,4}_{GT5}$	–
GT1-5	existence	–	–	$y^c_{GT1} - y^c_{GT5}$
AB1,2	operation	y^1_{AB1}, y^1_{AB2}	$y^{2,3,4}_{AB1}, y^{2,3,4}_{AB2}$	–
AB1,2	existence	–	–	y^c_{AB1}, y^c_{AB2}
HRSG1,2	operation	y^1_{HRSG1}, y^1_{HRSG2}	$y^{2,3,4}_{HRSG1}, y^{2,3,4}_{HRSG2}$	–
number of binary variables		9	27	7
GT1-5	power output	$\dot{W}^1_{GT1} - \dot{W}^1_{GT5}$	$\dot{W}^{2,3,4}_{GT1} - \dot{W}^{2,3,4}_{GT5}$	–
AB1,2	fuel flow rate	$\dot{m}^1_{f,AB1}, \dot{m}^1_{f,AB2}$	$\dot{m}^{2,3,4}_{f,AB1}, \dot{m}^{2,3,4}_{f,AB2}$	–
TMX	mass flow rate	$\dot{m}^1_{23}$	$\dot{m}^{2,3,4}_{23}$	–
	pressure	p^1_{14}	–	–
	pressure	p^1_8	–	–
SPHT	ΔT	$\Delta T_{SPHT,N}$	–	–
EVAP	ΔT	$\Delta T_{EVAP,N}$	–	–
ECON	ΔT	$\Delta T_{ECON,N}$	–	–
COND	ΔT	$\Delta T_{COND,N}$	–	–
number of cont. variables		17	24	0

partial-load performance, and cost equations. Additional constraints result from the limitation of temperatures, mass flow rates, and pressures in the partial-load cases compared to the design (full-load) case.

Due to the simultaneous solution of the optimization problem, *all* equations and constraints of the model have to be satisfied. Therefore, it is necessary that even if some gas turbines associated with one heat-recovery steam generator are not included in the actual design of the plant, the exhaust gas mass flow rate passing through the heat-recovery steam generator and its thermodynamic variables are not equal to zero. Otherwise the calculation of values like the logarithmic mean temperature difference ΔT_{log} in the heat exchangers yield function evaluation errors in the optimization process. Hence, e.g., the mixer M3 (Fig. 9.2) has to be modelled in an appropriate way:

$$y_{M3} \leq y_{94.3,b} + y_{94.2,b} \,, \tag{9.18}$$

$$y_{M3} \geq y_{94.3,b} \,, \tag{9.19}$$

$$y_{M3} \geq y_{94.2,b} \,. \tag{9.20}$$

Here, the binary variable y_{M3} indicates the operation of the heat-recovery steam generator in the respective load case ($y_{HRSG,2}$ in Table 9.2). The energy balance equation is formulated using a so-called big-M formulation incorporating the upper bounds on the variables $\dot{n}_1$ and h_1 ($\overline{\dot{n}}_1$ and $\overline{h}_1$, respectively):

$$\dot{n}_1\,h_1 \leq \dot{n}_{29}\,h_{29} + \dot{n}_{36}\,h_{36} + (1-y_{\mathrm{M3}})\,(-1)\,\overline{\dot{n}}_1\,\overline{h}_1\,, \tag{9.21}$$

$$\dot{n}_1\,h_1 \geq \dot{n}_{29}\,h_{29} + \dot{n}_{36}\,h_{36} - (1-y_{\mathrm{M3}})\,(-1)\,\overline{\dot{n}}_1\,\overline{h}_1\,. \tag{9.22}$$

Finally, the equations describing the chemical composition of the gas streams are formulated as

$$x_{1,\mathrm{N}_2}\,\dot{n}_1 \leq \dot{n}_{29}\,x_{29,\mathrm{N}_2} + \dot{n}_{36}\,x_{36,\mathrm{N}_2} + (1-y_{\mathrm{M3}})\,\overline{x}_{1,\mathrm{N}_2}\,\overline{\dot{n}}_1 \tag{9.23}$$

$$x_{1,\mathrm{N}_2}\,\dot{n}_1 \geq \dot{n}_{29}\,x_{29,\mathrm{N}_2} + \dot{n}_{36}\,x_{36,\mathrm{N}_2} - (1-y_{\mathrm{M3}})\,\overline{x}_{1,\mathrm{N}_2}\,\overline{\dot{n}}_1\,. \tag{9.24}$$

where $\dot{n}_i$ denote the mole flow rates and x_{N_2} the mole fraction of nitrogen. The equations for the other substances in the exhaust gas streams have to be formulated accordingly. Obviously, (9.21)–(9.24) can be satisfied for both cases, $y_{\mathrm{M3}} = 0$ and $y_{\mathrm{M3}} = 1$. For $y_{\mathrm{M3}} = 0$ the left hand side variables can be chosen according to the heat exchanger requirements.

9.3 Solution of the MINLP

As mentioned before, the presence of discrete decisions and nonlinear nonconvex equations describing the design, thermodynamic status, and behavior of the plant leads to a nonconvex MINLP whose solution requires sophisticated algorithms. Currently, there are only a few solvers available that can explicitly handle nonconvex MINLPs. To our best knowledge LaGO [45] is currently the only freely available one.

In this section we describe LaGOs enhanced Branch and Cut algorithm in more detail. At first the algorithm approximates nonconvex functions by convex underestimators, i.e., a convex function that underestimates the respective original function. Next, the obtained convex relaxation is linearized via the construction of supporting hyperplanes. These cutting planes are used to initialize and improve a linear relaxation of the problem. By means of this relaxation it is possible to efficiently compute reliable lower bounds to the global optimum and starting points for local searches for feasible solutions. A successive branching of the search space enables an improvement of the underestimators in the progression of the algorithm and thus a tightening of the linear relaxation.

In the following sections we focus on some components of LaGOs algorithm. The preprocessing routines include the investigation of the problem structure of a given MINLP (Sect. 9.3.1), the initialization of relaxations that lead to a linear outer approximation (Sect. 9.3.2), and methods that are used for the reduction of variables bounds (Sect. 9.3.3). Finally, we give a short overview of LaGOs Branch and Cut algorithm (Sect. 9.3.4). We use a general MINLP formulation to emphasize the wide applicability of the proposed method.

9.3.1 Problem Structure Analysis

Problem Formulation

A general MINLP can be formulated as

$$\begin{aligned} \min \quad & b_0^\top x && \text{(P)} \\ \text{such that} \quad & h(x) \leq 0, \\ & x \in [\underline{x}, \overline{x}], \\ & x_j \in \mathbb{Z}, \qquad j \in B, \end{aligned}$$

where $B \subseteq \{1, \ldots, n\}$, b_0, $\underline{x}$, $\overline{x} \in \mathbb{R}^n$, and $h : \mathbb{R}^n \to \mathbb{R}^m$ is twice-continuously differentiable. The set $[\underline{x}, \overline{x}] := \{x \in \mathbb{R}^n | \underline{x}_i \leq x_i \leq \overline{x}_i\}$ is referred as *box* and constitutes finite bounds on the variables. For the sake of simplicity we assume that the objective function is linear and equality constraints were replaced by two inequalities. Note, that to handle a nonlinear objective function $h_0(x)$, one can minimize a new variable y under the additional constraint $h_0(x) \leq y$.

LaGO requires procedures for the evaluation of function values, gradients, and Hessians. This restriction to "black-box functions" has the advantage that very general functions can be handled, but also the disadvantage that without insight into the algebraic structure of the functions $h_i(x)$ advanced reformulation and box reduction techniques (as in [1, 40, 53]) cannot be used and we are forced to use sampling methods in some components of LaGO.

Block Separability

At first, LaGO investigates the sparsity structure and block separability of the functions $h_i(x)$. A function is called *block-separable* if it can be represented as a sum of subfunctions, each depending only on a small number of variables. Block separability is a common property of real world applications where singular complex components are coupled by linear constraints. Also the model discussed in the previous section is highly block separable, since the power plant components, which are described by nonlinear equations, are linked only by linear equations for the working fluids (molar fraction x_i, mass flow rate $\dot{m}$, temperature T, pressure p, enthalpy h, and entropy s). Using a simple sampling technique for the recognition of sparsity patterns in the Hessian of the black-box functions $h_i(x)$ [45], LaGO automatically reformulates each function into the form

$$h_i(x) = c_i + b_i^\top x + \sum_{k=1}^{q_i} x_{Q_{i,k}}^\top A_{i,k} x_{Q_{i,k}} + \sum_{k=1}^{p_i} h_{i,k}(x_{N_{i,k}}), \tag{9.25}$$

where $c_i \in \mathbb{R}$, $b_i \in \mathbb{R}^n$, and the index sets $Q_{i,k}$ and $N_{i,k}$ denote quadratic and nonlinear nonquadratic variables, respectively. They are also referred as *blocks* of the function $h_i(x)$. The structure (9.25) allows us to distinguish between linear, quadratic, and nonquadratic parts of a function, and to treat each block separately if advantageous.

Convexity

Originating from the block separable formulation (9.25), LaGO checks for each quadratic block $x_{Q_{i,k}}^\top A_{i,k} x_{Q_{i,k}}$ and each nonquadratic block $h_{i,k}(x_{N_{i,k}})$ whether it represents a convex function or not. Therefore, for a quadratic function $x_{Q_{i,k}}^\top A_{i,k} x_{Q_{i,k}}$, it is sufficient to check whether the minimal eigenvalue of $A_{i,k}$ is nonnegative. For a nonquadratic function $h_{i,k}(x_{N_{i,k}})$, the minimal eigenvalue of the Hessian $\nabla^2 h_{i,k}(x_{N_{i,k}})$ is evaluated at sample points from the box $[\underline{x}_{N_{i,k}}, \overline{x}_{N_{i,k}}]$. Observe that only the sign of the eigenvalue is of interest, so that even for curvaceous functions a sufficiently rich set of sampling points yields correct results.

9.3.2 Relaxations

As a result of the structure analysis, LaGO knows for each block in the formulation (9.25) whether it is convex or not. For the computation of a convex underestimator $\breve{h}_i(x)$ of each nonconvex function $h_i(x)$, first nonconvex nonquadratic terms in $h_i(x)$ are underestimated by possibly nonconvex quadratic terms. Afterwards, each nonconvex quadratic term is convexified. Finally, a linear relaxation is generated by linearizing the convexified functions in reference points and dropping of integrality restrictions on the variables x_j, $j \in B$.

Quadratic Underestimators

Let $g : \mathbb{R}^r \to \mathbb{R}$ be a nonconvex function $h_{i,k}$ from (9.25), $r = |N_{i,k}|$. A quadratic underestimator $q(x) = x^\top Ax + b^\top x + c$ of $g(x)$ is computed by using a powerful sampling algorithm [41, 45]. This algorithm allows LaGO to determine quadratic underestimators of nonconvex functions for which only function and derivative evaluation methods are available. Starting with an initial sample set S consisting, among others, of vertices of the box $[\underline{x}, \overline{x}]$ and a distinguished sample point $\hat{x} \in S$ (often a local minimizer of $g(x)$), the following two steps are iterated:

1. Determine coefficients A, b, and c of $q(x)$ by solving the linear program

$$\begin{aligned} \min_{A,b,c} \quad & \sum_{x \in S} g(x) - q(x) && \text{(U)} \\ \text{such that} \quad & q(x) \le g(x), \qquad x \in S, \\ & q(\hat{x}) = g(\hat{x}). \end{aligned}$$

2. For points $\tilde{x} \in S$ with $q(\tilde{x}) = g(\tilde{x})$ maximize locally the error $q(x) - g(x)$ over the box $[\underline{x}, \overline{x}]$ by solving a nonlinear program starting from $\tilde{x}$. If this yields a point $\breve{x}$ with $q(\breve{x}) - g(\breve{x}) > \delta_{\text{tol}}$, add the inequality $q(\breve{x}) \le g(\breve{x})$ to (U) and go to step 1. Otherwise, i.e., the maximal error $\delta_{\max}$ is below the tolerance δ_{tol}, lower $q(x)$ by $\delta_{\max}$ (i.e., subtract $\delta_{\max}$ from c) and stop.

Figure 9.5 illustrates a quadratic underestimator for the logarithmic mean temperature difference obtained by this method.

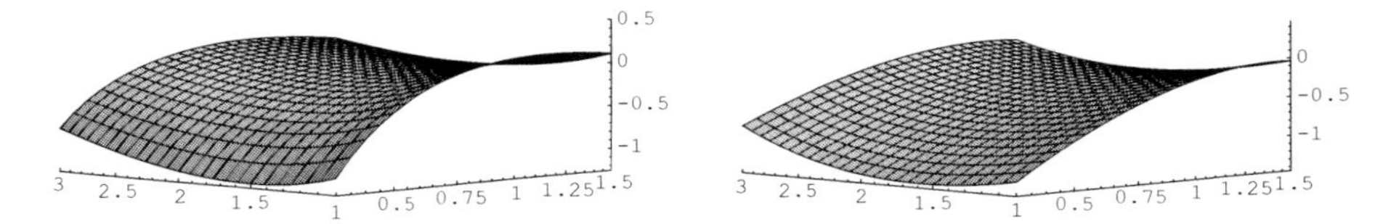

Fig. 9.5. The function $\Delta T_{\log} = (\Delta T_1 - \Delta T_2)/\left(\ln \Delta T_1 - \ln \Delta T_2\right)$ (logarithmic mean temperature difference) as used in the modelling of the partial-load performance (left, cf. (9.4)) and a corresponding quadratic underestimator (right). $\Delta T_{\log}$ has been fixed for visualization.

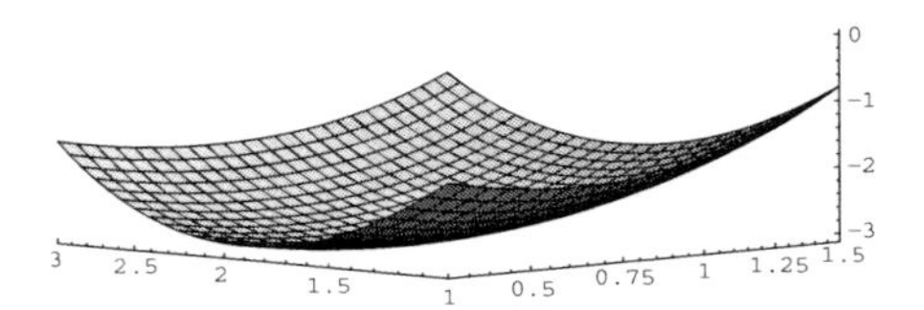

Fig. 9.6. Convex α-underestimator of the nonconvex quadratic underestimator in Figure 9.5

Convexification of Quadratic Terms

For the convexification of nonconvex quadratic terms $f(x) = x^\top Ax$, LaGO uses α-underestimators as introduced by Adjiman and Floudas [2]. An α-underestimator of $f(x)$ over the box $[\underline{x}, \overline{x}]$ is the function

$$\breve{f}(x) = f(x) + \sum_{i=1}^{r} \frac{\max\{0, -\lambda_1(WAW)\}}{(\overline{x}_i - \underline{x}_i)^2}(x_i - \underline{x}_i)(x_i - \overline{x}_i), \tag{9.26}$$

where $\lambda_1(D)$ denotes the minimal eigenvalue of a matrix D and the diagonal matrix W has the box-width $\overline{x} - \underline{x}$ on its diagonal and has been introduced for scaling reasons. It is clear that $\breve{f}$ is convex and $\breve{f}(x) \leq f(x)$ for all $x \in [\underline{x}, \overline{x}]$ [2]. Figure 9.6 illustrates a convexified quadratic underestimator.

Linear Relaxation

The linear relaxation (R) of (P) is generated by linearizing each nonlinear function $h_i(x)$ (if convex) or its convexification $\breve{h}_i(x)$ at a reference point $\hat{x}$. For initialization, the point $\hat{x}$ is chosen to be an optimal point of the nonlinear convex relaxation. During the Branch and Cut algorithm (Sect. 9.3.4), (R) is augmented by further linearizations at candidates for optimal points of (P) and optimal points of (R).

Further, to address the integrality restrictions on x_j, $j \in B$, in (R), mixed-integer-rounding cuts, which have their origin in mixed-integer linear programming [35, 38], are added to (R). These linear inequalities allow us to cut off nonintegral solutions from (R) [45].

As a third method for the computation of valid inequalities, LaGO can generate so called interval gradient cuts [9, 42], which are based on interval arithmetic calculations [39]. Assume that for a function $h_i(x)$ it is possible to compute an enclosure $[\underline{d}, \overline{d}] \subset \mathbb{R}^n$ of the gradient $\nabla h_i(x)$ over $[\underline{x}, \overline{x}]$, i.e., $\nabla h_i(x) \in [\underline{d}, \overline{d}]$ for all $x \in [\underline{x}, \overline{x}]$. Then, given a reference point $\hat{x} \in [\underline{x}, \overline{x}]$,

$$h_i(\hat{x}) + \nabla h_i(\hat{x})(x - \hat{x}) + \min_{d \in [\underline{d}, \overline{d}]} (d - \nabla h_i(\hat{x}))^\top (x - \hat{x}) \leq h_i(x) \quad \forall x \in [\underline{x}, \overline{x}].$$

Introducing new positive variables y^+ and y^- and writing $x - \hat{x} = y^+ - y^-$ with $y_j^+ = \max\{0, (x - \hat{x})_j\}$ and $y_j^- = \max\{0, -(x - \hat{x})_j\}$, $j = 1, \ldots, n$, we obtain (due to the inequality $h_i(x) \leq 0$ in (P)) the interval gradient cut

$$h_i(\hat{x}) + \nabla h_i(\hat{x})(x - \hat{x}) + (\underline{d} - \nabla h_i(\hat{x}))^\top y^+ - (\overline{d} - \nabla h_i(\hat{x}))^\top y^- \leq 0.$$

These cuts have the advantage that they can be derived directly from a nonlinear and nonconvex function and thus do not rely on a prior convexification step. On the other hand, the modeling of the conditions $y_j^+ = \max\{0, (x - \hat{x})_j\}$ and $y_j^- = \max\{0, -(x - \hat{x})_j\}$ would require additional discrete variables. Therefore, in order to fit into the linear relaxation (R), we instead add only the equations $x - \hat{x} = y^+ - y^-$, $y^+ \leq \overline{x} - \hat{x}$, $y^- \leq \hat{x} - \underline{x}$, and $y_i^+ + y_i^- \leq \max\{\overline{x}_i - \hat{x}_i, \hat{x}_i - \underline{x}_i\}$, $i = 1, \ldots, n$, to (R).

9.3.3 Methods to Tighten the Bounding Box

Since the quality of the underestimators and cuts depends strongly on the bounding box $[\underline{x}, \overline{x}]$, it can be advantageous to apply boxreduction procedures in the preprocessing. Also during the Branch and Cut algorithm, a branching operation might facilitate possible reductions of variable bounds, and even detect infeasibility for a subregion or fix binary variables.

In LaGO, two boxreduction techniques are currently implemented. The first method computes a new bounding box on the variables by enclosing the feasible set of (R), i.e., each or only some selected variable is minimized and maximized with respect to the constraints of (R) [45].

The second method is a simple constraint propagation method and thus utilizes only one constraint at a time, but works on the original formulation (P). Similar to the interval gradient cuts, this procedure relies on interval arithmetic operations (as they are available within the GAMS [24] interface): For a box $U \subseteq [\underline{x}, \overline{x}]$ assume that $h_i(x)$ can be written as $h_i(x) = g(x) + bx_j$ with x_j not appearing in $g(x)$ and $b \in \mathbb{R}$, $b \neq 0$. Denote by $g(U)$ an interval in $\mathbb{R} \cup \{\pm\infty\}$ s.t. $g(x) \in g(U)$ for all $x \in U$. Let $[\underline{y}_j, \overline{y}_j] = -g(U)/b$. If $b > 0$, $\overline{x}_j$ can be updated to $\min\{\overline{x}_j, \overline{y}_j\}$. Otherwise, if $b < 0$, $\underline{x}_j$ can be updated to $\max\{\underline{x}_j, \underline{y}_j\}$. In case that the new bounds define an empty box, infeasibility of the subproblem with box U is detected. After the bounds on x_j have been reduced, other constraints depending on x_j might yield further box reductions for other variables. Thus, the same procedure is applied to these constraints. This process iterates until the box stops to reduce significantly or infeasibility is detected.

9.3.4 Branch and Cut Algorithm

To search for a global optimum of (P), the algorithm follows a Branch and Bound scheme. Lower bounds on the global optimal value are computed by solving the linear relaxation, while upper bounds are given by the objective function value of incumbent solutions. These are points that are feasible for (P) and are found by a local search, that is the discrete variables are fixed in (P) and a descent algorithm is applied to the resulting nonlinear program. The fundamental idea of Branch and Bound is that partitioning the search space allows to improve the underestimating functions on each subregion and thus tightens the linear relaxation. An improved linear relaxation then results in higher lower bounds and new incumbent solutions due to better starting points for the local search.

The algorithm starts with considering the problem on its complete feasible region. This problem is also called the root problem. Solving (R) yields a lower bound and a starting point for a local search in (P). If the local search is successful, the first incumbent solution has been found and an upper bound can be computed. Otherwise, the upper bound is initialized with $+\infty$. If lower and upper bounds match, a globally optimal solution has been found and the procedure terminates. Otherwise, two new problems are constructed by dividing the feasible region of (P) using a subdivision of the box $[\underline{x}, \overline{x}]$ (*branching*). For each child the linear relaxation (R) is improved by adding further linearizations of nonlinear convex or convexified functions, interval gradient cuts for nonconvex constraints, and mixed-integer-rounding cuts. The new problems become children of the root problem, and the algorithm is applied recursively on each subproblem. This process constructs a tree of subproblems, the Branch and Bound tree. Since each node of the Branch and Bound tree has its own linear relaxation, the generated cutting planes need to be valid (i.e., underestimating the original functions) only on the corresponding subregion of the original feasible space. Hence, linearizations of convexified functions are generated with respect to α-underestimators that are valid for the corresponding subbox only, cf. (9.26).

The decision on how to subdivide a part of the search space (the *branching rule*) is based on the infeasibility of the solution of the linear relaxation, i.e., the fractionality of discrete variables and the distance between a quadratic function and its convexification [45]. Subdividing with respect to a variables x_j, $j \in B$, means to create two nodes with additional restrictions $x_j \leq \lfloor \hat{x}_j \rfloor$ and $x_j \geq \lceil \hat{x}_j \rceil$, respectively, where $\hat{x}_j \notin \mathbb{Z}$ is the value of x_j in a solution of (R). Subdividing with respect to a continuous variable x_j means to create one node with increased lower bound $\hat{x}_j$ for x_j and one node with decreased upper bound $\hat{x}_j$ for x_j. Thus, the α-underestimator (9.26) and the derived linearizations improve in the new nodes.

The choice of the next node to be processed (the *node selection rule*) is guided by the gap between the lower bounds of the nodes and the (uniform) upper bound [45].

Since the quadratic underestimators $q(x)$ are not updated during the algorithm and are computed by a heuristic method (Sect. 9.3.2), convergence of the gap between lower and upper bound to zero and locating a global optimum cannot be ensured for MINLPs with nonquadratic nonconvex terms. However, as our results in the next section show, LaGOs Branch and Cut algorithm is able to compute good local optimal points for difficult MINLPs. For some simplified models, LaGO is even able to reduce the gap below 1%.

9.4 Optimization Results

We consider the power plant model presented in Sect. 9.2. At full load, the plant has an output of $\dot{W} = 750\,\mathrm{MW}$ electric power and $\dot{m}_{32} = 133.1\,\mathrm{t/h}$ process steam. Further assumptions are made:

- The *existence* of two heat-recovery steam generators is provided. They can be operated independently.
- To simplify the model, the nominal isentropic efficiencies $\eta_{\mathrm{s,N}}$ of the steam turbines are fixed.
- The steam quality x_{32} must be within the steam vapor region and thus $0 \leq x_{32} \leq 1$. Due to this assumption there is no need for further binary variables that describe the thermodynamic properties of this stream.
- The lower heating value of methane is $LHV = 50.015\,\mathrm{kJ\,kg^{-1}}$. The fuel (natural gas) is approximated as methane.
- The fuel cost is $c_{\mathrm{f}} = 4$€ $/\mathrm{GJ_{LHV}}$, its real rate of increase is set to 1.0% per year, and the rate of inflation is set to $r_{\mathrm{i}} = 2.0\%$.

9.4.1 Design Optimization with Consideration of Partial-Load Behavior

The mathematical program that describes the plant design and plant operation at full and partial load consists of 2,204 continuous and 43 binary variables and 2,517 equations. LaGO computes 433 quadratic underestimators and 834 α-underestimators in the preprocessing (cf. Sect. 9.3.2).

The independent decision variables were given in Table 9.2, the respective optimization results for the design optimization considering the partial-load behavior are shown in the Tables 9.3 and 9.4. Note, that LaGO was not able to close the gap and thus prove global optimality for this model. Instead, we stopped the optimization after 24 h (approx. 8,000 Branch and Bound iterations on a Linux 2.6 AMD Athlon64 X2 6000+ computer with 3 GHz clock frequency and 4 GB RAM) at a gap between lower and upper bound of 15%. The presented design was found by LaGO after approx. 7 h.

Some of the results shown in Table 9.3 might be unexpected, since apparently the pinch temperature difference $\Delta T_{\mathrm{PINCH1}}$ and the temperature difference in the superheater $\Delta T_{\mathrm{SPHT1}}$ are rather high while the subcooling at the economizer outlet $\Delta T_{\mathrm{Sub,ECON1,N}}$ is low. The first two aspects are not required for the operation of the plant at full-load. However, since we included also three partial-load operation points into the model, we forced the solver to find a design that is flexible enough to operate at the considered partial-load operation points.

Although the requested electric power output can be satisfied by the use of only two gas turbines and a subsequent water–steam cycle, the existence of four gas turbines enables a more flexible partial-load operation. The additional burners AB1 and AB2, and the water injector TMX were incorporated in the plant structure but should not be operated in the full-load case OP1. The operation of the gas turbines, the additional burners and the water injector in the respective partial-load cases is shown in Table 9.4.

The use of the additional burners and the water injector enables a plant operation without evaporation of the working fluid in the economizer during partial-load operation. The relatively high pressure in the condenser is necessary to get acceptable steam qualities x_{13} at the steam turbine outlet at partial-load operation. Some of the four existing gas turbines are shut down

Table 9.3. Decision variables for the design optimization (nominal values, numbers refer to Fig. 9.2). Operating points are considered according to Table 9.1. The first part shows the binary variables and the respective power output of the gas turbines in the full-load case, the second part gives the continuous variables. $\Delta T_{\mathrm{SPHT1}} = T_{22} - T_{12}$, $\Delta T_{\mathrm{PINCH1}} = T_6 - T_{11}$, $\Delta T_{\mathrm{Sub,ECON1}} = T_{\mathrm{saturated}}(p_{10}) - T_{10}$, $\Delta T_{\mathrm{cw}} = T_{16} - T_{15}$

Decision variable	Value
V94.3A$_{\mathrm{a}}$	–
V94.3A$_{\mathrm{b}}$	1 (177.3 MW)
V94.3A$_{\mathrm{c}}$	1 (253.5 MW)
V94.2$_{\mathrm{a}}$	1 (93.1 MW)
V94.2$_{\mathrm{b}}$	1 (75.3 MW)
AB1 ($\dot{m}_{\mathrm{fuel}}$)	1 (0.0 kg s^{-1})
AB2 ($\dot{m}_{\mathrm{fuel}}$)	1 (0.0 kg s^{-1})
TMX1 ($\dot{m}_{\mathrm{water}}$)	1 (0.0 kg s^{-1})
$p_{14,\mathrm{N}}$ [bar]	53.5
$p_{8,\mathrm{N}}$ [bar]	0.12
$\Delta T_{\mathrm{SPHT1,N/SPHT2,N}}$ [K]	102.3/89.7
$\Delta T_{\mathrm{PINCH1,N/PINCH2,N}}$ [K]	32.9/22.4
$\Delta T_{\mathrm{Sub,ECON1,N/Sub,ECON2,N}}$ [K]	2.2/1.5
$\Delta T_{\mathrm{cw,N}}$ [K]	4.0

Table 9.4. Operation of the four gas turbines, the additional burners AB1 and AB2, and the water injector at the different operation points

Component	Variable	Unit	OP1	OP2	OP3	OP4
V94.3A$_a$	$\dot{W}$	MW	–	–	–	–
V94.3A$_b$	$\dot{W}$	MW	177.3	192.4	0.0	0.0
V94.3A$_c$	$\dot{W}$	MW	253.5	253.5	216.3	0.0
V94.2$_a$	$\dot{W}$	MW	93.1	149.3	149.3	149.3
V94.2$_b$	$\dot{W}$	MW	75.3	0.0	149.3	132.5
$\dot{m}_{\mathrm{f,AB1}}$	$\dot{m}$	kg s^{-1}	0.00	0.02	0.00	0.00
$\dot{m}_{\mathrm{f,AB2}}$	$\dot{m}$	kg s^{-1}	0.00	0.00	0.00	0.02
$\dot{m}_{23}$	$\dot{m}$	kg s^{-1}	0.00	0.00	1.80	1.80

Table 9.5. Operation costs for the first year with an optimized design. Operation according to Table 9.1. Fuel costs $\dot{C}_{\mathrm{f}}$ are calculated by (9.27)

Costs	Unit	Value
PEC	Mio. €	197.2
$\dot{C}_{\mathrm{f,OP1}}$	Mio. €/a	44.7
$\dot{C}_{\mathrm{f,OP2}}$	Mio. €/a	32.3
$\dot{C}_{\mathrm{f,OP3}}$	Mio. €/a	28.1
$\dot{C}_{\mathrm{f,OP4}}$	Mio. €/a	23.4
TRR_{lev}	Mio. €/a	236.2

successively as partial load decreases. Thus, each gas turbine is tried to be operated at conditions as close as possible to the respective full-load conditions.

Note that conventional optimization procedures (e.g., [3, 7, 13, 16, 54–57]) might easily have failed to find a design that is feasible for some of the considered partial-load cases due to a violation of the pinch and steam quality constraints. That is why conventional optimization approaches with consideration of only one operation point require a subsequent variation of some decision variables using heuristic methods. These methods do not guarantee cost optimality of the final design obtained after these corrections, e.g., the subcooling at the economizer outlet $\Delta T_{\mathrm{Sub,ECON,N}}$ would be too high.

The objective function value is the levelized total revenue requirement TRR_{lev}. The economic assumptions were presented at the beginning of this section. Table 9.5 shows the PEC, the levelized fuel cost flow rate $\dot{C}_{\mathrm{f,i}}$ for the respective load cases, and the TRR_{lev}. The fuel costs are calculated by

$$\dot{C}_{\mathrm{f,i}} = c_{\mathrm{f}}\,\mathrm{LHV}\,\dot{m}_{\mathrm{f,OP,i}}\,oh_{\mathrm{OP,i}}, \tag{9.27}$$

where $\dot{m}_{\mathrm{f,OP,i}}$ denotes the fuel mass flow rate and $oh_{\mathrm{OP,i}}$ the operating hours in the respective load case. The respective costs for the different products (process steam and electricity) can be calculated by using an exergy-based cost allocation method [8].

9.4.2 Sensitivity of an Optimal Design with Respect to Uncertain Investment Cost

In the following we present a first attempt to analyse the effect that uncertainty in the investment cost of a power plant component has on the optimal plant design. For this purpose we have considered only the design operation point (full-load case) with an operation of 8,000 h full load equivalent per year. We have chosen the two academic cost models shown in Fig. 9.4 (cf. Sect. 9.2.2) for the high-, intermediate-, and low-pressure sections of the steam turbine and compared the designs that are found by the solver when either one of the models is applied or when a weighted average (A/B) is used for calculating the *PEC* of a steam turbine. Table 9.6 presents the results.

The *PEC* and TRR_{lev} values depend on the cost function that was used. But the differences among the results (structure and operation variables) of the three cases (A, B, and A/B) are practically negligible. Due to this, the TRR_{lev} in this analysis is basically influenced by the *PEC* of the turbines. Altogether, in this example, the design is significantly influenced by the fuel cost, i.e., the cost optimal design tends towards a thermodynamic optimal design. Hence, the results of the optimized designs recalculated with different cost models are almost the same (last three lines of Table 9.6). Apparently the impact of the different steam turbine cost functions is not strong enough to influence the optimized structure or the operating parameters.

9.4.3 Comparison with Other MINLP Solvers

Finally, we depict our experience with other MINLP solvers on the problem instance from Sect. 9.4.1. This instance is available as model `chp_partload` in the MINLPLib [15]. We note that the results presented here do not allow conclusions about the performance of the considered solvers in general. Since the

Table 9.6. Comparison of optimization and simulation results with different cost models: A, B (cf. Sect. 9.2.2), and a weighted average of the cost functions (A/B).

			Optimized with cost approach		
	Variable	Unit	A	B	A/B
	PEC	Mio. €	225.42	159.96	192.70
	$\dot{C}_{\text{f}}$	Mio. €/a	170.74	170.62	170.68
	TRR_{lev}	Mio. €/a	300.19	278.57	289.38
	p_{14}	bar	38.17	38.17	38.17
	T_{21}	°C	415.64	415.64	415.64
	$\dot{W}_{\text{HPST}}$	MW	108.94	109.18	109.07
	$\dot{W}_{\text{LPST}}$	MW	85.21	85.40	85.31
*TRR*lev	A	Mio. €/a	300.19	300.20	300.20
calculated with	B	Mio. €/a	278.58	278.57	278.57
cost approach	A/B	Mio. €/a	289.39	289.39	289.38

power plant model has been developed with having the solver LaGO in mind, it is in some sense tailored for LaGO, e.g., a block-separable formulation where nonconvex functions have only low dimensionality has been emphasized. For other solvers, a different but equivalent formulation might be advantageous. Further, the choice of LaGOs parameters is based on a long experience with similar power plant optimization models, while for the other solvers we have made only small adjustments to the default setting.

For this study, we have chosen the MINLP solvers that are available with GAMS 22.6 [24]. These are AlphaECP, BARON, Bonmin, DICOPT, LINDOGlobal, OQNLP, and SBB. Note that only BARON and LINDOGlobal can explicitly handle nonconvex problems. Unfortunately, testing our model on LINDOGlobal was not possible because the model size exceeded the restrictions on the GAMS/LINDOGlobal license.

We have run each solver once without providing any feasible starting point and once with the solution discussed in Sect. 9.4.1 as starting point in order to see whether the solver is able to improve it further. Recall, that this solution has an objective function value of 236.24 Mio. €/a and LaGO reported a lower bound of 201.59 Mio. €/a when it was stopped. Each solver was run with a time limit of 3 h and a relative gap tolerance of 1%. Iteration, node, or memory limits were turned off. The computer was a Linux 2.6 Intel Core 2 Duo T7500 laptop with 2.2 GHz clock frequency and 2 GB RAM. The results are summarized in Table 9.7 and discussed in more detail in the following.

AlphaECP [62] implements an extended cutting plane algorithm which guarantees global optimality for pseudoconvex problems. The algorithm constructs and improves a MIP approximation of the problem by constructing a (possibly shifted) linearization of violated constraints. If a (partial) solution of the MIP approximation is feasible to the original problem, an upper bound might be updated. Otherwise AlphaECP can do a local search in the MINLP, i.e., discrete variables are fixed and a NLP solver is called using the MIP

Table 9.7. Best objective function values when running the power plant optimization instance from Sect. 9.4.1 with different MINLP solvers and a time limit of 3 h. The third column gives the best values when a solution with objective function value 236.24 Mio. €/a is provided as a starting point to the solver.

Solver	best solution Mio. €/a	best solution with starting point Mio. €/a
LaGO	244.17	236.24
AlphaECP	fail	fail
BARON	255.13	236.24
Bonmin	242.50	fail
DICOPT	fail	fail
OQNLP	fail	235.97
SBB	236.76	235.73

solution as starting point. When AlphaECP was run with default parameter values no feasible point was found. Also setting the option `callnlpiter` to increase the number of NLP subsolver calls did not improve the situation.

BARON [52, 53] implements a branch and reduce algorithm that is related to LaGOs methodology. Here, exact convex underestimators are constructed for nonconvex functions using a factorable reformulation of the model. These underestimators are used to generate a linear approximation that yields lower bounds. Further, constraint propagation and duality techniques are used to tighten the bounding box on the variables. On the power plant model, BARON with default parameters spend most of the time to do probing on variable bounds and did not find any feasible point. Thus, we rerun BARON with the restriction to do probing for at most ten variables (option `PDo 10`). Now a feasible point with objective function value 255.13 Mio. €/a was found. The lower bound at the end of the timeperiod coincides with LaGOs lower bound of 201.59 Mio. €/a. Providing BARON with LaGOs solution as starting point did not result in a better point.

COIN-OR/Bonmin [11] implements both outer approximation and Branch and Bound algorithms. We have decided for the Branch and Bound algorithm since it seems to be better suited for nonconvex problems. Here, lower bounds are computed from solutions of the relaxed MINLP, i.e., the MINLP with dropped integrality restrictions, and upper bounds from solving the MINLP with fixed discrete variables. Bonmin found a feasible solution with objective function value 242.50 Mio. €/a after 41 minutes (1,016 nodes). Unfortunately a further improvement was not possible because it stopped 10 min later due to a failing solve of a lower bounding problem by the NLP subsolver Ipopt. Running Bonmin with the provided starting point and with or without the options `num_resolve_at_root` and `num_resolve_at_node` set to 10 did not improve the situation.

DICOPT [18] implements an extended outer approximation algorithm. It iteratively solves MIP approximations generated by linearization of the MINLP and NLP subproblems obtained by fixing discrete variables in the MINLP. We have run DICOPT with stopping rule 0 and a very high value for `maxcycles`, so that it does not stop before the time limit is reached. Unfortunately no feasible point was found. Providing a feasible starting point did not improve the situation.

OQNLP [59] is a heuristic multistart algorithm. The solver generates starting points via a scatter search or by random and uses them as starting points for local searches. For the power plant optimization model, OQNLP did not find any feasible point within the timelimit when run without a starting point. However, if we provide OQNLP with the feasible starting point, it was able to find an improved point with objective function value 235.97 Mio. €/a after a few seconds, but then was not able to improve this point further.

SBB [6] is a Branch and Bound algorithm similar to the one used in Bonmin. Lower bounds are computed by solving NLP subproblems obtained by partly fixing and partly relaxing integrality restrictions in the MINLP. If the solution of such an NLP is feasible for the MINLP, a new upper bound has been found. We run SBB with the option `acceptnonopt 1` to increase tolerance with NLP solver failures. For the power plant optimization model, SBB found a feasible solution with objective function value 236.76 Mio. €/a and reported a lower bound of 202.49 Mio. €/a. When we provided LaGOs solution as starting point, it found a feasible solution with objective function value 235.73 Mio. €/a and reported a lower bound of 203.91 Mio. €/a. Note, that SBBs lower bounds are only guaranteed for convex models.

From the results presented here, it can be observed that algorithms based on a linearization of nonlinear constraints like AlphaECP and DICOPT are not suited for this problem. We presume, that this is due to the disregard of nonconvex behavior in the construction of the MIP approximation.

On the other hand, NLP based Branch and Bound algorithms like Bonmin and SBB seem to behave well on this problem instance. Even though the NLP relaxations solved in these algorithms constitute nonconvex problems, the NLP solver (CONOPT in the SBB run and Ipopt in the Bonmin run) seem to be able to find good solutions. The proper treatment of the 45 binary variables is then the remaining task of the MINLP solver. The moderate size of this combinatorial part (when compared to the 2,204 continuous variables) is probably advantageous for SBB and Bonmin.

The performance of BARON and LaGO indicate that the extra effort for the convexification of nonconvex functions in order to compute true lower bounds results in longer running times, either because the linear relaxation does not give useful information for branching decisions or because the starting points used for local searches are worse when compared with an NLP based Branch and Bound algorithm.

Finally, the slightly improved solution points found by OQNLP and SBB when started with LaGOs solution indicate that LaGO could benefit from a kind of local branching heuristic that searches for solutions that improve a recently found incumbent solution.

9.5 Conclusions

We presented a MINLP formulation of a model for a complex cogeneration plant and the Branch and Cut based solver LaGO that is used to solve the model. The operation under different load conditions and under uncertainties in the investment cost are discussed. It has been shown, that the consideration of only one typical design point is not sufficient, since the operation at some partial-load operation points may become impossible due to thermodynamic constraints of the plant so that heuristics methods would be required to satisfy the further requirements that are posed by partial-load operation. The presented method overcomes this problem by considering required partial-load conditions already in the design optimization and thus allows to find a cost optimal plant design that is feasible for the full-load and the considered partial-load cases. Of course, this enhancement comes in hand with a significantly more complex optimization model.

Further, we have seen that the optimized design is insensitive to changes in the steam turbine cost function. An approach to extend this sensitivity analysis towards the computation of a design that is robust with respect to changes in a cost function is to use a (weighted) average of cost function scenarios. However, in order to avoid that this approach collapses to the computation of a design that is just optimal for the averaged costs, one should also introduce recourse decisions into the model, i.e., allow the off-design variables to take different values in each considered cost scenario.

Due to the formulation of the problem as a system of equations, an extension of the model, such as the consideration of more partial-load operation points or more complex energy conversion plants is possible without any significant modeling effort. In particular there is no need to define a calculation order for the streams and components because the MINLP solver optimizes the entire problem simultaneously. We have seen that even though a typical engineering problem, when formulated as mathematical program, requires sophisticated solution algorithms, the investment in a complex MINLP model and powerful solver can be rewarded by realistic solutions of good quality.

From the algorithmic perspective, the main difficulty of the considered MINLP models are the nonlinearities and nonconvexities introduced by the equations for the thermodynamic behavior of the plant. LaGO handles nonconvex functions by computing quadratic underestimators that can be convexified using α-underestimators. While it is possible to compute good local minimal points, closing the gap between lower and upper bounds, and thus proving global optimality of the computed solutions, is still exceptional for complex MINLPs. To bring LaGO closer to this goal, ongoing research focuses on the additional generation of quadratic underestimators during the Branch and Bound process. Additional room for improvement is presented by the branching and node selection rules and a further exploration of block-separability. Also the consideration of a mixed-integer linear relaxation could help to improve convergence.

Acknowledgement. We are grateful to GAMS Development Corporation for providing us with an evaluation license for the GAMS MINLP solvers. Further, we thank two reviewers for their comments and helpful suggestions on an earlier version of this paper.

References

1. C. S. Adjiman, S. Dallwig, C. A. Floudas, and A. Neumaier. A global optimization method, αBB, for general twice-differentiable constrained NLPs – I. Theoretical advances. *Computers and Chemical Engineering*, 22:1137–1158, 1998.

2. C. S. Adjiman and C. A. Floudas. Rigorous convex underestimators for general twice-differentiable problems. *Journal of Global Optimization*, 9:23–40, 1997.
3. T. Ahadi-Oskui. Optimierung des Entwurfs komplexer Energieumwandlungsanlagen. In *Fortschritt-Berichte*, number 543 in Series 6. VDI-Verlag, Düsseldorf, Germany, 2006.
4. T. Ahadi-Oskui, H. Alperin, I. Nowak, F. Cziesla, and G. Tsatsaronis. A Relaxation-Based Heuristic for the Design of Cost-Effective Energy Conversion Systems. *Energy*, 31:1346–1357, 2006.
5. T. Ahadi-Oskui, S. Vigerske, I. Nowak, and G. Tsatsaronis. Optimizing the design of complex energy conversion systems by branch and cut. Preprint 07-11, Department of Mathematics, Humboldt-University Berlin. available at `http://www.math.hu-berlin.de/publ/pre/2007/P-07-11.pdf` and submitted, 2007.
6. ARKI Consulting and Development A/S and GAMS Inc. SBB. `http://www.gams.com/solvers/solvers.htm#SBB`, 2002.
7. J. Axmann, R. Dobrowolski, and R. Leithner. Evolutionäre Algorithmen zur Optimierung von Kraftwerkskonzepten und Anlagenbauteilen. In *Fortschritt-Berichte*, number 438 in Reihe 6, pages 251–265. VDI-Verlag, Düsseldorf, Germany, 1997.
8. A. Bejan, G. Tsatsaronis, and M. Moran. *Thermal Design and Optimization*. Wiley, New York, USA, 1996.
9. M.S. Boddy and D.P. Johnson. A new method for the global solution of large systems of continuous constraints. In Ch. Bliek, Ch. Jermann, and A. Neumaier, editors, *Global Optimization and Constraint Satisfaction*, volume 2861 of *Lecture Notes in Computer Science*, pages 143–156. Springer, Berlin, 2003.
10. C. G. E. Boender and H. E. Romeijn. Stochastic methods. In R. Horst and P. Pardalos, editors, *Handbook of Global Optimization*, pages 829–869. Kluwer, Dordrecht, 1995.
11. P. Bonami, L.T. Biegler, A.R. Conn, G. Cornuéjols, I.E. Grossmann, C.D. Laird, J. Lee, A. Lodi, F. Margot, N. Sawaya, and A. Wächter. An algorithmic framework for convex mixed integer nonlinear programs. *Discrete Optimization*, 5:186–204, 2008.
12. C. Bouvy. Kombinierte Struktur- und Einsatzoptimierung von Energieversorgungssystemen mit einer Evolutionsstrategie. Shaker Verlag, Aachen, 2007.
13. C. Bouvy and S. Herbergs. Mehrkriterielle Optimierung dezentraler Energieversorgungssysteme mit evolutionären Algorithmen. In *Optimierung in der Energiewirtschaft*, number 1908 in Fortschitt-Berichte, pages 265–277. VDI-Verlag, Düsseldorf, Germany, 2005.
14. J. Bruno, F. Fernandez, F. Castells, and I. Grossmann. A Rigorous MINLP Model for the Optimal Synthesis and Operation of Utility Plants. *Transactions of the Institution of Chemical Engineers*, 76:246–258, 1998.
15. M. R. Bussieck, A. S. Drud, and A. Meeraus. MINLPLib – A Collection of Test Models for Mixed-Integer Nonlinear Programming. *INFORMS Journal on Computing*, 15(1):114–119, 2003.
16. F. Cziesla. Produktkostenminimierung beim Entwurf komplexer Energieumwandlungsanlagen mit Hilfe von wissensbasierten Methoden. In *Fortschritt-Berichte*, number 438 in Reihe 6. VDI-Verlag, Düsseldorf, 2000.
17. DEWI, E.ON Netz, EWI, RWE Transportnetz Strom, and VE Transmission. Energiewirtschaftliche Planung für die Netzintegration von Windenergie in Deutschland an Land und Offshore bis zum Jahr 2020. `http://www.eon-netz.`

com/Ressources/downloads/dena_haupt_studie.pdf, February 2005. Studie im Auftrag der Deutschen Energie-Agentur GmbH (dena).
18. M. A. Duran and I. E. Grossmann. An outer-approximation algorithm for a class of mixed-integer nonlinear programs. *Mathematical Programming*, 36:307–339, 1986.
19. A. Epe, C. Küchler, W. Römisch, S. Vigerske, H.-J. Wagner, C. Weber, and O. Woll. Stochastic programming with recombining scenario trees – optimization of dispersed energy supply. Chapter 15 in this book.
20. M. Fenski. Prozessanalyse mittels selbstorganisierender neuronaler Karten am Beispiel eines kombinierten Gas- und Dampfturbinenwerks mit Fernwärmeauskopplung. Diplomarbeit, Technische Universität Berlin, 2006.
21. R. Fletcher and S. Leyffer. Solving Mixed Integer Nonlinear Programs by Outer Approximation. *Mathematical Programming*, 66(3(A)):327–349, 1994.
22. C. A. Floudas, I. G. Akrotirianakis, C. Caratzoulas, C. A. Meyer, and J. Kallrath. Global optimization in the 21st century: Advances and challenges. *Computers and Chemical Engineering*, 29(6):1185–1202, 2005.
23. C.A Frangopoulos. Optimal Synthesis and Operation of Thermal Systems by the Thermoeconomic Functional Approach. *Transactions of the ASME, Journal of Engineering for Gas Turbines and Power*, 114:707–714, 1992.
24. GAMS Development Corp. *GAMS – The Solver Manuals*. GAMS, Washington DC, 2007.
25. Gas Turbine World 2006 Handbook. Pequot, Fairfield, 2006.
26. M. Gebhardt, H. Kohl, and T. Steinrötter. Preisatlas, Ableitung von Kostenfunktionen für Komponenten der rationellen Energienutzung. http://www.iuta.de/thermodynamik/Preisatlas_Download.htm, June 2002.
27. R. Horst and P. Pardalos. *Handbook of Global Optimization*. Kluwer, Dordrecht, 1995.
28. K. Hüttenhofer and A. Lezuo. Cogeneration Power Plant Concepts Using Advanced Gas Turbines. *VGB PowerTech*, 81(6):50–56, 2001.
29. M. Jüdes, F. Cziesla, W. Ahrens, J. Petri, and G. Tsatsaronis. Neuronale Netze als Hilfsmittel zur Betriebsbewertung am Beispiel einer industriellen Kraft-Wärme-Kopplungsanlage. In *Optimierung in der Energieversorgung*, number 1792 in VDI-Berichte, pages 201–213. VDI-Verlag, 2003.
30. M. Jüdes and G. Tsatsaronis. Cost Effective Design Optimization and Maintenance Strategies with Consideration of the Partial Load Behavior of Power Plants. In A. Mirandola, Ö. Arnas, and A. Lazzaretto, editors, *Proceedings of ECOS 2007, Padova, Italy*, volume I, pages 251–258, June, 25-28 2007.
31. M. Jüdes and G. Tsatsaronis. Design Optimization of Power Plants by Considering Multiple Partial Load Operation Points. In *2007 ASME International Mechanical Engineering Congress and Exposition*, November 11-15, Seattle, Washington USA, IMECE2007. ASME, 2007.
32. O. Knacke, O. Kubaschewski, and K. Hesselmann. *Thermochemical Properties of Inorganic Substances*. Springer, Berlin, Germany, 1991.
33. T. Kohonen. *Self-Organizing Maps*. Springer Verlag, Heidelberg, 2001.
34. M. Krämer. Modellanalyse zur Optimierung der Stromerzeugung bei hoher Einspeisung von Windenergie. In *Fortschritt-Berichte*, number 492 in Reihe 6. VDI-Verlag, Düsseldorf, Germany, 2003.
35. H. Marchand and L.A. Wolsey. Aggregation and mixed integer rounding to solve MIPs. *Operations Research*, 49(3):363–371, 2001.

36. J.R. Muñoz and M.R. von Spakovsky. A Decomposition Approach for the Large Scale Synthesis/Design Optimization of Highly Coupled, Highly Dynamic Energy Systems. *International Journal of Applied Thermodynamics*, 4(1):19–33, 2001.
37. J.R. Muñoz and M.R. von Spakovsky. The Application of Decomposition to the Large Scale Synthesis/Design Optimization of Aircraft Energy Systems. *International Journal of Applied Thermodynamics*, 4(2):61–74, 2001.
38. G. L. Nemhauser and L. A. Wolsey. *Integer and Combinatorial Optimization*. Wiley, New York, 1988.
39. A. Neumaier. *Interval Methods for Systems of Equations*. Cambridge University Press, Cambridge, 1990.
40. A. Neumaier. Complete search in continuous global optimization and constraint satisfaction. In *Acta Numerica*, volume 13, chapter 4, pages 271–370. Cambridge University Press, Cambridge, 2004.
41. A. Neumaier and S. Vigerske. personal communication, October 2006.
42. I. Nowak. *Relaxation and Decomposition Methods for Mixed Integer Nonlinear Programming*. Birkhäuser Verlag, Basel, Schweiz, 2005.
43. I. Nowak, H. Alperin, and S. Vigerske. LaGO – an object oriented library for solving MINLPs. In Ch. Bliek, Ch. Jermann, and A. Neumaier, editors, *Global Optimization and Constraint Satisfaction*, volume 2861 of *Lecture Notes in Computer Science*, pages 31–43. Springer, Berlin, 2003.
44. I. Nowak and S. Vigerske. LaGO – Lagrangian Global Optimizer. `https://projects.coin-or.org/LaGO`.
45. I. Nowak and S. Vigerske. LaGO: a (heuristic) branch and cut algorithm for nonconvex MINLPs. *Central European Journal of Operations Research*, 16(2):127–138, 2008.
46. K. Papalexandri, E. Pistikopoulos, and B. Kalitventzeff. Modelling and Optimization Aspects in Energy Management and Plant Operation with Variable Energy Demands-Application to Industrial Problems. *Computers and Chemical Engineering*, 22(9):1319–1333, 1998.
47. D. Paulus. Single-Component Optimal Heat Exchanger Effectiveness using Specific Exergy Costs and Revenues. In S. Kjelstrup, J. E. Hustad, T. Gundersen, A. Røsjorde, and G. Tsatsaronis, editors, *Proceedings of ECOS 2005, Trondheim, Norway*, volume **III**, pages 1407–1414, June, 20-22 2005.
48. R. Peltier. TOP PLANTS: Tenaska Virginia Generating Station, Scottsville, Virginia. *POWER magazine*, 151(9):50–53, 2007.
49. R. Romero, A. Zobaa, E. Asada, and W. Freitas. Mathematical optimisation techniques applied to power systems operation and planning. *International Journal of energy technology and policy*, 5(4):393–403, 2007.
50. T. Savola. Simulation and Optimisation of Power Production in Biomass-Fuelled Small-Scale CHP plants. Licentiate's thesis, Helsinki University of Technology, March 2005.
51. SOFBID. EBSILONProfessional. `http://www.sofbid.com`.
52. M. Tawarmalani and N. V. Sahinidis. Global optimization of mixed-integer nonlinear programs: A theoretical and computational study. *Mathematical Programming*, 99:563–591, 2004.
53. M. Tawarmalani and N.V. Sahinidis. *Convexification and Global Optimization in Continuous and Mixed-Integer Nonlinear Programming: Theory, Algorithms, Software, and Applications*. Kluwer, Dordrecht, 2002.

54. G. Tsatsaronis, F. Cziesla, and Z. Gao. Avoidable Thermodynamic Inefficiencies and Costs in Energy Conversion Systems. Part 1: Methodology. In N. Houbak, B. Elmegaard, B. Qvale, and M.J. Moran, editors, *Proceedings of ECOS 2003, Copenhagen, Denmark* , volume **II**, pages 809–814, June 30–July 2, 2003.
55. G. Tsatsaronis, K. Kapanke, and A.M. Blanco Marigorta. Exergoeconomic estimates for a novel process with integrated CO_2 capture for the production of hydrogen and electric power. In C.A. Frangopoulos, C.D. Rakopoulos, and G. Tsatsaronis, editors, *Proceedings of ECOS 2006, July 12–14, Aghia Pelagia, Crete, Greece*, volume 3, pages 1581–1591, 2006.
56. G. Tsatsaronis, L. Lin, J. Pisa, and T. Tawfik. Thermoeconomic Design Optimization of a KRW-based IGCC power plant, Final Report submitted to Southern Company Services and the U.S. Department of Energy. DE-FC21-89MC26019, Center for Electric Power, Tennessee Technological University, 1991.
57. G. Tsatsaronis, T. Tawfik, L. Lin, and D.T. Gallaspy. Exergetic Comparison of two KRW-based IGCC Power Plants. *Journal of Engineering Gas Turbines and Power*, pages 219–299, 1994.
58. R. Turton, R. Bailie, W. Whiting, and J. Shaeiwitz. *Analysis, Synthesis and Desing of Chemical Processes*. Prentice Hall, New Jersey, USA, 1984.
59. Zsolt Ugray, Leon Lasdon, John Plummer, Fred Glover, Jim Kelly, and Rafael Martí. Scatter search and local NLP solvers: A multistart framework for global optimization. *INFORMS Journal on Computing*, 19(3):328–340, 2007.
60. G. Ulrich. *A guide to chemical engineering process design and economics*. Wiley, New York, USA, 1984.
61. W. Wagner. *Properties of Water and Steam*. Springer, Berlin, 1998.
62. T. Westerlund and R. Pörn. Solving pseudo-convex mixed integer optimization problems by cutting plane techniques. *Optimization and Engineering*, 3:253–280, 2002.

10

Optimally Running a Biomass-Based Energy Production Process

Maurizio Bruglieri and Leo Liberti

Summary. A multiplant biomass-based energy production process is able to extract the chemical energy from various agricultural products. Such a process consists of several plants that are able to deal with biomasses of different types. Each type of plant has distinct mass-to-energy yields for each particular product type. Since the scale of the process may be geographically wide, transportation costs also have an impact on the overall profitability. Biomasses have different unit costs, and end-products (electrical energy, refined bioethanol, but also several other cross-products of the biomasses that are not necessarily energy-related) have different selling prices; hence, deciding the amount of each different biomass to process in order to maximize revenues and minimize costs is a nontrivial task. In this paper we propose a mathematical programming formulation of this problem and discuss its application to a real-world example.

10.1 Introduction

This paper is concerned with a mathematical programming formulation of the problem of optimally running an energy production process based on biomasses. This model was developed for practical reasons arising in the establishment of a bioenergy production process in central Italy. Specifically, the involved chemical, agricultural and engineering enterprises needed to justify the profitability of the process to banks and funding agencies. This was carried out by employing sensitivity analysis around the optimum of the mathematical program describing the process.

The production of energy from biomasses is proving more and more popular what with the energy from fossil carbon-based fuels being costly to both the environment and society [15]. Mathematical programming is one of the main planning tools in this area. [13] examines the competitiveness of biomass-based fuel for electrical energy opposed to carbon-based fuel. In [6], a mathematical programming approach is proposed to localize both energy conversion plants and biomass catchments basins in provincial area. Among the advantages of this type of energy production, there is the potential for employing wasted materials of biological origin, like used alimentary fats and oils, agricultural waste and so on. A factory producing energy with such materials would benefit from both the sales of the energy and the gains obtained by servicing waste [2]. Other mathematical models for specific biomass discrete facility location problems are developed in [8] and [10].

The biomass-based energy production process considered in this paper (see Fig. 10.1) involves several processing plants of different types (for example, a solid biomass plant, a squeeze plant and a fermentation–distillation plant). Some of these plants (e.g., solid biomass plant) produce energy. Others (e.g., the fermentation–distillation plant) produce intermediate products which will then be routed to other plants for further processing. There are several possible input products (e.g., agricultural products, biological waste), obtained from different sources (e.g., direct farming or acquisition on the markets) at different unit costs. Apart from the energetic output, there may be other output products which are sold in different markets (e.g., molasses obtained from the fermentation–distillation plant and sold in the feed market). The optimization problem stemming from the process is that of modeling the production process as a net gain maximization supposing the type of plants involved and the end product demands are known.

Section 10.2 presents the mathematical programming formulation. In Sect. 10.3 we discuss a real-world application of our model. Section 10.4 concerns some realistic improvements to the model. Section 10.5 concludes the paper.

10.2 Modeling the Production Process

Modeling a flowsheet as that presented in Fig. 10.1 presents many difficulties. Notice that the products can be inputs, intermediate, outputs, or both (like alcohol, which is both an output product and an intermediate product). Likewise, processes can be intermediate or final or a combination (like the fermentation–distillation plant). Consider also that the decision maker may choose to buy an intermediate product from a different source to cover demand needs, thus making the product a combination of intermediate and input. Of course the input products may be acquired or produced at different locations and at different prices. Moreover, each flow arrow has an associated transportation cost. The time horizon for the optimization process is 1-year.

Because essentially this problem is connected with the transportation and processing of various materials through a network, we employ a model based on multicommodity flow, which is a standard and well-understood modeling technique in Operations Research (see for instance [1], [3] and [5]). The main concept in our model is the *process site*. A process site is a geographical location with at most one processing plant and/or various storage spaces for different types of goods (commodities). A place where production of a given commodity occurs is represented by a process site with a storage space. Thus,

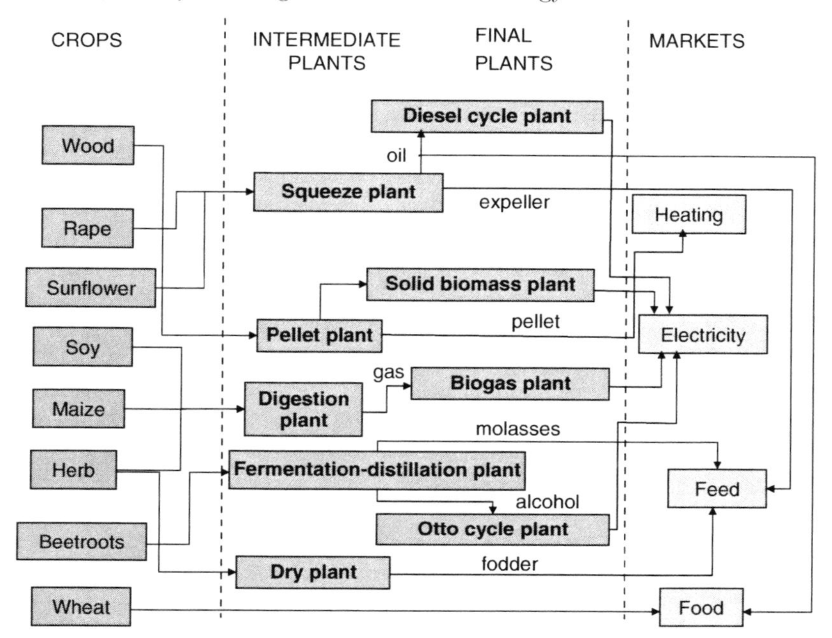

Fig. 10.1. A typical process flowsheet

for example, a geographical location with two fields producing rapes and sunflowers is a process site with two storage spaces and no processing plant. The fermentation–distillation plant is a process site with no storage spaces and one processing plant. Each output in Fig. 10.1 is represented by a process site with just one storage space for each output good. In this interpretation the concepts of input, output and intermediate products, and those of intermediate and final process, lose importance: this is appropriate because, as we have emphasized earlier, these distinctions are not always well-defined. Instead, we focus the attention on the material balance and on the transformation process in each process site. Furthermore, we are able to deal with the occurrence that a given commodity may be obtained at different costs depending on whether it is bought or produced directly.

We represent the process sites by a set V of vertices of a directed graph $G = (V, A)$ where the set of arcs A is given by the logistical connections among the locations. To each vertex $v \in V$ we associate a set of commodities $H^-(v)$ which may enter the process site, and a set of commodities $H^+(v)$ which may leave it. Thus, for example, the squeeze plant is a process site vertex where $H^-(\text{squeeze plant}) = \{\text{rape}, \text{sunflower}\}$ and $H^+(\text{squeeze plant}) = \{\text{oil, expeller}\}$. Furthermore, we let $H = \bigcup_{v \in V}(H^-(v) \cup H^+(v))$ be the set of all commodities involved in the production process, and we partition $V = V_0 \cup V_1$ into V_0, the set of process sites with an associated processing plant, and $V_1 = V \backslash V_0$. Figure 10.2 is the graph derived from the example in Fig. 10.1.

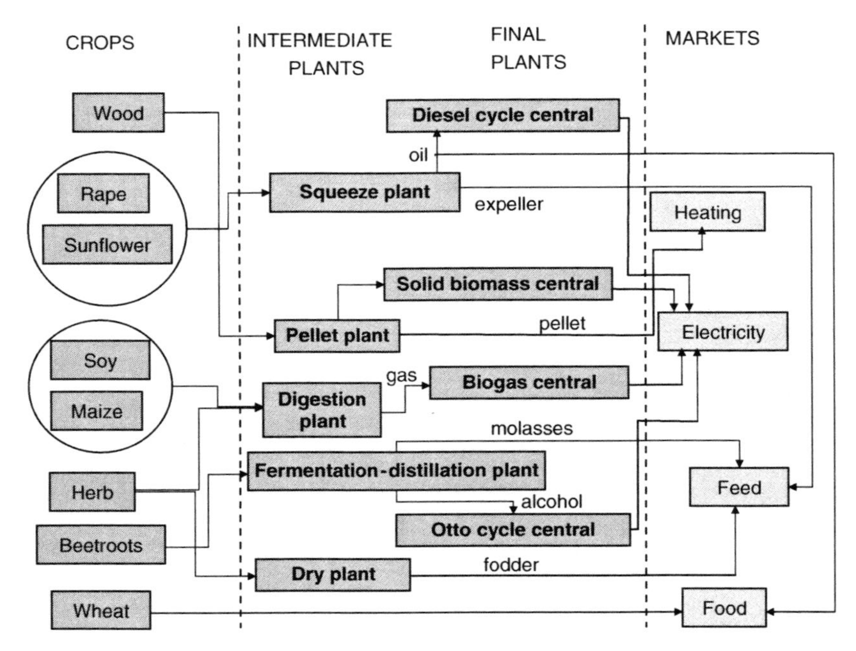

Fig. 10.2. The graph derived from the example in Fig. 10.1

We assume the following to be known parameters:

- c_{vk}: cost of supplying vertex v with a unit of commodity k (negative costs are associated with output nodes, as these represent selling prices; a negative cost may also be associated to the input node "waste," since waste disposal is a service commodity);
- C_{vk}: maximum quantity of commodity k in vertex v;
- τ_{uvk}: transportation cost for a unit of commodity k on the arc (u, v);
- T_{uvk}: transportation capacity for commodity k on arc (u, v);
- λ_{vkh}: cost of processing a unit of commodity k into commodity h in vertex v;
- π_{vkh}: yield of commodity h expressed as unit percentage of commodity k in vertex v;
- d_{vk}: demand of commodity k in vertex v.

It is clear that certain parameters make sense only when associated to a particular subset of vertices, like, e.g., the demands may only be applied to the vertices representing the outputs. In this case, the corresponding parameter should be set to 0 in all vertices for which it is not applicable.

The decision variables are:

- x_{vk}: quantity of commodity k in vertex v;
- y_{uvk}: quantity of commodity k on arc (u, v);
- z_{vkh}: quantity of commodity k processed into commodity h in vertex v.

Since the output demands are known *a priori*, we would like to minimize the total operation costs subject to demand satisfaction. There are three types of costs:

- cost of supplying vertices with commodities:

$$\gamma_1 = \sum_{k\in H}\sum_{v\in V} c_{vk}x_{vk};$$

- transportation costs:

$$\gamma_2 = \sum_{k\in H}\sum_{(u,v)\in A} \tau_{uvk}y_{uvk};$$

- processing costs:

$$\gamma_3 = \sum_{v\in V}\sum_{k\in H^-(v)}\sum_{h\in H^+(v)} \lambda_{vkh}z_{vkh},$$

so the objective function is

$$\min \sum_{i=1}^{3} \gamma_i(x,y,z). \tag{10.1}$$

We need to make sure that some material conservation equations are enforced in each process site where a plant is installed:

$$\sum_{k\in H^-(v)} \pi_{vkh}z_{vkh} = x_{vh}, \ \forall v \in V_0, h \in H^+(v). \tag{10.2}$$

Notice that these constraints do not actually enforce a conservation of mass, for in most processing plants a percentage of the input quantities goes to waste; but it is nonetheless a conservation law subject to the yield properties of the particular transformation process of the plant.

Secondly, the quantity of processed commodity must not exceed the quantity of input commodity in each vertex:

$$\sum_{h\in H^+(v)} z_{vkh} \le x_{vk}, \ \forall v \in V_0, k \in H^-(v). \tag{10.3}$$

Furthermore, we need the quantity of input commodity in each vertex to be consistent with the quantity of commodity in the vertex itself, and similarly for output commodities:

$$\sum_{u\in V:(u,v)\in A} y_{uvk} = x_{vk}, \ \forall v \in V, k \in H^-(v) \tag{10.4}$$

$$\sum_{u\in V:(v,u)\in A} y_{vuh} = x_{vh}, \ \forall v \in V, h \in H^+(v). \tag{10.5}$$

Finally, we have the bounds on the variables:

$$d_{vk} \leq x_{vk} \leq C_{vk}, \ \forall v \in V, k \in H \tag{10.6}$$
$$0 \leq y_{uvk} \leq T_{uvk}, \ \forall (u,v) \in A, k \in H \tag{10.7}$$
$$z_{vkh} \geq 0, \ \forall v \in V, k \in H^-(v), h \in H^+(v) \tag{10.8}$$

and some fixed variables for irrelevant vertices:

$$x_{vk} = 0, \ \forall v \in V_1, k \in H \backslash (H^-(v) \cup H^+(v)) \tag{10.9}$$
$$y_{uvk} = 0, \ \forall (u,v) \in A, k \in H \backslash H^-(v), \tag{10.10}$$
$$y_{uvk} = 0, \ \forall (u,v) \in A, k \in H \backslash H^+(u). \tag{10.11}$$

The main advantage to this model is that it can be easily extended to deal with more commodities and plants in a natural way, by adding appropriate vertices or changing the relevant $H^-(v), H^+(v)$ and related parameters.

10.3 A Real-World Application

The model described in Sect. 10.2 is a Linear Programming (LP) problem, which can be solved by using one of several LP solvers. Using our model we solved an instance derived from a real world application within the "Marche Bioenergia" project (the administrative Italian region where this project took place is called "Marche"). This project consists in the study of replacement/integration of the traditional crops (beetroots, wheat) with new crops exploitable by biomass-based energy production plants, as represented in Fig. 10.1. The target territory consists of some 40,000 ha of land around San Severino Marche, a small village in the center of Italy. One of the aims of the "Marche Bioenergia" project was that of estimating the real value of the national economical incentive to produce electric energy from agricultural products (so-called *green certificates*). With our model, we can do this by looking at the reduced cost attained at the optimum by the nonbasic variables z_{vkh} corresponding to unused power plants.

The processing costs λ_{vkh} and the transformation yields π_{vkh} take the values summarized in Tables 10.1, 10.2, 10.3 and 10.4. In particular Table 10.1 lists the yields and agricultural costs of the crops: such data have been obtained in collaboration with the regional farmers association "Coldiretti" of Ancona (the main city of the Marche region). Table 10.2 lists the yields and transformation costs of the intermediate plants (also supplied by "Coldiretti"), whereas Table 10.3 lists the yields and transformation costs of the power plants supplied directly by the "Marche Bioenergia" firm: large-scale solid biomass and Otto cycle plants (10 MW each) and small-scale Diesel cycle and biogas plants (1 MW each). Finally, Table 10.4 lists the prices $-c_{vk}$ of the final products obtained from "Sole 24 Ore" (1st June 2006 issue), the most important financial journal in Italy. The transportation cost τ_{uvk} have been set equal

Table 10.1. The agricultural costs and yields of considered crops

Crop	*Cost* (euro/ha)	*Yield* (ton/ha)
wood	1,000	130.00
rape	445	2.27
sunflower	697	2.25
soy	470	3.40
maize	704	6.00
herb	600	7.08
beetroots	1,360	33.70
wheat	473	4.00

Table 10.2. The processing costs and the yields of the intermediate plants

Plant	*Cost* (euro/ton)	*Output*	*Yield*
squeeze	18.00	oil	35%
		expeller	65%
pellet	70.00	pellet	95%
digestion	10.00	biogas	0.38%
fermentation–	5	alcohol	20%
distillation		molasses	80%
dry	7.30	fodder	75%

Table 10.3. The processing costs and the yields of the final power plants

Plant	*Cost* (euro/ton)	*Yield* (MWh/ton)
Diesel cycle	23.00	4.25%
Solid biomass	10.00	1.07%
Biogas	50.00	1.00%
Otto cycle	8.70	2.87%

Table 10.4. The prices of the final products sold in different markets

Product	*Market*	*Price* (euro/ton)
pellet	heating	150
electrical power	electricity	150 (euro/MWh)
molasses	feed	100
fodder	feed	115
rape oil	food	550
rape expeller	food	150
sunflower oil	food	650
sunflower expeller	food	125
wheat	food	135

to 10 Eur/ton for all products and 10 Eur/MWh for electric power since the territory considered is relatively small. All capacities, C_{vk} and T_{uvk}, and all demands d_{vk} are considered unbounded: the problem is bounded anyway by the total available land (40,000 ha).

We remark that most data in our model is financial and physical process related, and is thus subject to errors. However, as was mentioned in the introduction, the main application purpose of our study was to justify profitability of the enterprise to banks and funding agencies. It turned out that in practice an approximated cost estimate was enough to attain this purpose, even without considering randomness of data. On the other hand, obtaining robust solutions of LPs subject to uncertain data reduces to solving another LP [4], so it is computationally as tractable as solving the original one.

We solved (in rough 0.01 s) the LP model described in Sect. 10.2 to optimality on the instance presented here using the AMPL [7] modeling language and the CPLEX [9] solver. The obtained solution is shown graphically in Fig. 10.3.

In the proposed optimal solution, about half of the agricultural resources (21.728, 60 ha of land) is devoted to traditional market (rapes for food and feed markets). Slightly less than half of the total land is used to grow wood and beetroots for supplying the solid biomass and the Otto cycle plants, respectively. No other biomass-based energy plant is profitable: from postoptimality sensitivity analysis we infer that in order to produce electricity with a biogas

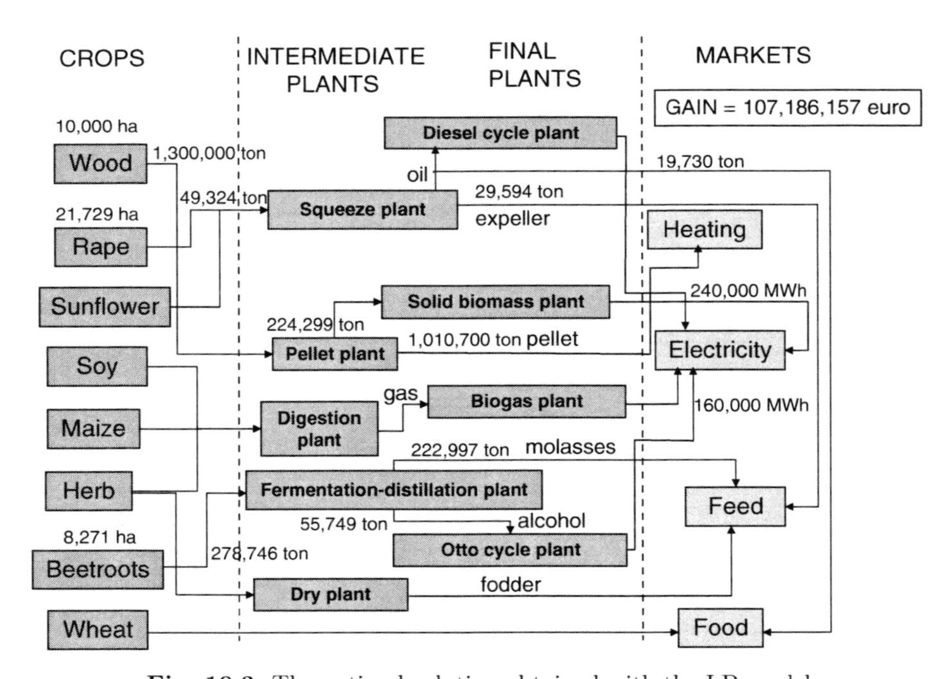

Fig. 10.3. The optimal solution obtained with the LP model

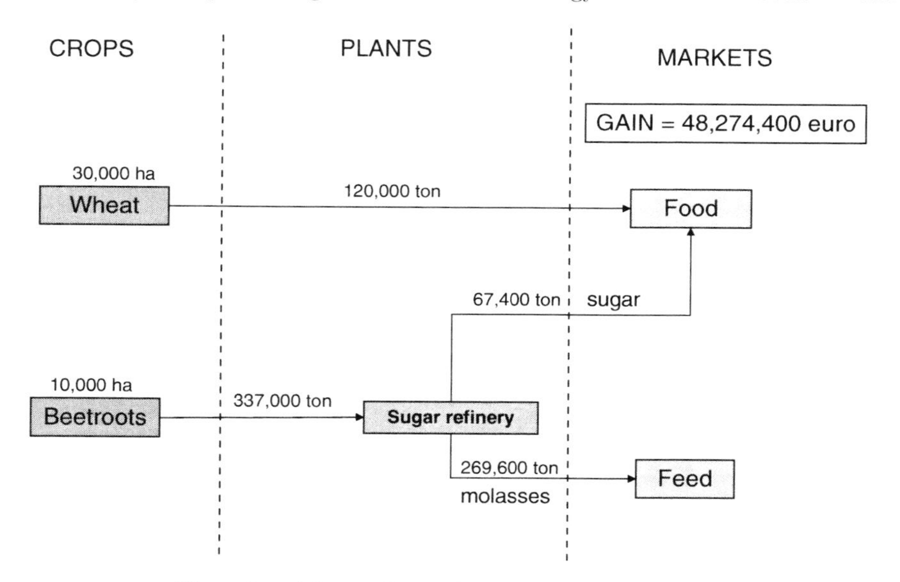

Fig. 10.4. The traditional agricultural production

plant, production costs should decrease by 317 Eur/MWh (reduced cost of variable z_{vkh}, where v =biogas production plant, k =gas, h =electricity); and in order to produce electricity with a Diesel cycle plant, production costs should decrease by 153 Eur/MWh. Finally, comparing our optimal solution and the solution associated to current traditional agricultural production (represented in Fig. 10.4), we notice that exploitation of crops providing biomass for power production more than doubles the total gain (from about 48 million of euro to about 107 million of euro).

10.4 Model Improvements

The model of Sect. 10.2 relies on some simplifications of real-world conditions. A more realistic model can be obtained as follows.

- Some of the plants considered in this paper produce electricity. These have very specific properties and behaviors [11, 12], among which:
  1. In a true market situation (i.e., no subsidization), electricity prices vary during the course of a single day, as demand rises and subsides
  2. Some electricity production plants are often designed to produce electricity *and* heat (which is either stored or conveyed directly into buildings in the area) – such plants are called Combined Heat Power (CHP)
  3. CHPs generate heat and electricity at the same hour and same location

- Transportation costs do not depend linearly on the distances due to the different means of transportation used [14]. For very short transportation distances, tractors may be used, which have higher transportation cost than lorries, used for medium to long distances; for very long transportation distances, trains or ships are used. More generally, the geography of the production process region deeply influences the costs of the single process activities; [14] suggests a methodology that combines Geographical Information Systems (GIS) software with process analysis to estimate these costs.

It turns out that the model in Sect. 10.2 can be extended to accommodate all of the above features. As regards the variability of energy prices during the course of a day and of the year, this can be dealt with in two ways: by employing storage space, or by explicitly adding a time dependency in the model. The former involves adding biomass storage space near the electricity plants (the energyPRO model [11] proposes an electricity plant planning methodology that locally optimizes each plant over a yearly time horizon with hourly time-steps). For the latter, consider the time set $T = \{1, \ldots, 8760\}$ of hours in a 365-day year. We reindex the cost parameters c_{vk} to c^t_{vk} for all $v \in V, k \in H, t \in T$, the decision variables x, y, z to x^t_{vk}, y^t_{uvk}, z^t_{vkh} for all appropriate $u, v \in V$, $k, h \in H$, $t \in T$. We rewrite $\gamma_1, \gamma_2, \gamma_3$ (terms of the objective function) as follows:

$$\gamma_1 = \sum_{t \in T} \sum_{k \in H} \sum_{v \in V} c^t_{vk} x^t_{vk}$$

$$\gamma_2 = \sum_{t \in T} \sum_{k \in H} \sum_{(u,v) \in A} \tau_{uvk} y^t_{uvk}$$

$$\gamma_3 = \sum_{t \in T} \sum_{v \in V} \sum_{k \in H^-(v)} \sum_{h \in H^+(v)} \lambda_{vkh} z^t_{vkh}.$$

All constraints (10.3)–(10.11) are changed accordingly (all occurrences of decision variables x, y, z and parameters c gain an index t and a quantifier $\forall t \in T$ is added to each constraint). This provides a model that can be exactly decomposed in $|T|$ separate LPs as that of Sect. 10.2, i.e., one for each hour $t \in T$. Although it may be possible to solve this problem every hour, it would not be reasonable to expect to change input or transported quantities every hour. Thus, we "connect" the decomposed LPs by means of equality constraints on input and transported quantities. We assume that for each vertex $v \in V$ and commodity $k \in K$ the input quantity x^t_{vk} can be changed every χ_{vk} hours, and that transporting commodity k on the arc (u, v) takes ξ_{uvk} hours (for simplicity we suppose that χ_{vk}, ξ_{uvk} are divisors of $|T|$). We then have:

$$\forall v \in V, k \in H, i \leq \frac{|T|}{\chi_{vk}},\ t_1 < t_2 \in \{\chi_{vk}(i-1)+1, \ldots, \chi_{vk}i\}\ x_{vk}^{t_1} = x_{vk}^{t_2} \tag{10.12}$$

$$\forall (u,v) \in A, k \in H, i \leq \frac{|T|}{\xi_{vk}},\ t_1 < t_2 \in \{\xi_{uvk}(i-1)+1, \ldots, \xi_{uvk}i\}\ y_{uvk}^{t_1} = y_{uvk}^{t_2}. \tag{10.13}$$

Constraints (10.12)–(10.13) simply state that input and transported quantities may only change at predetermined times. The solution of such a large-scale LP is not practically unreasonable using commercial-strength LP solvers such as CPLEX [9].

The combined production of electricity and heat (point 2 in the list above) can be dealt by our model by simply introducing an output process site representing heat, and adapting the λ and π parameters relative to the CHP, various inputs and heat output to reflect the situation. As a consequence of point 3 in the list above, this modeling is not wholly satisfactory, as generation of heat is time-dependent because it is linked to the generation of electricity: but again this time dependency can be dealt with by using process sites representing heat storage capacity or simply adding to the plant cost parameter.

Nonlinear transportation costs are already fully dealt with by our model, for with each arc we associate a transportation cost which is not unitary but rather depends on the vertices adjacent to the arc. Since the arc length is not used anywhere in the model, each arc can be assigned its proper cost.

10.5 Conclusion

In this paper we described a Linear Programming (LP) model for running a biomass-based energy production process, with a real-world application. Our model makes it possible to double the profit associated to traditional agricultural production. The financial benefit was so large that "Marche Bioenergia" was able to self-finance the project without having to seek economical incentives.

Acknowledgement. We wish to thank Prof. Antonio Roversi (Politecnico di Milano) and Ing. Valerio Bitetto (Tecnoplan) for suggesting the problem and kindly helping us to collect the input data for the real world application in the "Marche Bioenergia" project.

References

1. R. K. Ahuja, T. L. Magnanti, and J. B. Orlin. Network flows: Theory, algorithms, and applications, Prentice Hall, New Jersey, 1993.
2. R. Aringhieri, M. Bruglieri, F. Malucelli, and M. Nonato. An asymmetric vehicle routing problem arising in the collection and disposal of special waste. In L. Liberti and F. Maffioli, editors, *CTW04 Workshop on Graphs and Combinatorial Optimization*, volume 17 of *Electronic Notes in Discrete Mathematics*, pages 41–46, Elsevier, Amsterdam, 2004.
3. P. Belotti, L. Brunetta, and F. Malucelli. Multicommodity network design with discrete node costs. *Networks*, 49(1):100–115, 2007.
4. D. Bertsimas and M. Sym. The price of robustness. *Operations Research*, 52(1):35–53, 2004.
5. P. Cappanera and G. Gallo. A Multicommodity Flow Approach to the Crew Rostering Problem *Operations Research*, 52(4):583–596, 2004.
6. G. Fiorese, G. Marino, and G. Guariso. Environmental and economic evaluation of biomass as an energy resource: application to a farming provincial area. Technical report, DEI, Politecnico di Milano, 2005.
7. R. Fourer and D. Gay. *The AMPL Book.* Duxbury, Pacific Grove, 2002.
8. D. Freppaz, R. Minciardi, M. Robba, M. Rovatti, R. Sacile, and A. Taramasso. Optimizing forest biomass exploitation for energy supply at a regional level. *Biomass and Bioenergy*, 26:15–25, 2004.
9. ILOG. *ILOG CPLEX 8.0 User's Manual.* ILOG S.A., Gentilly, France, 2002.
10. E. G. Koukios, D. Voivontas, and D. Assimacopoulos. Assessment of biomass potential for power production: a GIS based method. *Biomass and Bionergy*, 20:101–112, 2001.
11. H. Lund and A. N. Andersen. Optimal designs of small CHP plants in a market with fluctuating electricity prices. *Energy Conversion and Management*, 46(6):893–904, 2005.
12. H. Lund and E. Münster. Integrated energy systems and local energy markets. *Energy Policy*, 34:1152–1160, 2006.
13. B. A. McCarl, D. M. Adams, R. J. Alig, and J. T. Chmelik. Competitiveness of biomass-fueled electrical power plants. *Annals of Operations Research*, 94:37–55, 2000.
14. B. Möller. The use of GIS in planning biomass industries. In *Biomass production conference "Energy from forestry and agriculture"*, Elgin, November 2005.
15. Regional Wood Energy Development Programme. *Proceedings of the regional expert consultation on modern applications of biomass energy.* Food and Agricultural Organization of the UN, Bangkok, Thailand, January 1998.

11

Mathematical Modeling of Batch, Single Stage, Leach Bed Anaerobic Digestion of Organic Fraction of Municipal Solid Waste

Takwai E. Lai, Abhay K. Koppar, Pratap C. Pullammanappallil, and William P. Clarke

Summary. Energy recovery can play an important role in municipal solid waste (MSW) management strategies by providing a saleable by-product and mitigating the environmental effects of the residue that requires disposal. Incineration, in-vessel anaerobic digestion processes and bioreactor landfills can all produce energy and can be used for pretreatment of MSW prior to eventual disposal. Organic fraction of municipal solid waste (OFMSW) is biochemically converted to methane and carbon dioxide in anaerobic digesters and bioreactor landfills. A lumped parameter mathematical model that describes this conversion process in a batch, single-stage, leach bed anaerobic digester under flooded conditions is developed and validated in this chapter. The model uses information such as mass of organic matter loaded in the vessel, amount of water used to flood the waste bed, headspace volume, alkalinity, pH and initial microbial concentrations to predict methane (or biogas) production rate, composition of biogas, residual concentration of organic matter, intermediate metabolites and alkalinity, and pH variations when the digester is operated at mesophilic (38°C) temperature. Most parameters of the model were obtained from literature and a sensitivity analysis used to identify those that required further refinement for improving model predictions. To improve numerical stability and rapid convergence of simulations, a novel solution procedure was developed to solve the charge balance equations in the differential algebraic equations set. Parameter estimation and model validation was carried out using data obtained from three pilot scale experiments conducted in 200 l vessels with 30 kg of OFMSW. Whereas parameter estimation was carried out using the results of one experiment, the model was validated using the results of the other two. The model was found to satisfactorily predict the experimental results and revealed that sufficient concentrations of microbial populations are present naturally in OFMSW and these can be activated rapidly by providing adequate alkalinity to prevent acidification. Such a start up procedure guarantees sustained and stable operation of the digester. Additional simulations determined that alkalinity and pH buffering capabilities provided by an initial concentration of $\approx 11\,\mathrm{g\,l^{-1}}$ of sodium bicarbonate ($NaHCO_3$) was sufficient to accomplish this.

11.1 Introduction

As defined by the United States Environmental Protection Agency (USEPA), municipal solid waste (MSW) is the waste that is generated by communities and includes wastes such as durable goods, nondurable goods, containers and packaging, food scraps, yard trimming and miscellaneous inorganic waste from residential commercial and institutional sources [56]. Excluded from MSW are mine wastes, construction waste, industrial process waste and hazardous waste. MSW management has become an environmental and social concern around the world not only because of the enormous quantities generated but also due to the adverse short and long term environmental effects resulting from improper management in the past. Landfilling is a commonly practiced method of disposal. Nowadays, to reduce the amount of waste that is eventually landfilled and to recover resources, it is taken through sorting, recycling and pretreatment processes. When placed in a landfill, the waste naturally undergoes biologically mediated degradation whereby a stable residue is produced which has a minimal impact on the environment. During this process landfill gas is also produced which can be extracted and used as a fuel because of its high methane content. But in conventional landfills due to the prolonged persistence of adverse conditions for microbial growth, it takes decades for the waste to degrade and yield the above quantities of methane. Moreover, the gas production is not sustained and is subjected to temporal and spatial variations across the landfill. Anaerobic digestion (or biogasification) technologies have been developed for accelerating the biological degradation of MSW either in bioreactor landfills or in-vessel systems.

In anaerobic digestion process, organic compounds like carbohydrates, fats and proteins are mineralized to biogas through the syntrophic action of several groups of microorganisms. The process occurs in nature in anaerobic environments (i.e., in the absence of molecular oxygen) like wetlands, rice fields, intestines of animals, aquatic sediments, and manures, and is responsible for carbon cycling in these environments. The engineered process is called anaerobic digestion which has been traditionally employed for waste treatment. A variety of feedstocks can be anaerobically digested including sewage, municipal solid waste, agricultural residues, wastewater from agro-processing industries, animal manures, and purpose-grown energy crops for biofuel production. As a process for waste treatment, it offers several advantages:

- Can handle high moisture feedstocks as it is mediated by microorganisms
- Reduces the polluting potential of wastes by converting the organic carbon to biogas
- Generates a renewable fuel in the form of biogas, primarily a mixture of methane (55–65%) and carbon dioxide (35–45%)
- In the case of solid wastes it reduces the amount (by weight and volume) of material that has to be disposed or landfilled

Table 11.1. Anaerobic digester design options for various feedstocks

Feedstocks	*Design Options*
Low solids (<2% total solids); Soluble industrial waste water [45], municipal sewage [30]	Anaerobic filter [61], Upflow anaerobic sludge blanket reactor (UASB) [45], fluidized bed reactor [41]
Medium solids (2–15% total solids); Sewage sludge [48], aquatic/marine plants [5], particulate industrial wastes, animal manures [1]	Continuous stirred tank reactor (CSTR) [1], solids concentrating reactor (SOLCON) [5]
High solids (>15% total solids); Municipal solid waste [11, 12, 22, 51], agricultural residue [53], energy crops [13]	Continuous stirred tank reactor (CSTR) [22], Leachbed reactor (e.g., SEBAC-sequential batch anaerobic composting) [11, 12, 51]

- Plant nutrients like nitrogen and phosphorus are conserved in the process, and these can be recycled by land application of the digested residue or effluent
- Can be operated at temperatures as high as 57°C, therefore inactivating human and plant pathogens
- Rates of biogas production are at least two to three orders of magnitude faster than that occurring in landfills

There are several reactor designs available for anaerobic digestion. The type of design that is employed for a particular feedstock depends on its solids content. Table 11.1 lists some anaerobic digester designs that are currently in use. Of particular interest for this chapter is the leach bed design that is used for digesting high-solids feedstocks like MSW. There are a few variations of this design and it can be operated in a batch or continuous mode using one or more vessels. In the one-vessel leach bed process studied here the feedstock is loaded into this vessel to form a bed of solids and sealed. Anaerobic digestion is initiated by flooding the bed with inoculum obtained from a previously digested bed of waste. Once digestion is completed the liquid is drained and stored to serve as inoculum to start up digestion in the next batch of feedstock. The residue is unloaded and the vessel is loaded with a fresh batch of feedstock. Some of the advantages of this process over other designs are:

- It does not require fine shredding of waste
- Does not require mixing or agitation of digester contents
- The process can be implemented in bioreactor landfills as well as in-vessel digesters

In this chapter the development and validation of a dynamic mathematical model to describe anaerobic digestion of OFMSW in a single stage, high solids, leach bed process is presented. Section 11.2 discusses proximate and ultimate characteristics of MSW along with quantities generated in various countries followed by a description of primary metabolic pathways in anaerobic digestion in Sect. 11.3. Section 11.4 addresses the construction of the model equations. Selection of parameters, stoichiometry of various metabolic processes and parameter estimation is discussed in Sect. 11.5. Section 11.6 is devoted to model implementation and simulation. The model validation procedures and outcomes are presented in Sect. 11.7, followed by a model application case study in Sect. 11.8 and conclusions in Sect. 11.9. All the model equations are listed in Appendix 11.9.

11.2 Characteristics of Municipal Solid Waste

Production and composition of municipal solid wastes vary from site to site and are influenced by various factors, including region, climate, extent of recycling, use of in-sink disposals, collection frequency, season and cultural practices. In considering MSW as a feedstock for anaerobic digestion, it is important to know the feed characteristics which are illustrated in Tables 11.2–11.4. In general, the components which can be biologically degraded are paper and putrescible fractions (yard and food wastes) and these typically comprise over 40% of the wet weight.

This organic matter can be effectively digested as unsorted or sorted MSW [42, 43], however the degree of separation of organics influences materials handling and the quality of the residue for use as compost [22]. The trend in sorting is toward source separation of the organic and nonorganic fractions [24]. This not only facilitates sorting of recyclables from the nonorganic fraction, but also results in digester feedstocks (and thus residues) that are relatively free of undesired components such as plastics, metals, glass, and heavy metals. Table 11.3 lists the constituents that make up the organic fraction of MSW. Also listed is the extent of conversion of this organic fraction in an anaerobic digester operated at a hydraulic retention time (HRT) of 20 days and temperature of 35°C [40]. An elemental composition of the organic fraction is given in Table 11.4.

Table 11.2. Composition of domestic waste in various countries

Component	India [49]	Port Harcourt, Nigeria, [25]	Australia [52]	United States [52]
Paper/Cardboard	5.7	12.4	9.9	33.9
Putrescibles (food and Yard wastes)	40.3	37.6	55.9	26.8
Plastics	3.9	9.9	7.3	11.7
Glass	2.1	13.5	6.8	5.3
Metals	1.9	17.2	7.1	7.6
Other	46.1	9.4	13	14.7

Table 11.3. Organic composition of municipal refuse

	[2] % dry wt.	[54] % volatile solids	[40] % dry wt.	[46] % dry wt.	[40] % Conversion
Volatile solids	78.6		73	88.5	58
Cellulose	51.2	40	32.9	15.5	75
Hemicellulose	11.9		5.2	9.5	94
Protein	4.2	5.6	9.6	6.87	10
Lignin	15.2	27.3	12.5	8.5	17
Lipids	–	6	5.9	8.5	66
Starch/sol. Sugars	0.5	3.3	–	–	-
Pectin	<3		–	–	-
Soluble Sugars	0.35	–	–	–	-

Table 11.4. Characteristics of the organic fraction of MSW [28]

Physical characteristics	
Moisture, %	21
Bulk density, $kg\,m^{-3}$	560
Chemical characteristics	
Carbon, %	46
Nitrogen, %	1.5
Phosphorus, %	0.08
Sulfur, %	0.2
C/N ratio	37
C/P ratio	575
Hydrogen, %	6
Oxygen, %	41
Residue, %	6
Cobalt, ppm	0.1
Iron, ppm	163
Molybdenum, ppm	1
Nickel, ppm	1.5
Selenium, ppm	<0.01
Tungsten, ppm	0.1

From a microbiological viewpoint, the organic fraction of MSW has a high solids content (>50%), limiting nitrogen content (C/N ratio >30), and limited surface area for degradation. The principal organic components are cellulose, hemicellulose and lignin. The biochemical methane potential of several MSW components was determined in order to compare the ultimate methane yield [39]. These data, summarized in Table 11.5, indicate that a typical

Table 11.5. Estimates of ultimate methane yield (Y_μ) for MSW components

Sample	Y_μ, $m^3\,kg^{-1}$ VS added
Controls	
Cellulose (a)	0.37
Cellulose (b)	0.37
Org. Fract. MSW	
Sumter County, Florida[1]	0.22
Levy County, Florida[1]	0.20
Yard waste	
Blend	0.14
Grass	0.21
Leaves	0.12
Branches	0.13
Paper	
Office	0.37
Corrugated	0.278
Newspaper (no ink)	0.084
Newspaper (with ink)	0.100
Cellophane	0.35
Food board (uncoated)	0.34
Food board (coated)	0.33
Milk carton	0.32
Wax paper	0.34

[1] Organic fraction of MSW prepared by hand/mechanical separations for composting conversion plants.

conversion efficiency of MSW is between 50 and 60% of volatile solids (VS) corresponding to a methane yield of approximately $0.2\,m^3\,kg^{-1}$ VS. The highest methane yields were observed for various types of paper, including office and food packaging. The lowest methane yield was observed for newspaper, and ink did not influence its biodegradability. As expected, the biodegradability of different types of yard wastes was quite variable. These data provide a basis for predicting potential methane production from wastes with known composition. The undegraded fraction consists of lignin, and cellulose that is tightly complexed with lignin which is refractory to anaerobic metabolism.

11.3 Metabolic Processes in Anaerobic Digestion

The conversion of organic matter to biogas does not occur in a single step. This mineralization process is accomplished through a series of biochemical reactions mediated by different species of microorganisms where the product of one conversion step serves as a substrate for the next step. Figure 11.1 depicts the major metabolic processes that occur during anaerobic digestion of solid

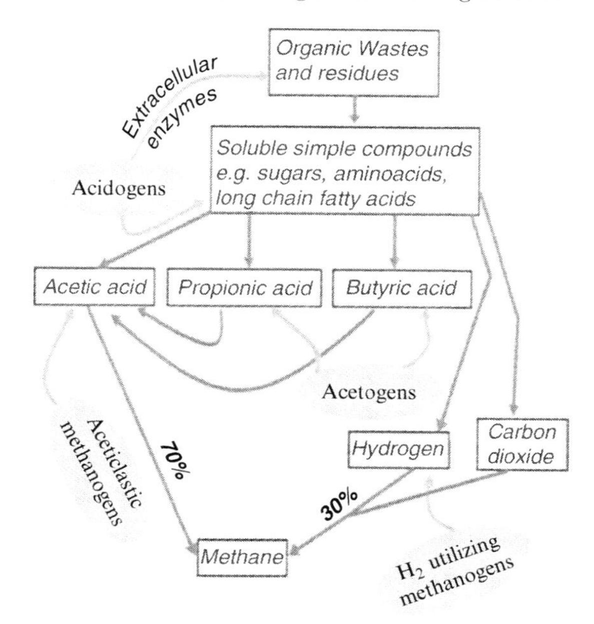

Fig. 11.1. Metabolic processes in anaerobic digestion

organic matter. Four primary conversion steps are distinguished: hydrolysis (or solubilization), acidogenesis, acetogenesis and methanogenesis.

When growing on solid substrates microorganisms cannot directly metabolize these particulates, so extracellular enzymes, commonly referred to as hydrolases are secreted to breakdown these macromolecules (polysaccharides, lipids, and proteins) to its constituent monomers (monosaccharides, long chain fatty acids, glycerol, and amino acids). This step is called hydrolysis (or solubilization). The monomers which are now in a soluble form can be transported into the cell for further metabolism.

Hydrolysis is followed by acidogenesis where the monomers are fermented to volatile organic acids – a mixture of acetic, propionic, butyric and valeric acids, hydrogen and carbon dioxide within the cell. Small amounts of alcohols and amines may also be produced. The ratio of volatile organic acids is regulated by the feedstock type, microbial species and the operating conditions. Microbial species mediating the acidogenesis reactions are grouped as acidogenic bacteria (or acidogens), although small populations of protozoa, fungi and yeasts have also been reported to carry out acidogenesis [55]. The acidogens accounts overall for approximately 90% of the total microbial population in a digester [62].

The third stage called acetogenesis involves the formation of acetic acid which is an important intermediate for the production of methane. In addition to acetic acid production during acidogenesis, there are two possible routes for the formation of acetic acid in this step. One is from the metabolism of higher molecular weight organic acids like propionic, butyric and valeric acids (acetogenic hydrogenation) and the other route is the reduction of carbon dioxide using hydrogen as electron donor (homoacetogenic fermentation). Homoacetogenic fermenters can also grow on monosaccharides. The microbial species responsible for these metabolic processes are classified as acetogenic bacteria (or acetogens). Measurements in anaerobic environments such as in swamps and sludges showed that the population of species responsible for acetogenic hydrogenations may be smaller by two orders of magnitude in comparison with the methanogens present in the same samples [62].

Methanogens use only a relatively small number of organic compounds as an energy source, these include carbon dioxide, formic acid, acetic acid, methanol, methylamines and dimethyl sulphide. Some methanogens may also use carbon monoxide [38]. Therefore for mineralization of organic matter to biogas it is necessary that these compounds be converted to one of the above intermediates before conversion to methane. There are two groups of methanogens the aceticlastic methanogens that utilize acetic acid (such as *Methanothrix soehngenii, Methanosarcina* TM-1, *Methanosarcina acetivorans*) and the hydrogen utilizing methanogens that convert hydrogen and carbon dioxide. Among hydrogen utilizing methanogens, there are quite a few species that metabolize formic acid (e.g., *Methanococcus thermolithotrophicus, Methanobacterium formicicum*) and carbon monoxide (e.g., *Methanobacterium thermoautotrophicum*). Finally, some methanogens are capable of metabolizing almost every substrate mentioned above. *Methanosarcina barkeri* and *Methanococcus mazei*, for example, utilize everything but formic acid [19,38,59]. In an anaerobic digester about 70% of the methane is produced from acetate and the rest from hydrogen.

Dynamic mathematical models of anaerobic treatment processes have been developed to improve understanding of the complex ecosystem within these systems and to predict the response of the system to changes in feedstock and operating conditions [4, 9, 36, 37, 45, 57, 58]. Mathematical models can serve as a tool to formulate operational and control strategies for the system. Good strategies will reduce operating costs, improve process stability, and enhance treatment efficiency and throughput. Other potential uses of a model include assessment of new reactor designs, diagnosis of poorly performing systems and as soft sensors in decision support systems for plant operation.

11.4 Model Description

A mathematical model for a batch, single stage (one-vessel), leach bed anaerobic digestion process was constructed based on nonsteady state mass balance equations and physico-chemical equilibrium expressions. Mass balance equations are developed for components in both the liquid phase and gas phase. The metabolic processes modeled were those shown in Fig. 11.1. Among the processes described in Sect. 11.3, homoacetogenesis was ignored, and acetic acid and hydrogen were assumed to be the sole substrates for methane production. Data summarized in Table 11.3 show that the principal organic components of MSW are lignocellulosic materials and that these materials are not completely degradable. Based on these observations, it was assumed that the organic fraction of MSW consists of a refractory portion that does not degrade and the degradable portion is comprised of polysaccharides. The polysaccharides upon hydrolysis produce soluble substrates (which were lumped together as glucose, a sugar molecule). Glucose is a product of hydrolysis of cellulose. The anaerobic digester biochemical system is represented in Fig. 11.2, showing the principal components in the liquid and gas phases following the metabolic pathway in Fig. 11.1. Some of the compounds, the weak acids and bases dissociates in water producing ionized species, hydronium (H_3O^+ or H^+) and hydroxide ion (OH^-). The gases, carbon dioxide and hydrogen dissolve in water and there is a transfer of theses gases from liquid to gas phase or vice versa to maintain equilibrium. Particulate organic matter and microbial components has not been separately distinguished as a solid phase in the model. It is assumed that these components are homogenously distributed within the liquid phase and mass transfer limitations to and from the solid surface has

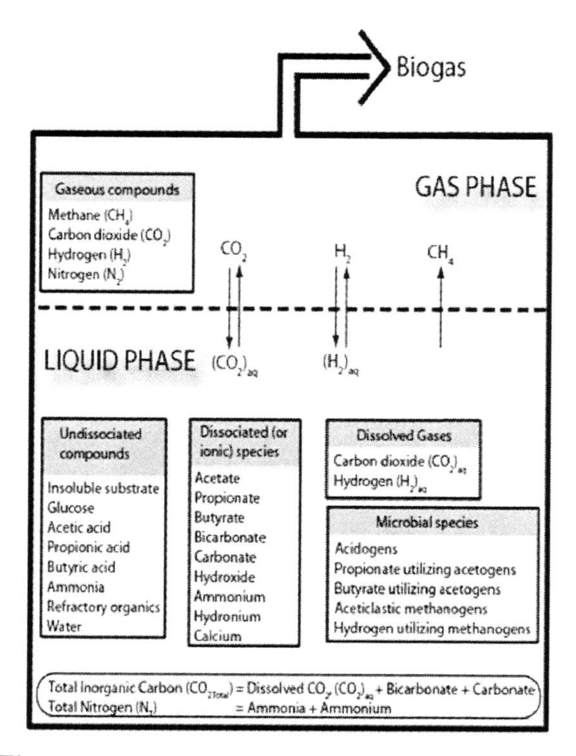

Fig. 11.2. Metabolic processes in anaerobic digestion

Table 11.6. List of state variables used in the model

State variables	Description	Unit
S_1	Insoluble substrate	$mg\ CODS\,l^{-1}$
S_2	Soluble substrate	$mg\ CODS\,l^{-1}$
S_3	Total acetic acid	$mg\ CODS\,l^{-1}$
S_4	Total propionic acid	$mg\ CODS\,l^{-1}$
S_5	Total butyric acid	$mg\ CODS\,l^{-1}$
S_6	Dissolved hydrogen $(H_2)_{aq}$	$mole\,l^{-1}$
S_7	Methane	$mole\,l^{-1}$
RS_1	Refractory component of S_1	$mg\ CODS\,l^{-1}$
X_1	Acidogens	$mg\ CODX\,l^{-1}$
X_2	Aceticlastic methanogens	$mg\ CODX\,l^{-1}$
X_3	Propionate utilizing acetogens	$mg\ CODX\,l^{-1}$
X_4	Butyrate utilizing acetogens	$mg\ CODX\,l^{-1}$
X_5	Hydrogen utilizing methanogens	$mg\ CODX\,l^{-1}$
y_{H2}	Hydrogen content in gas phase (volume fraction)	
y_{CH4}	Methane content in gas phase (volume fraction)	
y_{N2}	Nitrogen content in gas phase (volume fraction)	
y_{CO2}	Carbon dioxide content in gas phase (volume fraction)	
CO_{2Total}	Total inorganic carbon in liquid phase (i.e., sum of bicarbonate, carbonate and dissolved CO_2)	$mole\,l^{-1}$
Ca	Calcium ion	$mole\,l^{-1}$
N_T	Total dissolved nitrogen (i.e., sum of ammonia and ammonium ions)	$mole\,l^{-1}$

been ignored. The liquid and gas phase compounds listed in Table 11.6 are state variables in the model for which mass balance equations are written. The concentrations of the other components in the biochemical system can be determined from a knowledge of state variables and physico-chemical equilibrium relationships. In addition, ammonia has been included in the model as it is required for growth of microorganisms which is in turn supplied by the protein fraction of MSW. Ammonia is released during hydrolysis of proteins and as it is usually present in excess than that required for growth, it accumulates in the liquid. Other cations like alkali metals can be released during degradation of lignocellulosics in the MSW. These have been lumped together as calcium ions in the model (Ca).

11.4.1 Dynamic Mass Balance Equations

The mass balance equation for any compound (i) in the liquid phase (or active volume) of a batch reactor can be written as follows:

$$Rate\ of\ accumulation_i = Rate\ of\ formation_i - Rate\ of\ consumption\ (or\ loss)_i$$

Assuming uniform distribution of the compound within the reactor active volume or in other words a well mixed condition;

$$Rate\ of\ accumulation_i = \frac{d\,(V_L S_i)}{d\,t}$$

where V_L = active (or liquid) volume in reactor
$S_i = S_{1...7}$ or $X_{1...5}$ or CO_{2Total} or N_T or Ca in Table 11.6.

The rate at which substrate is consumed or product formed is proportional to the growth rate of the microbial species (or population) mediating the particular degradation, i.e.,

$$r_n = Y * Growth\ rate_i \quad \forall i = 1\,to\,5, n = i + 1$$

where

r_n = rate of consumption of substrates S_2 (soluble substrate), S_3 (acetic acid), S_4 (propionic acid), S_5 (butyric acid), S_6 (dissolved hydrogen)
Y = proportionality constant

The model includes five microbial species (X_1 to X_5) as carrying out degradation processes occurring in anaerobic digestion of OFMSW. The growth rate of these microbial species is assumed to follow Monod kinetics on limiting substrate and is written as:

$$Growth\ rate_i = \mu_i X_i$$

where

X_i = concentration of microbial biomass X_1, X_2, X_3, X_4 or X_5
μ_i = Specific growth of $X_i = \frac{\mu_{mi} S_n}{(K_{S,n} + S_n)} \quad \forall i = 1\ to\ 5,\ n = i + 1$
μ_{mi} = maximum specific growth rate of X_i
S_n = concentration of limiting substrate utilized by X_i for growth; X_1 grows on S_2, X_2 on S_3, X_3 on S_4, X_4 on S_5 and X_5 on S_6
$K_{s,n}$ = half saturation constant, i.e., the value of limiting substrate concentration where the specific growth rate equals ½ maximum specific growth rate.

The head space of the digester was comprised of hydrogen, carbon dioxide, methane and water vapor and these gases were assumed to be ideal and well mixed at a total pressure of 1 atm and 38°C. The partial pressure of water vapor was equal to the saturated vapor pressure at 38°C. Hydrogen, methane and carbon dioxide were generated from the microbial reactions. Methane was assumed to be insoluble in the liquid phase while dissolved hydrogen and carbon dioxide was transferred from the liquid phase to the gas phase to drive the system to equilibrium. The equilibrium dissolved concentration was determined by Henry's law:

$$g^* = \frac{y}{H}$$

where

g^* = concentration of dissolved gas (carbon dioxide or hydrogen) at equilibrium
y = partial pressure of hydrogen or carbon dioxide in head space
H = Henry law constant for hydrogen or carbon dioxide

The gas transfer rate is then expressed as

$$r_T = k_L a \cdot (\frac{y}{H} - g_{aq})$$

where

r_T = rate of transfer of dissolved gases (hydrogen or carbon dioxide), 11.14 and 11.23 in the Appendix
g_{aq} = concentration of gas (hydrogen or carbon dioxide) dissolved in liquid
$k_L a$ = mass transfer coefficient

A mass balance for a component (hydrogen, carbon dioxide or methane) in the headspace assuming ideal gas behavior for each gas component was written as

Rate of accumulation = Rate of flow from liquid phase – Rate of flow out of digester

$$Molar\ rate\ of\ accumulation\ of\ a\ gas\ i\ in\ headspace = \frac{d\,(n_i)}{dt}$$

where

n_i = moles of gas i in headspace (i = carbon dioxide, hydrogen, methane or nitrogen)

Since the gas composition is measured as a volume fraction, assuming ideal gas behavior, n_i can be written as

$$n_i = \frac{p_i\,V_G}{RT}$$

where

p_i = partial pressure of gas i (which is proportional of volume fraction, y_i)
V_G = head space volume
T = temperature of biogas (K)
R = universal gas constant

$$\frac{d\,(n_i)}{dt} = \frac{V_G}{RT}\,\frac{d\,(y_i)}{dt}$$

$$Molar\ rate\ of\ flow\ of\ gas\ i\ into\ headspace\ from\ liquid\ phase = r_i\,V_L$$

where

r_i = rate of transfer (per unit volume of liquid phase) of carbon dioxide or hydrogen from liquid, or the rate of production of methane in liquid phase, $r_i = 0$ for nitrogen.

$$Molar\ rate\ of\ flow\ of\ gas\ i\ out\ of\ the\ digester = \frac{q_G\, y_i\, P}{RT}$$

where

q_G = total volumetric biogas flow rate (Liters per unit time)
y_i = volume fraction of gas i in biogas
P = pressure of biogas (= 1 atm)

Substituting above expressions into the gas phase mass balance equation, yields equations 11.40, 11.41, 11.42 and 11.43 in the Appendix for hydrogen, methane, nitrogen and carbon dioxide respectively. Since methane is insoluble in liquid the rate of flow of methane in is equal to the rate of production of methane and since nitrogen is not produced in the reactions it is only expelled from the headspace.

The model also incorporated equilibrium relationships that described the dissolution of carbon dioxide, generation of alkalinity from breakdown of solids and the dissociation of the various weak organic acids (acetic, propionic and butyric acids). The physico-chemical reaction system was used to estimate the concentration of dissociated ionic species and thus the pH of the system. The implementation of charge balance in the liquid phase was used for the estimation of the pH. The rate of formation of calcium salts and total nitrogen was assumed to be directly proportional to the rate of hydrolysis of insoluble substrates. Other features of the model are described below.

11.4.2 Rate of Hydrolysis of Insoluble Substrates

Most models describe the hydrolysis step in the anaerobic digestion of organic matter using first order kinetics [4, 7, 9, 20, 50]. The first order relationship assumes that the rate of substrate removal is proportional to its concentration. The relationship requires only a substrate balance equation and one parameter, namely a hydrolysis rate constant. Therefore, the rate of hydrolysis neither depends on the biomass concentration nor the enzyme concentration. Moreover, in a batch leach bed process the concentration of insoluble substrate is highest at the beginning of the digestion step and if first order kinetics were true would result in a high rate of solubilization. This is also not observed in practice.

The hydrolysis reaction rate, i.e., the rate of solubilization of insoluble substrates was described by Contois function which is a modification of the Monod equation. Unlike Monod equation, the Contois expression incorporates the dependency on bacterial concentration in order to suppress possible over-estimation of the specific bacterial growth rate in a system like a batch digester loaded with MSW where substrate is not limiting initially. The expression for hydrolysis rate was written as

$$r_1 = \frac{k_h\, S_1}{(K_{x1}\, X_1 + S_1)} X_1$$

where

k_h = hydrolysis rate constant
K_{x1} = Contois constant
X_1 = concentration of acidogenic bacteria (which is responsible for hydrolysis of insoluble substrate)
S_1 = concentration of insoluble substrate

Justification for using the Contois equation is as follows. A kinetic model for the anaerobic digestion of organic feeds proposes that degradation takes place in three stages [3]. The organic matter is hydrolyzed into assimilable substrates, which are then transported into cells. The final stage is utilization of these substrates for cell growth and product formation. The kinetic equation describing the combined effect of these three steps was written as

$$\frac{\mu_m}{\mu} = \frac{K_S\, k\, Y}{K_h} \frac{(S_o - S)}{S} + \frac{K_S}{S} + 1$$

where

K_s = the half-saturation constant
k = hydrolyzed substrate transport first order rate coefficient
K_h = hydrolysis rate coefficient
Y = bacterial yield coefficient
S_o = substrate concentration in bulk liquid
S = substrate concentration.

During anaerobic digestion of complex feeds where $S \gg K_s$, the second term of the right hand side of above equation can be neglected and the equation becomes the Contois equation in which $K_{x1} = K_s k/K_h$. Accordingly, K_{x1} can be metabolically identified. Furthermore, in the case of readily degradable substrates in which K_h is very large value compared with the other values, the first term of the right hand side of above equation becomes negligible, and the equation reduces to the Monod equation. Several experimental results from the anaerobic digestion of complex organic wastes such as dairy manure, rice straw and cellulose show that the effluent substrate concentration depends on the influent substrate concentration in a continuous digester operating under steady state condition [8, 32, 34]. This is indicative of Contois type kinetics. Moreover, the above equation indicate that the digestion of insoluble substrate follows Contois-type kinetics whereas the digestion of soluble substrate such as acetic acid and propionic acid follows to a Monod kinetics.

11.4.3 Regulation and Inhibition by Molecular Hydrogen

Acidogenic fermentation produces several metabolic products like volatile and nonvolatile organic acids, alcohols and other chemicals. In an anaerobic digester treating carbohydrate feedstocks the primary products of acidogenic fermentation are volatile organic acids like acetic, propionic and butyric acids. Energy utilization by the acid forming anaerobic bacteria is maximized with the production of acetic acid along with hydrogen and carbon dioxide. In a properly operating anaerobic digester, the hydrogen is consumed by the hydrogen utilizing methanogen and converted to methane. If the hydrogen cannot be used, more reduced end products are produced to act as the electron sink. These reduced end products (usually propionic, butyric, and lactic acids) must then be removed from the system by being either washed out of the reactor or converted to other end products. It is thought that the ratios of formation of these volatile organic acids is regulated by the dissolved hydrogen concentration [6].

The regulation of acidogenesis has been linked to the presence of hydrogen generated by the acid forming bacteria. Hydrogen generated or consumed at the reaction sites within the bacteria is controlled by redox couples, the most common of which are the pyridine nucleotides, characterized by the nicotinamide adenine dinucleotide (NAD $\leftrightarrow$ NADH) couple. It has been estimated that for the anaerobic degradation of sewage sludge approximately 63% of the total hydrogen evolution would occur by electron transfer via pyridine nucleotides [26]. Hydrogen build up can inhibit acidogenesis.

The acetogenic bacteria produce acetic acid from the reduced end products of the acidogenic fermentations (propionic and butyric acids), with the release of electrons as hydrogen gas. These bacteria are also inhibited by the build-up of hydrogen. Two possible mechanisms of inhibition have been proposed: firstly, free-energy inhibition [27]; and secondly, electron transfer to produce hydrogen via the pyridine nucleotide redox system [26]. Unlike the acid-forming bacteria, the free energy for the metabolic reactions of the acetogenic bacteria are very small and are negative only at low hydrogen partial pressures. Hence, any significant build-up of hydrogen may stop the conversion of the propionic and butyric acids to acetic acid when the free energy for the conversion reactions becomes positive. There is also the possibility that the bacteria could find an alternative electron sink for the hydrogen, (such as inorganic electron acceptors) thereby allowing the substrate reaction to continue. Even if this was the case, the $NAD^+ \leftrightarrow NADH$ redox reaction would still remain an important inhibiting mechanism of the bacteria [60]. It was proposed that NADH could be oxidized with the corresponding production of the hydrogen gas via the following redox reaction:

$$NADH + H^+ \leftrightarrow NAD^+ + 2H_2$$

A model for hydrogen inhibition in anaerobic bacteria was proposed in which dissolved hydrogen concentration was related by the Nernst equation to the redox reactions of pyridine nucleotides (characterized by the $NAD^+ \leftrightarrow NADH$ couple) in the bacteria. Neglecting any resistance to mass transfer of hydrogen between gas and liquid phase, hydrogen in the gas phase of the reactor determines the ratio of oxidized to reduced NAD within the bacteria [33]. The ratio of NADH to NAD^+ was then simplified in terms of the partial pressure of hydrogen in the gas phase as follows [15]

$$zn = \frac{[NADH]}{[NAD^+]} = 2000 p_{H_2}$$

However, measurements of dissolved and gas phase hydrogen concentrations have shown that the dissolved hydrogen is not in equilibrium with partial pressure of hydrogen in the gas phase [18]. In the current model dissolved hydrogen concentration was used to calculate the ratio of NADH to NAD^+ as shown in 11.1 in the Appendix. A hydrogen inhibition term $\left(\frac{1}{1+zn}\right)$ was incorporated into rate expressions for acid formation (11.5), propionic acid utilization (11.7) and butyric acid utilization (11.8).

11.4.4 pH Inhibition

Operational failure of an anaerobic digester is usually associated with a fall in the pH and the subsequent death or inhibition of microbial consortia by the increasing hydrogen ion concentration. Therefore, anaerobic digesters typically incorporate process equipment for pH control. Process failure can also be caused by the rise in the pH to above 8.5; this usually only occurs if there are problems with the automatic dosing system and too much caustic is added to the reactor. A rise in pH will affect the more sensitive methanogens, causing an accumulation of volatile acids, making this problem somewhat self-regulatory and less likely to result in the complete failure of the process. Such a disturbance would, however, destabilize a reactor and in short term make it more susceptible to other modes of failure.

The most common type of failure in anaerobic treatment systems is caused by a substrate shock loading to the process resulting in the accumulation of acidic products and a fall in pH below 6.8. At this level most methanogenic bacteria are unable to effectively utilize acetic acid, causing it to accumulate in the reactor, resulting in a further fall in the pH. Specification of the hydrogen ion concentration at which pH inhibition occurs is very difficult because of the ability of anaerobic bacteria to adapt to their environmental conditions [14, 17]. Most studies on the inhibition of anaerobic treatment systems by hydrogen ions have concentrated on the methanogenic bacteria, specifically the aceticlastic bacteria. They are assumed to be most sensitive of all types of the anaerobic bacteria and are considered to play a pivotal role in maintaining the stability of an anaerobic reactor. Generally methane production has been shown to be maximum in a pH range of 7.0–7.2 and that methane production falls off on either side of this optimum range to a 50% level at plus or minus approximately 1 or 1.5 pH units [14,17]. It has also been shown that in mixed culture systems the optimum growth of a population of glucose consuming acidogenic bacteria occurred in a pH range of 5.8–6.2. There was a 50% decrease in the growth rate at a pH of 5.0, while for pH levels greater than 6.0 there was a gradual decrease in biological activity down to 25% at a pH of 8.0 [63]. A similar optimum pH range was found for an acidogenic population degrading a complex waste [16]. The optimum pH was found to be at 7.0, with significant acidification of the wastewater occurring within the range of 5.0–8.0.

In the model presented here, microbial growth was assumed to be inhibited by pH and two types of pH inhibition functions were used, one for acidic pH and another for alkaline pH [44].

$$pHin_i = \exp\left(-0.5\left(\frac{pH - pHm_i}{pHsd_i}\right)^2\right), \quad pH < 7, \text{ acidic range}$$

$$pHin_i a = \exp\left(-0.5\left(\frac{pHm_i a - pH}{pHsd_i a}\right)^2\right), \quad pH > 7, \text{ alkaline range}$$

where, i = 1, 2, 3, 4, 5 corresponding to X_1 to X_5.

pHm_i and $pHsd_i a$ are parameters that respectively determine the optimum pH for a bacterial group and the slope of the pH inhibition function. The pH inhibition function is incorporated into substrate utilization kinetics for acid formers (11.5), acetic acid (11.6), propionic acid (11.7), butyric acid (11.8) and hydrogen utilization (11.9).

11.4.5 Product Inhibition

Product inhibition can be characterized as an inhibitory response to accumulation of products excreted by the bacteria, which would depend upon the concentration of product relative to the amount of biomass present. In some cases a lowering of the free energy of the substrate reaction by the accumulating product is a major cause of the product inhibition. In this case the ratio of product to substrate would be more important than the absolute concentration of the product. Free-energy mitigated product inhibition is likely to occur in bacteria that utilize substrate reactions with a small physiological free energy. In an anaerobic ecosystem these bacteria would be the propionic and butyric acid utilizing acetogens. Hydrogen inhibition of these bacterial groups have been incorporated into this model. The acidogens and hydrogen-utilizing methanogens have large negative free energies for their substrate reactions and consequently any apparent product inhibition could only be caused by a general toxic or inhibitory effect of the product on the growth of these bacteria. Product inhibition for acidogens was included in this model to account for accumulation of organic acids at a neutral pH. A noncompetitive inhibition term was included to describe accumulation of acetic acid at high concentrations (11.5).

11.4.6 Model Assumptions

Other assumptions used in the construction of the model are:

1. The batch system operates at a constant temperature (38°C).
2. Active (liquid) volume of the digester is constant (V_L).
3. No physical losses.
4. No oxygen contamination.
5. No evaporative losses in the system (gas outlet fitted with cooling coil).
6. System at constant pressure (1 atm).
7. All physico-chemical equilibrium constants are fixed.
8. Gas in head space is well mixed and at constant volume (V_G), temperature (38°C) and pressure (1 atm).
9. All gases behave as ideal gas.
10. Partial pressure of water vapor is equal to saturated vapor pressure at 38°C.
11. Methane and hydrogen gas are insoluble in water.
12. The system described is well mixed and there is no spatial variation in the digester volume.
13. There are no aerobic reactions.
14. The degradation of proteins and lipids is not considered, i.e., only the degradation of carbohydrates is considered.

This model is a lumped parameter model. The aim of the model was to simulate the degradation of fresh MSW in flooded conditions. This model can be used as a tool to predict the degradation rate of MSW in terms of methane gas production, biogas content, volatile fatty acid production and the pH of the liquid phase. The model was used to study the sensitivity of initial conditions such as the initial bacterial concentrations and the initial buffer concentration of a flooded MSW bed.

11.5 Selection of Parameters

The parameters required for the model were classified into four groups and they are:

1. Yield coefficients (product and biomass)
2. Constants (experimental and equilibrium)
3. Fitting parameters (biological, yield coefficients of calcium and total nitrogen from insoluble substrates and mass transfer coefficients of carbon dioxide and hydrogen gas) and
4. Initial values of state variables

The proportionality constant, Y, that relates the growth rate of microbial species to the rate of substrate utilization or product formation, in other words the product and biomass yield coefficients, were calculated using the

Table 11.7. The stoichiometry of catabolic and anabolic reactions used in the model

Acidogens	
$C_6H_{12}O_6 + 2H_2O \rightarrow 2CH_3COOH + 2CO_2 + 4H_2$	$+$ 4ATP
$C_6H_{12}O_6 + 2H_2 \rightarrow 2CH_3CH_2COOH + 2H_2O$	$+$ 2ATP
$C_6H_{12}O_6 \rightarrow CH_3CH_2CH_2COOH + 2CO_2 + 2H_2$	$+$ 2ATP
$5C_6H_{12}O_6 + 6NH_3 \rightarrow 6C_5H_9O_3N + 12H_2O$	
Overall: $C_6H_{12}O_6 + 0.1740NH_3 \rightarrow 0.1740C_5H_9O_3N + 0.5700CH_3COOH + 0.5700CH_3CH_2COOH + 0.2850CH_3CH_2CH_2COOH + 1.140CO_2 + 1.140H_2 + 0.3481H_2O$	
Propionic acid bacteria	
$CH_3CH_2COOH + 2H_2O \rightarrow CH_3COOH + CO_2 + 3H_2$	0.5 ATP
$3CH_3CH_2COOH + CO_2 + 2NH_3 \rightarrow 2C_5H_9O_3N + 2H_2O + H_2$	
Overall: $CH_3CH_2COOH + 1.8556H_2O + 0.0361NH_3 \rightarrow 0.0361C_5H_9O_3N + 0.9459CH_3COOH + 0.9278CO_2 + 2.8556H_2$	
Butyric acid bacteria	
$CH_3CH_2CH_2COOH + 2H_2O \rightarrow 2CH_3COOH + 2H_2$	0.75 ATP
$CH_3CH_2CH_2COOH + CO_2 + NH_3 \rightarrow C_5H_9O_3N + H_2O$	
Overall: $CH_3CH_2CH_2COOH + 1.8376H_2O + 0.0542CO_2 + 0.0542NH3 \rightarrow 0.0542C_5H_9O_3N + 1.8917CH_3COOH + 1.8917H_2$	
Aceticlastic methanogen	
$CH_3COOH \rightarrow CH_4 + CO_2$	0.25 ATP
$5CH_3COOH + 2NH_3 \rightarrow 2C_5H_9O_3N + 4H_2O$	
Overall: $CH_3COOH + 0.0182NH_3 \rightarrow 0.0182C_5H_9O_3N + 0.9545CH_4 + 0.9545CO_2 + 0.0364H_2O$	
Hydrogen utilizing methanogen	
$4H_2 + CO_2 \rightarrow CH_4 + 2H_2O$	1.0 ATP
$5CO_2 + 10H_2 + NH_3 \rightarrow C_5H_9O_3N + 7H_2O$	
Overall: $H_2 + 0.290CO_2 + 0.016NH_3 \rightarrow 0.016C_5H_9O_3N + 0.532H_2O + 0.210CH_4$	

stoichiometry of catabolic and anabolic reactions of substrate consumption. The biomass yield coefficient for each consumption reaction was calculated based on the theoretical energy yield in which 10 g of biomass is produced per mole of ATP generated [15]. The empirical formula of the biomass was assumed to be $C_5H_9O_3N$ [33]. Table 11.7 lists the stoichiometric equations of substrate consumption for each bacterial group and the corresponding biomass formation stoichiometry. The overall stoichiometry for each reaction was arrived at by combining the catabolic and anabolic stoichiometries by replacing the moles of ATP produced with the equivalent mass of cellular biomass. To

Table 11.8. List of product and biomass yield coefficients used in the model

Yield	*Description*	*Value*
Y_{S3S2}	Acetic acid yield on glucose	0.1907 mg CODS/mg CODS
Y_{S4S2}	Propionic acid yield on glucose	0.3326 mg CODS/mg CODS
Y_{S5S2}	Butyric acid yield on glucose	0.2375 mg CODS/mg CODS
Y_{S3S4}	Acetic acid yield on propionic acid	0.5405 mg CODS/mg CODS
Y_{S3S5}	Acetic acid yield on butyric acid	0.7570 mg CODS/mg CODS
Y_{H2S2}	Hydrogen yield on glucose	5.937×10^{-6} mole H_2/mgCODS
Y_{H2S4}	Hydrogen yield on propionic acid	2.549×10^{-5} mole H_2/mgCODS
Y_{H2S5}	Hydrogen yield on butyric acid	1.182×10^{-5} mole H_2/mgCODS
Y_{CO2S2}	Carbon dioxide yield on glucose	5.937×10^{-6} mole CO_2/mgCODS
Y_{CO2S3}	Carbon dioxide yield on acetic acid	1.491×10^{-5} mole CO_2/mgCODS
Y_{CO2S4}	Carbon dioxide yield on propionic acid	8.281×10^{-6} mole CO_2/mgCODS
Y_{CO2S5}	Carbon dioxide yield on butyric acid	3.385×10^{-7} mole CO_2/mgCODS
Y_{CO2S6}	Carbon dioxide yield on hydrogen	0.290 mole CO_2/mole H_2
Y_{CH4S3}	Methane yield on acetic acid	1.491×10^{-5} mole CH_4/mgCODS
Y_{CH4S6}	Methane yield on hydrogen	0.210 mole CH_4/mole H_2
Y_{X1S2}	Acidogen yield on glucose	0.2333 mgCODX/mgCODS
Y_{X2S3}	Aceticlastic methanogen yield on acetic acid	0.0523 mgCODX/mgCODS
Y_{X3S4}	Propionic acid bacteria yield on propionic acid	0.0593 mgCODX/mgCODS
Y_{X4S5}	Butyric acid bacteria yield on butyric acid	0.0623 mgCODX/mgCODS
Y_{X5S6}	Hydrogen utilizing methanogen yield on hydrogen	2,926/mole H_2

derive the overall acidogenic catabolic stoichiometry the acetic, propionic and butyric acid reactions were combined on a 1:1:1 ratio.

Table 11.8 lists of biological yield coefficients (both product and biomass) calculated from the stoichiometries in Table 11.7. All the yield coefficients, including the yield on biomass were based on COD in terms of mass with the exception of hydrogen, carbon dioxide and methane gas which were expressed in terms of mole. The unit for the yield coefficients for substrate and biomass was mgCODS and mgCODX respectively. The values for the experimental constants (measured values) and the constants for all equilibrium expressions for the dissociated ionic species are shown in Table 11.9. All equilibrium constants for the acid–base pairs were based on literature values and the references are shown in the same table. The constants for the pH inhibition function were derived using a parameter estimation routine developed by Ramsay [44].

The kinetic parameters for the biological processes during anaerobic digestion were obtained from batch MSW digestion experiments [36]. These experiments also yielded refractory fraction of insoluble component (fr). Initial

Table 11.9. List of constants used in the model

Constant	*Description*	*Value*	*Ref.*
C1	Conversion factor (mgCOD to mole acetic acid)	64,020	
C2	Conversion factor (mgCOD to mole propionic acid)	112,036	
C3	Conversion factor (mgCOD to mole butyric acid)	159,984	
R	Universal gas constant	0.08206 l*atm/(mol*K)	
V_L	Liquid volume	80.5 l	
V_G	Gas head space	120 l	
T	Reactor temperature	311.2 K	
Y_{H2O}	Moisture content in gas phase	0.0562 atm	[23]
k2	Equilibrium constant for dissociation of acetate ion	3.2×10^{-5} mole l^{-1}	[31]
k3	Equilibrium constant for dissociation of bicarbonate ion	5.0×10^{-7} mole l^{-1}	[31]
k4	Equilibrium constant for dissociation of carbonate ion	5.0×10^{-11} mole l^{-1}	[31]
k5	Equilibrium constant for dissociation of ammonium ion	1.1×10^{-9} mole l^{-1}	[35]
kp	Equilibrium constant for dissociation of propionate ion	1.31×10^{-5} mole l^{-1}	[47]
kb	Equilibrium constant for dissociation of butyrate ion	1.44×10^{-5} mole l^{-1}	[47]
H_{CO2}	Henry's Law constant for CO_2	40.82 atm*l mol^{-1}	[35]
H_{H2}	Henry's Law constant for H_2	1334 atm*l mol^{-1}	[18]
$pHm_1/_1a$	Mean pH parameter for X_1	6.00	[44]
$pHm_2/_2a$	Mean pH parameter for X_2	6.00	[44]
$pHm_3/_3a$	Mean pH parameter for X_3	6.00	[44]
$pHm_4/_4a$	Mean pH parameter for X_4	6.00	[44]
$pHm_5/_5a$	Mean pH parameter for X_5	7.00	[44]
$pHsd_1/_1a$	Std. dev pH parameter for X_1	0.85	[44]
$pHsd_2/_2a$	Std. dev pH parameter for X_2	0.635	[44]
$pHsd_3/_3a$	Std. dev pH parameter for X_3	0.635	[44]
$pHsd_4/_4a$	Std. dev pH parameter for X_4	0.635	[44]
$pHsd_5/_5a$	Std. dev pH parameter for X_5	0.42	[44]

bacterial concentrations and the soluble substrate inhibition parameter were calibrated with methane production data from a sequencing leach bed experiment in which a leachate volume equal to 10% of the bed volume was sequenced [29]. The fitting parameters included the biological parameters, yield coefficients of calcium and total nitrogen from insoluble substrates and mass transfer coefficients of carbon dioxide and hydrogen gas. All the fitting parameters required for the model are listed in Table 11.10. The biological

Table 11.10. List of fitting parameters used in the model

Parameters	*Description*	*Value*	*Ref.*
fr	Refractory fraction of insoluble substrate	0.59–0.61	[29]
Y_{Ca} [1]	Calcium yield	1.45×10^{-6} mole/mgCODS	[35]
$Y_{N,T}$	Nitrogen yield	1.00×10^{-6} mole/mgCODS	[35]
k_h [1]	Hydrolysis rate constant	8.0 mgCODS/mgCODX/day	[35]
K_{x1}	Contois constant	35 mgCODS/mgCODX	[35]
μ_{m1}	Max growth rate of acidogens	2.75 1 per day	[35]
μ_{m2} [1]	Max growth rate of aceticlastic bacteria	0.36 1 per day	[35]
μ_{m3} [1]	Max growth rate of propionic acid bacteria	0.80 1 per day	[47]
μ_{m4} [1]	Max growth rate of butyric acid bacteria	6.90 1 per day	[47]
μ_{m5}	Max growth rate of H_2 utilizing bacteria	1.39 1 per day	[35]
$K_{S,2}$	Half-velocity constant for X_1	275 mgCODS/l	[35]
$K_{S,3}$	Half-velocity constant for X_2	360 mgCODS/l	[35]
$K_{S,4}$	Half-velocity constant for X_3	247 mgCODS/l	[47]
$K_{S,5}$	Half-velocity constant for X_4	154 mgCODS/l	[47]
$K_{S,6}$	Half-velocity constant for X_5	7.50×10^{-7} mole H_2/l	[35]
Kin [1]	Product inhibition factor	18000 mgCODS/l	[35]
k_{d1} [1]	Biomass death rate –X_1	0.048 1 per day	[35]
k_{d2}	Biomass death rate –X_2	0.101 1 per day	[35]
k_{d3} [1]	Biomass death rate –X_3	0.01 1 per day	[47]
k_{d4} [1]	Biomass death rate –X_4	0.03 1 per day	[47]
k_{d5} [1]	Biomass death rate –X_5	0.048 1 per day	[35]
kLa_{CO2}	CO_2 Transfer coefficient	100 1 per day	[35]
kLa_{H2} [1]	H_2 Transfer coefficient	12 1 per day	[18]

[1] parameters that required modification

parameters used in the model were the kinetic parameters used to describe the growth of each bacterial group. The Monod kinetic parameters include the maximum growth rate (μ_m), the half-velocity constant (K_S), the biomass yield coefficient ($Y_{X/S}$) and the first order death coefficient (k_d). The value of all other fitting parameters were derived from different literature sources (reference shown in Table 11.10). Some of the parameters were modified to improve model predictions of methane yield from the degradation of insoluble substrates. The modified parameters included Y_{Ca}, k_h, μ_{m2}, μ_{m3}, μ_{m4}, Kin,

k_{d1}, k_{d3}, k_{d4}, k_{d5} and kLa_{H2}. Some of these fitting parameters were more sensitive than the others and μ_{m2} was found to be the most sensitive parameter. The next section discusses the method used in parameter estimation and the outcomes of the sensitivity analysis. The values of the remaining fitting parameters (Y_{NH3}, K_{x1}, μ_{m1}, μ_{m5}, K_{S2}, K_{S3}, K_{S4}, K_{S5}, K_{S6}, k_{d2} and KLa_{CO2}) remained unchanged as they were found to be insensitive to the model outputs.

The initial conditions for some of the state variables were listed as a different set of parameters. The initial biomass concentrations ($X_{1,0}$ to $X_{5,0}$) could not be quantified experimentally and literature values on the indigenous bacterial concentration of MSW were not available, therefore these were estimated from experiments.

11.6 Model Implementation and Simulation

Anaerobic digestion process display behavior which can be characterized by two groups of time constants in the system. The growth rate of all bacterial groups are slow whilst the physico-chemical equilibrium reactions are very fast. The slow reactions are associated with large time constants while the fast reactions are associated with small time constants of the system. This physical behavior is good for propagation of error in the ODE set but not for a numerical method as the solution curves converges too quickly and this will eventually lead to an ill-conditioned behavior or a "*stiff*" problem [21]. The ODE set of this model was integrated using the *ODE15s* solver which is a built-in function of *MATLAB*. This *ODE15s* solver uses a backward differentiation formulae (BDF) method to solve for the "*stiffness*" problem. This BDF method is implemented in variable step-variable order mode so that the computational effort is optimized [21]. The main advantages of the BDF method are firstly, the high efficiency in solving complex ODE due to the lower number of evaluations of the equations compared to single step methods (e.g., Runge-Kutta technique which is used in other *MATLAB* solvers, e.g., *ODE45*) and secondly, the increase in demand of accuracy does not result in an increase in simulation time. The model equations (algebraic and state expressions) were coded in *MATLAB*. Minimum relative error tolerance was chosen in the driver routine to maximize the accuracy of the model predictions.

The pH of the leachate was an important parameter in this model as the growth rate of all bacterial groups were pH dependent. The rate of change of hydrogen ion concentration (mole) was monitored to model the pH of the liquid phase. A charge balance approach was initially applied to solve for the concentration of hydrogen ion which was assumed to be equal to the difference between the sum of the negatively charged ions and the positively charged ions. The physico-chemical reaction system incorporated equilibrium relationships that describe the dissolution of carbon dioxide, generation of alkalinity from breakdown of solids and the dissociation of volatile fatty acids.

Since all these equilibrium reactions could only be expressed in algebraic form, implementation of a subroutine along with the main program (which solves the differential equation sets) was required to solve for the charge balance numerically. Although this was a common technique used in solving this type of DAE system, it was not employed in this modeling exercise. This was because the implementation of a subroutine program would result in prolonged simulation time and possible numerical instability. An alternative method was developed to solve for the pH of the liquid phase of the system. The charge balance was first described in a differential form which is shown in 11.48 of Appendix. The terms on the right hand side of this equation are the derivatives of the explicit description of the dissociated ionic species which are shown in (11.15)–(11.21). To be able to integrate the charge balance in differential form directly, all the explicit descriptions of the dissociated ionic species were differentiated. The following equations show the differentiated form of the explicit description of the dissociated ionic species.

$$2\frac{dCO_3^{2-}}{dt}+\frac{dHCO_3^-}{dt}=\frac{dH^+}{dt}\left(\frac{k_3CO_{2Total}\left(\left(k_3k_4+k_3H^++\left(H^+\right)^2\right)-\left(H^++2k_4\right)\left(k_3+2H^+\right)\right)}{\left(k_3k_4+k_3H^++(H^+)^2\right)^2}\right)$$

$$+\frac{dCO_{2Total}}{dt}\left(\frac{k_3\left(H^++2k_4\right)}{\left(k_3k_4+k_3H^++(H^+)^2\right)^2}\right)$$

$$\frac{dOH^-}{dt}=\frac{dH^+}{dt}\left(-\frac{k_w}{(H^+)^2}\right)$$

$$\frac{dAc^-}{dt}=\frac{(k_2+H^+)\,k_2\frac{dS_3}{dt}-k_2S_3\frac{dH^+}{dt}}{(k_2+H^+)^2}$$

$$\frac{d\,\mathrm{Pr}^-}{dt}=\frac{(k_p+H^+)\,k_p\frac{dS_4}{dt}-k_pS_4\frac{dH^+}{dt}}{(k_p+H^+)^2}$$

$$\frac{dBu^-}{dt}=\frac{(k_b+H^+)\,k_b\frac{dS_5}{dt}-k_bS_5\frac{dH^+}{dt}}{(k_b+H^+)^2}$$

$$\frac{dNH_4^+}{dt}=\frac{H^+\,(k_5+H^+)\,\frac{dN_T}{dt}+k_5N_T\frac{dH^+}{dt}}{(k_5+H^+)^2}$$

where

CO_3^{2-} = carbonate ion
HCO_3^- = bicarbonate ion
OH^- = hydroxide ion
Ac^- = acetate ion
Pr^- = propionate ion
Bu^- = butyrate ion

By expressing the explicit description of the dissociated ionic species in the above format, the differential form of the hydrogen ion concentration expression (or conservation of charge balance) could be solved directly by separating the hydrogen ion differentials with all other terms. The differential form of charge balance became the last equation for the conservation balance after the rearrangement of terms within the equation. It should be noted that the right hand side of above equations only include state variables, the differential of state variables and equilibrium constants, this implies that once the initial condition of the state variables are provided, the charge balance conservation equation can be integrated directly. The time required to carry out the simulation was reduced significantly after the incorporation of this method together with the employment of the *MATLAB ODE15s* solver. The average simulation time was 2 min as opposed to hours required to solve a DAE system using the same personal computer.

11.7 Model Validation

The model was validated using data obtained from leach bed anaerobic digestion experiments carried out in 200 l stainless steel reactors (Fig. 11.3). The waste bed was supported on a stainless steel mesh screen with a net open area of 41%. A liquid distributor was installed on the bottom of the lid of the reactor. It consisted of four stainless steel arms which had regular openings for liquid discharge. The outer surfaces of the reactor were insulated using 50 mm, roll-faced glass wool which was covered by an aluminum casing. To reduce moisture from entering the gas meter, a stainless steel air-cooling coil was fitted as an extension to the gas outlet. Volumetric gas production from each reactor was measured using a positive displacement gas meter. The temperature of the MSW bed was controlled using a 450 W heating tape (KTeS series type, *ISOPAD GmbH*, Germany). The heating tape was mounted on the outer surface of the reactor. Each reactor was equipped with a Type 'T' thermocouple which had a miniature head consisting of 3 mm stainless steel insertion sheath. The thermocouple was interfaced to a personal computer with *Labtech Pro* and *Realtime vision* software (Laboratory Technologies Corporation, USA) installed. The temperature was measured through the thermocouple and the data was recorded using the software. A proportional–integral control algorithm was set up to control the temperature of the waste bed at $38 \pm 2°C$ using the heating tape. Each reactor was connected to an external perspex tank with a holding capacity of 24 l. This tank was pressure compensated with the reactor. Both the reactor and the tank were fitted with two polypropylene submersible pumps (Tauchpumpe, Nr.511.0412, *Mocar GmbH*, Germany) in which one served as a spare pump. During the transfer process, leachate was pumped from the reactor to the external collection tank and the volume was recorded. The leachate was then pumped from the collection tank

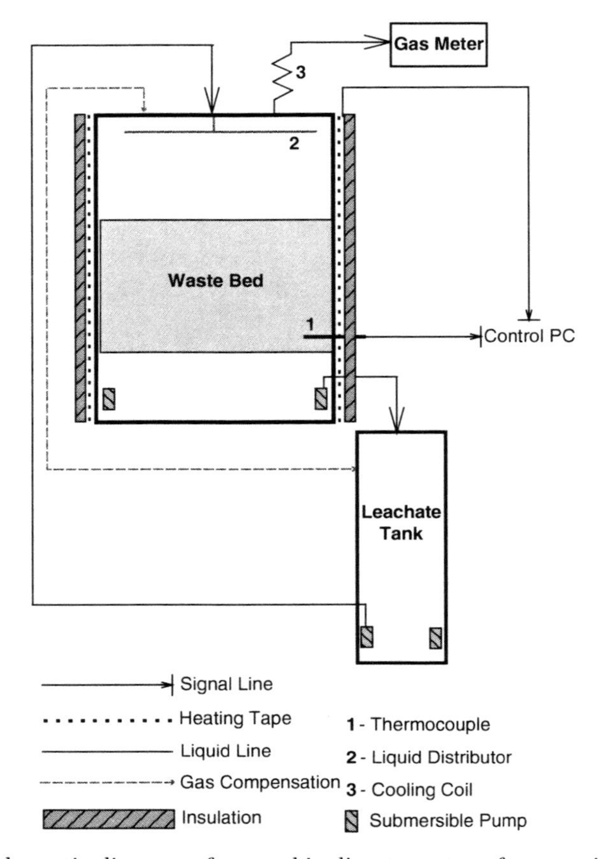

Fig. 11.3. Schematic diagram of anaerobic digester set up for experimental validation of model

into the reactor through the liquid distributor. Complete details of the design of these reactors can be found in [10].

Experiments were carried out using unsorted coarsely shredded municipal solid waste collected from a municipal transfer station. To minimize variations in feedstock between experiments, 1.5 tonnes of MSW was collected, shredded to average size of 5–10 cm and stored in polypropylene drums in a commercial freezer. Each drum held about 30 kg of feedstock. When required a drum was removed from the freezer, thawed overnight and loaded into the digester.

Three experiments were conducted to collect data for model validation. It was previously shown that anaerobic digestion can be initiated in a bed of MSW by simply flooding the bed with a solution containing pH buffer [29]. Therefore experiments carried out here employed this concept. Prior to loading the waste 40 l of sodium bicarbonate ($NaHCO_3$) solution was pumped into the reactor so as to fill up the leachate collection area below the screen.

The concentration of $NaHCO_3$ used was the same in each experiment. In experiment 1, 27.2 kg of waste was packed in the digester at a bulk density of 441 kg m^{-3}. Thirty liters of 11.21 g l^{-1} (equivalent to an alkalinity of 6,000 mg $CaCO_3$ l^{-1}) $NaHCO_3$ (sodium bicarbonate) solution was pumped into the waste bed so as to flood it (total buffer solution volume is 70 l). Liquid was not exchanged for the rest of the duration of experiment 1. In experiment 2, 30.4 kg of waste was packed in the digester at a bulk density of 433 kg m^{-3}. Thirty liters of 4.67 g l^{-1} (equivalent to an alkalinity of 2,500 mg $CaCO_3$ l^{-1}) $NaHCO_3$ (sodium bicarbonate) solution was pumped into the waste bed so as to flood it. This yielded a total of 70 l of buffer solution in reactor. On day 1, 10 l of liquid from digester was replaced with 10 l of $NaHCO_3$ solution. On day 2, 19 l of liquid in digester was replaced with 19 l of $NaHCO_3$ solution. This was repeated for 11 more days until day 13. The liquid in the digester was not exchanged for the rest of the duration of experiment 2. Experiment 3 verified the capability of the model in predicting recovery of an imbalanced digestion process (i.e., situations where rate of acidogenesis exceeds methanogenesis). In this experiment, 26.3 kg of waste was packed at a bulk density of 441 kg m^{-3}. The bed was moistened with tap water and left undisturbed for 90 days. On day 91, 75 l of 11.21 g l^{-1} $NaHCO_3$ solution was pumped into the reactor. At the beginning of the experiment there was a total of 85 l of free liquid in the reactor. Liquid was not exchanged for the rest of the duration of experiment 3. Both liquid and gas samples were withdrawn daily during the experiments. The liquid samples were analyzed for pH, and acetic, propionic and butyric acids [29]. These acid concentrations was summed up and reported as total VFA (volatile fatty acid) concentration in units of mg COD/l. The gas samples were analyzed for methane, carbon dioxide and hydrogen [29]. In addition, gas production rate was monitored on-line. Methane production was then determined as the product of total biogas produced daily multiplied by methane concentration. The model predictions of cumulative methane production, methane composition, pH of leachate and total VFA was compared to experimental data.

One set of experimental results was used for parameter estimation and two separate sets of experimental results were used for model validation. Parameter estimation was carried out using the results of Experiment 1. The model was initially run using the set of parameters listed in Tables 11.8, 11.9 and 11.10. The model parameters were then tuned using a sensitivity analysis approach. During the investigation of the sensitivity of each parameter, model predictions and experimental results were compared qualitatively. Model predictions on daily and cumulative methane production, daily biogas content, total VFAs production and pH of the leachate were all compared with experimental data. Each fitting parameter (those listed in Table 11.10) was initially tuned by doubling or halving its magnitude manually, the effects of increasing and decreasing the fitting parameter were then monitored by comparing the model predictions with the experimental data. Different results collected during the course of experiment 1 that were used for the estimation of parameters

Table 11.11. The estimation of parameters using different results from experiment 1

Parameter	*Experimental results*
Y_{Ca} (yield of calcium salts)	Carbon dioxide gas content trend
k_h (hydrolysis rate constant)	Lag phase of methane production
Kin (glucose inhibition constant)	curve
k_{d1} (acidogens death rate constant)	
KLa_{H2} (mass transfer coefficient of H_2)	Hydrogen gas content trend
μ_{m3} (maximum growth rate of propionic acid bacteria)	Daily methane rate trend, cumulative methane production profile and VFAs production profile
μ_{m4} (maximum growth rate of butyric acid bacteria)	
$k_{d3} - k_{d5}$ (death rate constant of propionic acid bacteria, butyric acid bacteria and hydrogen-utilizing bacteria)	
X_{i0} (initial concentration of all five bacterial groups)	

is summarized in Table 11.11. The tuning of each fitting parameter was continued until the fitting of model outputs on experimental data was qualitatively satisfied. The values of some fitting parameters were increased while some were decreased depending on their effects on the model predictions. Fine tuning on some fitting parameters was required as they were relatively sensitive. Based on the findings of the parameter estimation exercise, it was concluded that some model parameters could be applied for simulation without any alteration and they were Y_{NH3}, K_{x1}, μ_{m1}, μ_{m5}, K_{S2}, K_{S3}, K_{S4}, K_{S5}, K_{S6}, k_{d2} and KLa_{CO2} while the rest became the new fitting parameters. These fitting parameters included Y_{Ca}, k_h, μ_{m2}, μ_{m3}, μ_{m4}, Kin, k_{d1}, k_{d3}, k_{d4}, k_{d5} and KLa_{H2}. Some of these fitting parameters were more sensitive than the others and μ_{m2}, was found to be the most sensitive parameter. The initial concentrations of all bacterial groups (acidogens, aceticlastic methane bacteria, propionic acid and butyric acid bacteria and hydrogen utilizing bacteria) were also categorized as fitting parameters because actual biomass concentrations could not be determined experimentally and literature values were not available.

The tuned parameters are listed in Table 11.12 and it should be noted that the initial concentration of all five bacterial groups were found to be sensitive. The estimation of the maximum specific growth rate of aceticlastic methane bacteria (μ_{m2}) was found to be very crucial as the predictions of the degradation of insoluble carbohydrates (in terms of methane gas and VFAs production and leachate pH) depended heavily on this parameter. The maximum specific growth rate of butyric acid bacteria (μ_{m4}) was reduced from 6.90 d^{-1} to 0.95

Table 11.12. The fitting parameters estimated using data from experiment 1

Parameters	*Description*	*Estimated value*
Y_{Ca}	Calcium yield	4.0×10^{-7} mole/mgCODS
k_h	Hydrolysis rate constant	3.4 mgCODS/ mgCODX·day
μ_{m2}	Max growth rate – aceticlactic bacteria	0.65 1/day
μ_{m3} [1]	Max growth rate – propionic acid bacteria	0.95 1/day
μ_{m4} [1]	Max growth rate – butyric acid bacteria	0.95 1/day
Kin	Product inhibition factor	9,000 mgCODS/l
k_{d1}	Biomass death rate -X1	0.07 1/day
k_{d3} [1]	Biomass death rate -X3	0.20 1/day
k_{d4} [1]	Biomass death rate -X4	0.07 1/day
k_{d5}	Biomass death rate -X5	0.07 1/day
KLa_{H2}	H_2 transfer coefficient	72 1/day
$X_{1,0}$	Initial conc. – acidogen	3.0 mg/l
$X_{2,0}$	Initial conc. – aceticlastic methanogen	6.0 mg/l
$X_{3,0}$	Initial conc. – propionic acid bacteria	0.5 mg/l
$X_{4,0}$	Initial conc. – butyric acid bacteria	0.5 mg/l
$X_{5,0}$	Initial conc. – hydrogen utilizing methanogen	3.0 mg/l

[1] Reference: [47]

d^{-1} and the first-order biomass death rate constant of propionic acid bacteria (k_{d3}) was increased from $0.01d^{-1}$ to $0.20\ d^{-1}$ in order to improve the model predictions. A high maximum specific growth rate for butyric acid bacteria and an insignificant first order biomass death rate for propionic acid bacteria would result in rapid utilization of butyric and propionic acid respectively and this would eventually result in an over prediction of methane gas. To be able to predict the lag phase of methane production accurately, the hydrolysis rate constant (k_h) and glucose inhibition constant (Kin) were reduced while the first order death constant of acidogens was increased to slow down the rate of hydrolysis of insoluble carbohydrate. The yield coefficient of calcium salts (Y_{Ca}) was reduced by almost four times to improve the prediction of carbon dioxide in the gas phase. This reduction was necessary because an unjustified increase of positively charged dissociated ions (i.e., Ca^{2+}) would result in an increase in the production of negatively charged dissociated ions (e.g., HCO_3^- and CO_3^{2-}) which would eventually lead to a tendency to dissolve more carbon dioxide from the gas phase. A sixfold increase in mass transfer coefficient of hydrogen gas (KLa_{H2}) was required to promote the dissolution of hydrogen gas into liquid phase which would then improve the prediction of hydrogen gas content. The value for the maximum specific growth rate of propionic acid bacteria (μ_{m3}) and the first order biomass death rate constant of butyric acid bacteria and hydrogen utilizing bacteria (k_{d4} and k_{d5}) were slightly adjusted to improve the model predictions on daily and cumulative methane production and VFAs production. Figure 11.4 shows the model fit to data from experiment 1 using the estimated parameters.

The model was then used to simulate experiment 2. Slight modifications of the model structure was required as in this experiment a volume of liquid was replaced with fresh buffer solution daily for 19 days. It was assumed that the digester contents were well mixed and removing liquid would remove the soluble components in the liquid. The model predictions were compared to experimental data and the corresponding plots are shown in Fig. 11.5. As can be seen the predictions of all data except pH of leachate was satisfactory. The model overpredicts the pH from day 25 until the end of the experiment.

As shown in Fig. 11.6, for experiment 3 reasonable predictions over the trends of cumulative methane production, methane gas content, total VFAs production and leachate pH were achieved using the same set of model parameters. The model under predicted the production of methane and total VFAs from day 10 to 20, however, towards the end of the simulation, the model began to over predict the methane production as a result of rapid utilization of VFAs. This methane and VFAs trend might be related to the initial biomass concentrations used in the model. For validation purpose, the initial biomass concentrations of experiment 3 were assumed to be the same as a fresh bed of MSW (for e.g. in experiment 1) despite the fact that the bed had been flushed daily for 90 days with water. It was hard to justify the validity of this assumption because it could be either an over-estimation or an under-estimation. It was reasonable to argue that after 90 days of flushing, most of the bacteria had been washed out and the remaining ones did not survive the acidic conditions. This argument was not totally true as experimental results had shown that there was still sufficient level of bacteria to initiate the start up of the degradation process. Based on those experimental results, it was then reasonable to argue that there was actually more biomass in the 90 days old "sour" bed than any fresh bed of MSW. The bacteria might have survived the acidic conditions by turning themselves into spores and they remained dormant until the physical conditions became favorable again. It was very likely that either one or both of the above arguments were valid depending on the time of the digestion process and as a result, no conclusions regarding these two arguments should be made unless the concentrations were actually quantified and monitored. The importance of the initial bacterial concentrations on model predictions should be reiterated because when the model was validated using the experimental results of experiment 3, the initial values of all state variables (apart from initial biomass concentrations) were actually experimental measurements. It was anticipated that the errors associated with those measurements were within some kind of limits and as a result, the only unknown initial conditions were the biomass concentrations. As discussed before, the initial concentration of all five bacterial groups were found to be sensitive and consequently, any modifications on these initial conditions would result in different model predictions in terms of methane production, total VFAs production and leachate pH profile.

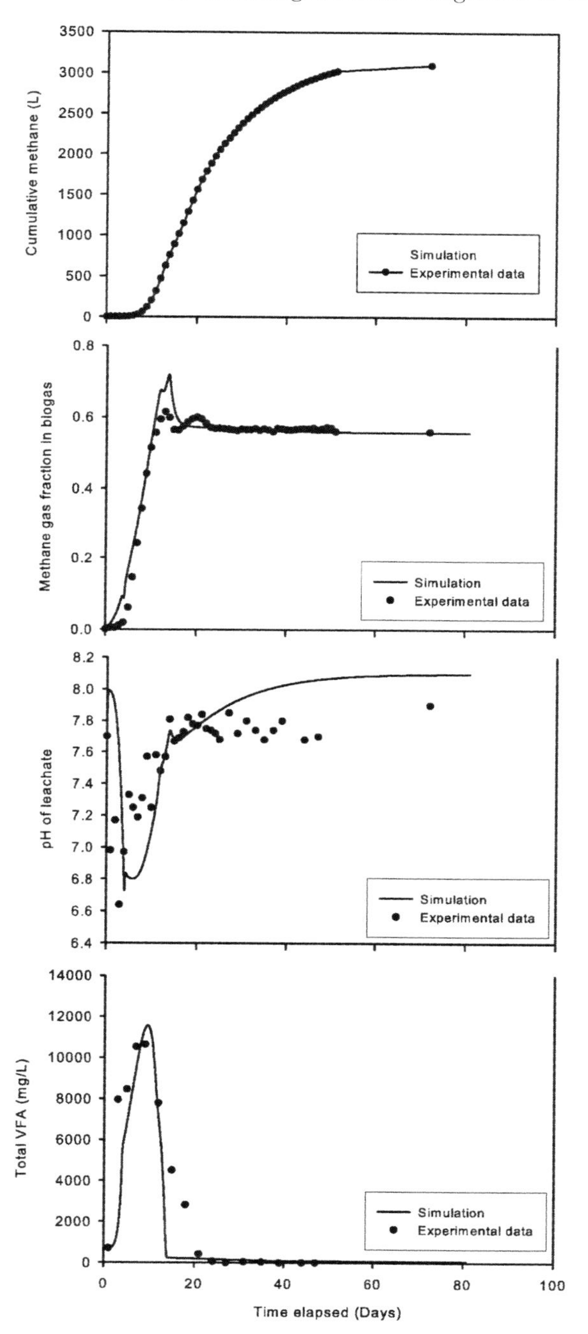

Fig. 11.4. Model fit to data from experiment 1

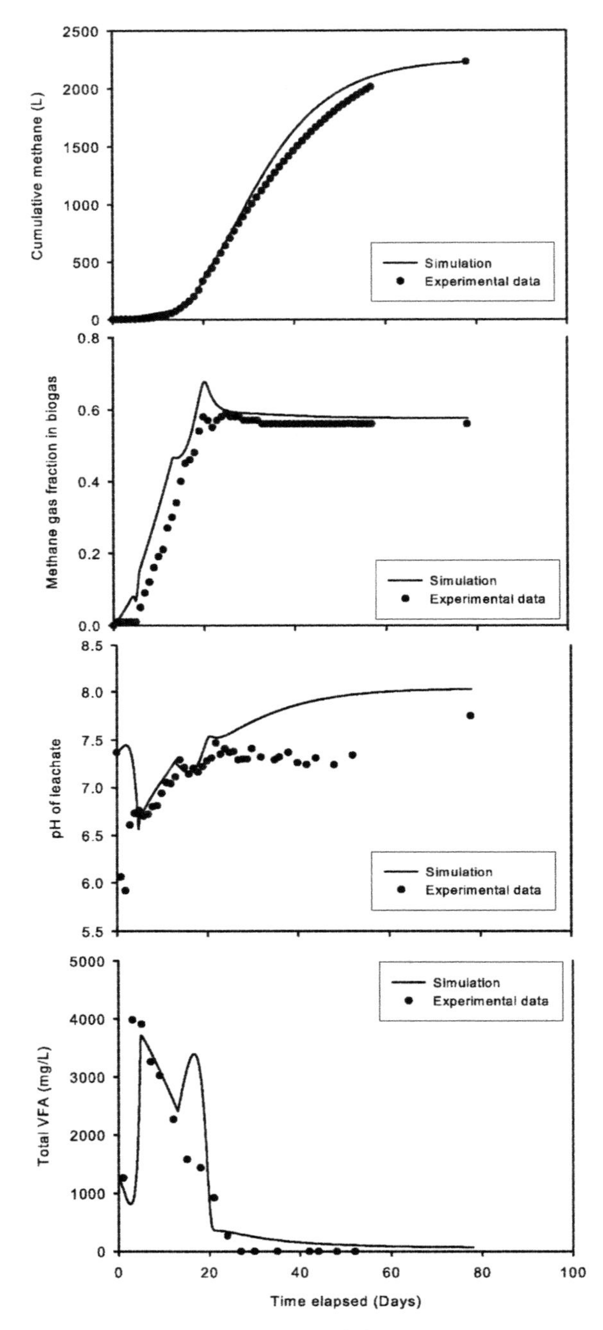

Fig. 11.5. Model prediction of data from experiment 2

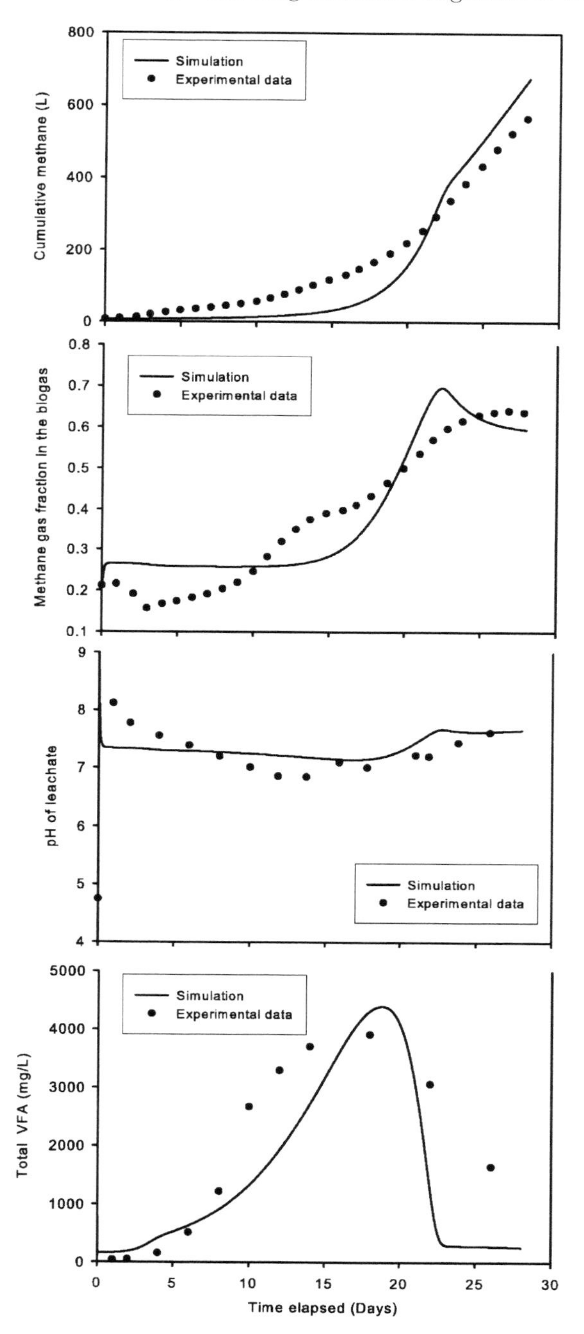

Fig. 11.6. Model prediction of data from experiment 3

11.8 Model Application

The validated model was used to predict the optimal initial sodium bicarbonate concentration for anaerobic digestion of MSW under flooded conditions. Six buffer concentrations were investigated and two of them were the same as the experimental values in experiments 1 and 2. The six concentrations are known as case 1 to case 6 and they were 580, 2,300, 2,800, 4,670, 11,200 and 17,000 mg $NaHCO_3\,l^{-1}$ respectively. It should be noted that the initial gas content and the initial concentrations of total inorganic carbon dioxide and calcium ion will change according to the initial buffer concentration. These changes were estimated using the physico-chemical equilibrium relationships and the charge balance equation. Table 11.13 shows the initial value for the nitrogen and carbon dioxide gas content and the initial concentration of total inorganic carbon dioxide and calcium ion for all six cases.

The simulation time for all six cases was 1-year and the output variables shown in Fig. 11.7 are the cumulative methane production and leachate pH. When the initial buffer concentration was 580 mg $NaHCO_3\,l^{-1}$, the model predicted very little methane production as the pH of the bed of MSW was always below 5.0 which is inhibitory to all five bacterial groups involved in the digestion process. When considering the buffer concentration of 2300 and 2,800 mg$NaHCO_3\,l^{-1}$, the model predicted more methane production as the pH of the system eventually returned to a non acidic condition. The predicted cumulative methane was higher when the bed of MSW was flooded with 2,800 instead of 2,300 mg $NaHCO_3\,l^{-1}$ buffer solution because the model predicted that the pH of the system could return to a neutral level faster due to a higher capacity to neutralize the acids. When considering the remaining three initial buffer concentrations (4,670, 11,200 and 17,000 mg $NaHCO_3\,l^{-1}$), it can be seen from Fig. 11.7 that there were basically no differences in model predictions in terms of cumulative methane production. The model predicted that the same amount of methane was produced in the same period of time if the bed of MSW was initially flooded with buffer solution at either of those three concentrations. However, when the bed was initially flooded with 4,670 mg

Table 11.13. The initial conditions for different initial sodium bicarbonate concentration

Case	*Sodium bicarbonate (mg/l)*	*N_2 content*	*CO_2 content*	*CO_{2Total} (mole/l)*	*Ca (mole/l)*
1	580	0.9216	0.0221	0.00595	0.00521
2	2,300	0.9065	0.0372	0.02381	0.01398
3	2,800	0.8972	0.0465	0.02976	0.01685
4	4,670[1]	0.8682	0.0755	0.04832	0.02580
5	11,200[2]	0.7626	0.1811	0.1159	0.05838
6	17,000	0.3714	0.5723	0.1786	0.08934

[1] Experiment 2
[2] Experiment 1

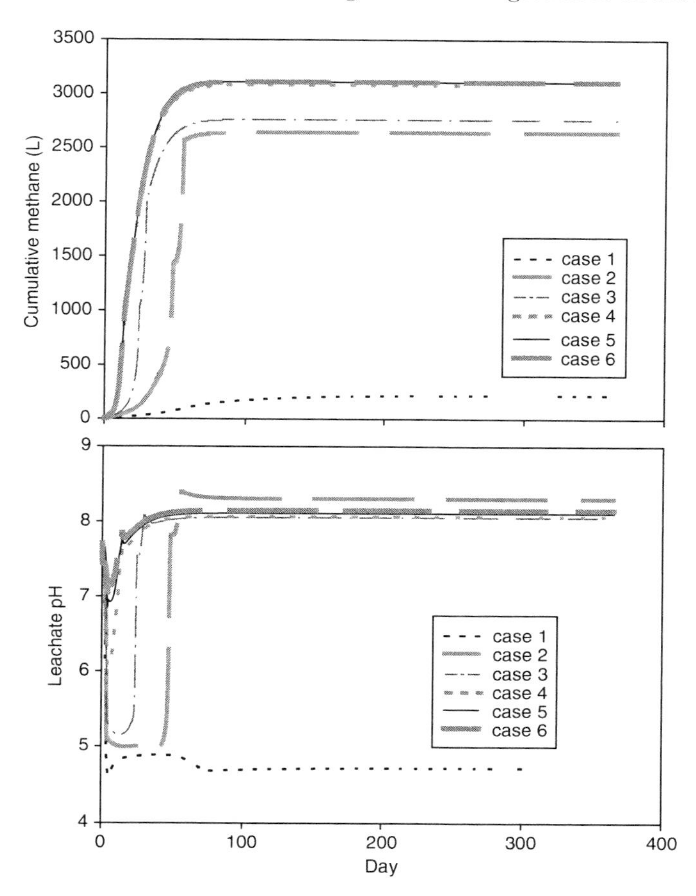

Fig. 11.7. Model application: Comparisons of methane production and system pH among six different initial buffer conditions

$NaHCO_3\,l^{-1}$ (same as experiment 2), the model predicted that the pH of the system would drop to as low as 6.00 as opposed to neutral for the other two cases. Based on the model predictions of a flooded MSW bed with an initial buffer concentration of 4,670 mg $NaHCO_3\,l^{-1}$ (case 4), it is not unreasonable to speculate that if the fresh MSW bed of experiment 2 was left unattended, sufficient level of buffering capacity might have been generated through the natural carbonate equilibrium relationships and the pH of the system might have returned to a neutral level within a short period of time. But since exchanging of free liquid with fresh buffer solution during the first 13 days of experiment 2 did seem to be necessary, it is therefore not feasible to conclude that an initial buffer concentration of 4,670 mg $NaHCO_3\,l^{-1}$ is optimal. However, it is reasonable to conclude that any further increase in initial buffer concentration beyond 11,200 mg $NaHCO_3\,l^{-1}$ will not result in any improvement on digestion efficiency. This is because the model predicted two very similar methane production profiles for case 5 and case 6 where the bed of MSW was initially flooded with 11,200 and 17,000 mg $NaHCO_3\,l^{-1}$ respectively. Based on the findings of this model analysis exercise, it can be concluded that rapid start up of the anaerobic digestion of fresh MSW was achieved by flooding the bed with buffer solution at a concentration above 4,670 mg $NaHCO_3\,l^{-1}$ initially. However, in order to ensure that there is a sufficient level of buffering capacity for the neutralization of acidic conditions due to the accumulation of VFAs, it is preferable that the fresh MSW bed be flooded with buffer solution at concentration higher than 4,670 mg $NaHCO_3\,l^{-1}$ initially. It was proven experimentally that an initial buffer concentration of 11,200 mg $NaHCO_3\,l^{-1}$ was sufficient to maintain a non acidic system pH for the entire duration of the digestion process.

11.9 Conclusions

Energy recovery can serve an important role in sustainably managing municipal solid waste. Anaerobic digestion of organic fraction of municipal solid waste (OFMSW) not only produces a fuel but also reduces the amount of residue that requires disposal and conserves plant nutrients (nitrogen and phosphorus) that can be recycled through land application of digested residue. Among various technologies available for anaerobic digestion, the high-solids, leach bed process offers flexibility for in-vessel or bioreactor implementation. A mathematical model to describe anaerobic digestion of organic fraction of municipal solid waste was developed and validated using data from pilot scale experiments. The model uses information on the amount of waste, initial alkalinity and pH, and digester operating conditions (liquid and gas volumes and temperature) to predict the progression of methane generation, gas composition and quality of digester liquid (volatile organic acids concentration and pH). The model consists of unsteady state mass balance equations, physicochemical equilibrium expressions and charge balance to enable computation of the above output parameters. The model was implemented in MATLAB for simulation. A novel approach of writing the charge balance in a differential form (i.e., as a derivative with respect to time) enabled simulations to numerically converge rapidly. It was found that the model satisfactorily predicted cumulative methane production, methane composition in gas phase, accumulation and depletion of total volatile organic acids and pH. Initial values of bacterial concentrations were estimated to provide best fit for one set of data and these values were satisfactory in simulating another set of experiments. Sufficient concentrations of microbial populations are present naturally in OFMSW and these can be activated rapidly by providing adequate alkalinity to prevent acidification by acidogenesis. Such a start up procedure guarantees sustained and stable operation of the digester. Simulations showed that an initial concentration of $\approx 11\,g\,l^{-1}$ of $NaHCO_3$ was sufficient to accomplish this.

In an experiment, a digester loaded with OFMSW was deliberately acidified initially and then reactivated by addition of alkalinity. There was a mismatch with model predictions of this experiment indicating that the model was not adequate in simulating bacterial population changes during inhibitory conditions such as acidification. Since parameters of the model were estimated by only a visual fit to experimental data, perhaps estimation of parameters through statistical regression procedures would have provided better fits for a range of operating conditions.

References

1. F. Alatriste-, P. Samar, H. H. J. Cox, B. K. Ahring, and R. Iranpour. Anaerobic codigestion of municipal, farm, and industrial organic wastes: A survey of recent literature. *Water Environment Research*, 78(6):607–636, 2006.
2. M. A. Barlaz, R. K. Ham, and D. M. Schaefer. Methane production from municipal refuse: a review of enhancement techniques and microbial dynamics. *Critical Review in Environmental Control*, 19:557–584, 1990.
3. A. Barthakur, M. Bora, and H. D. Singh. Kinetic model for substrate utilization and methane production in the anaerobic digestion of organic feeds. *Biotechnology Progress*, 7:369–376, 1991.
4. D. J. Batstone, J. Keller, I. Angelidaki, S. V. Kalyuzhnyi, S. G. Pavlostathis, A. Rozzi, W. T. M. Sanders, H. Siegrist, and V. A. Vavilin. *Anaerobic Digestion Model No.1 (ADM No.1)*. IWA Task Group for Mathematical Modelling of Anaerobic Digestion Processes, London, 2002.
5. R. Biljetina, V. J. Srivastava, D. P. Chynoweth, and T. D. Hayes. *Anaerobic digestion of water hyacinth and sludge, Aquatic Plants for Water Treatment and Resource Recovery*. Magnolia, Orlando, FL, pages 725–738, 1987. W. H. Smith and K. R. Reddy (eds).
6. D. Boone and R. Mah. *Anaerobic Digestion of Biomass*, chapter Transitional bacteria. Elsevier Applied Science, London, 1987. D. P. Chynoweth and R. Isaacson (eds).
7. J. D. Bryers. Structured modeling of the anaerobic digestion of biomass particulates. *Biotechnology and Bioengineering*, 27:638–649, 1985.
8. Y. R. Chen and A. G. Hashimoto. Substrate utilization kinetic model for biological treatment processes. *Biotechnology and Bioengineering*, 22:2081–2095, 1980.
9. O. Christ, P. A. Wilderer, R. Angerhofer, and M. Faulstich. Mathematical modeling of the hydrolysis of anaerobic processes. *Water Science and Technology*, 41(3):61–65, 2000.
10. S. Chugh. *Enhanced degradation of municipal solid waste*. PhD thesis, Department of Chemical Engineering, The University of Queensland, Brisbane, Australia, 1996.
11. S. Chugh, D. P. Chynoweth, W. Clarke, P. Pullammanappallil, and V. Rudolph. Degradation of unsorted municipal solid waste by a leach-bed process. *Bioresource Technology*, 69(2):103–115, 1999.
12. D. P. Chynoweth, G. Bosch, J. F. K. Earle, J. Owens, and R. Legrand. Sequential batch anaerobic composting of the organic fraction of municipal solid waste. *Water Science Technology*, 24:327–339, 1992.

13. D. P. Chynoweth, J. M. Owens, and R. L. Legrand. Renewable methane from anaerobic digestion of biomass. *Renewable Energy*, 22:1–8, 2001.
14. R. H. Clark and R. E. Speece. The pH tolerance of anaerobic digestion advances in water pollution research. In *5th International Conference on Water Pollution Research*, number 2, pages 27/1–27/14, 1971.
15. D. J. Costello, P. F. Greenfield, and P. L. Lee. Dynamic modelling of a single-stage high-rate anaerobic reactor I model derivation. *Water Research*, 25:847–858, 1991.
16. G. Dinopoulou, T. Rudd, and J. N. Lester. Anaerobic acidogenesis of a complex wastewater: I. The influence of operational parameters on reactor performance. *Biotechnology and Bioengineering*, 31:958–968, 1988.
17. A. C. Duarte and G. K. Anderson. Inhibition modeling in anaerobic digestion. *Water Science and Technology*, 14:749–763, 1982.
18. J. C. Frigon and S. R. Guiot. Impact of liquid to gas hydrogen mass transfer on substrate conversion efficiency of an upflow anaerobic sludge bed and filter reactor. *Enzyme and Microbial Technology*, 17:1080–1086, 1995.
19. G. Gottschalk. *Bacterial metabolism.* Springer, New York, pages 209–282, 1986.
20. W. Gujer and A. J. B. Zehnder. Conversion processes in anaerobic digestion. *Water Science and Technology*, 15:127–167, 1983.
21. K. M. Hangos and I. T. Cameron. *Process modeling and model analysis.* Academic, San Diego, 2001.
22. H. Hartmann and B. K. Ahring. Strategies for the anaerobic digestion of the organic fraction of municipal solid waste: an overview. *Water Science and Technology*, 53(8):7–22, 2006.
23. D. M. Himmelblau. *Basic principles and calculations in chemical engineering.* Prentice Hall, New Jersey, 5 edition, 1989.
24. IEA (International Energy Agency). *Biogas from municipal solid waste: overview of systems and markets for anaerobic digestion of MSW*, 1994. Report of International Energy Agency Task XI: Conversion of MSW to Energy; Activity 4: Anaerobic Digestion of MSW. IEA, Paris
25. A. H. Ignoni, M. J. Ayotamuno, S. O. T. Ogaji, and S. D. Probert. Municipal solid waste in Port Harcourt, Nigeria. *Applied Energy*, 84:664–670, 2007.
26. H. F. Kaspar and K. Wuhrmann. Kinetic parameters and relative turnovers of some important catabolic reactions in digesting sludge. *Applied and Environmental Microbiology*, 36(1):1–7, 1978.
27. H. F. Kaspar and K. Wuhrmann. Product inhibition in sludge digestion. *Microbial Ecology*, 4:241–248, 1978.
28. M. Kayhanian and S. Hardy. The impact of four design parameters on the performance of a high-solids anaerobic digestion of municipal solid waste for fuel gas production. *Environmental Technology*, 15:557–567, 1994.
29. T. E. Lai. *Rate limiting factors of the anaerobic digestion of municipal solid waste in bioreactor landfills.* PhD thesis, Department of Chemical Engineering, The University of Queensland, Brisbane, Australia, 2001.
30. E. J. La Motta, E. Silva, A. Bustillos, H. Padron, and J. Luque. Combined anaerobic/aerobic secondary municipal wastewater treatment: pilot-plant demonstration of the UASB/aerobic solids contact system. *Journal of Environmental Engineering*, 133(4):397–403, 2007.
31. F. M. Morel and J. G. Hering. *Principles and applications of aquatic chemistry.* Wiley, New York, 1993.

32. G. R. Morris. Anaerobic fermentation of animal wastes: A kinetic and empirical design evaluation. Master's thesis, Cornell University, Ithaca, New York, 1976.
33. F. Mosey. Mathematical modeling of the anaerobic digestion process: Regulatory mechanisms for the formation of short chain volatile acids from glucose. *Water Science and Technology*, 15:209–232, 1983.
34. T. Noike, G. Endo, J. -E. Chang, J. L. Yaguchi, and J. I. Matsumoto. Characteristics of carbohydrate degradation and the rate limiting step in anaerobic digestion. *Biotechnology and Bioengineering*, 27:1482–1489, 1985.
35. A. Nopharatana. *Modelling leach-bed anaerobic digestion of municipal solid waste*. PhD thesis, Department of Chemical Engineering, The University of Queensland, Brisbane, Australia, 2000.
36. A. Nopharatana, P. C. Pullammanappallil, and W. P. Clarke. A dynamic mathematical model for sequential leach-bed anaerobic digestion of organic fraction of municipal solid waste. *Biochemical Engineering Journal*, 13(1):21–33, 2003.
37. A. Nopharatana, P. C. Pullammanappallil, and W. P. Clarke. Kinetics and dynamic modelling of batch anaerobic digestion of municipal solid waste in a stirred reactor. *Waste Management*, 27(5):595–603, 2007.
38. R. S. Orelmand. *Biology of anaerobic microorganisms*. Wiley, New York, pages 641–705, 1988. A. J. B. Zehnder (ed).
39. J. M. Owens and D. P. Chynoweth. Biochemical methane potential of MSW components. *Water Science and Technology*, 27:1–14, 1993.
40. C. S. Peres, C. R. Sanchez, C. Matumoto, and W. Schimidell. Anaerobic biodegradability of the organic fraction of the organic componens of municipal solid wastes (OFMSW). *Water Science and Technology*, 25:285–294, 1992.
41. M. Perez-Garcia, L. I. Rornero-Garcia, R. Rodriguez-Cano, and D. Sales-Marquez. High rate anaerobic thermophilic technologies for distillery wastewater treatment. *Water Science and Technology*, 51(1):191–198, 2005.
42. J. T. Pfeffer. Temperature effects on anaerobic fermentation of domestic refuse. *Biotechnology and Bioengineering*, 16:771–787, 1974.
43. J. T. Pfeffer and K. A. Kahn. Microbial production of methane from municipal refuse. *Biotechnology and Bioengineering*, 18:1179–1191, 1976.
44. I. R. Ramsay. *Modeling and control of high rate anaerobic wastewater treatment systems*. PhD thesis, Department of Chemical Engineering, University of Queensland, Brisbane, Australia, 1997.
45. I. R. Ramsay and P. C. Pullammanappallil. Full-scale validation of a dynamic mathematical model for a two-stage, high-rate anaerobic brewery wastewater treatment system. *Journal of Environmental Engineering*, 131(7):1030–1036, 2005.
46. M. S. Rao and S. P. Singh. Bioenergy conversion studies of organic fraction of MSW: Kinetic studies and gas yield organic loading relationships for process optimization. *Bioresource Technology*, 95:173–185, 2004.
47. M. Romli. *Modelling and verification of a two-stage, high-rate, anaerobic wastewater treatment system*. PhD thesis, Department of Chemical Engineering, The University of Queensland, Brisbane, Australia, 1993.
48. W. Rulkens. Sewage sludge as a biomass resource for the production of energy: Overview and assessment of the various options. *Energy and Fuels*, 22(1):9–15, 2008.
49. M. Sharholy, K. Ahmad, G. Mahmood, and R. C. Trivedi. Municipal solid waste management in Indian cities a review. *Waste Management*, 28:459–467, 2008.

50. H. Siegrist, D. Renggli, and W. Gujer. Mathematical modeling of anaerobic mesophilic sewage sludge treatment. *Water Science and Technology*, 27:25–36, 1993.
51. W. Six and L. de Baere. Dry anaerobic conversion of municipal solid-waste by means of the DRANCO process. *Water Science and Technology*, 25(7):295–300, 1992.
52. K. Sormunen, M. Ettala, and J. Rintala. Detailed internal characterization of two Finnish landfills by waste sampling. *Waste Management*, 28:151–163, 2008.
53. L. M. Svensson, L. Bjornsson, and B. Mattiasson. Enhancing performance in anaerobic high-solids stratified bed digesters by straw bed implementation. *Bioresource Technology*, 98(1):46–52, 2007.
54. E. Ten Brummeler, H. C. J. M. Horbach, and I. W. Koster. Dry anaerobic batch digestion of the organic fraction of municipal solid waste. *Journal of Chemical Technology and Biotechnology*, 50:191–209, 1991.
55. D. F. Toerien and W. H. J. Hattingh. Anaerobic digestion i. the microbiology of anaerobic digestion. *Water Research*, 3(6):385–416, 1969.
56. USEPA. *Municipal solid waste generation, recycling and disposal in the United States: Facts and figures for 2006*. EPA-530-F-07-030, 2007. Available at `http://www.epa.gov/epaoswer/non-hw/muncpl/pubs/msw06.pdf`.
57. V. A. Vavilin, L. Y. Lokshina, X. Flotats, and I. Angelidaki. Anaerobic digestion of solid material: Multidimensional modeling of continuous-flow reactor with non-uniform influent concentration distributions. *Biotechnology and Bioengineering*, 97(2):354–366, 2007.
58. V. A. Vavilin, L. Y. Lokshina, J. P. Y. Jokela, and J. A. Rintala. Modeling solid waste decomposition. *Bioresource Technology*, 94(1):69–81, 2004.
59. G. D. Vogels, J. T. Keltjens, and C. van der Drift. *Biology of anaerobic microorganisms*, pages 707–770. John Wiley and Sons Inc., New York, 1988. A. J. B. Zehnder (ed).
60. M. J. Wolin. Metabolic interactions among intestinal microorganisms. *The American Journal of Clinical Nutrition*, 27:1320–1328, 1974.
61. T. Yilmaz, A. Yuceer, and M. Basibuyuk. A comparison of the performance of mesophilic and thermophilic anaerobic filters treating papermill wastewater. *Bioresource Technology*, 99(1):156–163, 2008.
62. T. C. Zeikus. *Anaerobic Digestion*. Applied Science Publishers, London, pages 61–89, 1980. D. A. Stafford, B. I. Wheatley and D. E. Hungus (eds).
63. R. J. Zoetmeyer, J. C. van den Heuvel, and A. Cohen. pH influence on acidogenic dissimilation of glucose in an anaerobic digester. *Water Research*, 16:303-311, 1982.

Appendix

The complete set of equations that represent the model is listed below. For nomenclature please refer to text.

Constitutive Relations

$$zn = (H_2)_{aq} \cdot \left(10^{\left(pH - \frac{1139}{T+273}\right)}\right) \qquad (11.1)$$

$$pHin_i = \exp\left(-0.5 \cdot \left(\frac{pH - pHm_i}{pHsd_i}\right)^2\right),$$
$$\text{where } i = 1,2,3,4,5 \quad (pH < 7) \tag{11.2}$$

$$pHin_i a = \exp\left(-0.5 \cdot \left(\frac{pHm_i, a - pH}{pHsd_i, a}\right)^2\right),$$
$$\text{where } i = 1,2,3,4,5 \quad (pH > 7) \tag{11.3}$$

$$r_1 = \frac{k_h \cdot S_1}{(K_{x1} \cdot X_1 + S_1)} \cdot X_1 \tag{11.4}$$

$$r_2 = \left(\frac{\mu_{m1} \cdot S_2}{Y_{X1/S2} \cdot (K_{S,2} + S_2)}\right) \cdot X_1 \cdot pHin_1 \cdot pHin_1 a \cdot \left(\frac{1}{1 + \frac{S_2}{Kin}}\right) \cdot \left(\frac{1}{1 + zn}\right) \tag{11.5}$$

$$r_3 = \left(\frac{\mu_{m2} \cdot S_3}{Y_{X2/S3} \cdot (K_{S,3} + S_3)}\right) \cdot X_2 \cdot pHin_2 \cdot pHin_2 a \tag{11.6}$$

$$r_4 = \left(\frac{\mu_{m3} \cdot S_4}{Y_{X3/S4} \cdot (K_{S,4} + S_4)}\right) \cdot X_3 \cdot pHin_3 \cdot pHin_3 a \cdot \left(\frac{1}{1 + zn}\right) \tag{11.7}$$

$$r_5 = \left(\frac{\mu_{m4} \cdot S_5}{Y_{X4/S5} \cdot (K_{S,5} + S_5)}\right) \cdot X_4 \cdot pHin_4 \cdot pHin_4 a \cdot \left(\frac{1}{1 + zn}\right) \tag{11.8}$$

$$r_6 = \left(\frac{\mu_{m5} \cdot S_6}{Y_{X5/S6} \cdot (K_{S,6} + S_6)}\right) \cdot X_5 \cdot pHin_5 \cdot pHin_5 a \tag{11.9}$$

$$r_{di} = k_{di} \cdot X_i \qquad \forall i = 1,2,3,4,5 \tag{11.10}$$

$$r_{CO2} = Y_{CO2S2} \cdot r_2 + Y_{CO2S3} \cdot r_3 + Y_{CO2S4} \cdot r_4 - Y_{CO2S5} \cdot r_5 - Y_{CO2S6} \cdot r_6 \tag{11.11}$$

$$r_{CH4} = Y_{CH4S3} \cdot r_3 + Y_{CH4S6} \cdot r_6 \tag{11.12}$$

$$r_{H2} = Y_{H2S2} \cdot r_2 + Y_{H2S4} . r_4 + Y_{H2S5} \cdot r_5 - r_6 \tag{11.13}$$

$$r_{H2,T} = KLa_{H2} \cdot \left(\frac{y_{H2}}{H_{H2}} - (H_2)_{aq}\right) \tag{11.14}$$

$$Ac^- = \frac{k2 \cdot \left(\frac{S_3}{C1}\right)}{k2 + [H^+]} \tag{11.15}$$

$$\bar{\text{Pr}} = \frac{kp \cdot \left(\frac{S_4}{C2}\right)}{kp + [H^+]} \tag{11.16}$$

$$Bu^- = \frac{kb \cdot \left(\frac{S_5}{C3}\right)}{kb + [H^+]} \tag{11.17}$$

$$NH_3 = \frac{k5 \cdot [N_T]}{k5 + [H^+]} \tag{11.18}$$

$$CO_3^{2-} = \frac{(CO_{2Total})}{1 + \frac{[H^+]}{k4} + \frac{[H^+]^2}{k3 \cdot k4}} \tag{11.19}$$

$$HCO_3^- = \frac{[CO_{2Total}]}{1 + \frac{k4}{[H^+]} + \frac{[H^+]}{k3}} \tag{11.20}$$

$$OH^- = \frac{1e - 14}{[H^+]} \tag{11.21}$$

$$(CO_2)_{aq} = CO_{2Total} - HCO_3^- - CO_3^{2-} \tag{11.22}$$

$$r_{CO2,T} = KLa_{CO2} \cdot \left(\frac{y_{CO2}}{H_{CO2}} - (CO_2)_{aq} \right) \tag{11.23}$$

$$q_G = \frac{R \cdot T \cdot V_L}{(1 - Y_{H2O})} \cdot (r_{CH4} - r_{H2,T} - r_{CO2,T}) \tag{11.24}$$

$$r_{Ca} = Y_{Ca} \cdot r_1 \tag{11.25}$$

$$r_{N,T} = Y_{N,T} \cdot r_1 \tag{11.26}$$

Conservation Balances

$$\frac{dS_1}{dt} = -r_1 \tag{11.27}$$

$$\frac{dS_2}{dt} = (1 - fr) \cdot r_1 - r_2 \tag{11.28}$$

$$\frac{dS_3}{dt} = Y_{S3S2} \cdot r_2 - r_3 + Y_{S3S4} \cdot r_4 + Y_{S3S5} \cdot r_5 \tag{11.29}$$

$$\frac{dS_4}{dt} = Y_{S4S2} \cdot r_2 - r_4 \tag{11.30}$$

$$\frac{dS_5}{dt} = Y_{S5S2} \cdot r_2 - r_5 \tag{11.31}$$

$$\frac{dS_6}{dt} = r_{H2} + r_{H2,T} \tag{11.32}$$

$$\frac{dS_7}{dt} = r_{CH4} \tag{11.33}$$

$$\frac{dRS_1}{dt} = fr \cdot r_1 \tag{11.34}$$

$$\frac{dX_1}{dt} = Y_{X1S2} \cdot r_2 - r_{d1} \tag{11.35}$$

$$\frac{dX_2}{dt} = Y_{X2S3} \cdot r_3 - r_{d2} \tag{11.36}$$

$$\frac{dX_3}{dt} = Y_{X3S4} \cdot r_4 - r_{d3} \tag{11.37}$$

$$\frac{dX_4}{dt} = Y_{X4S5} \cdot r_5 - r_{d4} \tag{11.38}$$

$$\frac{dX_5}{dt} = Y_{X5S6} \cdot r_6 - r_{d5} \tag{11.39}$$

$$\frac{dy_{H2}}{dt} = -\frac{q_G}{V_G} \cdot y_{H2} - \frac{V_L}{V_G} \cdot R \cdot T \cdot r_{H2,T} \tag{11.40}$$

$$\frac{dy_{CH4}}{dt} = -\frac{q_G}{V_G} \cdot y_{CH4} + \frac{V_L}{V_G} \cdot R \cdot T \cdot r_{CH4} \tag{11.41}$$

$$\frac{dy_{N2}}{dt} = -\frac{q_G}{V_G} \cdot y_{N2} \tag{11.42}$$

$$\frac{dy_{CO2}}{dt} = -\frac{q_G}{V_G} \cdot y_{CO2} - \frac{V_L}{V_G} \cdot R \cdot T \cdot r_{CO2,T} \tag{11.43}$$

$$= -\left(\frac{dY_{CH4}}{dt} + \frac{dY_{H2}}{dt} + \frac{dY_{N2}}{dt}\right) \tag{11.44}$$

$$\frac{dCO_{2Total}}{dt} = r_{CO2} + r_{CO2,T} \tag{11.45}$$

$$\frac{dCa}{dt} = r_{Ca} \tag{11.46}$$

$$\frac{dN_T}{dt} = r_{N,T} \tag{11.47}$$

$$\frac{dH^+}{dt} = 2 \cdot \frac{dCO_3^{2-}}{dt} + \frac{dHCO_3^-}{dt} + \frac{dOH^-}{dt} + \frac{dAc^-}{dt} + \frac{d\,\mathrm{Pr}^-}{dt} + \frac{dBu^-}{dt} - \frac{dNH_4^+}{dt} - 2 \cdot \frac{dCa}{dt} \tag{11.48}$$

12

Spatially Differentiated Trade of Permits for Multipollutant Electric Power Supply Chains

Trisha Woolley, Anna Nagurney, and John Stranlund

Summary. In this paper, we consider electric power supply chain networks in which the power generators have distinct power plants and associated technologies and we develop a model of tradable pollution permits in the case of multiple pollutants and spatially distinct receptor points. We formulate the governing equilibrium conditions as a finite-dimensional variational inequality and demonstrate that, under the proposed multipollutant permit trading scheme, the environmental standards are achieved. Finally, we describe a computational procedure that exploits the structure of the problem. We also present numerical examples.

12.1 Introduction

Electric power plants emit several different air pollutants, such as carbon dioxide (CO_2), sulfur dioxide (SO_2), nitrous oxide (NO_x), and mercury (Hg) with differing environmental impacts. For example, carbon dioxide is a major cause of global climate change; sulfur dioxide and nitrous oxide are responsible for acid rain and fine particle concentrations in the atmosphere; nitrous oxide also contributes to ground-level ozone, and mercury may travel vast distances before deposited in, for example, waterways, bioaccumulating in the food chain resulting in impaired neurological development [2, 7]. Moreover, SO_2, NO_x, and Hg have important spatial characteristics; that is, the impacts of these pollutants depend critically on the location of their sources and where their impacts are realized.

Although most environmental regulations attempt to control one pollutant at a time, integrated multipollutant regulations have advantages over the standard piecemeal approach. Multipollutant approaches can account for the substitutability or complementarity effects of emissions from power plants. As one pollutant is reduced, another may rise, as in, for example, if an electric power generating firm invests in low sulfur coal to reduce SO_2 emissions, this will result in an increased amount of NO_x and Hg emissions [29, 30]. However, to exploit the complementarity effects of pollutants, firms may invest in electrostatic precipitators (EPSs) that will reduce SO_2 and NO_x together. Thus, a generator will choose a technology that is not the cheapest, but reduces multiple pollutants while meeting the current pollutant standard [31]. Furthermore, the relationship between pollutants may vary between seasons, across regions, and, possibly, over time as the composition of the atmosphere changes [34].

Because of such advantages, there have been several existing and proposed regulations to control multiple pollutants. The Regional Clean Air Incentives Market (RECLAIM) program was implemented in California to control NO_x and SO_2 pollutants; the proposed but not enacted Clear Skies was a national cap to reduce SO_2, NO_x, and Hg; and the US Environmental Protection Agency's Clean Air Interstate Rule (CAIR) capped emissions of SO_2 and NO_x in a large region covering more than 20 states, mostly east of the Mississippi, and the District of Columbia [28].

Pollution by electric power entities can be controlled by price, in the form, for example, of a carbon tax that is imposed for emissions that exceed a predetermined bound, or by quantity, as in the case of emission trading schemes (cf. [5, 21, 35], and the references therein). There are two types of emission trading policies, project-based (generators purchase credits from a project aimed to reduce emissions) and an allowance market (also known as cap and trade programs). In the latter type, electric power generators are given credits (or allowances) by a central environmental authority. The advantage of emissions trading is that credit trading generates pollution prices that distribute emissions control in a cost-effective manner. For additional background on tradable pollution permits, see [12, 13, 19, 24, 32, 33], and [5].

In this paper, we model the trading of emission rights by electric power producers who emit multiple pollutants with impacts that depend on the spatial dispersion of sources and receptors (for additional background on the electric power industry and associated modeling issues, see [1, 8–10, 22, 36], and [23]). The control of multiple, spatially differentiated, pollutants via emission trading calls for multiple pollution permit markets. Moreover, unlike the previous literature, we emphasize the use of alternative power production technologies as well as the underlying supply chain aspects of electric power generation and distribution. The results in this paper are particularly relevant given the current trends in environmental policies governing emissions in the electric power industry. The new model allows for the determination of the equilibrium numbers and prices of the various tradable pollution permits simultaneously with the equilibrium electric power flows and prices. The model builds upon the electric power supply chain model with alternative power plant technologies developed by Wu et al. (2006), which, however, only considered a single pollutant (and, in effect, a single receptor point). The model developed by [35] was further transformed into a transportation network equilibrium model (see also, e.g., [20]).

This paper is organized as follows. In Sect. 12.2, we present the model of the electric supply chain network with different power plant technologies and with the inclusion of multipollutant tradable permits with multiple receptor points. We also discuss a special case of the model and demonstrate that the environmental standards are achieved. In Sect. 12.3, we describe the computational procedure which exploits the structure of the problem. We also present examples. Section 12.4 summarizes the results in this paper and presents our conclusions.

12.2 The Electric Power Supply Chain Network Model with Multipollutant Tradable Permits

We now develop the model that captures the behavior of the electric power supply chain network decision-makers in the presence of a multipollutant permit trading scheme. The decision-makers in the electric power supply chain are the electric power generators, with their associated power plants, the suppliers, the transmission service providers, and the consumers at the demand markets. The equilibrium conditions of the electric power supply chain network will be given as well as the equivalent variational inequality formulation.

The electric power supply chain network is represented in Fig. 12.1 with the top tier of nodes consisting of the G power generators (also referred to as "gencos"), enumerated by $1, \dots, g, \dots, G$. Power generators are the decision-makers who own and operate the M power plants, with a typical power plant technology denoted by m, and depicted in the second tier of nodes in Fig. 12.1. Such nodes are enumerated as $11, \dots, GM$ with node gm denoting the m-th power plant of genco g. The gencos produce electric power using the different power plants, which are powered, for example, by different forms of technology such as coal, natural gas, uranium, oil, sun, wind, etc., and with different associated costs and environmental impacts. The gencos sell the electric power to the power suppliers in the third tier of nodes in the electric power supply chain, as depicted in Fig. 12.1.

In Fig. 12.1, we also represent the R receptor points, with a typical receptor point denoted by r, associated with the pollutants generated by the power plants. These receptor points are spatially separated. We also assume that there are J pollutants with a typical pollutant denoted by j.

The suppliers do not physically handle the electricity, but function as intermediaries who only hold and trade the right for the electric power. The nodes corresponding to the power suppliers are enumerated as: $1, \dots, s, \dots, S$ with node s corresponding to supplier s. Suppliers sell the electric power to the consumers at the different demand markets via the V transmission service providers, who are the entities who own and operate the electric power transmission and distribution systems. We denote a typical transmission service provider by v.

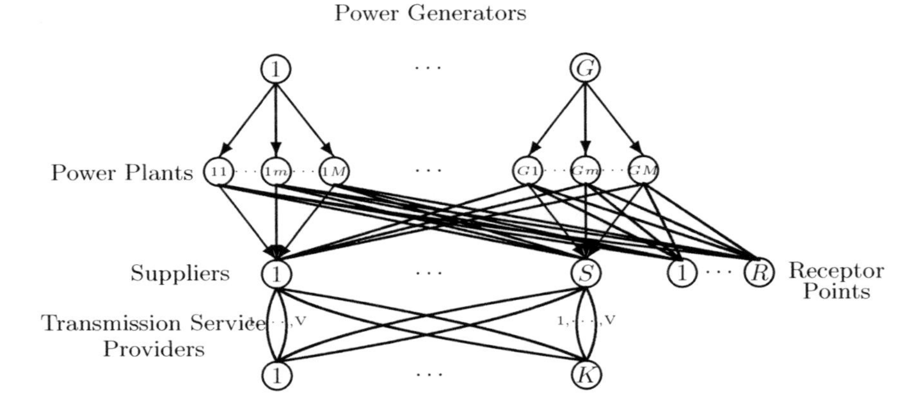

Fig. 12.1. The electric power supply chain network with power plants and associated technologies and with pollutant receptor points

Transmission service providers are not represented as nodes in the network model, since they do not make decisions such as to where or from whom the electric power will be delivered (see also [23] and [35]). The bottom-tiered nodes in Fig. 12.1 represent the demand markets, which can differ by their geographic location or the type of associated consumers; for example, whether they correspond to businesses or households. The nodes corresponding to the demand markets are enumerated as: $1, \dots, k, \dots, K$ with node k corresponding to demand market k. The majority of the notation needed for the model is given in Table 12.1. An equilibrium solution is denoted by "*." All vectors are assumed to be column vectors, except where noted otherwise.

We now focus on the notation for the permits. Similar to the discussion in [17, 18] and [14], let l^j_{gmr}; $j = 1, \dots, J$; $g = 1, \dots, G$; $m = 1, \dots, M$; $r = 1, \dots, R$ denote the number of permits/licenses for pollutant of type j held by genco g that uses power plant m, and which affects receptor point r with l^{j0}_{gmr} denoting the initial allocation. Group the former permits into the $JGMR$-dimensional vector l.

Let e^j_{gmr}; $j = 1, \dots, J$; $g = 1, \dots, G$; $m = 1, \dots, M$; $r = 1, \dots, R$ denote the unit contribution of the ambient concentration of pollutant type j affecting the receptor point r generated per unit of electric power produced by genco g using his power plant m. Hence, the total amount of ambient concentration of pollutant j at receptor point r associated with genco g and power plant m is $e^j_{gmr} q_{gm}$.

Table 12.1. Notation for the electric power supply chain network model with power plants (cf. [35])

Notation	*Definition*
q_{gm}	Quantity of electricity produced by generator g using power plant m, where $g = 1, \ldots, G; m = 1, \ldots, M$
q_m	G-dimensional vector of electric power generated by the gencos using power plant technology m with components: $q_{1m}, \ldots, q_{Gm}$
q	GM-dimensional vector of all the electric power outputs generated by the gencos at the power plants
Q^1	GMS-dimensional vector of electric power flows between the power plants of the power generators and the power suppliers with component gms denoted by q_{gms}
Q^2	SVK-dimensional vector of power flows between suppliers and demand markets with component svk denoted by q^v_{sk} and denoting the flow between supplier s and demand market k via transmission service provider v
d	K-dimensional vector of market demands with component k denoted by d_k
$f_{gm}(q_m)$	Power generating cost function of power generator g using power plant m with marginal power generating cost with respect to q_{gm} denoted by $\frac{\partial f_{gm}}{\partial q_{gm}}$
$c_{gms}(q_{gms})$	Transaction cost incurred by power generator g using power plant m in transacting with power supplier s with marginal transaction cost denoted by $\frac{\partial c_{gms}(q_{gms})}{\partial q_{gms}}$
h	S-dimensional vector of the power suppliers' supplies of the electric power with component s denoted by h_s, with $h_s \equiv \sum_{g=1}^{G} \sum_{m=1}^{M} q_{gms}$
$c_s(h) \equiv c_s(Q^1)$	Operating cost of power supplier s with marginal operating cost with respect to h_s denoted by $\frac{\partial c_s}{\partial h_s}$ and the marginal operating cost with respect to q_{gms} denoted by $\frac{\partial c_s(Q^1)}{\partial q_{gms}}$
$c^v_{sk}(q^v_{sk})$	Transaction cost incurred by power supplier s in transacting with demand market k via transmission service provider v with marginal transaction cost with respect to q^v_{sk} denoted by $\frac{\partial c^v_{sk}(q^v_{sk})}{\partial q^v_{sk}}$
$\hat{c}_{gms}(q_{gms})$	Transaction cost incurred by power supplier s in transacting with power generator g for power generated by plant m with marginal transaction cost denoted by $\frac{\partial \hat{c}_{gms}(q_{gms})}{\partial q_{gms}}$
$\hat{c}^v_{sk}(Q^2)$	Unit transaction cost incurred by consumers at demand market k in transacting with power supplier s via transmission service provider v
$\rho_{3k}(d)$	Demand market price function at demand market k

12.2.1 The Behavior of the Power Generators and their Optimality Conditions

Let ρ^*_{1gms} denote the unit price charged by power generator g for the transaction with power supplier s for electric power produced at plant m with $g = 1, \ldots, G$; $m = 1, \ldots, M$, and $s = 1, \ldots, S$. ρ^*_{1gms} is an endogenous variable and can be determined once the complete electric power supply chain network equilibrium model is solved. Let τ^{j*}_r; $j = 1, \ldots, J$; $r = 1, \ldots, R$ denote the price of the permit at equilibrium for pollutant of type j of emission affecting receptor point r. These prices are also endogenous to the model and will be determined once the complete model is solved.

We assume that each electric power generator seeks to determine his optimal production portfolio across his power plants and his sales allocations of the electric power to the suppliers as well as the optimal holdings of pollution permits in order to maximize his own profit. Since we have assumed that each individual power generator is a profit-maximizer, the objective function of power generator g can be expressed as follows:

$$\text{Maximize} \quad \sum_{m=1}^{M}\sum_{s=1}^{S} \rho^*_{1gms} q_{gms} - \sum_{m=1}^{M} f_{gm}(q_m) - \sum_{m=1}^{M}\sum_{s=1}^{S} c_{gms}(q_{gms})$$

$$- \sum_{j=1}^{J}\sum_{m=1}^{M}\sum_{r=1}^{R} \tau^{j*}_r (l^j_{gmr} - l^{j0}_{gmr}). \tag{12.1}$$

The first term in the objective function (12.1) represents the revenue of power generator g and the next two terms represent his power generation cost and transaction costs, respectively. The last term denotes the expenditure or revenue from transacting permits for the generator based on the total pollutants by his power plants affecting the ambient concentrations at the receptor points.

The structure of the network in Fig. 12.1 guarantees that the conservation of flow equations associated with the electric power production and distribution are satisfied. Conservation of flow equation (12.2) below states that the amount of power generated at a particular power plant (and corresponding to a particular genco) is equal to the electric power transacted by the genco from that power plant with all the suppliers and this holds for each of the power plants, subject to:

$$\sum_{s=1}^{S} q_{gms} = q_{gm}, \quad m = 1, \ldots, M. \tag{12.2}$$

Equation (12.3) below states that each power plant cannot pollute at an amount greater than the plant is licensed to at that receptor point.

$$l^j_{gmr} \geq e^j_{gmr} q_{gm}, \quad j = 1 \ldots, J; m = 1, \ldots, M; r = 1, \ldots, R. \tag{12.3}$$

The following nonnegativity conditions must also hold:

$$q_{gms} \geq 0, \quad m = 1, \ldots, M; s = 1, \ldots, S,$$

$$l^j_{gmr} \geq 0, \quad j = 1, \ldots, J; m = 1, \ldots, M; r = 1, \ldots, R. \tag{12.4}$$

Hence, the optimization problem of power generator g; $g = 1, \ldots, G$ consists of (12.1), subject to constraints: (12.2) and (12.3), with the nonnegativity assumption on the electric power outputs at the power plants and the number of permits (cf. following (12.1)). Assume now, as was done in [21] and [35], that the generating cost and the transaction cost functions for each power generator are continuously differentiable and convex, and that the power generators compete in a noncooperative manner in the sense of Nash ([26,27]). The optimality conditions for all power generators, under the above assumptions (cf. [15]), coincide with the solution of the following variational inequality: determine $(q^*, Q^{1*}, l^*, \lambda^*) \in \mathcal{K}^1$ satisfying

$$\sum_{g=1}^{G}\sum_{m=1}^{M}\left[\frac{\partial f_{gm}(q^*_m)}{\partial q_{gm}} + \sum_{j=1}^{J}\sum_{r=1}^{R}\lambda^{j*}_{gmr}e^j_{gmr}\right] \times [q_{gm} - q^*_{gm}]$$
$$+\sum_{g=1}^{G}\sum_{m=1}^{M}\sum_{s=1}^{S}\left[\frac{\partial c_{gms}(q^*_{gms})}{\partial q_{gms}} - \rho^*_{1gms}\right] \times [q_{gms} - q^*_{gms}]$$
$$+\sum_{j=1}^{J}\sum_{g=1}^{G}\sum_{m=1}^{M}\sum_{r=1}^{R}[\tau^{j*}_r - \lambda^{j*}_{gmr}] \times [l^j_{gmr} - l^{j*}_{gmr}]$$
$$+\sum_{j=1}^{J}\sum_{g=1}^{G}\sum_{m=1}^{M}\sum_{r=1}^{R}[l^{j*}_{gmr} - e^j_{gmr}q^*_{gm}] \times [\lambda^j_{gmr} - \lambda^{j*}_{gmr}] \geq 0, \quad \forall (q, Q^1, l, \lambda) \in \mathcal{K}^1, \tag{12.5}$$

where $\mathcal{K}^1 \equiv \{(q, Q^1, l, \lambda)|(q, Q^1, l, \lambda) \in R^{GM+GMS+2JGMR}_+$ and (12.2) holds$\}$.

Note that λ^j_{gmr} is the Lagrange multiplier associated with the (jmr)-th constraint (12.3), which we refer to as a shadow price.

Equilibrium Conditions for the Permits

Furthermore, we know that (cf. [5]) the multipollutant permit market is also subject to equilibrium conditions given by the following. For each pollution permit of type j; $j = 1, \ldots, J$ and receptor point r; $r = 1, \ldots, R$, a multipollutant tradable permit scheme is said to be in equilibrium if:

$$\sum_{g=1}^{G}\sum_{m=1}^{M}[l^{j0}_{gmr} - l^{j*}_{gmr}]\begin{cases} = 0, \text{ if } & \tau^{j*}_r > 0, \\ \geq 0, \text{ if } & \tau^{j*}_r = 0. \end{cases} \tag{12.6}$$

Expression (12.6) states that if the market price of a permit for pollutant of type j and receptor point r is positive, then there is no excess of permits for that pollutant at that receptor point; if the price is zero, then there can be an excess of such permits. Clearly, these equilibrium conditions guarantee that the total number of required permits cannot exceed the initial allocation of permits by the regulatory agency for each receptor point and pollutant.

The optimality conditions for all power generators simultaneously (cf. (12.5)), under the above assumptions (cf. [15]), coupled with the equilibrium conditions (12.6) for all pollutant types and receptor points, coincide, in turn, with the solution of the following variational inequality: determine $(q^*, Q^{1*}, l^*, \lambda^*, \tau^*) \in \mathcal{K}^2$ satisfying

$$\sum_{g=1}^{G}\sum_{m=1}^{M}\left[\frac{\partial f_{gm}(q_m^*)}{\partial q_{gm}} + \sum_{j=1}^{J}\sum_{r=1}^{R}\lambda_{gmr}^{j*}e_{gmr}^{j}\right] \times [q_{gm} - q_{gm}^*]$$

$$+\sum_{g=1}^{G}\sum_{m=1}^{M}\sum_{s=1}^{S}\left[\frac{\partial c_{gms}(q_{gms}^*)}{\partial q_{gms}} - \rho_{1gms}^*\right] \times [q_{gms} - q_{gms}^*]$$

$$+\sum_{j=1}^{J}\sum_{g=1}^{G}\sum_{m=1}^{M}\sum_{r=1}^{R}[\tau_r^{j*} - \lambda_{gmr}^{j*}] \times [l_{gmr}^{j} - l_{gmr}^{j*}]$$

$$+\sum_{j=1}^{J}\sum_{g=1}^{G}\sum_{m=1}^{M}\sum_{r=1}^{R}[l_{gmr}^{j*} - e_{gmr}^{j}q_{gm}^*] \times [\lambda_{gmr}^{j} - \lambda_{gmr}^{j*}]$$

$$+\sum_{j=1}^{J}\sum_{r=1}^{R}\left[\sum_{g=1}^{G}\sum_{m=1}^{M}(l_{gmr}^{j0} - l_{gmr}^{j*})\right] \times [\tau_r^{j} - \tau_r^{j*}] \geq 0, \quad \forall (q, Q^1, l, \lambda, \tau) \in \mathcal{K}^2, \tag{12.7}$$

where $\mathcal{K}^2 \equiv \{(q, Q^1, l, \lambda, \tau) | (q, Q^1, l, \lambda, \tau) \in R_+^{GM+GMS+2JGMR+JR}$ and (12.2) holds$\}$.

The Behavior of Power Suppliers and their Optimality Conditions

The power suppliers transact with the power generators and with the consumers at the demand markets through the transmission service providers. Suppliers are aware as to the types of power plants used and associated costs when purchasing electric power from the power generators. Analogous to the gencos, we assume that the power suppliers compete with one another in a noncooperative manner.

Since electric power cannot be stored, the following conservation of flow constraint states that the total amount of electricity sold by a power supplier is equal to the total electric power that he purchased from the generators and produced via the different power plants available to the generators, that is:

$$\sum_{k=1}^{K}\sum_{v=1}^{V} q_{sk}^{v} = \sum_{g=1}^{G}\sum_{m=1}^{M} q_{gms}, \quad s = 1, \dots, S. \tag{12.8}$$

Let ρ_{2sk}^{v*} denote the price charged by power supplier s to demand market k via transmission service provider v. This price is determined endogenously in the model once the entire network equilibrium problem is solved. It is assumed that each power supplier seeks to maximize his own profit, hence the optimization problem faced by supplier s may be expressed as follows:

$$\text{Maximize} \quad \sum_{k=1}^{K}\sum_{v=1}^{V} \rho_{2sk}^{v*} q_{sk}^{v} - c_s(Q^1) - \sum_{g=1}^{G}\sum_{m=1}^{M} \rho_{1gms}^{*} q_{gms} - \sum_{g=1}^{G}\sum_{m=1}^{M} \hat{c}_{gms}(q_{gms})$$

$$- \sum_{k=1}^{K}\sum_{v=1}^{V} c_{sk}^{v}(q_{sk}^{v}) \tag{12.9}$$

subject to:

$$\sum_{k=1}^{K}\sum_{v=1}^{V} q_{sk}^{v} = \sum_{g=1}^{G}\sum_{m=1}^{M} q_{gms}, \tag{12.10}$$

$$q_{gms} \geq 0, \quad g = 1, \dots, G; m = 1, \dots, M,$$

$$q_{sk}^{v} \geq 0; \quad k = 1, \dots, K; v = 1, \dots, V. \tag{12.11}$$

The first term in (12.9) denotes the revenue of supplier s from the sale of electricity to the demand market k via transmission service provider v, with the associated operating cost in the second term. The third term denotes the cost to purchase electricity for each supplier from each genco, and the last two terms represent the associated transaction costs for transactions with each genco and each demand market, respectively.

We assume that the transaction costs and the operating costs in (12.9) are all continuously differentiable and convex, and that the power suppliers compete in a noncooperative manner. Hence, the optimality conditions for all suppliers, simultaneously, under the above assumptions, can be expressed as the following variational inequality: determine $(Q^{2*}, Q^{1*}) \in \mathcal{K}^3$ such that

$$\sum_{s=1}^{S}\sum_{k=1}^{K}\sum_{v=1}^{V} \left[\frac{\partial c_{sk}^{v}(q_{sk}^{v*})}{\partial q_{sk}^{v}} - \rho_{2sk}^{v*} \right] \times [q_{sk}^{v} - q_{sk}^{v*}]$$

$$+ \sum_{g=1}^{G}\sum_{m=1}^{M}\sum_{s=1}^{S} \left[\frac{\partial c_s(Q^{1*})}{\partial q_{gms}} + \frac{\partial \hat{c}_{gms}(q_{gms}^{*})}{\partial q_{gms}} + \rho_{1gms}^{*} \right] \times [q_{gms} - q_{gms}^{*}] \geq 0, \tag{12.12}$$

$\forall (Q^2, Q^1) \in \mathcal{K}^3$, where $\mathcal{K}^3 \equiv \{(Q^2, Q^1) | (Q^2, Q^1) \in R_{+}^{SVK+GMS}$ and (12.10); equivalently (12.8) holds$\}$.

For notational convenience, and as was done in [35], we let

$$h_s \equiv \sum_{g=1}^{G} \sum_{m=1}^{M} q_{gms}, \quad s = 1, \ldots, S. \tag{12.13}$$

As defined in Table 12.1, the operating cost of power supplier s, c_s, is a function of the total electricity inflows to the power supplier, that is:

$$c_s(h) \equiv c_s(Q^1), \quad s = 1, \ldots, S. \tag{12.14}$$

Hence, his marginal cost with respect to h_s is equal to the marginal cost with respect to q_{gms}:

$$\frac{\partial c_s(h)}{\partial h_s} \equiv \frac{\partial c_s(Q^1)}{\partial q_{gms}}, \quad s = 1, \ldots, S; m = 1, \ldots, M; g = 1, \ldots, G. \tag{12.15}$$

After the substitution of (12.13) and (12.15) into (12.12), and algebraic simplification, we obtain a variational inequality equivalent to (12.12), as follows: determine $(h^*, Q^{2*}, Q^{1*}) \in \mathcal{K}^4$ such that

$$\sum_{s=1}^{S} \frac{\partial c_s(h^*)}{\partial h_s} \times [h_s - h_s^*] + \sum_{s=1}^{S} \sum_{k=1}^{K} \sum_{v=1}^{V} \left[\frac{\partial c_{sk}^v(q_{sk}^{v*})}{\partial q_{sk}^v} - \rho_{2sk}^{v*} \right] \times [q_{sk}^v - q_{sk}^{v*}]$$

$$+ \sum_{g=1}^{G} \sum_{m=1}^{M} \sum_{s=1}^{S} \left[\frac{\partial \hat{c}_{gms}(q_{gms}^*)}{\partial q_{gms}} + \rho_{1gms}^* \right] \times [q_{gms} - q_{gms}^*] \geq 0, \tag{12.16}$$

$\forall (h, Q^2, Q^1,) \in \mathcal{K}^4$,where $\mathcal{K}^4 \equiv \{(h, Q^2, Q^1) | (h, Q^2, Q^1) \in R_+^{S(1+VK+GM)}$ and (12.10) and (12.13) hold$\}$.

Equilibrium Conditions for the Demand Markets

At each demand market k the following conservation of flow equation must be satisfied:

$$d_k = \sum_{s=1}^{S} \sum_{v=1}^{V} q_{sk}^v, \quad k = 1, \ldots, K. \tag{12.17}$$

For each power supplier s; $s = 1, \ldots, S$ and transaction mode v; $v = 1, ..., V$, the market equilibrium conditions at demand market k take the form:

$$\rho_{2sk}^{v*} + \hat{c}_{sk}^v(Q^{2*}) \begin{cases} = \rho_{3k}(d^*), \text{ if } & q_{sk}^{v*} > 0, \\ \geq \rho_{3k}(d^*), \text{ if } & q_{sk}^{v*} = 0. \end{cases} \tag{12.18}$$

According to [21, 23], and [35], consumers at the demand market will purchase electricity from a supplier via a transmission service provider if the price that the consumer at the demand market is willing to pay is equal to the price charged by the power supplier plus the unit transaction cost. However, if the purchase price plus the unit transaction cost exceeds the purchase price that the consumer is willing to pay, then no transaction will take place. The equivalent variational inequality, given that, in equilibrium, condition (12.18) must hold simultaneously for all demand markets: $k = 1, \ldots, K$, takes the form: determine $(Q^{2*}, d^*) \in \mathcal{K}^5$, such that

$$\sum_{s=1}^{S} \sum_{k=1}^{K} \sum_{v=1}^{V} \left[\rho_{2sk}^{v*} + \hat{c}_{sk}^{v}(Q^{2*})\right] \times \left[q_{sk}^{v} - q_{sk}^{v*}\right] - \sum_{k=1}^{K} \rho_{3k}(d^*) \times \left[d_k - d_k^*\right] \geq 0, \quad (12.19)$$

$\forall (Q^2, d) \in \mathcal{K}^5$, where $\mathcal{K}^5 \equiv \{(Q^2, d) | (Q^2, d) \in R_+^{KSV+K} \text{and (12.17) holds}\}$.

The Equilibrium Conditions for the Electric Power Supply Chain Network with Multipollutant Permits

In equilibrium, the optimality conditions for all the power generators, the optimality conditions for all the power suppliers, and the equilibrium conditions for all the demand markets as well as the equilibrium conditions for the permits must be simultaneously satisfied so that no decision-maker has any incentive to alter his transactions. We now formally state the equilibrium conditions for the entire electric power supply chain network along with the variational inequality formulation, which follows directly from the definition.

Definition 12.1: Electric Power Supply Chain Network Equilibrium with Multipollutant Permits

The equilibrium state of the electric power supply chain network with power plants and multipollutant permits is one where the electric power flows between the tiers of the network coincide and the electric power flows and the multipollutant tradable permits and prices satisfy the sum of conditions (12.5), (12.16), and (12.19).

Variational Inequality Formulation of the Electric Power Supply Chain Network Equilibrium with Multipollutant Permits

The equilibrium conditions governing the electric power supply chain network according to Definition 12.1 coincide with the solution of the variational inequality given by: determine the vector of equilibrium electric power production quantities and flows, the demands, the number of permits, the shadow prices, and the permit prices $(q^*, h^*, Q^{1*}, Q^{2*}, d^*, l^*, \lambda^*, \tau^*) \in \mathcal{K}^6$ *satisfying:*

$$\sum_{g=1}^{G} \sum_{m=1}^{M} \left[\frac{\partial f_{gm}(q_m^*)}{\partial q_{gm}} + \sum_{j=1}^{J} \sum_{r=1}^{R} \lambda_{gmr}^{j*} e_{gmr}^{j}\right] \times \left[q_{gm} - q_{gm}^*\right] + \sum_{s=1}^{S} \frac{\partial c_s(h^*)}{\partial h_s} \times \left[h_s - h_s^*\right]$$

$$+\sum_{g=1}^{G}\sum_{m=1}^{M}\sum_{s=1}^{S}\left[\frac{\partial c_{gms}(q^*_{gms})}{\partial q_{gms}}+\frac{\partial \hat{c}_{gms}(q^*_{gms})}{\partial q_{gms}}\right]\times[q_{gms}-q^*_{gms}]$$

$$+\sum_{s=1}^{S}\sum_{k=1}^{K}\sum_{v=1}^{V}\left[\frac{\partial c^v_{sk}(q^{v*}_{sk})}{\partial q^v_{sk}}+\hat{c}^v_{sk}(Q^{2*})\right]\times[q^v_{sk}-q^{v*}_{sk}]-\sum_{k=1}^{K}\rho_{3k}(d^*)\times[d_k-d^*_k]$$

$$+\sum_{j=1}^{J}\sum_{g=1}^{G}\sum_{m=1}^{M}\sum_{r=1}^{R}\left[\tau^{j*}_r-\lambda^{j*}_{gmr}\right]\times\left[l^j_{gmr}-l^{j*}_{gmr}\right]$$

$$+\sum_{j=1}^{J}\sum_{g=1}^{G}\sum_{m=1}^{M}\sum_{r=1}^{R}\left[l^{j*}_{gmr}-e^j_{gmr}q^*_{gm}\right]\times\left[\lambda^j_{gmr}-\lambda^{j*}_{gmr}\right]$$

$$+\sum_{j=1}^{J}\sum_{r=1}^{R}\left[\sum_{g=1}^{G}\sum_{m=1}^{M}(l^{j0}_{gmr}-l^{j*}_{gmr})\right]\times[\tau^j_r-\tau^{j*}_r]\geq 0, \forall(q,h,Q^1,Q^2,d,l,\lambda,\tau)\in\mathcal{K}^6, \tag{12.20}$$

where $\mathcal{K}^6 \equiv \{(q,h,Q^1,Q^2,d,l,\lambda,\tau)|(q,h,Q^1,Q^2,d,l,\lambda,\tau) \in R_+^{GM+S+GMS+SKV+K+2JGMR+JR}$ and $(12.2), (12.10), (12.13),$ *and* (12.17) hold$\}$.

We now put variational inequality (12.20) into standard form (cf. [15]), which can be expressed as:

$$\langle F(X^*), X-X^*\rangle \geq 0, \quad \forall X\in\mathcal{K}, \tag{12.21}$$

where $X \equiv (q,h,Q^1,Q^2,d,l,\lambda,\tau) \in R_+^{GM+S+GMS+SKV+K+2GMRJ+RJ}$ and $F(X)$ as a column vector consisting of the column vectors $(P_{gm}, H_s, \Lambda_{gms}, G_{skv}, D_k, L_{jgmr}, C_{jgmr}, T_{jr})$ with indices: $g = 1,\dots,G$; $m = 1,\dots,M$; $s = 1,\dots,S$; $k = 1,\dots,K$; $v = 1,\dots,V$; $j = 1,\dots,J$; $r = 1,\dots,R$, and the specific components of F given by the functional terms preceding the multiplication signs in (12.20), respectively. The term $\langle\cdot,\cdot\rangle$ denotes the inner product in N-dimensional Euclidean space R^N.

We now identify a special case of the above model which will correspond to a particular pollution permit trading scheme. Chen and Hobbs ([3]; also see [4]) considered a single pollutant and single receptor point tradable permit market scheme. We now provide additional theoretical results which are important for environmental decision-making and policy-making. Similar results can be found in [5], but not generalized to the electric power industry with multiple power plants. Let $\bar{E}^j_r$; $j = 1,\dots,J$; $r = 1,\dots,R$, denote the imposed environmental standard for receptor r and emission type j. We now state the following.

Theorem 12.1 (Equilibrium Pattern Independence from Initial Permit Allocation)

If $l_{gmr}^{j0} \geq 0$, *for all* $j = 1, \ldots, J$; $g = 1, \ldots, G$; $m = 1, \ldots, M$, *and* $r = 1, \ldots, R$, *and* $\sum_{g=1}^{G} \sum_{m=1}^{M} l_{gmr}^{j0} = \bar{E}_r^j$, *for* $j = 1, \ldots, J$; $r = 1, \ldots, R$ *with each* $\bar{E}_r^j$ *positive and fixed, then the equilibrium pattern* $(q^*, h^*, Q^{1*}, Q^{2*}, d^*, l^*, \lambda^*, \tau^*)$ *is independent of* $\{l_{gmr}^{j0}\}$.

Proof. The last term in (12.20) (unlike the first seven in (12.20) which are independent of l_{gmr}^{j0}) depends only on the sum $\sum_{g=1}^{G} \sum_{m=1}^{M} l_{gmr}^{j0}$, for a fixed receptor point j and a fixed pollutant of type j. □

In the next Theorem, we provide a means for the selection of the sums of the initial permit/license allocation so that the imposed environmental standards are achieved.

Theorem 12.2 (Attainment of Environmental Standards)

An equilibrium vector, satisfying variational inequality (12.20), attains the environmental quality standards represented by vector $\bar{E} = (\bar{E}_1, \ldots, \bar{E}_R)$ *where* $\bar{E}_r = (\bar{E}_r^1, \ldots, \bar{E}_r^J)$ *for* r; $r = 1, \ldots, R$, *provided that the following is satisfied:*

$$\sum_{g=1}^{G} \sum_{m=1}^{M} l_{gmr}^{j0} = \bar{E}_r^j, \quad \forall r, \forall j. \tag{12.22}$$

Proof. From constraint (12.3) we have that

$$l_{gmr}^{j*} \geq e_{gmr}^j q_{gm}^*, \quad j = 1 \ldots J; m = 1 \ldots, M; r = 1, \ldots, R. \tag{12.23}$$

It is then clear from the assumption and variational inequality (12.20) that

$$\bar{E}_r^j = \sum_{g=1}^{G} \sum_{m=1}^{M} l_{gmr}^{j0} \geq \sum_{g=1}^{G} \sum_{m=1}^{M} l_{gmr}^{j*} \geq \sum_{g=1}^{G} \sum_{m=1}^{M} e_{gmr}^j q_{gm}^* \tag{12.24}$$

for all $j = 1, \ldots, J$; $r = 1, \ldots, R$.

Theorem 12.2 provides a mechanism for the determination of the sums of the initial permit/license allocations so that the environmental standards are attained. Indeed, all one needs to do is to set the initial permit allocation so that (12.22) is satisfied. We will illustrate this with examples in the next section.

12.3 Algorithm and Examples

Clearly, there are distinct variational inequality algorithms that may be applied to solve variational inequalities (12.20), and, in particular, we note the modified projection method (see [15]) which has been successfully applied to solve variational inequality problems in which the function F (cf. (12.21)) is monotone and Lipschitz continuous.

Wu et al. ([35]) in turn, proposed an Euler method for the electric power supply chain network equilibrium problem with power plants and reassigned carbon taxes. That Euler method was introduced by Dupuis and Nagurney ([6]), and is a special case of a general iterative scheme for the solution of variational inequalities as well as projected dynamical systems. [35] showed that the electric power supply chain problem with preassigned taxes could be transformed into a transportation network equilibrium problem over an appropriately constructed abstract network or supernetwork.

In the model and special case developed in this paper, we can no longer transform the variational inequalities (12.20) *directly* into transportation network equilibrium problems as was also done by [16] for supply chain network equilibrium problems. However, we can still exploit the connection by noticing that the variational inequality problems in this paper are defined over feasible sets that are, in effect, decomposable into subproblems in the flows and subproblems in the licenses, the shadow prices, and the license prices. Furthermore, the former subproblems retain the transportation network structure identified in [35] and this can be exploited algorithmically. Hence, we can apply the Euler method, whose general statement to solve a variational inequality is given by: determine $X^* \in \mathcal{K}$ such that

$$\langle F(X^*), X - X^* \rangle \geq 0, \quad \forall X \in \mathcal{K}, \tag{12.25}$$

and is given immediately following.

The Euler Method

The Euler method (see [6]) has been applied by [25] to solve the variational inequality governing elastic demand transportation network equilibrium problems in path flows. Convergence results can be found in the above references. For the solution of (12.25), the Euler method takes the form: at iteration l compute X^{l+1} by solving the variational inequality problem:

$$X^{l+1} = P_{\mathcal{K}}(X^l - a_l F(X^l)), \tag{12.26}$$

where $P_{\mathcal{K}}$ is the projection operator, and the sequence $\{a_l\}$ must satisfy the conditions: $\sum_{l=0}^{\infty} a_l = \infty$, $a_l > 0$, for all l, and $a_l \to 0$, as $l \to \infty$.

For completeness, we now present several examples. The examples consisted of two power generators, each of which had two power plants. There were two power suppliers and two demand markets with a single transmission service provider. We also assumed that there was a single pollutant and a single receptor point, as shown in Fig. 12.2.

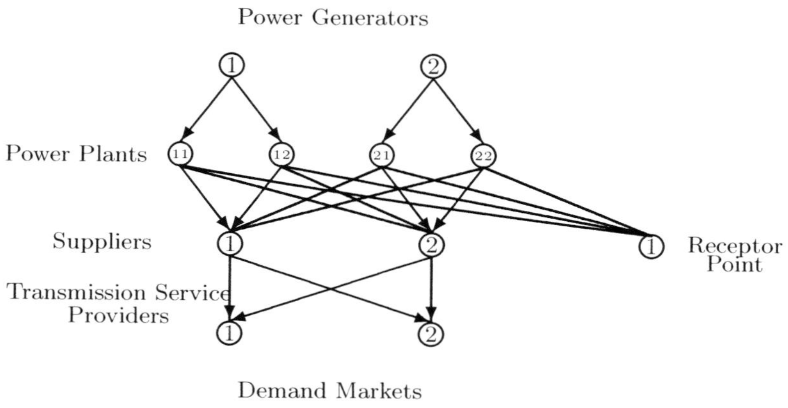

Fig. 12.2. Electric power supply chain network with a single receptor point for the examples

Example 12.1

The data for the first example are given below. The functional forms of the power generating cost functions, the transaction cost functions, the operating cost functions, and the demand price functions are identical to those in Example 12.1 in [35].

The emission terms: e_{gm}; $g = 1, 2$; $m = 1, 2$ were all equal to 1. The power generating cost functions for the power generators were given by:

$$f_{11}(q_1) = 2.5q_{11}^2 + q_{11}q_{21} + 2q_{11}, \quad f_{12}(q_2) = 2.5q_{12}^2 + q_{11}q_{12} + 2q_{22},$$

$$f_{21}(q_1) = .5q_{21}^2 + .5q_{11}q_{21} + 2q_{21}, \quad f_{22}(q_2) = .5q_{22}^2 + q_{12}q_{22} + 2q_{22}.$$

The transaction cost functions faced by the power generators and associated with transacting with the power suppliers were given by:

$$c_{111}(q_{111}) = .5q_{111}^2 + 3.5q_{111}, \quad c_{112}(q_{112}) = .5q_{112}^2 + 3.5q_{112},$$

$$c_{121}(q_{121}) = .5q_{121}^2 + 3.5q_{121}, \quad c_{122}(q_{122}) = .5q_{122}^2 + 3.5q_{122},$$

$$c_{211}(q_{211}) = .5q_{211}^2 + 2q_{211}, \quad c_{212}(q_{212}) = .5q_{212}^2 + 2q_{212},$$

$$c_{221}(q_{221}) = .5q_{221}^2 + 2q_{221}, \quad c_{222}(q_{222}) = .5q_{222}^2 + 2q_{222}.$$

The operating costs of the power generators, in turn, were given by:

$$c_1(Q^1) = .5\left(\sum_{i=1}^{2} q_{i1}\right)^2, \quad c_2(Q^1) = .5\left(\sum_{i=1}^{2} q_{i2}\right)^2.$$

The demand market price functions at the demand markets were:

$$\rho_{31}(d) = -1.33d_1 + 366.6, \quad \rho_{32} = -1.33d_2 + 366.6,$$

and the transaction costs between the power suppliers and the consumers at the demand markets were given by: $\hat{c}^1_{sk}(q^1_{sk}) = q^1_{sk} + 5, s = 1, 2; k = 1, 2$. All other transaction costs were assumed to be equal to zero.

In Example 12.1, the emissions standard $\bar{E} = 100$ with the initial license allocation given by: $l^0_{11} = l^0_{12} = l^0_{21} = l^0_{22} = 25$. The equilibrium electric power flows and demands and the equilibrium licenses and prices are given in Table 12.2. The demand was 50.00 at each demand market and the demand market price at each market for electric power was 300.10.

Example 12.2

Example 12.2 had the same data as Example 12.1, but we now tightened the emissions standard so that $\bar{E} = 50$. The initial license allocation was now given by: $l^0_{11} = l^0_{12} = l^0_{21} = l^0_{22} = 12.5$. The equilibrium solution is given in Table 12.2. It is clear that, as predicted by the theory, the environmental standard is achieved.

Example 12.3

Example 12.3 had the identical data to that in Examples 12.1 and 12.2, except that the environmental standard was further tightened to $\bar{E} = 20$ with the new initial license allocation given by: $l^0_{11} = l^0_{12} = l^0_{21} = l^0_{22} = 5$. The new equilibrium pattern is reported in Table 12.2. In this example, it is also clear that the equilibrium license numbers are such that the environmental standard is attained.

Example 12.4

Example 12.4 had the same data as Example 12.3 except that we modified the second demand market price function for electric power to:

$$\rho_{32}(d) = -1.33d_2 + 733.30.$$

The new equilibrium electric power flow, license, and price pattern is also reported in Table 12.2. In this example, there is zero demand for electric power at the first demand market. As in the preceding examples, the environmental standard is achieved. Note that as the equilibrium price of the permits increases, as expected, as the environmental standard is tightened for each successive example.

Table 12.2. Solutions to examples 1, 2, 3, and 4

Equilibrium solution	Example 1	Example 2	Example 3	Example 4
Equilibrium electric power flows				
q_{11}^*	15.20	7.48	2.85	2.87
q_{12}^*	6.63	3.17	1.10	1.10
q_{21}^*	15.53	7.82	3.19	3.20
q_{22}^*	62.65	31.53	12.86	12.91
q_{111}^*	7.60	3.74	1.43	1.43
q_{112}^*	7.60	3.74	1.43	1.43
q_{121}^*	3.31	1.59	0.55	0.55
q_{122}^*	3.31	1.59	0.55	0.55
q_{211}^*	7.76	3.91	1.59	1.60
q_{212}^*	7.76	3.91	1.59	1.60
q_{221}^*	31.32	15.77	6.43	6.46
q_{222}^*	31.32	15.77	6.43	6.46
h_1^*	50.00	25.00	10.00	10.00
h_2^*	50.00	25.00	10.00	10.00
q_{11}^{1*}	25.00	12.50	5.00	0.00
q_{12}^{1*}	25.00	12.50	5.00	10.00
q_{21}^{1*}	25.00	12.50	5.00	0.00
q_{22}^{1*}	25.00	12.50	5.00	10.00
Equilibrium demands				
d_1^*	50.00	25.00	10.00	0.00
d_2^*	50.00	25.00	10.00	20.00
Equilibrium pollution permit price and shadow prices				
$\tau^* = \lambda_{11}^* = \lambda_{12}^* = \lambda_{21}^* = \lambda_{22}^*$	115.50	236.38	308.91	656.96
Equilibrium permits/licenses				
l_{11}^*	15.20	7.48	2.85	2.87
l_{12}^*	6.63	3.17	1.10	1.10
l_{21}^*	15.53	7.82	3.19	3.20
l_{22}^*	62.65	31.53	12.86	12.91

12.4 Summary and Conclusions

As noted in the Introduction, pollution by electric power entities can be controlled by price, in the form, for example, of a carbon tax that is imposed for emissions that exceed a predetermined bound (and as modeled in [35] and [21]) or by quantity, as in the case of emission trading schemes. In this paper, we developed a multipollutant permit trading model in the case of electric power supply chains in which there are different technologies associated with electric power production. We derived the governing equilibrium conditions of the model and showed that it satisfies a finite-dimensional variational inequality problem. Moreover, we demonstrated that the model guarantees that the environmental standards are achieved, provided that the initial license allocation is set accordingly. Finally, we described how the equilibrium electric power flows and the pollution permits/licenses, along with their prices could be computed. For completeness, we also provided several numerical examples. Future research will include the identification of efficient computational procedures for large-scale electric power supply chains with tradable pollution permits.

The research in this chapter is the first to incorporate the substitutability and complementarity effects of multiple pollutants. This research can aid a regulatory agency in the determination of the number of permits required to achieve the reduction of emissions below a predetermined bound. Moreover, this model focuses specifically on electric power supply chains and the effects of governmental mandates regarding environmental standards on the associated prices and quantities. The importance of environmental-energy modeling to address market failures in energy is growing as awareness of pollution effects, emission abatement technologies, and government policies are changing. A limitation of the model is the requirement of the electric power industry to report accurate and true data regarding the costs of producing electricity. A future application of this model could include the empirical implementation of a tradable permit system, such as, for example, for the electric power supply chain of New England (see [11]).

Acknowledgement. The authors are indebted to the two anonymous reviewers for their helpful comments and suggestions on an earlier version of this chapter. The research of the first two authors was supported, in part, by NSF Grant. No.: IIS 00026471 and, in part, by the John F. Smith Memorial Fund at the Isenberg School of Management. This support is gratefully acknowledged and appreciated.

References

1. J. Boucher and Y. Smeers. Studies in the economics of transportation. *Operations Research*, 49:821–838, 2002.
2. D. Burtraw, D.A. Evans, A. Krupnick, K. Palmer, and R. Toth. Alternative models of restructured electricity system, part 1: No market power. *Annual Review of Environment and Resources*, 30:253–289, 2005.
3. Y.H. Chen and B.F. Hobbs. An oligopolistic electricity market model with tradable nox permits. *IEEE Transactions on Power Systems*, 20:119–129, 2005.
4. Y.H. Chen, B.F. Hobbs, S. Leyffer, and T.S. Munson. Leader-follow equilibria for electric power and no_x allowances markets. *Computational Management Science Online*, pages 1619–6988, 2006.
5. K.K. Dhanda, A. Nagurney, and P. Ramanujam. *Environmental Networks: A Framework for Economic Decision-Making and Policy Analysis.* Edward Elgar, Cheltenham, England, 1999.
6. P. Dupuis and A. Nagurney. Dynamical systems and variational inequalities. *Annals of Operations Research*, 44:9–42, 1993.

7. C. Hanisch. Where is mercury deposition coming from? uncertainties about the roles of different natural and synthetic sources are fueling the debate on how to regulate emissions. *Environmental Science and Technology*, 32:176A–179A, 1998.
8. W.W. Hogan. Contract networks for electric power transmission. *Journal of Regulatory Economics*, 4:211–242, 1992.
9. W. Jing-Yuan and Smeers Y. patial oligopolistic electricity models with cournot generators and regulated transmission prices. *Operations Research*, 47:102–112, 1999.
10. E.P. Kahn. Numerical techniques for analyzing market power in electricity. *The Electricity Journal*, 11:34–43, 1998.
11. Z. Liu and A. Nagurney. An integrated electric power supply chain and fuel market network framework: Theoretical modeling with empirical analysis for new england. Isenberg School of Management, University of Massachusetts, Amherst, USA, 2008.
12. J.P. Montero. Marketable pollution permits with uncertainty and transaction costs. *Resource and Energy Economics*, 20:27–50, 1997.
13. J.P. Montero. Multipollutant markets. *The RAND Journal of Economics*, 32:762–774, 2001.
14. W.D. Montgomery. Markets in licenses and efficient pollution control programs. *Journal of Economic Theory*, 5:395–418, 1972.
15. A. Nagurney. *Network Economics: A Variational Inequality Approach Second Edition*. Kluwer, Dordrecht, The Netherlands, 1999.
16. A. Nagurney. On the relationship between supply chain and transportation network equilibria: A supernetwork equivalence with computations. *Transportation Research E*, 42:293–316, 2006.
17. A. Nagurney and K.K Dhanda. Variational inequality approach for marketable pollution permits. *Computational Economics*, 9:363–384, 1996.
18. A. Nagurney and K.K Dhanda. Marketable pollution permits in oligopolistic markets with transaction costs. *Operations Research*, 48:424–435, 2000.
19. A. Nagurney, K.K Dhanda, and J.K. Stranlund. A general multiproduct, multipollutant market pollution permit model: A variational inequality approach. *Energy Economics*, 19:57–76, 1997.
20. A. Nagurney and Z. Liu. Transportation network equilibrium reformulations of electric power networks with computations. Isenberg School of Management, University of Massachusetts, Amherst, USA, 2005.
21. A. Nagurney, Z. Liu, and T. Woolley. Optimal endogenous carbon taxes for electric power supply chains with power plants. *Mathematical and Computer Modelling*, 44:899–916, 2006.
22. A. Nagurney and D. Matsypura. A supply chain network perspective for electric power generation, supply, transmission, and consumption. In *Proceedings of the International Conference in Computing, Communications and Control Technologies*, volume VI, pages 127–134, Austin, Texas, 2004.
23. A. Nagurney and D. Matsypura. A supply chain network perspective for electric power generation, supply, transmission, and consumption. In *Advances in Computational Economics, Finance and Management Science*. Springer, Berlin, Germany, 2005.
24. A. Nagurney, P. Ramanujam, and K.K Dhanda. A multimodal traffic network equilibrium model with emission pollution permits: Compliance versus noncompliance. *Transportation Research D*, 3:349–374, 1998.

25. A. Nagurney and D. Zhang. *Projected Dynamical Systems and Variational Inequalities with Applications.* Kluwer, Boston, Massachusetts, 1996.
26. J.F. Nash. Equilibrium points in n-person games. *Proceedings of the National Academy of Sciences*, 36:48–49, 1950.
27. J.F. Nash. Noncooperative games. *Annals of Mathematics*, 54:286–298, 1951.
28. K. Palmer, D. Burtraw, and J.S. Shih. The benefits and costs of reducing emissions from the electricity sector. *Journal of Environmental Management*, 83:115–130, 2007.
29. E.S. Rubin, M.B. Berkenpas, A. Farrel, G.A. Gibbon, and D.N. Smith. Multi-pollutant emission control of electric power plants. In *Proceedings of EPA-DOE-EPRI Mega Symposium*, Chicago, Illinois, 2001.
30. E.S. Rubin, J.R. Kalagnanam, H.C. Frey, and M.B. Barkenpas. Integrated environmental control modeling of coal-fired power systems. *Journal of Air and Waste Management Association*, 47:1180–1188, 1997.
31. P. Schwarz. Multipollutant efficiency standards for electricity production. *Contemporary Economic Policy*, 23:341–356, 2005.
32. T.H. Tietenberg. *Emissions trading: An exercise in reforming pollution policy.* Resources for the Future, Washington, DC, 1985.
33. J. Tschirhart and S. Wen. Tradable allowances in a restructuring electric industry. *Journal of Environmental Economics and Management*, 38:195–214, 1999.
34. Western Regional Air Partnership. Draft for wrap market trading forum review stationary source nox and pm emissions in the wrap region: An initial assessment of emissions, controls, and air quality impacts. Technical report, 2005.
35. K. Wu, A. Nagurney, Z. Liu, and J.K. Stranlund. Modelling generator power plant portfolios and pollution taxes in electric power supply chain networks: A transportation network equilibrium transformation. *Transportation Research D*, 11:171–190, 2006.
36. G. Zaccour. *Deregulation of Electric Utilities.* Kluwer, Boston, Massachusetts, 2001.

13

Applications of TRUST-TECH Methodology in Optimal Power Flow of Power Systems

Hsiao-Dong Chiang, Bin Wang, and Quan-Yuan Jiang

Summary. The main objective of the optimal power flow (OPF) problem is to determine the optimal steady-state operation of an electric power system while satisfying engineering and economic constraints. With the structural deregulation of electric power systems, OPF is becoming a basic tool in the power market. In this paper, a two-stage solution algorithm developed for solving OPF problems has several distinguished features: it numerically detects the existence of feasible solutions and quickly locates them. The theoretical basis of stage I is that the set of stable equilibrium manifolds of the quotient gradient system (QGS) is a set of feasible components of the original OPF problem. The first stage of this algorithm is a fast, globally convergent method for obtaining feasible solutions to the OPF problem. Starting from the feasible initial point obtained by stage I, an interior point method (IPM) at stage II is used to solve the original OPF problem to quickly locate a local optimal solution. This two-stage solution algorithm can quickly obtain a feasible solution and robustly solve OPF problems. Numerical test systems include a 2,383-bus power system.

13.1 Introduction

Since the early 1960s, the optimal power flow (OPF) problem has been one of the most widely studied topics in power system analysis and computation [1–3]. This problem is relevant in power system operations, scheduling, and planning [4–6]. The main objective of the OPF problem is to determine the optimal steady-state operation of an electric power system while satisfying engineering and economic constraints. With the structural deregulation of electric power systems, OPF is becoming a basic tool in the power market.

Mathematically, OPF is modeled as a nonlinear programming (NLP) problem, which usually minimizes the total generation dispatch cost, transmission loss, or their combination subject to a set of equality and inequality constraints. From a computational viewpoint, the OPF problem is a large-scale nonconvex NLP problem, in which both the objective function and the constraint functions are nonlinear. The OPF problem becomes a mixed-integer NLP problem when discrete control variables such as transformer taps, shunt capacitor banks and FACTS devices are taken into account. Furthermore, if transient stability constraints are considered, then it is expressed as a set of large-scale differential-algorithm equations (DAE).

Numerous NLP optimization methods have been proposed to solve the OPF problem, such as mathematical programming methods, stochastic global optimization approaches and metaheuristics, see for example, [7–24]. The mathematical programming methods include the generalized reduced gradient technique, successive quadratic programming (SQP), Lagrangian Newton approaches, successive linear programming (SLP) and interior point methods (IPM). Metaheuristics include simulated annealing type methods, genetic algorithms, evolutionary programming methods, particle swarm optimization (PSO) and the immune algorithm (IA). These metaheuristical methods may be well-suited to nonmonotonic solution surfaces, where many local optima exist. One basic disadvantage of these methods is poor computational speed. OPF problems with stability constraints are still under development.

The emergence of a deregulated electricity market has posed new challenges to the task of solving power market related OPF problems. The Independent System Operator (ISO) uses OPF tools to provide timely power market settlements and to ensure market fairness and efficiency [25, 26]. To ensure market efficiency and fairness, the global optimal solution of the underlying OPF problem is needed. However, it is extremely difficult to obtain the global optimal solution in a timely manner. Hence, some compromise of these two basic requirements may be necessary.

These challenges are thus translated into several requirements for OPF problem formulations and solution algorithms. Some of these requirements are listed below:

(1) Problem formulation includes realistic problem modeling such as AC power flow equations, smooth and nonsmooth objective functions.
(2) Solution algorithm finds a solution point in a timely manner.
(3) Solution algorithm finds a local optimal solution in a deterministic and robust manner.
(4) Problem formulation includes the stability constraints under a list of credible contingencies.
(5) Solution algorithm is preferred to find the global optimal solution.
(6) Solution algorithm is robust in varied operating conditions.

Interior point methods (IPMs) have been widely studied for solving the optimal power flow (OPF) problem in the last decade, see for example, [27–38]. Unlike Newton OPF methods, which use penalty functions to handle inequality constraints, IPMs convert the inequality constraints to equalities by introducing nonnegative slack variables. The improvement has shown better convergence performance and computational efficiency. Among the many variants of IPMs, the predictor–corrector primal-dual IPM has been the most efficient one for solving large-scale OPF problems. One attractive feature of IPMs is their ability to handle nonlinear inequalities without requiring an active set identification. Additionally, a strictly feasible starting point is not mandatory; only the strict positive conditions on a subset of variables must be satisfied at the initial point and subsequent iterations. Another advantage is their fast convergence performance and lower sensitivity to system dimension. However, if the step size is not chosen properly, the sublinear problem may lead to a solution that is infeasible for the original nonlinear domain. Furthermore, IPMs may suffer from issues such as divergence, initialization, termination, and optimality criteria. Among them, how to improve convergence performance is an important one.

In this paper, a two-stage solution algorithm is developed for solving OPF problems. This two-stage solution algorithm has several distinguished features: it numerically detects the existence of feasible solutions and quickly locates them. To quickly locate a local optimal solution, an interior point method is used in the second stage. The first stage is a fast, globally convergent method for obtaining a feasible solution to the OPF problem. To this end, TRUST-TECH methodology [39–43] is extended to an OPF problem to quickly locate feasible solutions. Starting from the feasible initial point obtained by stage I, IPM is used to solve the original OPF problem. The proposed AS-QGS (active-set QGS) at stage I is improved based on the active-set strategy. From a numerical viewpoint, the improvement is reflected in the system dimension which is only 20–25% of the dimension of QGS. Hence, the numerical implementation of stage I of the TRUST-TECH methodology can greatly reduce the dimension of QGS and dramatically improves its computation efficiency for large-scale OPF problems. AS-QGS is represented as a set of ordinary differential equations (ODE), which can be solved by effective ODE solvers. The theoretical basis of stage I is that the set of stable equilibrium manifolds of QGS is a set of feasible components of original OPF problems. In stage II, based on the feasible point obtained by stage I, an IPM algorithm is used to obtain a local optimal solution. This two-stage solution algorithm can not only quickly obtain a feasible solution, but it also greatly improves the robustness of IPMs in solving OPF problems.

This chapter is organized as follows: Sect. 13.2 explains the OPF problem in terms of a nonlinear constrained optimization formulation and Sect. 13.3 gives an overview of the TRUST-TECH methodology. In Sect. 13.4, the computational and analytical basis for the two-stage TRUST-TECH implementation is discussed and conceptual algorithms are presented. Section 13.5 and 13.6 cover the preferred implementations for the two stages, including the AS-QGS for stage I and IPMs for stage II, respectively, when the methodology is applied to solve large-scale systems. Numerical studies have been carried out on different sized systems and the results are presented in Sect. 13.7.

13.2 Optimal Power Flow

For purposes of description, we consider a generalized OPF problem:

$$\begin{aligned} \min\ & f(x_s, x_c) \\ s.t.\ & g(x) = 0 \\ & h(x) \leq 0 \\ & x^l \leq x \leq x^u \end{aligned} \tag{13.1}$$

where,

- $x_s = [\delta_1, \delta_2, \cdots, \delta_n, V_1, V_2, \cdots, V_n]^T$ is the state vector (bus voltage amplitude and angle)
- $x_c = [P_{g1}, P_{g2}, \cdots, P_{gm}, V_{g1}, V_{g2}, \cdots, V_{gm}, tap_i, \cdots]^T$ is the control vector, which includes control variables such as generator output power and voltage, transformer taps, FACTS control variables, shunt capacitor banks, and so on
- $x = [x_s, x_c]$
- $g(x_s, x_c)$ represents power flower equations
- $h(x_s, x_c)$ represents functional inequality constraints such as power flow limits on transmission lines and transformers, limits on VAR injections for reactive control buses and real power injection for the slack bus
- x^u and x^l are the upper and lower bounds of x

In the OPF problem, the objective function $f(x)$ may be fuel cost generation, active and/or reactive power transmission loss, reactive power reserve margin, security margin index, emission, and environmental index.

Without loss of generality, a typical OPF model in the rectangular coordination is considered in this paper:

$$\begin{aligned} \min\ & f(P_g, Q_g, V_e, V_f) \\ s.t.\ & P(V_e, V_f) + P_d - P_g = 0 \\ & Q(V_e, V_f) + Q_d - Q_g = 0 \\ & P_g^m \leq P_g \leq P_g^M \\ & Q_g^m \leq Q_g \leq Q_g^M \\ & V^m \leq |V| \leq V^M \\ & S^m \leq S(V_e, V_f) \leq S^M \end{aligned} \tag{13.2}$$

where, $P_g = \{P_{gi}\}$ and $Q_g = \{Q_{gi}\}$, $i = 1, \cdots, n_g$, are the real and reactive power output of the generators, respectively; $V_e = \{V_{ej}\}$ and $V_f = \{V_{fj}\}$, $j = 1, \cdots, n_b$, are the real and imagery parts of the bus voltages, respectively; P_d, Q_d are the active and reactive load powers of all buses; "m" means the lower limits, "M" means the upper limits. The OPF problem can be converted into a standard OPF problem of (13.1), in which the objective function, for example, is the minimization of the total cost:

$$\min_x\ f(x) = \sum_{i=1}^{n_g} \left(a_{2i} P_{gi}^2 + a_{1i} P_{gi} + a_{0i} \right) \tag{13.3}$$

where a_{2i}, a_{1i} and a_{0i} are the cost parameters of the i-th generator. The equality constraint functions can be reformulated as follows:

$$H(x) = \begin{pmatrix} P(V_e, V_f) + P_d - P_g \\ Q(V_e, V_f) + Q_d - Q_g \end{pmatrix} \tag{13.4}$$

where

$$P_i(V_e, V_f) = V_{ei} \sum_{j \in i} (G_{ij} V_{ej} - B_{ij} V_{fj}) + V_{fi} \sum_{j \in i} (G_{ij} V_{fj} + B_{ij} V_{ej})$$
$$Q_i(V_e, V_f) = V_{fi} \sum_{j \in i} (G_{ij} V_{ej} - B_{ij} V_{fj}) + V_{ei} \sum_{j \in i} (G_{ij} V_{fj} + B_{ij} V_{ej})$$

The inequality constraint functions become

$$G(\mathbf{x}) = \begin{pmatrix} P_g - P_g^M \\ P_g^m - P_g \\ Q_g - Q_g^M \\ Q_g^m - Q_g \\ (V_e^2 + V_f^2) - V^{M2} \\ V^{m2} - (V_e^2 + V_f^2) \\ S(V_e, V_f) - S^M \\ S^m - S(V_e, V_f) \end{pmatrix} \tag{13.5}$$

where

$$S_{ij}(V_e, V_f) = G_{ii}(V_{e_i}^2 + V_{f_i}^2) + G_{ij}(V_{e_i} V_{e_j} + V_{f_i} V_{f_j}) - B_{ij}(V_{e_i} V_{f_j} - V_{f_i} V_{e_j})$$

13.3 Overview of TRUST-TECH Methodology

The focus of this section is on an overview of the TRUST-TECH methodology for constrained nonlinear optimization problems and its numerical implementation.

13.3.1 Mathematical Preliminary

Some concepts of general nonlinear dynamical systems and a certain class of nonhyperbolic dynamical systems studied in [39] are briefly reviewed. Consider the following nonhyperbolic dynamical system:

$$\dot{x}(t) = F(x) = M(x) \cdot H(x) \tag{13.6}$$

where $H : R^n \to R^{m \times 1}$ and $H : R^n \to R^{n \times m}$ with $n > m$. It is assumed that both $H(x)$ and $M(x)$ are C^2 functions and that $F(x)$ satisfies the sufficient conditions for the existence and uniqueness of the solutions of (13.6).

The solution curve of (13.6) starting from x at $t = 0$ is called a trajectory and is denoted by $\Phi(x, :) : R \to R^n$.

A path-connected component of $F^{-1}(0)$ is called an equilibrium manifold of (13.6) and denoted by Σ. An equilibrium manifold Σ is stable if for a given $\epsilon > 0$, there exists $\delta > 0$ such that $x \in B_\delta(\Sigma)$ implies $\Phi(x,t) \in B_\epsilon(\Sigma), \forall t \in R$. Furthermore, an equilibrium manifold Σ is asymptotically stable if for a given $\epsilon > 0$, there exists $\delta > 0$ such that $x \in B_\delta(\Sigma)$ implies $\lim_{t\to\infty} \Phi(x,t) \in \Sigma$. If an equilibrium manifold Σ is not stable, it is unstable.

A p-dimensional equilibrium manifold Σ of (13.6) is hyperbolic if for any $x_0 \in \Sigma$, the Jacobian of $F(x)$ evaluated at x_0, denoted by $J_F(x_0)$, has no eigenvalues with zero real part on the normal space $N_{x_0}(\Sigma)$, which is the orthogonal complement of the tangent space $T_{x_0}(\Sigma)$. Furthermore, a hyperbolic equilibrium manifold Σ is (asymptotically) stable if for each $x_0 \in \Sigma$, all eigenvalues of its corresponding Jacobian $J_F(x_0)$ on $N_{x_0}(\Sigma)$ have negative real part. Conversely, a hyperbolic equilibrium manifold Σ is unstable if for each $x_0 \in \Sigma$, some eigenvalues of its corresponding Jacobian $J_F(x_0)$ on $N_{x_0}(\Sigma)$ have positive real part.

A p-dimensional equilibrium manifold Σ is called a *type* $-$ k equilibrium manifold if for each $x_0 \in \Sigma$, its corresponding Jacobian $J_F(x_0)$ has exactly k eigenvalues with positive real part on $N_{x_0}(\Sigma)$. For a hyperbolic equilibrium manifold Σ, its stable manifold $W^s(\Sigma)$ and unstable manifold $W^u(\Sigma)$ can be defined as follows:

$$W^s(\Sigma) = \left\{ x \in R^n : \lim_{t\to\infty} \Phi(x,t) \in \Sigma \right\}$$

$$W^u(\Sigma) = \left\{ x \in R^n : \lim_{t\to-\infty} \Phi(x,t) \in \Sigma \right\}.$$

The transversality condition can be similarly defined for two hyperbolic equilibrium manifolds through their stable and unstable manifolds. Two p-dimensional hyperbolic equilibrium manifolds Σ_i and Σ_j satisfy the transversality condition if either (1) $W^u(\Sigma_i) \cap W^s(\Sigma_j) = \emptyset$ or (2) $dim(W^u(\Sigma_i) \cap W^s(\Sigma_j)) \geq p + 1$ and $W^u(\Sigma_i) \oplus W^s(\Sigma_j) = R^n$.

It can be shown that for a (asymptotically) stable hyperbolic equilibrium manifold Σ_s there exists $\delta > 0$ such that for any $x \in B_\delta(\Sigma_s)$, the trajectory starting from x will converge to Σ_s. If δ can be arbitrarily large, the stable equilibrium manifold is said to be globally stable. A hyperbolic equilibrium manifold Σ_s may be stable but not necessarily globally stable. This leads to the concept of the stability region $A(\Sigma_s)$ of a stable equilibrium manifold Σ_s:

$$A(\Sigma_s) = \left\{ x \in R^n : \lim_{t\to\infty} \Phi(x,t) \in \Sigma_s \right\} \tag{13.7}$$

The boundary of $A(\Sigma_s)$ is defined as the stability boundary of Σ_s and is denoted $\partial A(\Sigma_s)$. From a topological viewpoint, $A(\Sigma_s)$ is an open, invariant, and connected set while the stability boundary $\partial A(\Sigma_s)$ is an $(n-1)$-dimensional closed and invariant set.

13.3.2 Overview of TRUST-TECH Constrained Methodology

The TRUST-TECH constrained methodology considers the following continuous constrained nonlinear programming problem:

$$\begin{aligned} &\min\ f(x) \\ &s.t.\ \ h_i(x) = 0,\ i \in I = \{1, 2, \cdots, e_c\}, \\ &\qquad g_j(x) \le 0,\ j \in J = \{1, 2, \cdots, i_c\}, \\ &\qquad x \in R^n \text{ and } f, h_i, g_j \in C^2. \end{aligned} \tag{13.8}$$

If auxiliary variables $s = \{s_j\}$ with $j = 1, \cdots, i_c$ are introduced to the inequality constraints, problem (13.8) can be transformed into a nonlinear optimization problem with only equality constraints:

$$\begin{aligned} &\min\ f(x) \\ &s.t.\ \ h_i(x) = 0,\ i \in I = \{1, 2, \cdots, e_c\}, \\ &\qquad g_j(x) + s_j^2 = 0,\ j \in J = \{1, 2, \cdots, i_c\}, \\ &\qquad x \in R^n \text{ and } f, h_i, g_j \in C^2. \end{aligned} \tag{13.9}$$

or equivalently,

$$\begin{aligned} &\min\ f(X) \\ &s.t.\ \ H(X) = 0 \\ &\qquad X \in R^{(n+i_c)\times 1} \end{aligned} \tag{13.10}$$

where $H(X) = [h_1(X), h_2(X), \cdots, h_m(X)]^T$ with $m = e_c + i_c$ and $X = (x, s)^T = (x_1, \cdots, x_n, s_1, \cdots, s_{i_c})^T$.

To facilitate the presentation, (13.10) is simplified as the following problem:

$$\begin{aligned} &\min\ f(x) \\ &s.t.\ \ H(x) = 0 \\ &\qquad x \in R^{n\times 1} \end{aligned} \tag{13.11}$$

where $H(x) = [h_1(x), h_2(x), \cdots, h_m(x)]^T$. Problem (13.11) is used to introduce the TRUST-TECH methodology in the following presentation.

Define the Lagrangian function associated with problem (13.11):

$$L(x, \lambda) = f(x) + \sum_{k=1}^{m} \lambda_k h_k(x) \tag{13.12}$$

with Lagrangian-multipliers $\lambda = (\lambda_1, \lambda_2, \cdots, \lambda_m)$. $\overline{x}$ is a *critical point* of the Lagrangian function if

$$\nabla_x L(\overline{x}, \overline{\lambda}) = \nabla f(\overline{x}) + \sum_{k=1}^{m} \overline{\lambda}_k \nabla h_k(\overline{x}) = 0$$

or in compact form

$$\nabla_x L(\overline{x}, \overline{\lambda}) = H(\overline{x}) = 0$$

with Lagrangian-multipliers $\lambda = (\overline{\lambda}_1, \overline{\lambda}_2, \cdots, \overline{\lambda}_m)$. Moreover, the constraint set, i.e., the feasible set, for problem (13.11) is defined as:

$$M = \{x \in R^n : H(x) = 0\}. \quad (13.13)$$

The following assumptions [C1] to [C3] are needed for problem (13.11):

[C1] At each $x \in M$, $\{\nabla h_k(x), k = 1, \cdots, m\}$ are linearly independent.
[C2] At each critical point $\overline{x} \in M$, $d^T \nabla^2_{xx} L(\overline{x}, \overline{\lambda}) d \neq 0$ for all $d \neq 0$ satisfying $\nabla h_k(\overline{x})^T d = 0$ for all $k = 1, \cdots, m$.
[C3] The objective function has only finitely many critical points at which it attains different values of f.

It can be shown that assumptions [C1]–[C3] are generically true. Therefore, problem (13.11) is structurally stable and its solutions persist under small perturbations of $f, h_1, \cdots, h_m$.

13.4 Computational and Analytical Basis

The difficulties of solving problem (13.11) are well recognized. First, the feasible region may be composed of several disconnected feasible components in the entire search space. Second, there may exist multiple local optimal solutions inside each feasible component. Hence, the computational challenges are the following:

- How to compute each feasible component of the feasible region, and
- How to compute each local optimal solution lying in each feasible component

Obviously, any effective algorithm for solving problem (13.11) should possess the ability to identify these disconnected components and to locate all the local optimal solutions lying within each feasible component.

The TRUST-TECH methodology overcomes these two challenges in two stages: Instead of directly solving the constrained optimization problem (13.11), TRUST-TECH defines two nonlinear dynamical systems and explores some trajectories of these systems: the "Quotient Gradient System" (QGS) and the "Projected Gradient System" (PGS). TRUST-TECH methodology explores some trajectories of QGS to locate multiple feasible components and explores PGS to locate multiple local optimal solutions lying within each feasible component. By exploring some trajectories of these two dynamical systems, TRUST-TECH can locate multiple local optimal solutions in each disconnected feasible component. These explorations will be explained in the following.

Since the proposed two-stage solution algorithm is composed of only Phase I of TRUST-TECH to locate feasible solutions and an IPM to locate a local OPF, we only discuss Phase I of TRUST-TECH.

13.4.1 Quotient Gradient Systems (QGS)

We consider the following QGS which is designed to dynamically characterize the feasible set of the constrained optimization problem (13.11).

$$\dot{x}(x) \;=\; -\,DH(x)^T \cdot H(x) \tag{13.14}$$

where $DH(x)$ is the Jacobian matrix of the vector function $H(x)$ and can be expressed as

$$DH(x) \;=\; \begin{bmatrix} \frac{\partial h_1}{\partial x_1} & \frac{\partial h_1}{\partial x_2} & \cdots & \frac{\partial h_1}{\partial x_n} \\ \frac{\partial h_2}{\partial x_1} & \frac{\partial h_2}{\partial x_2} & \cdots & \frac{\partial h_2}{\partial x_n} \\ \cdots & \cdots & \cdots & \cdots \\ \frac{\partial h_m}{\partial x_1} & \frac{\partial h_m}{\partial x_2} & \cdots & \frac{\partial h_m}{\partial x_n} \end{bmatrix}. \tag{13.15}$$

The following two assumptions are required for QGS:

[A1] All equilibrium manifolds of system (13.14) are hyperbolic and finite in number.
[A2] Either (1) $\|H(x)\|$ is a proper map (i.e. the preimage of a compact set is a compact set); or (2) for any $\phi > 0$, define the set $\Phi = \{x \in R^n : \|H(x)\| \le \phi, \|DH(x)^T \cdot H(x)\| \ne 0\}$; then, for any closed set $\Delta \subseteq \Phi$, $\inf\{\|DH(x)^T H(x)\| : x \in \Delta\} > 0$.

13.4.2 Theoretical Basis of QGS

Theorem 1 establishes a relationship between stable equilibrium manifolds and feasible components. This theorem implies that it suffices to locate the stable equilibrium manifolds of QGS in order to identify feasible components of the constrained optimization problem (13.11).

Theorem 1. (Feasible components and equilibrium manifolds) [39]

(1) Each feasible component of the constrained optimization problem (13.11) is a stable equilibrium manifold of QGS (13.14).
(2) If Σ is a stable equilibrium manifold of QGS (13.14), then each $x_\Sigma \in \Sigma$ is a local optimal solution of the following optimization problem:

$$\min \; E(x) \tag{13.16}$$

with $x \in R^n$, where

$$E(x) = \frac{1}{2}\|H(x)\|^2.$$

$E(x)$ is called the energy function of QGS.

Theorem 1 asserts that each feasible component of constrained optimization problem (13.11) corresponds to one stable equilibrium manifold of QGS (13.14). The set of all feasible components is contained in the set of all stable equilibrium manifolds. Therefore, all or multiple disconnected feasible components of the constrained optimization problem (13.11) can be identified by locating all or multiple stable equilibrium manifolds of QGS (13.14).

Theorem 2. (Completely stable) [39]
Under assumptions [A1] and [A2], QGS is completely stable.

To characterize the stability boundary of a stable equilibrium manifold of QGS (13.14), the following assumptions are needed:

[A3] Let Σ_s be a stable equilibrium manifold. If Σ is an equilibrium manifold and $\Sigma \cap \partial A(\Sigma_s) \neq \emptyset$, then $\Sigma \subset \partial A(\Sigma_s)$.
[A4] The stable and unstable manifolds of equilibrium manifolds on $\partial A(\Sigma_s)$ satisfy the transversality condition.
[A5] All equilibrium manifolds on $\partial A(\Sigma_s)$ have the same dimension.

Theorem 2 above characterizes the global behaviors of QGS (13.14). In combination with Theorem 1, they assert that the search procedure via following trajectories of QGS (13.14) can lead to feasible components of the constrained optimization problem (13.11).

Theorem 3. (Characterization of stability boundary)
Under assumptions [A3]–[A5], the stability boundary $\partial A(\Sigma_s)$ of the stable equilibrium manifold Σ_s of QGS can be characterized by the following relationship:

$$\partial A(\Sigma_s) = \bigcup_i W^s(\Sigma_i) \tag{13.17}$$

where Σ_i denotes the equilibrium manifold on the stability boundary $\partial A(\Sigma_s)$.

Theorem 4. (Unstable equilibrium manifold on the stability boundary)
If $\Sigma \subset \partial A(\Sigma_s)$ is a hyperbolic, isolated equilibrium manifold and satisfies assumptions [A3]–[A5], then $W^u(\Sigma) \cap A(\Sigma_s) \neq \emptyset$.

Remark: The characterization in Theorem 3 can be used to identify the unstable equilibrium manifold on the boundary. Theorem 4 asserts that the unstable equilibrium manifold on the stability boundary of a stable equilibrium manifold links two adjacent stable equilibrium manifolds. These theorems provide a theoretical basis for the TRUST-TECH constrained methodology in locating all the feasible components of the constrained optimization problem. The search procedure first detects the stability boundary and then identifies the unstable equilibrium manifold on the boundary (cf. Theorem 3). With the aid of the found unstable equilibrium manifold, the unique adjacent stable equilibrium manifold (cf. Theorem 4) can be subsequently located. Hence another feasible component is found (cf. Theorem 1).

A mechanism to move from one stable equilibrium manifold and reach its adjacent stable equilibrium manifolds is devised in the TRUST-TECH constrained methodology. Such a mechanism allows TRUST-TECH to identify multiple stable equilibrium manifolds of (13.14) (or equivalently, multiple feasible components)

13.4.3 Conceptual Algorithms

The TRUST-TECH constrained methodology develops computational mechanisms to escape from a stable equilibrium manifold (i.e., a feasible component) to adjacent stable equilibrium manifolds (i.e., adjacent feasible components) by exploring trajectories of QGS. In this section, a conceptual algorithm of Stage I of the TRUST-TECH constrained methodology is presented.

Feasible-Set Phase

The basis of the feasible-set phase can be briefly described as follows: The entire state space is decomposed into multiple stability regions of stable equilibrium manifolds. Two adjacent stability regions are separated by the intersection of their stability boundaries. To search for the target stable equilibrium manifold from the initial stable equilibrium manifold, it suffices to develop a computational mechanism to move toward the stability boundary of the initial stability region and reach the target stability region with the aid of an unstable equilibrium manifold on the stability boundary of the initial stability region. This leads to the following conceptual algorithm:

S1. Generate a search path moving away from the initial stable equilibrium manifold and toward the stability boundary of the initial stable equilibrium manifold.
S2. Move along the stability boundary to identify the corresponding unstable equilibrium manifold separating the initial stable equilibrium manifold and a target stable equilibrium manifold.
S3. Move along the unstable manifold of the identified unstable equilibrium manifold to locate the adjacent stable equilibrium manifold.

13.5 Active-Set Quotient Gradient System

A traditional implementation of TRUST-TECH handles the inequality constraints by converting them all to equality constraints by adding slack variables, that is,

$$\tilde{G}(x,s) \;=\; g(x) + s^2 \;=\; 0 \tag{13.18}$$

Consequently, the dimension of the QGS system becomes larger than the original one and the number of control variables also increases dramatically. In OPF problems, the number of inequality constraints is much larger than the number of equality constraints. As a result, the dimensionality of the QGS becomes significantly larger than that of the initial problem. Furthermore, all the inequality constraints have to be evaluated during each iteration. Hence, using this QGS formulation to obtain a feasible point is computationally expensive.

In order to reduce the computational burden, an active set-QGS (AS-QGS) formulation is proposed. During the iterative procedure, the inequality constraints can be divided into two categories: the active (i.e., violated) inequality constraints, and the inactive (i.e., satisfied) inequality constraints. During each step in the search procedure, we only need to consider the active inequality constraints and find a search direction for improving both equality violations and inequality violations. By exploring this property of active violations, an effective numerical implementation for handling the inequality constraints in TRUST-TECH is proposed in this paper.

In this new numerical implementation, the AS-QGS for quickly locating a feasible solution point from an initial point is formulated as the following unconstrained optimization problem:

$$\min\ E(x) = \frac{1}{2}\{\|H(x)\|^2 + \alpha \cdot \|G_t(x)\|^2\} \tag{13.19}$$

where $\alpha > 0$ is a weighting constant, commonly $\alpha = 1$, and $G_t(x)$ is the set of current violated constraints with $\{G_t(x) < 0\}$. This set of inequality constraints is also called the active (inequality) constraint set. Instead of adding slack variables to convert the whole set of the inequality constraints into equality ones, the active-set strategy removes all slack variables in the original implementation of the TRUST-TECH methodology.

Theoretically, the active set of inequality constraints should be selected as those satisfying $\{G_t(x) < 0\}$. In the real implementation of the proposed AS-QGS in Stage I, the active set is chosen to be a collection of the near-binding [44] plus the violated inequality constraints, for the following two reasons: (1) because of the finite numerical precision provided by the digital computer, the exact inequality $G_t(x) < 0$ would be difficult to get; and (2) in order to provide more room to move before reaching other inactive boundaries of the feasible region. More precisely, the active set implemented in this paper is taken as $\{G_t(x) < \epsilon\}$ rather than $\{G_t(x) < 0\}$, where $\epsilon > 0$ is a suitably small value.

The optimization problem is solved via the following nonlinear dynamical system, or the active-set QGS:

$$\dot{x} \;=\; -\left[J_H^T(x) \cdot H(x) + \alpha \cdot J_{G_t}^T(x) \cdot G_t(x)\right] \tag{13.20}$$

Since no auxiliary slack variables are involved, the dimension of the AS-QGS is the same as that of the original control variables. Compared to original QGS, the dimensions of Jacobian matrices in AS-QGS are significantly smaller. Furthermore, because only the active set of inequality constraints is considered in AS-QGS, the computational memory and CPU seconds will reduce dramatically.

To examine the computational performance of AS-QGS, a comparison is given in Table 13.1 in which the results were obtained in C++ code, 1.77 GHz CPU, 1 GB RAM Memory. We choose one effective ODE software-"CVODE"

Table 13.1. Comparison of QGS and AS-QGS

Bus number	Dimensions of ODEs			CPU time (seconds)		
	QGS	AS-QGS	(%)	QGS	AS-QGS	(%)
30	320	72	22.5	0.89	0.03	3.4
39	400	98	24.5	2.02	0.06	2.9
57	590	128	21.7	12.3	0.11	0.9
118	1,540	344	22.3	68.3	0.75	1.1
300	3,258	738	22.7	∞	12.6	0
678	7,408	1,696	22.9	∞	243.7	0
2,383	23,078	5,420	23.5	∞	∞	/

Table 13.2. Fast feasible solutions based on the numerical implementation of pseudotransient method

Cases	Iterations	CPU seconds	Max equality mismatch	Max violated inequality	Energy of AS-QGS
30	6	0.11	4.2×10^{-10}	8.8×10^{-10}	2.7×10^{-18}
39	14	0.18	5.4×10^{-12}	6.9×10^{-11}	2.6×10^{-21}
57	23	0.21	5.5×10^{-9}	7.8×10^{-9}	1.6×10^{-16}
118	16	0.52	4.9×10^{-10}	1.5×10^{-10}	4.7×10^{-19}
300	28	0.8	3.1×10^{-5}	9.7×10^{-5}	1.2×10^{-7}
678	28	1.2	9.7×10^{-4}	1.4×10^{-5}	1.1×10^{-3}
2052	39	3.5	4.8×10^{-7}	1.8×10^{-9}	4.0×10^{-13}
2383	20	2.82	5.0×10^{-11}	7.5×10^{-11}	1.1×10^{-20}

as the ODE solver. In this Table, "∞" means the CPU time is larger than 7,200 s. It is observed from this table that the dimension of AS-QGS is only 20–25% of the original QGS, and the CPU times of AS-QGS is less than 2% of those of QGS for large-scale OPF problems.

Although AS-QGS with CVODE solver improves the computation speed to nearly 100 times that of QGS with CVODE for large-scale cases, it is still time-consuming. As can be seen from Table 13.1, AS-QGS still takes more than 200 s to compute the 678-bus case, and fails to give a stable equilibrium manifold for the 2,383-bus case in 2 h. Instead of using an ODE solver to obtain a stable equilibrium manifold of AS-QGS, we apply the pseudotransient method to improve the computational speed. The pseudotransient method [45, 46] is a way to implement this category of inexact Newton's methods. This method was originally designed as a method for finding steady-state solutions to time-dependent differential equations without computing accurate trajectories.

In order to verify the efficiency of the pseudotransient method in solving AS-QGS, the numerical results are summarized in Table 13.2. Comparing Table 13.1 with Table 13.2, we observe that the numerical implementation based on the pseudotransient method is much faster than that based on ODE solver to obtain a stable equilibrium manifold of AS-QGS. In Table 13.1, one of the best ODE solvers, CVODE, needs 12.6 CPU seconds for the 300-bus case, 243.7 CPU seconds for the 678-bus case, and fails to obtain a stable equilibrium manifold in 2 h for the 2,383-bus case, while it only took the numerical implementation based on the pseudotransient method 2.82 s to obtain a feasible solution for the 2,383-bus case.

13.6 Stage II – IPM

Stage I of the proposed two-stage solution algorithm aims to obtain a feasible point of the OPF problem. In Stage II, one effective interior point method is used to obtain a local optimal solution. The IPM consists of three crucial elements: (1) the barrier method to handle inequality constraints (2) Lagrange's method to handle equality constraints and (3) improved Newton method to solve the set of nonlinear equations which come from the KKT conditions.

In order to solve NLP (13.1), IPM first applies the Fiacco–McCormick barrier method and adds slack variables to transform the OPF problem (13.1) into the following equality-constrained optimization problem:

$$\begin{aligned} \min_x \; & f(x) - \mu \textstyle\sum_{i=1}^{r} \ln u_i \\ s.t. \; & h(x) = 0 \\ & g(x) + u = 0 \\ & u > 0 \end{aligned} \tag{13.21}$$

We construct the following augmented Lagrangian function:

$$L_g = f(x) - y^T h(x) - w^T [g(x) + u] - \mu \sum_{i=1}^{r} \ln u_i$$

where y, w are Lagrangian multipliers for equality and inequality constraints, respectively. The Karush–Kuhn–Tucker first-order necessary conditions for the Lagrangian function L_g are given as follows:

$$\begin{aligned} L_x &= \nabla_x f(x) - \nabla_x h(x) y - \nabla_x g(x) w = 0 \\ L_y &= h(x) = 0 \\ L_w &= g(x) + u = 0 \\ L_u &= UWE + \mu E = 0 \end{aligned} \tag{13.22}$$

where $U = diag(u_1, u_2, \cdots, u_r)$, $W = diag(w_1, w_2, \cdots, w_r)$, $E = [1, 1, \cdots, 1]^T$. Applying the Newton method to solve the above nonlinear equations, we can get the following two decomposed linear equations:

$$\begin{bmatrix} H & \nabla_x h(x) \\ \nabla_x^T h(x) & 0 \end{bmatrix} \begin{bmatrix} \Delta x \\ \Delta y \end{bmatrix} = \begin{bmatrix} L'_x \\ -L_y \end{bmatrix} \tag{13.23}$$

and

$$\begin{bmatrix} U & W \\ 0 & I \end{bmatrix} \begin{bmatrix} \Delta w \\ \Delta u \end{bmatrix} = \begin{bmatrix} L'_u \\ -L_w - \nabla_x^T g(x) \Delta x \end{bmatrix} \tag{13.24}$$

where

$$\begin{aligned} H &= \nabla_x^2 f(x) + \nabla_x^2 h(x) y + \nabla_x^2 g(x) U^{-1} W \nabla_x^T g(x), \\ L'_x &= L_x + \nabla_x g(x)[U^{-1}(L'_u - W L_w)], \\ L'_u &= UWE + \mu E + \Delta u \Delta w \end{aligned}$$

Observing equations (13.23)–(13.24), it is impossible to solve (13.23)–(13.24) directly because the right-hand side includes unknown high-order deviations $\Delta u \Delta w$. In order to solve this problem, the predictor–corrector interior point method was proposed, in which a predictor step and a corrector step are needed at each iteration. Because the predictor step and the corrector step share the same coefficient matrix with two different right-hand sides, only one LU factorization is needed. The predictor–corrector interior point method is composed of the following steps:

(1) Initialization: set iteration number $k = 0$, give the initial values of state variables x^0, slack variables u^0, Lagrange multipliers y^0, w^0
(2) Let $\mu = 0$, $\Delta u \Delta w = 0$, solve the linear equations (13.23) to obtain the affine direction Δx_{of}, Δy_{of}, then obtain Δu_{of}, Δw_{of} by back substitution of (13.24)
(3) Compute step sizes α_{ofp} and α_{ofd}, and modify complementary gaps: GAP and GAP_{of}, update the barrier parameter:

$$\begin{aligned} \alpha_{ofp} &= \min\left\{0.9995 \min_i \left(\frac{-u_i}{\Delta u_{ofi}}, \Delta u_{ofi} < 0\right), 1\right\} \\ \alpha_{ofd} &= \min\left\{0.9995 \min_i \left(\frac{-w_i}{\Delta w_{ofi}}, \Delta w_{ofi} > 0\right), 1\right\} \\ GAP &= -u^T w \\ GAP_{of} &= -(u + \alpha_{ofp} \Delta u_{of})^T (w + \alpha_{ofd} \Delta w_{of}) \\ \mu_{of} &= \min\left\{\left(\frac{GAP_{of}}{GAP}\right)^2, 0.1\right\} \frac{GAP_{of}}{2r} \end{aligned}$$

(4) Set $\mu = \mu_{of}$, $\Delta u \Delta w = \Delta u_{of} \Delta w_{of}$, and resolve the linear equations (13.22) using the same LU factorization matrix obtained in step 2 to obtain centering-corrector direction Δx, Δy. Then obtain Δu, Δw by back substitution of (13.23), and update all the variables:

$$\begin{aligned} \alpha_p &= \min\left\{0.9995 \min_i \left(\frac{-u_i}{\Delta u_i}, \Delta u_i < 0\right), 1\right\} \\ \alpha_d &= \min\left\{0.9995 \min_i \left(\frac{-w_i}{\Delta w_i}, \Delta w_i > 0\right), 1\right\} \end{aligned}$$

$$\begin{aligned} x &= x + \alpha_p \Delta x \\ I &= I + \alpha_p \Delta I \\ u &= u + \alpha_p \Delta u \\ y &= y + \alpha_d \Delta y \\ z &= z + \alpha_d \Delta z \\ w &= w + \alpha_d \Delta w \end{aligned}$$

(5) Compute complementary gap $GAP = -u^T w$. If GAP and maximal absolute power flow mismatch are less than the given precision, or the maximal iteration is reached, then stop; otherwise, go to step 2

The proposed two-stage solution algorithm for solving OPF problem is summarized as follows:

Fast Obtaining a feasible solution

Stage: Run AS-QGS system to obtain a feasible solution of the OPF problem.

Obtaining a Local OPF solution

S1. Run the predictor–corrector interior point method for solving the OPF problem; if it converges, then stop; otherwise, go to Stage S2.
S2. Run AS-QGS system to obtain a feasible solution of the OPF problem, and go to Stage S1.

Remark: One of the most important merits of the TRUST-TECH methodology is that it is designed to cooperate with and to take full advantage of the existing (local) optimization methods to find local optima efficiently. Consequently, the actual implementation of the TRUST-TECH is not confined to any fixed solution methodology. In fact, different (local) methods would be preferred to be incorporated into TRUST-TECH for specific applications. The merit of the TRUST-TECH methodology is its ability to embrace them seamlessly when tackling different challenging problems. For the OPF problem studied in this paper, the IPMs have been proven to be one category of the best and most efficient methods in finding local optimal solutions. Hence, it is natural for us to incorporate IPMs into the TRUST-TECH framework to rapidly find OPF solutions, and the benefits thus acquired have been validated via the numerical results shown in this paper.

13.7 Numerical Studies

IPM has good convergence performance when solving OPF problems with a flat starting point. However, IPM can be very sensitive to initial points for some classes of OPF problems. In this section, three difficult cases are given to demonstrate the efficiency of the proposed TRUST-TECH based methodology. Table 13.3 shows a summary of the test problems for the 678-bus, 2,052-bus, and 2,383-bus cases. In order to solve these large-scale power systems,

Table 13.3. Test cases

Number of bus	Number of generators	Number of transformers	Number of branches	Number of variables	Number of inequalities constraints	Number of equalities constraints
678	170	266	919	1,696	2,856	1,356
2,052	212	858	2,533	4,526	7,540	4,104
2,383	327	170	2,896	5,420	8,829	4,766

Table 13.4. The summary of stage I and II

Number of Bus	Iterations of Stage I	CPU Seconds of Stage I	Iterations of Stage II	CPU Seconds of Stage II
678	286	12	18	5.2
2,052	39	38	16	8.4
2,383	20	28	25	12.5

advanced sparse matrix technology is used. In these numerical studies, the maximum iteration is set to be 100, and the convergence tolerance was set to 10^{-6} for both the complementary gap and the maximal absolute power flow mismatch. All the cases are tested in Matlab 7 with Intel 1.77 GHz CPU, 1 GB RAM memory. The objective function of the OPF problem is minimization of the fuel cost.

In these numerical studies, the flat starting point is determined as follows:

(a) Initialization of control variables:

$$\begin{aligned} P_{gi} &= (P_{gi\,\min} + P_{gi\,\max})/2 \\ Q_{gi} &= (Q_{gi\,\min} + Q_{gi\,\max})/2 \\ V_{ei} &= 1.0 \\ V_{fi} &= 0.0 \end{aligned}$$

(b) Initialization of slack variables:

$$\begin{aligned} I_0 &= \gamma(g_+ - g_-) \\ u_0 &= (1-\gamma)(g_+ - g_-) \\ z_0 &= \mu_0/I_0 \\ w_0 &= -\mu_0/u_0 \\ y_0 &= 1 \end{aligned}$$

where $\gamma = 0.64$, $\mu_0 = n_b$.

When the above flat starting is used, IPM diverges for three test cases. In order to improve the convergence performance of IPM in solving these cases, Stage I of the proposed solution algorithm is used. The numerical results are summarized in Table 13.4 and in Fig. 13.1–13.3.

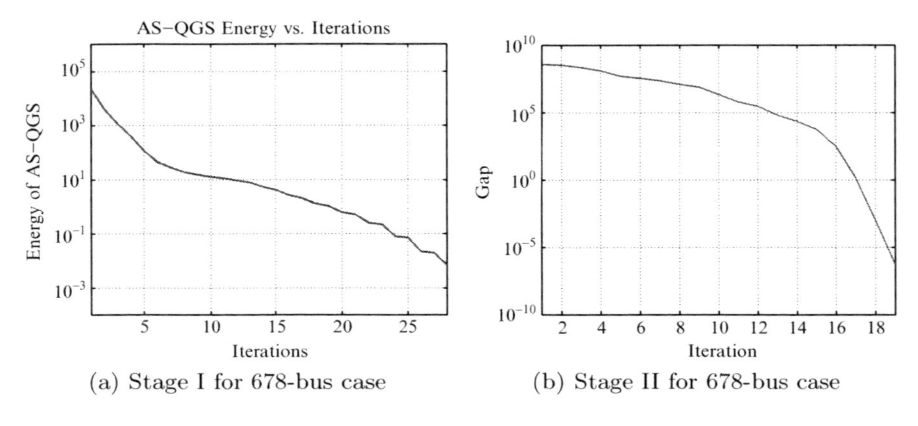

(a) Stage I for 678-bus case (b) Stage II for 678-bus case

Fig. 13.1. Convergence performance for 678-bus case

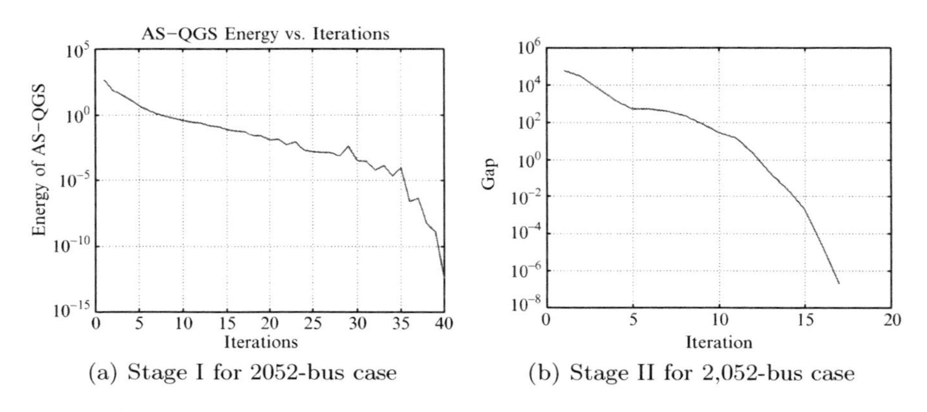

(a) Stage I for 2052-bus case (b) Stage II for 2,052-bus case

Fig. 13.2. Convergence performance for 2,052-bus case

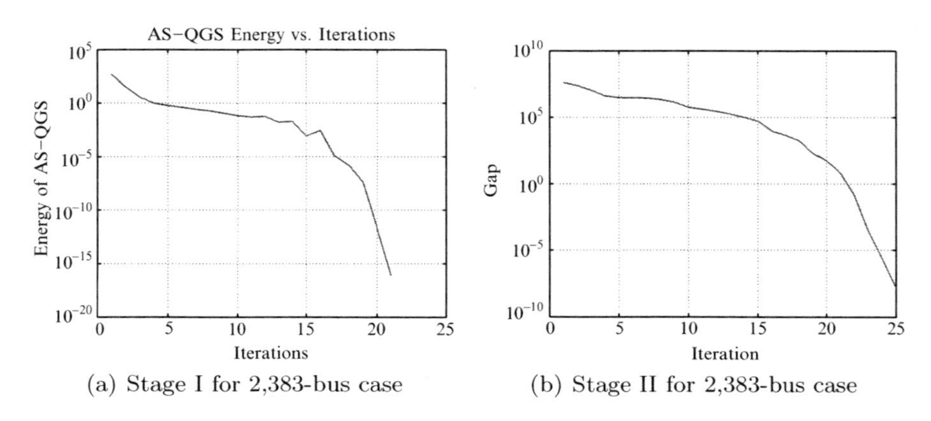

(a) Stage I for 2,383-bus case (b) Stage II for 2,383-bus case

Fig. 13.3. Convergence performance for 2,383-bus case

Please note that the numbers of iterations shown in Table 13.4 are different to that in Table 13.2, since the tolerances (on the AS-QGS energy value) for convergence are different in the two experiments. In the experiments carried out in this section, Stage I is considered to converge when the AS-QGS energy is lower than 10^{-10} for all the cases. From Table 13.2, we can see that the 678-bus system is a difficult case whose AS-QGS energy decreases much slower than other cases. Thus, it is reasonable to see that for the given tolerance, this case requires significantly more iterations to converge.

Furthermore, in carrying out experiments on the test cases presented in this paper, only one feasible component was found. Consequently, in the experimental results shown in this section, the result by Stage I is a single feasible point in the only feasible component.

It is observed from Table 13.4 that when Stage I is used to produce a feasible initial point for Stage II (i.e., IPMs), KKT points are obtained for all these difficult cases. Although Stage I takes some CPU seconds, it significantly improves the convergence performance of the Stage II (i.e., IPMs). The same observations can also be made from Fig. 13.1–13.3. In other words, Stage I is important for some cases in which IPM diverges. When IPM diverges, Stage I of the proposed solution algorithm can obtain a feasible point which can serve as the initial point for the IPM to rapidly converge to KKT points.

13.8 Concluding Remarks

Although IPMs exhibit excellent computational efficiency for large-scale OPF problems, its convergence performance heavily depends on a good starting strategy. While a feasible initial point is not required for IPMs, a feasible initial point can improve the convergence performance of IPMs especially for some difficult OPF problems. In this paper, a two-stage solution algorithm has been developed for solving OPF problems. This two-stage solution algorithm has several distinguished features: it numerically detects the existence of feasible solutions and is fast to locate them. To quickly locate a local optimal solution, an interior point method is used in the second stage. The first stage is a fast, globally convergent method for obtaining a feasible solution of the OPF problem. In stage I, unlike the traditional QGS, the proposed active-set QGS is computationally efficient. In stage II, based on the feasible point obtained by stage I, an IPM algorithm is used to obtain a local optimal solution. This two-stage solution algorithm not only rapidly obtains a feasible solution, but also greatly improves the robustness of the IPMs. The proposed two-stage solution algorithm can meet the following requirements presented in the introduction:

(1) The problem formulation includes realistic problem modeling such as AC power flow equations, smooth and nonsmooth objective functions.
(2) The solution algorithm finds a solution point in a timely manner.
(3) The solution algorithm finds a local optimal solution in a deterministic and robust manner.

References

1. J. Carpentier. Contribution to the economic dispatch problem. *Bulletin de la Societ Franhise dElectricit* , 8(1):431–437, 1962.
2. H.W. Dommel and W.F. Tinney. Optimal power flow solutions. *IEEE Transactions on Power Apparatus and Systems*, PAS-87(10):1866–1876, 1968.
3. O. Alsac and B. Stott. Optimal load flow with steady-state security. *IEEE Transactions on Power Apparatus and Systems*, PAS-93(3):745–751, 1974.
4. S.N. Talukdar and F.F. Wu. Computer-aided dispatch for electric power systems. *Proceedings of IEEE*, 69(10):1212–1231, 1981.
5. M. Huneault and F.D. Galiana. A survey of the optimal power flow literature. *IEEE Transactions on Power Systems*, 6(2):762–770, 1991.
6. J.A. Momoh, M.E. El-Hawary, and R. Adapa. A review of selected optimal power flow literature to 1993: Part-i and part-ii. *IEEE Transactions on Power Systems*, 14(1):96–111, 1999.
7. N. Grudinin. Combined quadratic-separable programming opf algorithm for economic dispatch and security control. *IEEE Transactions on Power Systems*, 12(4):1682–1688, 1997.
8. R.C. Burchett, H.H. Happ, and D.R. Vierath. Quadratically convergent optimal power flow. *IEEE Transactions on Power Apparatus and Systems*, PAS-103(11):3267–3276, 1984.
9. J. Nanda. New optimal power-dispatch algorithm using fletcher's quadratic programming method. *IEE Proceedings C*, 136(3):153–161, 1989.
10. K.C. Almeida and R. Salgado. Optimal power flow solutions under variable load conditions. *IEEE Transactions on Power Systems*, 15(4):1204–1211, 2000.
11. S.A. Pudjianto and G. Strbac. Allocation of var support using lp and nlp based optimal power flows. *IEE Proceedings – Generation, Transmission and Distribution*, 149(4):377–383, 2002.
12. T.N. Saha and A. Maitra. Optimal power flow using the reduced newton approach in rectangular coordinates. *International Journal of Electrical Power and Energy Systems*, 20(6):383–389, 1998.
13. Y.Y. Hong, C.M. Liao, and T.G. Lu. Application of newton optimal power flow to assessment of var control sequences on voltage security: case studies for a practical power system. *IEE Proceedings C*, 140(6):539–544, 1993.
14. J.A. Momoh. Improved interior point method for opf problems. *IEEE Transactions on Power Systems*, 14(3):1114–1120, 1999.
15. Y.C. Wu and A.S. Debs. Initialization, decoupling, hot start, and warm start in direct nonlinear interior point algorithm for optimal power flows. *IEE Proceedings – Generation, Transmission and Distribution*, 148(1):67–75, 2001.
16. K.C. Almeida, F.D. Galiana, and S. Soares. A general parametric optimal power flow. *IEEE Transactions on Power Systems*, 9(1):540–547, 1994.
17. P. Ristanovic. Successive linear programming based opf solution. Technical report, IEEE Tutorial Course Manual #96 TP 111-0, Piscataway, NJ, 1996.
18. N. Grudinin. Reactive power optimization using successive quadratic programming method. *IEEE Transactions on Power Systems*, 13(4):1219–1225, 1998.
19. D. Sun, B. Ashley, B. Brewer, A. Hughes, and W. Tinney. Optimal power flow by newton approach. *IEEE Transactions on Power Apparatus and Systems*, PAS-103(10):2864–2880, 1984.

20. Y.C. Wu, A.S. Debs, and R.E. Marsten. A direct nonlinear predictor-corrector primal-dual interior point algorithm for optimal power flows. *IEEE Transactions on Power Systems*, 9(2):876–883, 1994.
21. F.F. Wu, G. Gross, J.F. Luini, and P.M. Lock. A two-stage approach to solving large scale optimal power flow. In *IEEE Proceedings of Power Industry Computer Applications Conference (PICA'79)*, pages 126–136, Cleveland, OH, May 1979.
22. V.H. Quintana, G.L. Torres, and J. Medina-Palomo. Interior-point methods and their applications to power systems: a classification of publications and software codes. *IEEE Transactions on Power Systems*, 15(1):170–176, 2000.
23. G.L. Torres and V.H. Quintana. An interior-point methods for nonlinear optimal power flow using voltage rectangular coordinates. *IEEE Transactions on Power Systems*, 13(4):1211–1218, 1998.
24. S. Granville. Optimal reactive dispatch through interior point methods. *IEEE Transactions on Power Systems*, 9(1):136–146, 1994.
25. Pjm manual 06, 11, 12: Scheduling operations, 2006. Available at: `http://www.pjm.com/contributions/pjm-manuals/manuals.html`.
26. Federal Energy regulatory Commission. Principles for efficient and reliable reactive power supply and consumption. *FERC Staff reports, Docket No. AD05-1-000*, pages 161–162, February 2005. Available at: `http://www/ferc.gov/legal/staff-reports.asp`.
27. H. Wei, H. Sasaki, J. Kubokawa, and R. Yohoyama. An interior point methods for power systems weighted nonlinear l1 norm static state estimation. *IEEE Transactions on Power Systems*, 13(2):617–623, 1998.
28. G.D. Irisarri, X. Wang, J. Tong, and S. Mokhtari. Maximum loadability of power systems using interior point method nonlinear optimization. *IEEE Transactions on Power Systems*, 12(1):162–172, 1997.
29. S. Granville, J.C.O. Mello, and A.C.G. Melo. Application of interior point methods to power flow unsolvability. *IEEE Transactions on Power Systems*, 11(2):1096–1103, 1996.
30. X. Wang, G.C. Ejebe, J. Tong, and J.G. Waight. Preventive/corrective control for voltage stability using direct interior point method. *IEEE Transactions on Power Systems*, 13(3):878–883, 1998.
31. J. Medina, V.H. Quintana, A.J. Conejo, and F.P. Thoden. A comparison of interior-point codes for medium-term hydrothermal coordination. *IEEE Transactions on Power Systems*, 13(3):836–843, 1998.
32. X. Yan and V.H. Quintana. An efficient predictor - corrector interior point algorithm for security-constrained economic dispatch. *IEEE Transactions on Power Systems*, 12(2):803–810, 1997.
33. S. Mehrotra. On the implementation of a primal-dual interior point method. *SIAM Journal on Optimization*, 2(4):575–601, 1992.
34. J. Gondzio. Multiple centrality corrections in a primal-dual method for linear programming. *Computational Optimization and Applications*, 6(2):137–156, 1996.
35. G.L. Torres and V.H. Quintana. On a nonlinear multiple-centrality corrections interior-point method for optimal power flow. *IEEE Transactions on Power Systems*, 16(2):222–228, 2001.
36. Y.C. Wu and A.S. Debs. Initialisation, decoupling, hot start, and warm start in direct nonlinear interior point algorithm for optimal power flows. *IEE-Proceeding*, 148(1):67–75, 2001.

37. R.A. Jabr, A.H. Coonick, and B.J. Cory. A primal-dual interior point method for optimal power flow dispatching. *IEEE Transactions on Power Systems*, 17(3):654–662, 2002.
38. H. Wei, H. Sasaki, J. Kubakawa, and R. Yokoyama. An interior point nonlinear programming for optimal power flow problems with a novel data structure. *IEEE Transactions on Power Systems*, 13(3):870–877, 1998.
39. J. Lee and H.D. Chiang. A dynamical trajectory-based methodology for systematically computing multiple optimal solutions of general nonlinear programming problems. *IEEE Transactions on Automatic Control*, 49(6):888–899, 2004.
40. H.D. Chiang and C.C. Chu. A systematic search method for obtaining multiple local optimal solutions of nonlinear programming problems. *IEEE Transactions on Circuits and Systems*, 43(2):99–109, 1996.
41. H.D. Chiang and J. Lee. *TRUST-TECH Paradigm for computing high-quality optimal solutions: method and theory*, pages 209–234. Wiley-IEEE, New Jersey, February 2006.
42. C.R. Karrem. *TRUST-TECH Based Methods for Optimization and Learning.* PhD thesis, Cornell University, Ithaca, NY, 2007.
43. J.H. Chen. *Hybrid TRUST-TECH Algorithms and Their Applications to Mixed Integer and Mini-Max Optimization Problems.* PhD thesis, Cornell University, Ithaca, NY, 2007.
44. M.S. Bazaraa, H.D. Sherali, and C.M. Shetty. *Nonlinear Programming, Theory and Algorithms.* Wiley, Hoboken, NJ, third edition, 2006.
45. C.T. Kelley and D.E. Keyes. Convergence analysis of pseudo-transient continuation. *SIAM Journal on Numerical Analysis*, 35(2):508–523, 1998.
46. T.S. Coffey, C.T. Kelley, and D.E. Keyes. Pseudo-transient continuation and differential-algebraic equations. *SIAM Journal of Scientific Computing*, 25(2):553–569, 2003.

Part III

Stochastic Programming: Methods and Applications

14

Scenario Tree Approximation and Risk Aversion Strategies for Stochastic Optimization of Electricity Production and Trading

Andreas Eichhorn, Holger Heitsch, and Werner Römisch

Summary. Dynamic stochastic optimization techniques are highly relevant for applications in electricity production and trading since there are uncertainty factors at different time stages (e.g., demand, spot prices) that can be described reasonably by statistical models. In this paper, two aspects of this approach are highlighted: scenario tree approximation and risk aversion. The former is a procedure to replace a general statistical model (probability distribution), which makes the optimization problem intractable, suitably by a finite discrete distribution. Our methods rest upon suitable stability results for stochastic optimization problems. With regard to risk aversion we present the approach of polyhedral risk measures. For stochastic optimization problems minimizing risk measures from this class it has been shown that numerical tractability as well as stability results known for classical (nonrisk-averse) stochastic programs remain valid. In particular, the same scenario approximation methods can be used. Finally, we present illustrative numerical results from an electricity portfolio optimization model for a municipal power utility.

14.1 Introduction

The deregulation of energy markets has led to several new challenges for electric power utilities. Electric power has to be generated in a competitive environment and, in addition, coordinated with several trading activities. Electricity portfolios for spot and derivative markets become important, and the electrical load as well as electricity prices become increasingly unpredictable. Hence, the number of uncertainty sources and the financial risk for electric utilities have increased. These facts initiated the development of stochastic optimization models for producing and trading electricity. We mention, for example, stochastic hydro-electric and trading models [13, 31] and stochastic hydro-thermal production and trading models [12, 18, 19, 28, 36–38]. For an overview on stochastic programming models in energy we refer to [40].

Typical stochastic optimization models for producing and trading electricity, however, are focused on (expected) profit maximization while *risk management* is considered as an extra task. Power utilities often separate the planning of their hydro-thermal electricity production versus a preliminary and simplified trading model from the risk management. However, alternatively, risk management may be integrated into the (hydro-thermal) power production and trading planning by maximizing expected profit and minimizing (or bounding) a certain risk functional simultaneously [3, 9, 26]. Such *integrated risk management* strategies promise additional overall efficiency for power utilities.

Mathematical modeling of integrated risk management of an electricity producing and trading utility leads to multistage stochastic programs with risk objectives or risk constraints. In the present paper, we discuss two basic aspects of implementing such models: (1) the approximate representation of the underlying probability distribution by a finite discrete distribution, i.e., by a finite number of scenarios with their probabilities, and (2) modeling and minimization of risk.

The first is typically an indispensable first step towards a solution of a stochastic optimization model. On the other hand, this is a highly sensitive concern, in particular, if dynamic decision structures are involved (multistage stochastic programming [35]). Then, the scenarios of the approximate distribution must exhibit *tree structure*. Moreover, it is of interest to get by with a moderate number of scenarios to have the resulting problem tractable. We refer to the overview [6] and to several different approaches [4,5,20,23,25,27,30] for scenario tree generation.

In Sect. 14.4 we assume that scenarios of the underlying stochastic load–price process are available, e.g., by sampling from a properly developed stochastic (time series) model or by some other approximation scheme. We describe a methodology based on clustering and scenario reduction that produces a tree of scenarios and represents a good approximation of the stochastic process. The approach is based on suitable stability results ensuring that the obtained approximate problems are indeed related to the original (infinite dimensional) ones. For interested readers these stability results are presented in Sect. 14.3. The methodology as well as the stability arguments are based on distances of random vectors that allow to decide about their closeness. Moreover, since multistage stochastic programs look for decisions that do not anticipate, but depend at each time period t only on information that is available at t, a distance measure for the information flow is needed. It is expressed by a distance of filtrations, since the information increase over time is modeled by σ-fields forming a filtration that is associated to the stochastic process.

The second topic requires the selection of appropriate risk functionals that allow to quantify risk in a meaningful way and preserve tractability of the optimization model. We argue that *polyhedral risk functionals* satisfy both demands. These are given as (the optimal values of) certain simple linear stochastic programs. Well-known risk functionals such as Average Value-at-Risk AVaR and expected polyhedral utility belong to this class and, moreover, multi-period risk functionals for multistage stochastic programs are suggested. For stochastic programs incorporating polyhedral risk functionals it has been shown that numerical tractability as well as stability results known for classical (nonrisk-averse) stochastic programs remain valid. In particular, the same scenario tree approximation methods can be used.

In a case study, we present illustrative numerical results from an electricity portfolio optimization model for a municipal power utility. In particular, it is shown that the use of different risk objectives leads to different risk aversion strategies by trading at derivative markets. They require less than additional 1% of the optimal expected revenue.

14.2 Mathematical Framework

Let a finite number of time steps $T \in \mathbb{N}$ as well as a multivariate discrete-time *stochastic process* $\xi = (\xi_1, ..., \xi_T)$ be given. This means that each ξ_t is a d-dimensional random vector (with some fixed dimension $d \in \mathbb{N}$) whose realization can be observed at time step $t = 1, ..., T$, respectively. Since $t = 1$ represents the present we require that ξ_1 is deterministic, i.e., $\xi_1 \in \mathbb{R}^d$. For $t \geq 2$ we require that each ξ_t has *statistical moments* of order r with some number $r \geq 1$ (that will be specified later on), i.e., $\mathbb{E}[|\xi_t|^r] < \infty$ for $t = 1, ..., T$ where $\mathbb{E}[\,.\,]$ denotes the expected value functional and $|\,.\,|$ refers to the Euclidean norm in $\mathbb{R}^d$.

Mathematically, these requirements are typically expressed by means of the so-called L_r-spaces: $\xi_t \in L_r(\Omega, \mathcal{F}, \mathbb{P}; \mathbb{R}^d)$ where $(\Omega, \mathcal{F}, \mathbb{P})$ is a given *probability space*. Now, in multistage stochastic programming, decisions x_t can be made at each time step $t = 1, ..., T$ based on the observations until time t, respectively. This means that x_t may depend and may only depend on (the concrete realization of) $\xi^t := (\xi_1, ..., \xi_t)$, respectively. This *nonanticipativity* requirement can be expressed by $x_t \in L_{r'}(\Omega, \sigma(\xi^t), \mathbb{P}; \mathbb{R}^{m_t})$ with some moment order $r' \geq 1$ (specified later on) and some dimensions $m_t \in \mathbb{N}$ $(t = 1, ..., T)$. In other words: x_t must be a $\sigma(\xi^t)$-measurable random element where $\sigma(\xi^t)$ is the sub-σ-field of the original σ-field $\mathcal{F}$ generated by $\xi_1, ..., \xi_t$. The sequence of all σ-fields is increasing, i.e., $\{\emptyset, \Omega\} = \sigma(\xi^1) \subseteq \sigma(\xi^2) \subseteq ... \subseteq \sigma(\xi^T) = \mathcal{F}$ and thus forms a so-called *filtration*. Assume for the moment that the input random vector ξ is represented in the form of a scenario tree, where d real variables are associated to each node of the tree. Then the $\sigma(\xi^t)$-measurability of x_t for every $t \in \{1, \ldots, T\}$ means that the decision vector x is represented by the same tree (as ξ), but with m_t real variables associated to each node at time t.

In this presentation, we consider *linear programming* multistage stochastic programs of the form

$$\min_{x_1,\ldots,x_T} \left\{ \mathbb{E}\left[\sum_{t=1}^{T} \langle b_t(\xi_t), x_t \rangle\right] \;\middle|\; \begin{array}{l} x_t \in L_{r'}(\Omega, \sigma(\xi^t), \mathbb{P}; \mathbb{R}^{m_t}), \\ x_t \in X_t \;\; \mathbb{P}\text{-almost surely (a.s.)}, \\ A_{t,0}x_t + A_{t,1}(\xi_t)x_{t-1} = h_t(\xi_t) \text{ a.s.} \\ (t = 1, ..., T) \end{array} \right\} \qquad (14.1)$$

with some numbers $m_t, n_t \in \mathbb{N}$, given polyhedral sets $X_t \subseteq \mathbb{R}^{m_t}$, recourse matrices $A_{t,0} \in \mathbb{R}^{n_t \times m_t}$, technology matrices $A_{t,1} \in \mathbb{R}^{n_t \times m_{t-1}}$ (where we assume $A_{1,1} \equiv 0$), and vectors $h_t \in \mathbb{R}^{n_t}$ and $b_t \in \mathbb{R}^{m_t}$ (cost factors). The items $A_{t,1}$, h_t, and b_t may depend on ξ_t ($t = 1, ..., T$). It is assumed that this dependence is affinely linear. This allows, for example, to model that some components of b_t, h_t and/or some elements of the matrix $A_{t,1}$ are stochastic and ξ denotes the vector of all such stochastic inputs.

Note that in (14.1) optimality of the stochastic costs $\langle b_t(\xi_t), x_t \rangle$ is determined in terms of the expected value, i.e., the objective is linear (risk-neutral). In Sects. 14.5 and 14.6 we will consider the risk–averse alternative

$$\min_{x_1,\ldots,x_T} \left\{ \begin{array}{c} \gamma \cdot \rho(z_{t_1}, ..., z_{t_J}) \\ -(1-\gamma) \cdot \mathbb{E}\left[z_T\right] \end{array} \middle| \begin{array}{l} x_t \in L_{r'}(\Omega, \sigma(\xi^t), \mathbb{P}; \mathbb{R}^{m_t}), \\ x_t \in X_t \text{ a.s.}, \\ A_{t,0}x_t + A_{t,1}(\xi_t)x_{t-1} = h_t(\xi_t) \text{ a.s.} \\ z_t := -\sum_{\tau=1}^{t} \langle b_\tau(\xi_\tau), x_\tau \rangle \text{ a.s.} \\ (t = 1, ..., T) \end{array} \right\} \tag{14.2}$$

where the objective is supplemented with a (multiperiod) risk functional ρ (risk measure). The number $\gamma \in [0, 1]$ is a fixed weighting parameter. The random values z_t represent the accumulated revenues at each time t. Clearly, it holds that $z_t \in L_p(\Omega, \sigma(\xi^t), \mathbb{P})$ with $p \in [1, \infty]$ given by

$$\frac{1}{p} = \begin{cases} \frac{1}{r'} \;, & \text{if all } b_t \text{ are nonrandom} \\ \frac{1}{r} + \frac{1}{r'} \;, & \text{otherwise.} \end{cases}$$

The risk functional ρ is applied to a subset of J time steps $1 < t_1 < t_2 < ... < t_J = T$. Note that, since risk functionals are essentially nonlinear by nature, problem (14.2) is no longer linear. However, we will concentrate on the employment of risk functionals from the class of *polyhedral risk functionals* which exhibit a favorable sort of nonlinearity; cf. Sect. 14.5.

14.3 Stability of Multistage Problems

Studying stability of the multistage stochastic program (14.1) consists in regarding it as an optimization problem in the infinite dimensional linear space $\times_{t=1}^{T} L_{r'}(\Omega, \mathcal{F}, \mathbb{P}; \mathbb{R}^{m_t})$. This is a Banach space when endowed with the norm

$$\begin{aligned} \|x\|_{r'} &:= \left(\textstyle\sum_{t=1}^{T} \mathbb{E}\left[|x_t|^{r'}\right] \right)^{1/r'} \quad \text{for } r' \in [1, \infty), \\ \|x\|_\infty &:= \max_{t=1,\ldots,T} \operatorname{ess\,sup} |x_t|, \end{aligned}$$

where $|\,.\,|$ denotes some norm on the relevant Euclidean spaces and $\operatorname{ess\,sup} |x_t|$ denotes the essential supremum of $|x_t|$, i.e., the smallest constant C such that $|x_t| \le C$ holds $\mathbb{P}$-almost surely. Analogously, ξ can be understood as an element of the Banach space $\times_{t=1}^{T} L_r(\Omega, \mathcal{F}, \mathbb{P}; \mathbb{R}^d)$ with norm $\|\xi\|_r$. For the integrability numbers r and r' it will be imposed that

$$r := \begin{cases} \in [1,\infty) \ , & \text{if only costs or only right-hand sides are random} \\ 2 \ , & \text{if only costs and right-hand sides are random} \\ T \ , & \text{if all technology matrices are random} \end{cases}$$

$$r' := \begin{cases} \frac{r}{r-1} \ , & \text{if only costs are random} \\ r \ , & \text{if only right-hand sides are random} \\ \infty \ , & \text{if all technology matrices are random} \end{cases} \tag{14.3}$$

with regard to problem (14.1). The choice of r and the definition of r' are motivated by the knowledge of existing moments of the input process ξ, by having the stochastic program well defined (in particular, such that $\langle b_t(\xi_t), x_t\rangle$ is integrable for every decision x_t and $t = 1, ..., T$), and by satisfying the conditions (A2) and (A3) (see below).

Since r' depends on r and our assumptions will depend on both r and r', we will add some comments on the choice of r and its interplay with the structure of the underlying stochastic programming model. To have the stochastic program well defined, the existence of certain moments of ξ has to be required. This fact is well known for the two-stage situation (see, e.g., [35, Chap. 2]). If either right-hand sides or costs in a multistage model (14.1) are random, it is sufficient to require $r \geq 1$. The flexibility in case that the stochastic process ξ has moments of order $r > 1$ may be used to choose r' as small as possible in order to weaken the condition (A3) (see below) on the feasible set. If the linear stochastic program is fully random (i.e., costs, right-hand sides and technology matrices are random), one needs $r \geq T$ to have the model well defined and no flexibility with respect to r' remains.

14.3.1 Assumptions

Next we introduce some notation. We set $s := Td$ and $m := \sum_{t=1}^{T} m_t$. Let

$$F(\xi, x) := \mathbb{E}\big[\textstyle\sum_{t=1}^{T}\langle b_t(\xi_t), x_t\rangle\big]$$

denote the objective function defined on $L_r(\Omega,\mathcal{F},\mathbb{P};\mathbb{R}^s) \times L_{r'}(\Omega,\mathcal{F},\mathbb{P};\mathbb{R}^m)$ and let

$$\mathcal{X}(\xi) := \big\{x \in \times_{t=1}^{T} L_{r'}(\Omega, \sigma(\xi^t), \mathbb{P}; \mathbb{R}^{m_t}) \,|\, x_t \in \mathcal{X}_t(x_{t-1};\xi_t) \text{ a.s. } \ (t = 1, ..., T)\big\}$$

denote the set of feasible elements of (14.1) with $x_0 \equiv 0$ and

$$\mathcal{X}_t(x_{t-1};\xi_t) := \big\{x_t \in \mathbb{R}^{m_t} : x_t \in X_t, A_{t,0}x_t + A_{t,1}(\xi_t)x_{t-1} = h_t(\xi_t)\big\}$$

denoting the t-th feasibility set for every $t = 1, ..., T$. That allows to rewrite the stochastic program (14.1) in the short form

$$\min\big\{F(\xi, x) : x \in \mathcal{X}(\xi)\big\}. \tag{14.4}$$

In the following, we need the optimal value

$$v(\xi) = \inf \{F(\xi, x) : x \in \mathcal{X}(\xi)\}$$

for every $\xi \in L_r(\Omega, \mathcal{F}, \mathbb{P}; \mathbb{R}^s)$ and, for any $\varepsilon \geq 0$, the *ε-approximate solution set* (level-set)

$$S_\varepsilon(\xi) := \{x \in \mathcal{X}(\xi) : F(\xi, x) \leq v(\xi) + \varepsilon\}$$

of the stochastic program (14.4). Since, for $\varepsilon = 0$, the set $S_\varepsilon(\xi)$ coincides with the set of solutions to (14.4), we will also use the notation $S(\xi) := S_0(\xi)$. The following conditions will be imposed on (14.4):

(A1) The numbers r, r' are chosen according to (14.3) and $\xi \in L_r(\Omega, \mathcal{F}, \mathbb{P}; \mathbb{R}^s)$.

(A2) There exists a $\delta > 0$ such that for any $\tilde{\xi} \in L_r(\Omega, \mathcal{F}, \mathbb{P}; \mathbb{R}^s)$ satisfying $\|\tilde{\xi} - \xi\|_r \leq \delta$, any $t = 2, ..., T$ and any $x_\tau \in L_{r'}(\Omega, \sigma(\tilde{\xi}^\tau), \mathbb{P}; \mathbb{R}^{m_\tau})$ $(\tau = 1, ..., t-1)$ satisfying $x_\tau \in \mathcal{X}_\tau(x_{\tau-1}; \tilde{\xi}_\tau)$ a.s. (where $x_0 = 0$), there exists $x_t \in L_{r'}(\Omega, \sigma(\tilde{\xi}^t), \mathbb{P}; \mathbb{R}^{m_t})$ satisfying $x_t \in \mathcal{X}_t(x_{t-1}; \tilde{\xi}_t)$ a.s. (*relatively complete recourse locally around* ξ).

(A3) The optimal values $v(\tilde{\xi})$ of (14.4) with input $\tilde{\xi}$ are finite for all $\tilde{\xi}$ in a neighborhood of ξ and the objective function F is *level-bounded locally uniformly at* ξ, i.e., for some $\varepsilon_0 > 0$ there exists a $\delta > 0$ and a bounded subset B of $L_{r'}(\Omega, \mathcal{F}, \mathbb{P}; \mathbb{R}^m)$ such that $S_{\varepsilon_0}(\tilde{\xi})$ is contained in B for all $\tilde{\xi} \in L_r(\Omega, \mathcal{F}, \mathbb{P}; \mathbb{R}^s)$ with $\|\tilde{\xi} - \xi\|_r \leq \delta$.

For any $\tilde{\xi} \in L_r(\Omega, \mathcal{F}, \mathbb{P}; \mathbb{R}^s)$ sufficiently close to ξ in L_r, condition (A2) implies the existence of some feasible $\tilde{x}$ in $\mathcal{X}(\tilde{\xi})$ and (14.3) implies the finiteness of the objective $F(\tilde{\xi}, .)$ at any feasible $\tilde{x}$. A sufficient condition for (A2) to hold is the *complete recourse condition* on every recourse matrix $A_{t,0}$, i.e., $A_{t,0}X_t = \mathbb{R}^{n_t}$, $t = 1, ..., T$. The locally uniform level-boundedness of the objective function F is quite standard in perturbation results for optimization problems (see, e.g., [34, Theorem 1.17]). The finiteness condition on the optimal value $v(\xi)$ is not implied by the level-boundedness of F for all relevant pairs (r, r'). In general, the conditions (A2) and (A3) get weaker for increasing r and decreasing r', respectively.

14.3.2 Optimal Values

The first stability result for multistage stochastic programs represents a quantitative continuity property of the optimal values. Its main observation is that multistage models behave stable at some stochastic input process if both its probability distribution and its filtration are approximated with respect to the L_r-distance and the filtration distance

$$D_{\mathrm{f}}(\xi, \tilde{\xi}) := \sup_{\varepsilon > 0} \inf_{\substack{x \in S_\varepsilon(\xi) \\ \tilde{x} \in S_\varepsilon(\tilde{\xi})}} \sum_{t=2}^{T-1} \max\Big\{\big\|x_t - \mathbb{E}[x_t|\sigma(\tilde{\xi}^t)]\big\|_{r'}, \big\|\tilde{x}_t - \mathbb{E}[\tilde{x}_t|\sigma(\xi^t)]\big\|_{r'}\Big\} \qquad (14.5)$$

where $\mathbb{E}[\,.\,|\sigma(\xi^t)]$ and $\mathbb{E}[\,.\,|\sigma(\tilde{\xi}^t)]$ $(t = 1, ..., T)$ are the corresponding conditional expectations, respectively. Note that for the supremum in (14.5) only small ε's are relevant and that the approximate solution sets are bounded for $\varepsilon \in (0, \varepsilon_0]$ according to (A3).

The following stability result for optimal values of program (14.4) is taken from [24, Theorem 2.1].

Theorem 1. *Let (A1), (A2), and (A3) be satisfied and the sets X_1 be nonempty and bounded. Then there exist positive constants L and δ such that the estimate*

$$|v(\xi) - v(\tilde{\xi})| \leq L \left(\|\xi - \tilde{\xi}\|_r + D_{\mathrm{f}}(\xi, \tilde{\xi}) \right) \tag{14.6}$$

holds for all random elements $\tilde{\xi} \in L_r(\Omega, \mathcal{F}, \mathbb{P}; \mathbb{R}^s)$ with $\|\tilde{\xi} - \xi\|_r \leq \delta$.

The result states that the changes of optimal values are at most proportional to the errors in terms of L_r- and filtration distance when approximating ξ. The corresponding constant L depends on $\|\xi\|_r$ (i.e., the r-th moment of ξ), but is not known in general.

14.3.3 Approximate Solutions

To prove a stability result for (approximate) solutions of (14.4) a stronger version of the filtration distance D_{f} is needed, namely,

$$D_{\mathrm{f}}^*(\xi, \tilde{\xi}) := \sup_{x \in \mathcal{B}_\infty} \sum_{t=2}^{T} \left\| \mathbb{E}[x_t|\sigma(\xi^t)] - \mathbb{E}[x_t|\sigma(\tilde{\xi}^t)] \right\|_{r'}, \tag{14.7}$$

where $\mathcal{B}_\infty := \{x : \Omega \to \mathbb{R}^m : x \text{ is } \mathcal{F}\text{-measurable}, |x(\omega)| \leq 1, \mathbb{P}\text{-almost surely}\}$. Notice that the sum is extended by the additional summand for $t = T$ and that the former infimum is replaced by a supremum with respect to a sufficiently large bounded set. If we require, in addition to assumption (A3), that for some $\varepsilon_0 > 0$ there exist constants $\delta > 0$ and $C > 0$ such that $|\tilde{x}(\omega)| \leq C$ for $\mathbb{P}$-almost every $\omega \in \Omega$ and all $\tilde{x} \in S_{\varepsilon_0}(\tilde{\xi})$ with $\tilde{\xi} \in L_r(\Omega, \mathcal{F}, \mathbb{P}; \mathbb{R}^s)$ and $\|\tilde{\xi} - \xi\|_r \leq \delta$, we have

$$D_{\mathrm{f}}(\xi, \tilde{\xi}) \leq C\, D_{\mathrm{f}}^*(\xi, \tilde{\xi}). \tag{14.8}$$

Sometimes it is sufficient to consider the unit ball in $L_{r'}$ rather than $\mathcal{B}$ (cf. [22, 23]). However, in contrast to D_{f} the distance D_{f}^* always satisfies the triangle inequality.

Now, we state the second stability result that represents a Lipschitz property of approximate solution sets ([22] Theorem 2.4).

Theorem 2. *Let (A1), (A2) and (A3) be satisfied with $r' \in [1, \infty)$ and the set X_1 be nonempty and bounded. Assume that the solution set $S(\xi)$ is nonempty. Then there exist $\bar{L} > 0$ and $\bar{\varepsilon} > 0$ such that*

$$d\!\!l_\infty\left(S_\varepsilon(\xi), S_\varepsilon(\tilde{\xi})\right) \leq \frac{\bar{L}}{\varepsilon} \left(\|\xi - \tilde{\xi}\|_r + D_{\mathrm{f}}^*(\xi, \tilde{\xi}) \right) \tag{14.9}$$

holds for every $\tilde{\xi} \in L_r(\Omega, \mathcal{F}, \mathbb{P}; \mathbb{R}^s)$ *with* $\|\xi - \tilde{\xi}\|_r \leq \delta$ *(with* $\delta > 0$ *from (A3)) and* $S(\tilde{\xi}) \neq \emptyset$*, and for any* $\varepsilon \in (0, \bar{\varepsilon})$*. Here,* $\mathbb{d}_\infty$ *denotes the Pompeiu–Hausdorff distance of closed bounded subsets of* $L_{r'} = L_{r'}(\Omega.\mathcal{F}, \mathbb{P}; \mathbb{R}^m)$*, which is given by*

$$\mathbb{d}_\infty(B, \tilde{B}) = \sup_{x \in L_{r'}} \left| d_B(x) - d_{\tilde{B}}(x) \right|$$

with $d_B(x)$ *denoting the distance of* x *to* B*, i.e.,* $d_B(x) = \inf_{y \in B} \|x - y\|_{r'}$*.*

The most restrictive assumption in Theorem 2 is the existence of solutions to both problems. Notice that solutions always exist if the underlying random vector has a finite number of scenarios or if $r' \in (1, \infty)$. For a more thorough discussion we refer to [22, Sect. 2]. Notice that the constant $\frac{\bar{L}}{\varepsilon}$ gets larger for decreasing ε and that, indeed, Theorem 2 does not remain true for the Pompeiu–Hausdorff distance of solution sets $S(\xi) = S_0(\xi)$ and $S(\tilde{\xi}) = S_0(\tilde{\xi})$, respectively.

14.4 Construction of Scenario Trees

In this section we want to introduce a general approach to generate appropriate scenario trees by making use of the stability theory of the previous section. To this end we assume that $r \geq 1$ and r' are selected such that ξ has a finite r-th moment and according to (14.3), respectively. Then we aim at generating a scenario tree ξ_{tr} such that the distances

$$\|\xi - \xi_{\text{tr}}\|_r \quad \text{and} \quad D_{\text{f}}^*(\xi, \xi_{\text{tr}}) \tag{14.10}$$

are small, where the latter is given by (14.7). We conclude that the optimal values $v(\xi)$ and $v(\xi_{\text{tr}})$, and the approximate solution sets $S_\varepsilon(\xi)$ and $S_\varepsilon(\xi_{\text{tr}})$ are close to each other according to Theorem 1 and Theorem 2, respectively.

14.4.1 General Approach

The scenario tree construction method starts with a good initial scenario approximation consisting of a finite number of scenarios. These scenarios might be obtained by quantization techniques [16] or by sampling or resampling techniques based on parametric or nonparametric stochastic models of the input process ξ. Let us denote the initial approximation of ξ by $\hat{\xi}$ having scenarios $\xi^i = (\xi_1^i, \ldots, \xi_T^i) \in \mathbb{R}^{Td}$ with probabilities $p_i > 0$, $i = 1, \ldots, N$, and a common root, i.e., $\xi_1^1 = \ldots = \xi_1^N =: \xi_1^*$.

In the following we assume that

$$\|\xi - \hat{\xi}\|_r + D_{\text{f}}^*(\xi, \hat{\xi}) \leq \varepsilon \tag{14.11}$$

holds for some given (initial) tolerance $\varepsilon > 0$. For example, condition (14.11) may be satisfied for D_{f}^* and any tolerance $\varepsilon > 0$ if $\hat{\xi}$ is obtained by sampling from a finite set with sufficiently large sample size (see [23, Example 5.3]). A more general case is discussed in [20], where the only assumption is that the initial set of scenarios provides a good approximation with respect to the L_r-distance.

Next we describe an algorithmic procedure that starts from $\hat{\xi}$ and ends up with a scenario tree process ξ_{tr} having the same root node ξ_1^*, less nodes than $\hat{\xi}$ and allowing for constructive estimates of $\|\hat{\xi} - \xi_{\text{tr}}\|_r$. The idea of the algorithm consists in forming clusters of scenarios based on scenario reduction on the time horizon $\{1, ..., t\}$ recursively for increasing time t. To this end, the seminorm $\| \,.\, \|_{r,t}$ on $L_r(\Omega, \mathcal{F}, \mathbb{P}; \mathbb{R}^s)$ (with $s = Td$) given by

$$\|\xi\|_{r,t} := \left(\mathbb{E}\left[|\xi|_t^r\right]\right)^{1/r} \tag{14.12}$$

is used at step t, where $|\,.\,|_t$ is a seminorm on $\mathbb{R}^s$ which, for each $\xi = (\xi_1, ..., \xi_T) \in \mathbb{R}^s$, is given by $|\xi|_t := |(\xi_1, ..., \xi_t, 0, ..., 0)|$.

The scenario tree construction algorithm determines recursively stochastic processes $\hat{\xi}^t$ having scenarios $\hat{\xi}^{t,i}$ endowed with probabilities p_i, $i \in I := \{1, ..., N\}$, and, in addition, partitions $\mathcal{C}_t = \{C_t^1, ..., C_t^{K_t}\}$ of the index set I, i.e.,

$$C_t^k \cap C_t^{k'} = \emptyset \quad (k \neq k') \quad \text{and} \quad \bigcup_{k=1}^{K_t} C_t^k = I. \tag{14.13}$$

The index sets $C_t^k \in \mathcal{C}_t$, $k = 1, ..., K_t$, represent clusters of scenarios (see Fig. 14.1 for an illustration). To define these clusters we aim at aggregating similar scenarios at each time step.

The initialization of the scenario tree generation procedure consists in setting $\hat{\xi}^1 := \hat{\xi}$, i.e., $\hat{\xi}^{1,i} = \xi^i$, $i \in I$, and $\mathcal{C}_1 = \{I\}$. At step t (with $t > 1$) we consider each cluster C_{t-1}^k, i.e., each scenario subset $\{\hat{\xi}^{t-1,i}\}_{i \in C_{t-1}^k}$, separately and delete scenarios $\{\hat{\xi}^{t-1,j}\}_{j \in J_t^k}$ by the forward selection algorithm of [21] (see also [23, Sect. 2]) such that

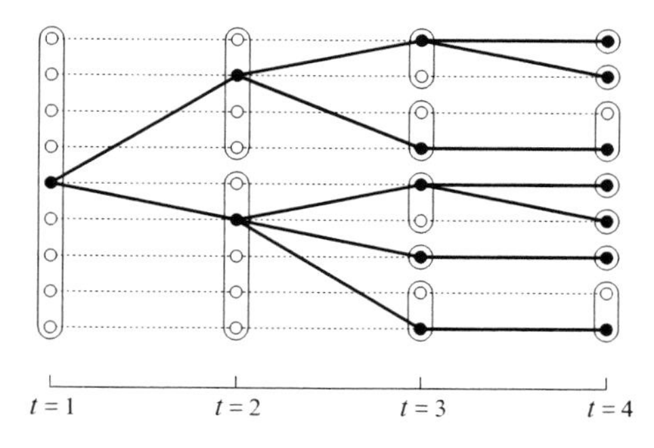

Fig. 14.1. Illustration of the tree construction by recursive scenario clustering

$$\left(\sum_{k=1}^{K_{t-1}} \sum_{j \in J_t^k} p_j \min_{i \in I_t^k} \left| \hat{\xi}^{t-1,i} - \hat{\xi}^{t-1,j} \right|_t^r \right)^{1/r}$$

is bounded from above by some prescribed tolerance. Here, the index set I_t^k of remaining scenarios is given by

$$I_t^k = C_{t-1}^k \setminus J_t^k.$$

As in the general scenario reduction procedure in [21], the index set J_t^k is subdivided into index sets $J_{t,i}^k$, $i \in I_t^k$ such that

$$J_t^k = \bigcup_{i \in I_t^k} J_{t,i}^k \quad \text{and} \quad J_{t,i}^k := \{j \in J_t^k : i = i_t^k(j)\}$$
$$\text{with} \quad i_t^k(j) \in \arg\min_{i \in I_t^k} |\hat{\xi}^{t-1,i} - \hat{\xi}^{t-1,j}|_t^r.$$

Next we define a mapping $\alpha_t : I \to I$ such that

$$\alpha_t(j) = \begin{cases} i_t^k(j) \ , \ j \in J_t^k, \ k = 1, ..., K_{t-1} \\ j \quad , \text{ otherwise.} \end{cases} \tag{14.14}$$

Then the scenarios of the stochastic process $\hat{\xi}^t = \{\hat{\xi}_\tau^t\}_{\tau=1}^T$ are defined by

$$\hat{\xi}_\tau^{t,i} = \begin{cases} \xi_\tau^{\alpha_\tau(i)} \ , \ \tau \leq t \\ \xi_\tau^i \quad , \text{ otherwise} \end{cases} \tag{14.15}$$

with probabilities p_i for each $i \in I$. The processes $\hat{\xi}^t$ are illustrated in Fig. 14.2, where $\hat{\xi}^t$ corresponds to the t-th picture for $t = 1, ..., T$. The partition $\mathcal{C}_t$ at t is defined by

$$\mathcal{C}_t = \{\alpha_t^{-1}(i) : i \in I_t^k, \ k = 1, ..., K_{t-1}\}, \tag{14.16}$$

i.e., each element of the index set I_t^k defines a new cluster and the new partition $\mathcal{C}_t$ is a refinement of the former partition $\mathcal{C}_{t-1}$.

The scenarios of the final scenario tree $\xi_{\text{tr}} := \hat{\xi}^T$ and their probabilities are given by the structure of the final partition $\mathcal{C}_T$, i.e., they have the form

$$\xi_{\text{tr}}^k = \left(\xi_1^*, \xi_2^{\alpha_2(i)}, ..., \xi_t^{\alpha_t(i)}, ..., \xi_T^{\alpha_T(i)}\right) \quad \text{and} \quad \pi_T^k = \sum_{j \in C_T^k} p_j \quad \text{if } i \in C_T^k \tag{14.17}$$

for each $k = 1, ..., K_T$. The index set I_t of realizations of ξ_t^{tr} is given by

$$I_t := \bigcup_{k=1}^{K_{t-1}} I_t^k.$$

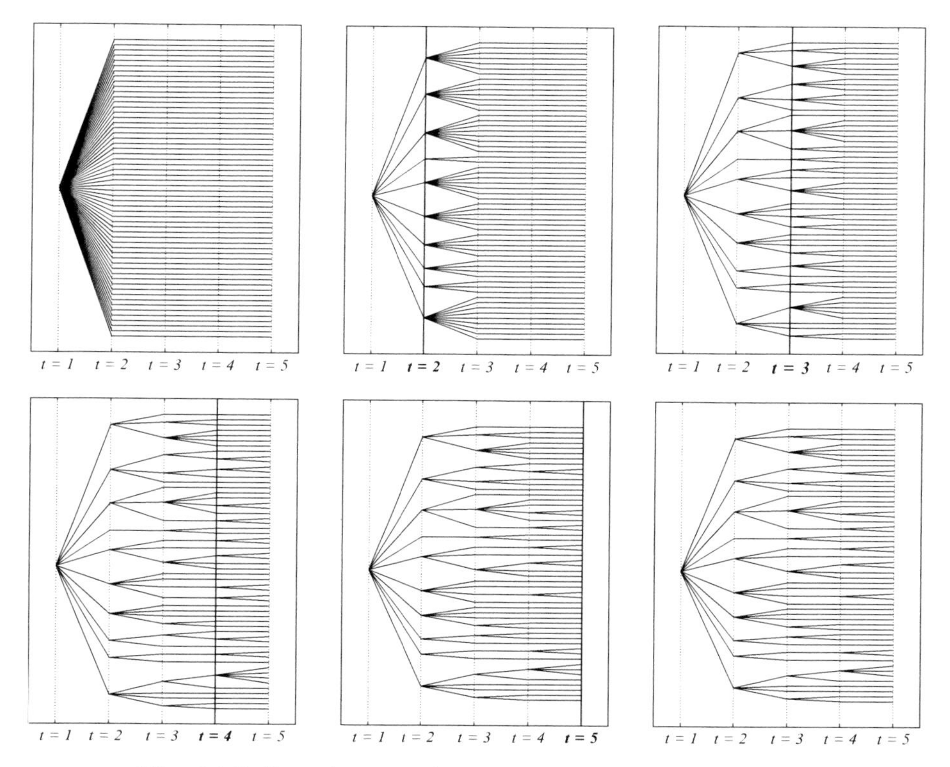

Fig. 14.2. Stepwise scenario tree construction for an example

For each $t \in \{1, ..., T\}$ and each $i \in I$ there exists an unique index $k_t(i) \in \{1, ..., K_t\}$ such that $i \in C_t^{k_t(i)}$. Moreover, we have $C_t^{k_t(i)} = \{i\} \cup J_{t,i}^{k_{t-1}(i)}$ for each $i \in I_t$. The probability of the i-th realization of ξ_t^{tr} is $\pi_t^i = \sum_{j \in C_t^{k_t(i)}} p_j$. The branching degree of scenario $i \in I_{t-1}$ coincides with the cardinality of $I_t^{k_t(i)}$.

The next result quantifies the relative error of the t-th construction step and is proved in [23, Theorem 3.4].

Theorem 3. *Let the stochastic process $\hat{\xi}$ with fixed initial node ξ_1^*, scenarios ξ^i and probabilities p_i, $i = 1, ..., N$, be given. Let ξ_{tr} be the stochastic process with scenarios $\xi_{\mathrm{tr}}^k = (\xi_1^*, \xi_2^{\alpha_2(i)}, ..., \xi_t^{\alpha_t(i)}, ..., \xi_T^{\alpha_T(i)})$ and probabilities π_T^k for $i \in C_T^k$, $k = 1, ..., K_T$. Then we have*

$$\|\hat{\xi} - \xi_{\mathrm{tr}}\|_r \leq \sum_{t=2}^{T} \left(\sum_{k=1}^{K_{t-1}} \sum_{j \in J_t^k} p_j \min_{i \in I_t^k} |\xi_t^i - \xi_t^j|^r \right)^{1/r} . \tag{14.18}$$

14.4.2 Flexible Algorithm

Summarizing the above ideas yields the following scenario tree construction algorithm that allows to control the tree structure as well as the approximation tolerance with respect to the L_r-distance.

Algorithm 1 (forward tree construction)
Let N scenarios ξ^i with probabilities p_i, $i = 1, ..., N$, fixed root $\xi_1^* \in \mathbb{R}^d$, $r \geq 1$ and tolerances ε_r, ε_t, $t = 2, ..., T$, be given such that $\sum_{t=2}^{T} \varepsilon_t \leq \varepsilon_r$.

Step 1: Set $\hat{\xi}^1 := \hat{\xi}$ and $\mathcal{C}_1 = \{\{1, ..., N\}\}$.
Step t: Let $\mathcal{C}_{t-1} = \{C_{t-1}^1, ..., C_{t-1}^{K_{t-1}}\}$. Determine disjoint index sets I_t^k and J_t^k such that $I_t^k \cup J_t^k = C_{t-1}^k$, the mapping $\alpha_t(\,.\,)$ according to (14.14) and a stochastic process $\hat{\xi}^t$ having N scenarios $\hat{\xi}^{t,i}$ with probabilities p_i according to (14.15) and such that

$$\left\|\hat{\xi}^t - \hat{\xi}^{t-1}\right\|_{r,t}^r = \sum_{k=1}^{K_{t-1}} \sum_{j \in J_t^k} p_j \min_{i \in I_t^k} |\xi_t^i - \xi_t^j|^r \leq \varepsilon_t^r.$$

Set $\mathcal{C}_t = \{\alpha_t^{-1}(i) : i \in I_t^k,\ k = 1, ..., K_{t-1}\}$.
Step T+1: Let $\mathcal{C}_T = \{C_T^1, ..., C_T^{K_T}\}$. Construct a stochastic process ξ_{tr} having K_T scenarios ξ_{tr}^k such that $\xi_{\text{tr},t}^k := \xi_t^{\alpha_t(i)}$, $t = 1, ..., T$, if $i \in C_T^k$ with probabilities π_T^k according to (14.17), $k = 1, ..., K_T$.

While the first picture in Fig. 14.2 illustrates the process $\hat{\xi}$, the t-th picture corresponds to the situation after Step t, $t = 2, 3, 4, 5$ of the algorithm. The final picture corresponds to Step 6 and illustrates the final scenario tree ξ_{tr}. The proof of the following corollary is also given in [23].

Corollary 1. *Let a stochastic process $\hat{\xi}$ with fixed initial node ξ_1^*, scenarios ξ^i and probabilities p_i, $i = 1, ..., N$, be given. If ξ_{tr} is constructed by Algorithm 1, we have*

$$\|\hat{\xi} - \xi_{\text{tr}}\|_r \leq \sum_{t=2}^{T} \varepsilon_t \leq \varepsilon_r.$$

The next results states that the distance $|v(\xi) - v(\xi_{\text{tr}})|$ of optimal values gets small if the initial tolerance ε in (14.11) as well as ε_r are small (cf. [22, Theorem 3.4]).

Theorem 4. *Let (A1), (A2), and (A3) be satisfied with $r' \in [1, \infty)$ and the set X_1 be nonempty and bounded. Let $L > 0$, $\delta > 0$ and $C > 0$ be the constants appearing in Theorem 1 and (14.8), respectively. If $(\varepsilon_r^{(n)})$ is a sequence tending to 0 such that the corresponding tolerances $\varepsilon_t^{(n)}$ in Algorithm 1 are nonincreasing for all $t = 2, ..., T$, the corresponding sequence $(\xi_{\text{tr}}^{(n)})$ has the property*

$$\limsup_{n\to\infty} |v(\xi) - v(\xi_{\mathrm{tr}}^{(n)})| \leq L \max\{1, C\}\varepsilon, \tag{14.19}$$

where $\varepsilon > 0$ is the initial tolerance in (14.11).

14.5 Polyhedral Risk Functionals

The results and methods from Sect. 14.3 and Sect. 14.4 rest upon the linearity of problem (14.1) to some extent. Hence, in general they are not valid for the risk–averse problem (14.2) incorporating a general (nonlinear) risk functional ρ such as, e.g., Value-at-Risk ($\rho = \mathrm{VaR}_\alpha$) or standard deviation. Also algorithmic approaches for (14.1) might be destroyed by the incorporation of general risk functionals. However, in this section we consider the risk–averse problem (14.2) with ρ being chosen as a so-called *polyhedral risk functional.* This class of risk functionals has been introduced in [7,8]. The key feature of these functionals is that they, though being nonlinear, do not destroy mathematical structures of stochastic programs such as linearity or convexity.

14.5.1 Definition

The reason for the favorable behavior of polyhedral risk functionals in (14.2) is obvious from their definition: a polyhedral risk functional ρ is given by (the optimal value of) a linear stochastic minimization problem of the form

$$\rho(z) = \inf\left\{ \mathbb{E}\left[\sum_{j=0}^{J} \langle c_j, y_j\rangle\right] \middle| \begin{array}{l} y \in \times_{j=0}^{J} L_p(\Omega, \sigma(\xi^{t_j}), \mathbb{P}; \mathbb{R}^{k_j}) \\ y_j \in Y_j \text{ } \mathbb{P}\text{-almost surely (a.s.) } (j = 0, ..., J), \\ \sum_{\tau=0}^{j} \langle w_{j,\tau}, y_{j-\tau}\rangle = z_{t_j} \text{ a.s. } (j = 1, ..., J), \\ \sum_{\tau=0}^{j} V_{j,\tau} y_{j-\tau} = r_j \text{ a.s. } (j = 0, ..., J) \end{array} \right\} \tag{14.20}$$

for every $z = (z_{t_1}, ..., z_{t_J}) \in \times_{j=1}^{J} L_p(\Omega, \sigma(\xi^{t_j}), \mathbb{P})$ with some $p \in [1, \infty)$. The numbers $k_j \in \mathbb{N}_0$, $d_j \in \mathbb{N}_0$ $(j = 0, ..., J)$, vectors $c_j \in \mathbb{R}^{k_j}$, $r_j \in \mathbb{R}^{d_j}$ $(j = 0, ..., J)$, $w_{j,\tau} \in \mathbb{R}^{k_{j-\tau}}$ $(j = 1, ..., J,\ \tau = 0, ..., j)$, matrices $V_{j,\tau} \in \mathbb{R}^{d_j \times k_{j-\tau}}$ $(j = 0, ..., J,\ \tau = 0, ..., j)$, and polyhedral cones $Y_j \subseteq \mathbb{R}^{k_j}$ $(j = 0, ..., J)$ have to be chosen in advance such that the resulting functional exhibits suitable risk functional properties. Clearly, if definition (14.20) is inserted into (14.2) with[1] $\gamma = 1$, one ends up with the problem

[1] The choice $\gamma = 1$ is not restrictive at all since the so-called *mean-risk objective* $\gamma \cdot \rho(z_{t_1}, ..., z_{t_J}) - (1 - \gamma) \cdot \mathbb{E}[z_T]$ can be expressed as another polyhedral risk functional of the form (14.20); cf. [7,8].

$$\min\left\{\mathbb{E}\left[\sum_{j=0}^{J}\langle c_j, y_j\rangle\right] \left| \begin{array}{l} x \in \times_{t=1}^{T} L_{r'}(\Omega, \mathcal{A}_t, \mathbb{P}; \mathbb{R}^{m_t}),\ x_t \in X_t \text{ a.s. } (t \geq 1), \\ y \in \times_{j=1}^{J} L_p(\Omega, \mathcal{A}_{t_j}, \mathbb{P}; \mathbb{R}^{k_j}),\ y_j \in Y_j \text{ a.s. } (j \geq 0), \\ A_{t,0}x_t + A_{t,1}(\xi_t)x_{t-1} = h_t(\xi_t) \text{ a.s. } (t = 2, ..., T), \\ z_t = z_t(x, \xi) := -\sum_{\tau=1}^{t}\langle b_\tau(\xi_\tau), x_\tau\rangle \ (t = 1, ..., T), \\ \sum_{\tau=0}^{j}\langle w_{j,\tau}, y_{j-\tau}\rangle = z_{t_j} \text{ a.s. } (j = 1, ..., J), \\ \sum_{\tau=0}^{j} V_{j,\tau}y_{j-\tau} = r_j \text{ a.s. } (j = 0, ..., J) \end{array}\right.\right\} \tag{14.21}$$

i.e., the nonlinearity of the functional ρ is transformed into a problem of the form (14.1) with additional decision variables y_j and additional linear constraints. This fact is not only useful for stability analysis (see below), it is also appreciated with regard to algorithmic issues. Note that this transformation is also possible if integer variables are incorporated into (14.1).

Most well-known risk functionals (e.g., VaR_α and standard deviation which are both not polyhedral) depend on a single random variable z only rather than on a finite sequence $z_{t_1}, ..., z_{t_J}$. In the framework of (14.2) this means $J = 1$ and $t_1 = T$. Several coherence axioms for such single-period risk functionals have been suggested in [1, 14, 29] which are broadly accepted. For medium- and long-term economic activities (such as the model in Sect. 14.6) one may want to use multiperiod risk functionals ($J > 1$) that take into account the temporal development of profits and losses, e.g., to avoid liquidity problems at intermediate time steps. Also for this case coherence axioms are suggested [2, 15,32]. In both the single- and the multiperiod case such axioms give directions for the choice of the vectors and matrices in (14.20).

14.5.2 Properties

Because the arguments z_{t_j} in (14.20) appear on the right-hand sides of the constraints, it can be concluded that the functional ρ is always *convex* [7,8]. Hence, the theory of convex duality can be applied. This yields dual representations for ρ which can be useful for interpretation and verification of coherence axioms, and for algorithmic approaches, too.

Theorem 5. *([7, 8, 32]) Let ρ be a polyhedral risk functional of the form (14.20) and let the following conditions be satisfied for Y_j, c_j, $w_{j,\tau}$, and $V_{j,\tau}$:*

- Complete recourse*:* $\begin{pmatrix} V_{j,0} \\ w'_{j,0} \end{pmatrix} Y_j = \mathbb{R}^{d_j+1}$ $(j = 1, ..., J)$,
- Dual feasibility*:* $\bigcap_{j=0}^{J} \mathcal{D}_{\rho,j} \neq \emptyset$ *with*

$$\mathcal{D}_{\rho,j} := \left\{\begin{array}{l} (u_v, u_w) \in \mathbb{R}^J \times \mathbb{R}^{\sum d_j} : \\ c_j + \sum_{\nu=\max\{1,j\}}^{J} u_{v,\nu}w_{\nu,\nu-j} + \sum_{\nu=j}^{J} V^*_{\nu,\nu-j}u_{w,\nu} \in -Y_j^* \end{array}\right\}.$$

Then the functional ρ is finite, convex, and continuous on $\times_{j=1}^{J} L_p(\Omega, \sigma(\xi^{t_j}), \mathbb{P})$ and it is representable by

$$\rho(z) = \sup \left\{ -\mathbb{E}\left[\sum_{j=1}^{J} (\lambda_j z_{t_j} + \langle \mu_j, r_j \rangle)\right] \;\middle|\; \begin{array}{l} \lambda_j \in L_{p'}(\Omega, \sigma(\xi^{t_j}), \mathbb{P}), \\ \mu_j \in L_{p'}(\Omega, \sigma(\xi^{t_j}), \mathbb{P}; \mathbb{R}^{d_j}), \\ (\mathbb{E}[\lambda|\xi^{t_j}], \mathbb{E}[\mu|\xi^{t_j}]) \in \mathcal{D}_{\rho,j} \;\; a.s. \\ (j = 0, ..., J) \end{array} \right\}$$

with $p' \in (1, \infty]$ *being defined by* $1/p + 1/p' = 1$.

The above dual representation can be read as follows: the supremum operator aims at making λ large where z is small (in compliance with the respective constraints). Hence, $\rho(z)$ can be understood as a *worst case weighted expectation* of z (possibly biased by $\langle \mu_j, r_j \rangle$). If ρ satisfies the coherence axioms from [2], then (and only then) the constraints in the dual representation are such that all the λ multipliers are probability densities and $\langle \mu_j, r_j \rangle$ is always zero.

14.5.3 Single-Period Examples

For $J = 1$ and $t_1 = T$, i.e., for the single-period situation, polyhedral risk functionals can be found in economic literature.

Example 1. The *Conditional* or *Average Value-at-Risk* at level $\alpha \in (0, 1)$ (CVaR_α or AVaR_α, cf. [33] and [14, Sect. 4.4 in Chap. 4]) is given by

$$\text{AVaR}_\alpha(z) := \tfrac{1}{\alpha} \int_0^\alpha \text{VaR}_{\bar{\alpha}}(z) d\bar{\alpha} = \inf_{y_0 \in \mathbb{R}} \left\{ y_0 + \tfrac{1}{\alpha} \mathbb{E}\left[(y_0 + z)^- \right] \right\} \tag{14.22}$$

where the representation on the right is due to [33]. By introducing variables for positive and negative parts of $y_0 + z$, respectively, AVaR_α can be rewritten in the form (14.20) with $J = 1$, $d_0 = d_1 = 0$, $k_0 = 1$, $k_1 = 2$, $c_0 = 1$, $c_1 = (0, \frac{1}{\alpha})$, $w_{1,0} = (1, -1)$, $w_{1,1} = -1$, $Y_0 = \mathbb{R}$, and $Y_1 = \mathbb{R}^2_+$. Hence, AVaR_α is a polyhedral risk functional. Moreover, complete recourse and dual feasibility are satisfied and the dual representation of Theorem 5 reads

$$\text{AVaR}_\alpha(z) = \sup \left\{ -\mathbb{E}[\lambda z] : \lambda \in L_{p'}(\Omega, \mathcal{F}, \mathbb{P}),\ \lambda \in [0, \tfrac{1}{\alpha}] \text{ a.s.}, \mathbb{E}[\lambda] = 1 \right\}$$

where the λ multipliers can be interpreted as densities. We note that AVaR_α is known to be a convex risk functional in the sense of [14], a coherent risk functional in the sense of [1], and it is first- and second-order stochastic dominance consistent [29].

Example 2. Consider *expected utility* as a risk functional, i.e., $\rho_u(z) = -\mathbb{E}[u(z)]$ with some concave and nondecreasing utility function $u : \mathbb{R} \to \mathbb{R}$. This approach goes back to [39]. Typically, nonlinear utility functions $u : \mathbb{R} \to \mathbb{R}$ are used within this framework. Of course, in this case ρ_u cannot be represented by a linear stochastic program. However, in cases when the domain of the outcome z can be bounded a priori, it makes sense to consider piecewise linear utility functions u. In that case, $-u$ is convex and piecewise linear, hence, according to [34, Example 3.54] there exist $k \in \mathbb{N}$, $w \in \mathbb{R}^k$, $c \in \mathbb{R}^k$, and $v \in \{0,1\}^k$ such that

$$-u(\mu) = \inf\left\{ \langle c, y\rangle \,\middle|\, y \in \mathbb{R}^k,\ y \geq 0\ \langle w, y\rangle = \mu,\ \langle v, y\rangle = 1 \right\}$$

for all $\mu \in \mathbb{R}$. For this case, the expected utility risk functional reads

$$\rho_u(z) = \inf\left\{ \mathbb{E}\left[\langle c, y_1\rangle\right] \,\middle|\, \begin{array}{l} y_1 \in L_p(\Omega, \mathcal{A}, \mathbb{P}; \mathbb{R}^k),\ y_1 \geq 0 \ \text{ a.s.} \\ \langle w, y_1\rangle = z \ \text{ a.s.}, \langle v, y_1\rangle = 1 \ \text{ a.s.} \end{array} \right\}$$

where [34, Theorem 14.60] is used to justify the interchange of infimum and expectation. Hence, ρ_u is a polyhedral risk functional with $k_0 = d_0 = 0$, $k_1 = k$, $d_1 = 1$, $c_1 = c$, $w_{1,0} = w$, $V_{1,0} = v'$, and $Y_1 = \mathbb{R}^{k_1}_+$. The special case of the *expected regret* (expected loss), i.e., the case that $\rho(z) = \mathbb{E}[(z-\gamma)^-]$ with some target $\gamma \in \mathbb{R}$, is obtained by setting $k = 3$, $w = (\gamma, 1, -1)$, $v = (1, 0, 0)$, and $c = (0, 0, -1)$.

14.5.4 Multiperiod Examples

For $J > 1$, i.e., for the multiperiod situation, only few (polyhedral) risk functionals are suggested in economic literature. However, the framework of polyhedral risk functionals is *constructive*: various multiperiod polyhedral risk functionals have been proposed in [7, 8, 32] that can be understood as multiperiod extentions of AVaR_α. They all satisfy the basic risk coherence axioms from [2], but they differ with respect to the incorporation of the information dynamics. We present a selection of those in the following (keeping the original index numbers). It is assumed that the random variables z_t represent *accumulated* revenues as in problem (14.2).

Example 3. The functional

$$\rho_2(z_{t_1}, ..., z_{t_J}) := \inf_{y_0 \in \mathbb{R}} \left\{ y_0 + \tfrac{1}{\alpha}\tfrac{1}{J} \textstyle\sum_{j=1}^J \mathbb{E}\left[\left(z_{t_j} + y_0\right)^- \right] \right\}$$

from [8] can be understood as AVaR_α applied to a compound lottery, i.e., applied to z_0 given by $z_0(\omega) := z_{\iota(\omega)}(\omega)$ with ι being uniformly distributed on $\{t_1, ..., t_J\}$ and independent of $z_{t_1}, ..., z_{t_J}$. Clearly, ρ_2 can be represented through (14.20) by introducing (stochastic) variables for the positive and the negative part of $z_{t_j} + y_0$, respectively, for $j = 1, ..., J$. Hence, it is a polyhedral risk functional. It satisfies complete recourse and dual feasibility. The dual representation according to Theorem 5 given by

$$\rho_2(z) = \sup\left\{ -\mathbb{E}\left[\textstyle\sum_{j=1}^J \lambda_j z_{t_j} \right] \,\middle|\, \begin{array}{l} \lambda \in \times_{j=1}^J L_p(\Omega, \sigma(\xi^{t_j}), \mathbb{P}),\ \sum_{j=1}^J \mathbb{E}\left[\lambda_j\right] = 1 \\ \lambda_j \in [0, \frac{1}{\alpha}] \text{ a.s. } (j = 1, ..., J), \end{array} \right\}$$

aims at placing the available probability mass of λ to stages where $z = (z_{t_1}, ..., z_{t_J})$ attains low values.

Example 4. The polyhedral risk functional ρ_4 from [8], though being defined via an infimum representation of the form (14.20), is easier to catch by its dual representation according to Theorem 5 given by

$$\rho_4(z) = \sup \left\{ -\mathbb{E}\left[\sum_{j=1}^{J} \lambda_j z_{t_j}\right] \middle| \begin{array}{l} \lambda \in \times_{j=1}^{J} L_p(\Omega, \sigma(\xi^{t_j}), \mathbb{P}), \\ \lambda_j \in [0, \frac{1}{\alpha}] \text{ a.s., } \mathbb{E}[\lambda_j] = \frac{1}{J} \ (j = 1, ..., J) \\ \lambda_j = \mathbb{E}[\lambda_{j+1} | \sigma(\xi^{t_j})] \text{ a.s. } (j = 1, ..., J-1) \end{array} \right\}$$

with $z = (z_{t_1}, ..., z_{t_J})$. Here, the multiplier process λ has to be a martingale and, hence, all time steps are weighted equally.

Example 5. In [2] it was suggested to apply a single-period risk functional to the pointwise minimum of $z = (z_{t_1}, ..., z_{t_J})$, i.e., to z_0 given by $z_0(\omega) := \min\{z_{t_1}(\omega), ..., z_{t_J}(\omega)\}$. Doing so by using AVaR_α yields the functional

$$\begin{aligned} \rho_6(z) &= \inf_{y_0 \in \mathbb{R}} \left(y_0 + \tfrac{1}{\alpha}\mathbb{E}\left[(y_0 + z_0)^-\right]\right) \\ &= \inf_{y_0 \in \mathbb{R}} \left(y_0 + \tfrac{1}{\alpha}\mathbb{E}\left[\max\{0, -y_0 - z_{t_1}, ..., -y_0 - z_{t_J}\}\right]\right) \end{aligned}$$

which can also be represented in the form (14.20) by introducing (stochastic) variables $y_{j,2} = \max\{0, -y_0 - z_{t_1}, ..., -y_0 - z_{t_j}\} = \max\{y_{j-1,2}, -y_0 - z_{t_j}\}$ for $j = 1, ..., J$; cf. [7]. Then, complete recourse and dual feasibility are satisfied and there is also a dual representation according to Theorem 5.

14.5.5 Stability

At the first glance it seems as if stability of problem (14.2) with ρ being chosen as a polyhedral risk functional (14.20) were covered by the results from Sect. 14.3 due to the reformulation (14.21). However, a closer look to the latter problem reveals that it is not completely of the form (14.1): the resulting recourse matrices become stochastic when the dynamic constraints in (14.21) are integrated. Hence, Theorem 1 and Theorem 2 are not valid for problem (14.21) and cannot be suitably modified easily.

For this reason, stability of (14.2) is analyzed in [7, 10] systematically. Starting with the finding of further continuity properties of ρ (stronger than plain continuity as stated in Theorem 5), a stability theorem for the optimal values (corresponding to Theorem 1) can be proven. However, the filtration distance there is even more involved than D_{f} in (14.5) from Theorem 1.

For the justification of the scenario tree generation methods in Sect. 14.4, it is necessary to estimate these problem dependent objects by problem independent ones as in (14.8). In order to get a similar estimate for the involved filtration distance for problem (14.2), it turns out to be necessary to impose further technical conditions on ρ (beside complete recourse and dual feasibility). However, these conditions can be shown to be satisfied for all known polyhedral risk functionals from [7,8,32] as long as the integrability number p is set to 1. We conclude that there is a theoretical basis for the scenario tree approximation methods from Sect. 14.4 also in the situation of the risk–averse problem (14.2) if ρ is chosen as a suitable polyhedral risk functional.

14.6 Case Study

In this final section we demonstrate the use of the above theoretical results by presenting some simulation results from a power portolio optimization model; cf. Fig. 14.3. For motivation and for a detailed technical description of this model see [9, 11]; in the following, we describe its components on a more abstract level only. Its numerical output shall then illustrate the usage of scenario trees as well as the effect of different polyhedral risk functionals.

14.6.1 Model

Taking into account uncertainties in power portfolio optimization yields quite automatically to stochastic programming; see, e.g., [40]. The optimization model here is a mean-risk multistage stochastic program of the form (14.2). It is tailored to the 1-year planning situation of a certain (German) municipal power utility serving an electricity demand and a heat demand for certain customers; see Fig. 14.3. The (German) power market induces an hourly time discretization, hence, we have $T = 365 \cdot 24 = 8{,}760$ time steps. Energy demands as well as market prices for each hour in the future are unknown at previous time steps. These uncertainties can be described reasonably by stochastic time series models; cf. [11]. It is assumed that the power utility is sufficiently small such that it can be considered as a *price-taker*, i.e., its decisions do not affect market prices or demands.

The concrete situation of the power utility is supposed to be as follows: It features a *combined heat and power* (CHP) production plant that can serve the heat demand completely but the electricity demand only in part. Hence,

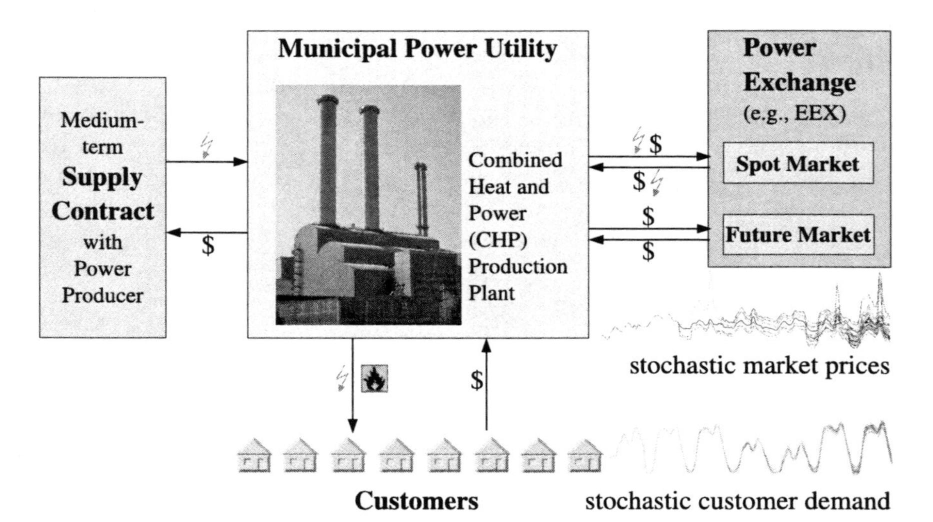

Fig. 14.3. Schematic diagram for the optimization model components

additional sources of electricity have to be used. Electricity can be obtained from the spot market of a power exchange (such as the European Energy Exchange EEX in Germany), or by purchasing a bilateral supply contract from a larger power producer. The latter possibility is suspected to be more expensive, but relying on the spot market only is known to be extremely risky. Spot price risk, however, may be reduced (hedged) by means of derivative products. Here, we consider futures from EEX (Phelix-futures, purely financial contracts).

The original practical purpose of this model was to evaluate given supply contracts in comparison with the possibility of relying on spot and future market only [9,11]. In the presentation here, however, we focus on the qualitative output with respect to the effect of the different polyhedral risk functionals from Sect. 14.5. Therefore, no such supply contracts are considered in the portfolio here.

The stochastic input process $\xi = (\xi_1, ..., \xi_T)$, modeled by an appropriate time series model (cf. [11]), is approximated by a scenario tree (cf. Fig. 14.4) according to the methods from Sect. 14.4. Each random vector ξ_t consists of 27 components: electricity demand ξ_t^e, heat demand ξ_t^h, EEX spot prices ξ_t^s, as well as base and peak future prices ξ_t^{fbm} and ξ_t^{fpm} (for each month $m = 1, ..., 12$). However, to avoid technical problems related to arbitrage, the future prices are calculated as *fair prices* from the spot prices in the scenario tree, i.e., the methods from Sect. 14.4 are applied only to the first three components ξ_t^e, ξ_t^h, and ξ_t^s $(t = 1, ..., T)$.

The decisions at each time t consist of CHP production amounts, EEX spot market volumes (electricity may be bought or sold), future stock, and contract flexibility (if there is any). The CHP production is subject to several technical (dynamic) constraints which are slightly simplified such that

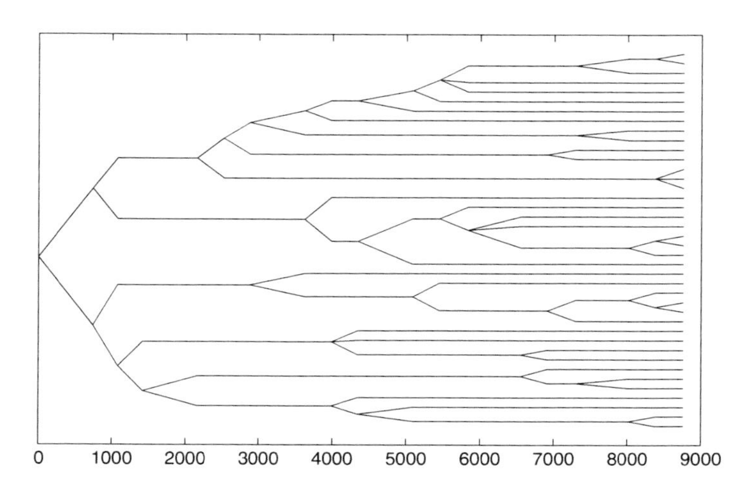

Fig. 14.4. Branching structure of the input scenario tree of 40 scenarios ($T = 8760$)

no integer variables come into play, i.e., everything is linear. There are no particular constraints for spot and future trading, but the pricing rules for EEX futures (initial margin, variation margin, transaction costs) make it necessary to introduce some auxiliary variables and constraints. Finally, there are the demand satisfaction constraints requiring that electricity demand and heat demand are always met. For further details we refer to [9]. The overall model (incorporating a polyhedral risk functional) is linear, i.e., it is of the form (14.2) resp. (14.21). Of course, the latter formulation is used for implementation.

14.6.2 Simulations Results

Together with a fixed scenario tree (cf. Fig. 14.4) the overall optimization model is a (large-scale) linear program. For the simulation results presented here, we used a scenario tree of 40 scenarios and approx. 150,000 nodes. The decision variables are defined on the nodes of the tree. For solving the linear program the ILOG CPLEX 9.1 software was employed. We restrict the presentation here to the case that no additional supply contracts are involved (beside EEX futures). Then, the different effects of the polyhedral risk functionals from Sect. 14.5 can be observed best.

In Fig. 14.5 the accumulated revenues z_t over time for each scenario, i.e., the temporal developments of the company's wealth, are shown after optimization with different polyhedral risk functionals. Of course, the tree structure of the input scenario tree can also be found in these outputs since the (optimal) revenues are stochastic in the same manner as the inputs. Optimizing the expected overall revenue $\mathbb{E}[z_T]$ only (without any risk functional) yields large dispersion (spread) at time T (cf. top of Fig. 14.5). The incorporation of the (single-period) AVaR applied to z_T reduces this spread considerably, but yields high spread and very low values for z_t at earlier time steps $t < T$. Clearly, this behavior is not acceptable for a (small) power utility. The multiperiod polyhedral risk functionals from Sect. 14.5 are effective such that dispersion is somehow better distributed over all time steps.

The graphs in Fig. 14.5 suggest that the effect of ρ_2, ρ_4, and ρ_6 is more or less the same. However, Fig. 14.6 reveals that there are further differences among these multiperiod risk functionals. For the calculation of these graphs, the fuel costs for the CHP plant have been slightly augmented in order to give the cash value curves a different direction. The difference between the multiperiod functionals is, roughly speaking, that ρ_4 aims at equal spread at all times, whereas ρ_2 and ρ_6 try to find a maximal level that is rarely underrun.

The different shapes of the cash value curves are achieved by different policies of future trading. Future trading is revealed through the jumps in the cash value curves and is explicitly shown in Fig. 14.7. These graphs display the overall future stock volumes (in Euro) at each time step. If no risk is considered then there is no future trading at all since, due to the fair-price assumption, there is no benefit from futures in terms of the expected revenue.

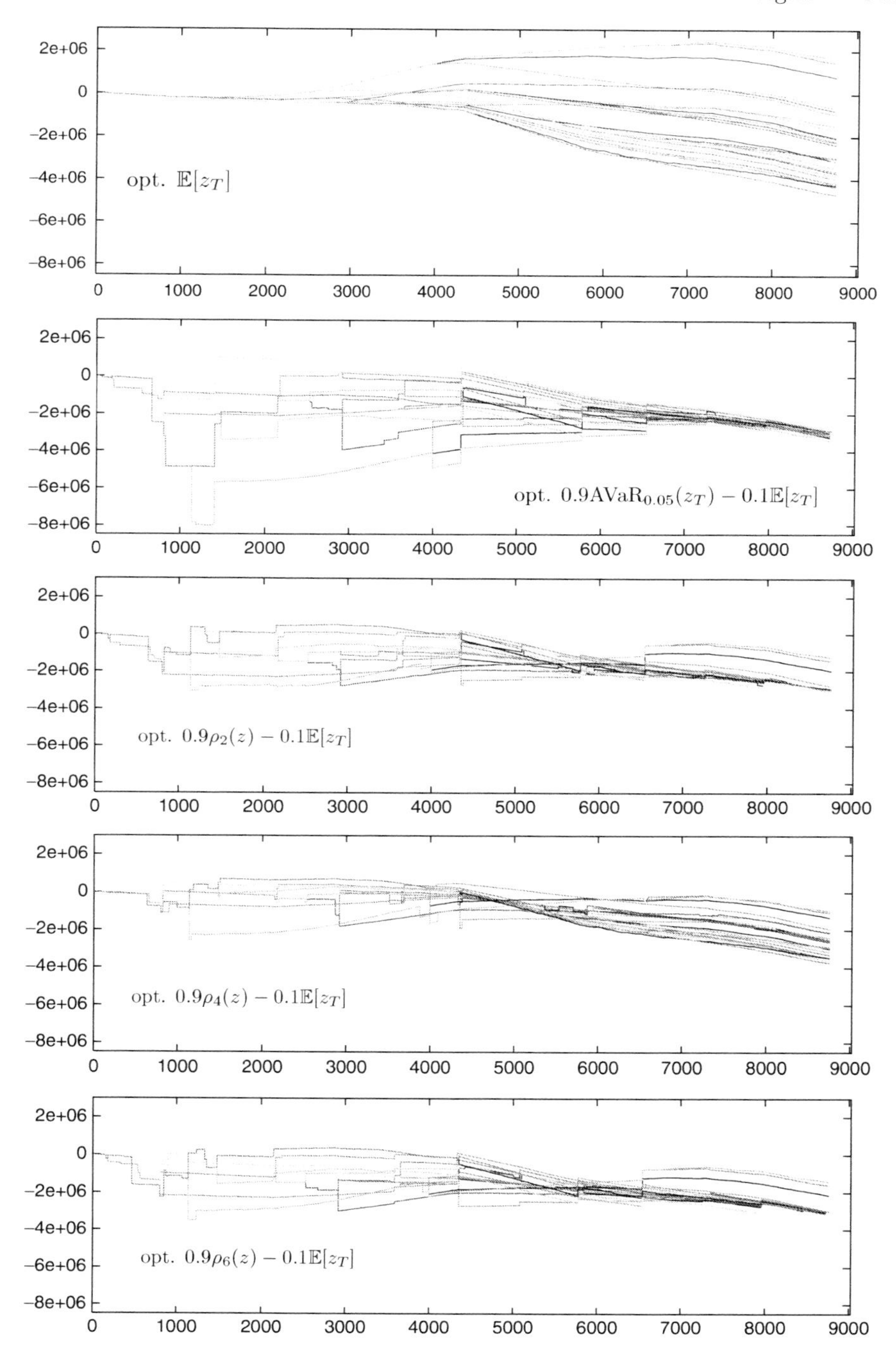

Fig. 14.5. Optimal cash values z_t (wealth) over time ($t = 1, ..., T$) for each scenario

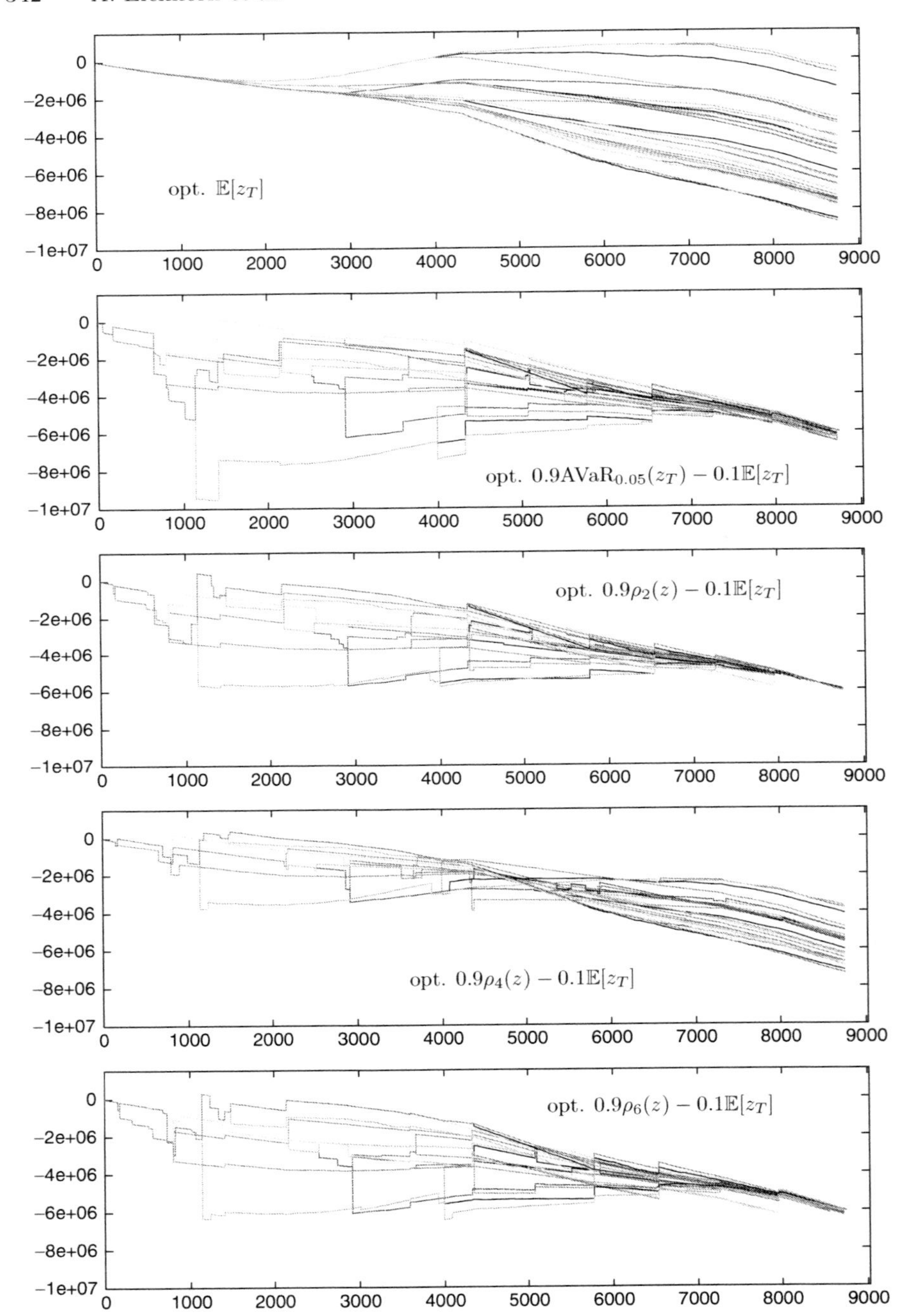

Fig. 14.6. Optimal cash values z_t (wealth) over time ($t = 1, ..., T$), **high fuel costs**

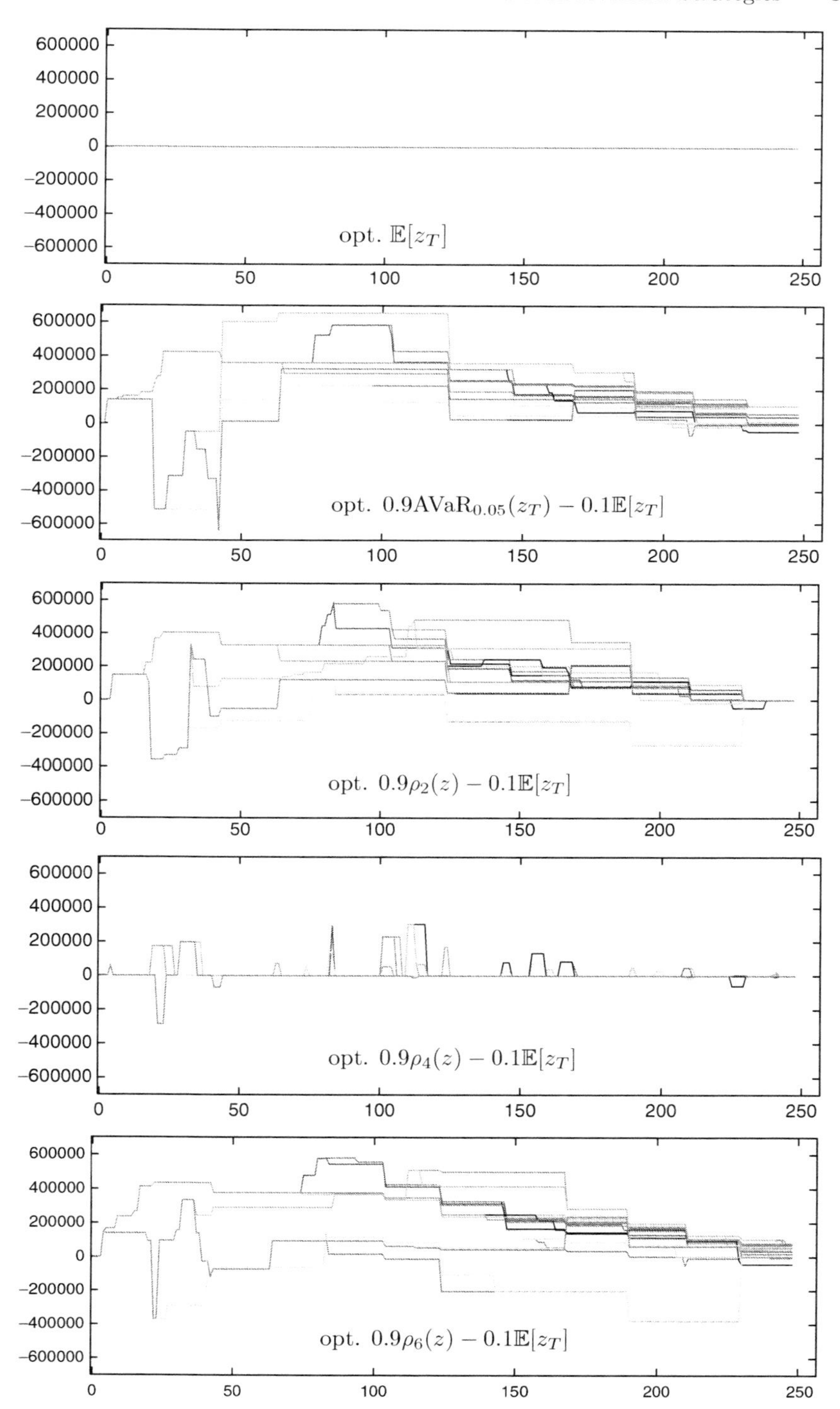

Fig. 14.7. Overall future stock over time (248 trading days), **high fuel costs**

Using AVaR, ρ_2, or ρ_6 leads to extensive future trading activity, whereas the application of ρ_4 yields more moderate future trading activity.

Finally, we mention that, within this application model, the incorporation of a polyhedral risk functional into the objective reduces the expected overall revenue $\mathbb{E}[z_T]$ only by approx. 1%. The additional computational effort arising from the risk measure is also very moderate.

14.7 Conclusion

We have presented a capacious theory for the framework of multistage stochastic programming. Though appearing rather technical and abstract at the first glance, these results are highly relevant in practice: Problems become numerically tractable by finite scenario tree approximation of the underlying stochastic input data. Moreover, risk–aversion requirements can be incorporated without significant increase of complexity by means of polyhedral risk functionals. In particular, there is a theoretical basis for the scenario tree approximation methods in both cases, the risk-neutral and the risk–averse case. For illustration, we have presented an exemplary model for mean-risk optimization of an electricity portfolio.

Acknowledgement. The presented work was supported by the DFG Research Center MATHEON "Mathematics for Key Technologies" in Berlin (`http://www.matheon.de`) and by the "Wiener Wissenschafts-, Forschungs- und Technologiefonds" in Vienna (`http://www.univie.ac.at/crm/simopt`).

References

1. P. Artzner, F. Delbaen, J.-M. Eber, and D. Heath. Coherent measures of risk. *Mathematical Finance*, 9:203–228, 1999.
2. P. Artzner, F. Delbaen, J.-M. Eber, D. Heath, and H. Ku. Coherent multiperiod risk adjusted values and Bellman's principle. *Annals of Operations Research*, 152:5–22, 2007.
3. B. Blaesig. *Risikomanagement in der Stromerzeugungs- und Handelsplanung*, volume 113 of *Aachener Beitrge zur Energieversorgung*. Klinkenberg, Aachen, Germany, 2007. PhD Thesis.
4. M.S. Casey and S. Sen. The scenario generation algorithm for multistage stochastic linear programming. *Mathematics of Operations Research*, 30:615–631, 2005.
5. M.A.H. Dempster. Sequential importance sampling algorithms for dynamic stochastic programming. *Zapiski Nauchnykh Seminarov POMI*, 312:94–129, 2004.
6. J. Dupačová, G. Consigli, and S.W. Wallace. Scenarios for multistage stochastic programs. *Annals of Operations Research*, 100:25–53, 2000.

7. A. Eichhorn. *Stochastic Programming Recourse Models: Approximation, Risk aversion, Applications in Energy.* PhD thesis, Department of Mathematics, Humboldt University, Berlin, 2007.
8. A. Eichhorn and W. Römisch. Polyhedral risk measures in stochastic programming. *SIAM Journal on Optimization*, 16:69–95, 2005.
9. A. Eichhorn and W. Römisch. Mean-risk optimization models for electricity portfolio management. In *Proceedings of the 9th International Conference on Probabilistic Methods Applied to Power Systems (PMAPS)*, Stockholm, Sweden, 2006.
10. A. Eichhorn and W. Römisch. Stability of multistage stochastic programs incorporating polyhedral risk measures. *Optimization*, 57:295–318, 2008.
11. A. Eichhorn, W. Römisch, and I. Wegner. Mean-risk optimization of electricity portfolios using multiperiod polyhedral risk measures. In *IEEE St. Petersburg PowerTech Proceedings*, 2005.
12. S.-E. Fleten and T.K. Kristoffersen. Short-term hydropower production planning by stochastic programming. *Computers and Operations Research*, 35:2656–2671, 2008.
13. S.-E. Fleten, S.W. Wallace, and W.T. Ziemba. Hedging electricity portfolios via stochastic programming. In C. Greengard and A. Ruszczyński, editors, *Decision Making under Uncertainty: Energy and Power*, volume 128 of *IMA Volumes in Mathematics and its Applications.* Springer, New York, pages 71–93, 2002.
14. H. Föllmer and A. Schied. *Stochastic Finance. An Introduction in Discrete Time*, volume 27 of *De Gruyter Studies in Mathematics.* Walter de Gruyter, Berlin, 2nd edition, 2004.
15. M. Frittelli and G. Scandolo. Risk measures and capital requirements for processes. *Mathematical Finance*, 16:589–612, 2005.
16. S. Graf and H. Luschgy. *Foundations of Quantization for Probability Distributions*, volume 1730 of *Lecture Notes in Mathematics.* Springer, Berlin, 2000.
17. C. Greengard and A. Ruszczyński, editors, *Decision Making under Uncertainty: Energy and Power*, volume 128 of *IMA Volumes in Mathematics and its Applications.* Springer, New York, 2002.
18. N. Gröwe-Kuska, K.C. Kiwiel, M.P. Nowak, W. Römisch, and I. Wegner. Power management in a hydro-thermal system under uncertainty by Lagrangian relaxation. In C. Greengard and A. Ruszczyński, editors, *Decision Making under Uncertainty: Energy and Power*, volume 128 of *IMA Volumes in Mathematics and its Applications.* Springer, New York, pages 39–70, 2002.
19. N. Gröwe-Kuska and W. Römisch. Stochastic unit commitment in hydrothermal power production planning. In S. W. Wallace and W. T. Ziemba, editors, *Applications of Stochastic Programming*, MPS/SIAM Series on Optimization, pages 633–653. SIAM, Philadelphia, PA, USA, 2005.
20. H. Heitsch. *Stabilität und Approximation stochastischer Optimierungsprobleme.* PhD thesis, Department of Mathematics, Humboldt University, Berlin, 2007.
21. H. Heitsch and W. Römisch. Scenario reduction algorithms in stochastic programming. *Computational Optimization and Applications*, 24:187–206, 2003.
22. H. Heitsch and W. Römisch. Stability and scenario trees for multistage stochastic programs. In G. Infanger, editor, *Stochastic Programming – The State of the Art.* 2009. to appear.
23. H. Heitsch and W. Römisch. Scenario tree modeling for multistage stochastic programs. *Mathematical Programming*, to appear, 2009.

24. H. Heitsch, W. Römisch, and C. Strugarek. Stability of multistage stochastic programs. *SIAM Journal on Optimization*, 17:511–525, 2006.
25. R. Hochreiter and G. Ch. Pflug. Financial scenario generation for stochastic multi-stage decision processes as facility location problems. *Annals of Operations Research*, 152:257–272, 2007.
26. R. Hochreiter, G. Ch. Pflug, and D. Wozabal. Multi-stage stochastic electricity portfolio optimization in liberalized energy markets. In *System Modeling and Optimization*, IFIP International Federation for Information Processing. Springer, Boston, MA, USA, pages 219–226, 2006.
27. K. Høyland, M. Kaut, and S.W. Wallace. A heuristic for moment-matching scenario generation. *Computational Optimization and Applications*, 24:169–185, 2003.
28. B. Krasenbrink. *Integrierte Jahresplanung von Stromerzeugung und -handel*, volume 81 of *Aachener Beiträge zur Energieversorgung*. Klinkenberg, Aachen, Germany, 2002. PhD Thesis.
29. W. Ogryczak and A. Ruszczyński. On consistency of stochastic dominance and mean-semideviation models. *Mathematical Programming*, 89:217–232, 2001.
30. T. Pennanen. Epi-convergent discretizations of multistage stochastic programs via integration quadratures. *Mathematical Programming*, 116:461–479, 2008.
31. M.V.F. Pereira and L.M.V.G. Pinto. Multi-stage stochastic optimization applied to energy planning. *Mathematical Programming*, 52:359–375, 1991.
32. G. Ch. Pflug and W. Römisch. *Modeling, Measuring, and Managing Risk*. World Scientific, Singapore, 2007.
33. R.T. Rockafellar and S. Uryasev. Conditional value-at-risk for general loss distributions. *Journal of Banking and Finance*, 26:1443–1471, 2002.
34. R.T. Rockafellar and R.J-B. Wets. *Variational Analysis*, volume 317 of *Grundlehren der mathematischen Wissenschaften*. Springer, Berlin, 1st edition, 1998. (Corr. 2nd printing 2004).
35. A. Ruszczyński and A. Shapiro, editors, *Stochastic Programming*, volume 10 of *Handbooks in Operations Research and Management Science*. Elsevier, Amsterdam, 1st edition, 2003.
36. H.K. Schmöller. *Modellierung von Unsicherheiten bei der mittelfristigen Stromerzeugungs- und Handelsplanung*, volume 103 of *Aachener Beitrge zur Energieversorgung*. Klinkenberg, Aachen, Germany, 2005. PhD Thesis.
37. S. Sen, L. Yu, and T. Genc. A stochastic programming approach to power portfolio optimization. *Operations Research*, 54:55–72, 2006.
38. S. Takriti, B. Krasenbrink, and L.S.-Y. Wu. Incorporating fuel constraints and electricity spot prices into the stochastic unit commitment problem. *Operations Research*, 48:268–280, 2000.
39. J. von Neumann and O. Morgenstern. *Theory of Games and Economic Behavior*. Princeton University Press, Princeton, NJ, USA, 1944.
40. S.W. Wallace and S.-E. Fleten. Stochastic programming models in energy. In Ruszczyński and Shapiro [35], Chap. 10, vol. 10 of Handbooks in Operations Research and Management Science. North-Holland, The Netherlands, pp. 637-677, 2003.

15

Optimization of Dispersed Energy Supply – Stochastic Programming with Recombining Scenario Trees

Alexa Epe, Christian Küchler, Werner Römisch, Stefan Vigerske, Hermann-Josef Wagner, Christoph Weber, and Oliver Woll

Summary. The steadily increasing share of wind energy within many power generating systems leads to strong and unpredictable fluctuations of the electricity supply and is thus a challenge with regard to power generation and transmission. We investigate the potential of energy storages to contribute to a cost optimal electricity supply by decoupling the supply and the demand. For this purpose we study a stochastic programming model of a regional power generating system consisting of thermal power units, wind energy, different energy storage systems, and the possibility for energy import. The identification of a cost optimal operation plan allows to evaluate the economical possibilities of the considered storage technologies.

On the one hand the optimization of energy storages requires the consideration of long-term planning horizons. On the other hand the highly fluctuating wind energy input requires a detailed temporal resolution. Consequently, the resulting optimization problem can, due to its dimension, not be tackled by standard solution approaches. We thus reduce the complexity by employing recombining scenario trees and apply a decomposition technique that exploits the special structure of those trees.

15.1 Introduction

Electric power, one of the most important fields within energy supply, has two main characteristics: on the one hand supply and demand have to be balanced at every time, on the other hand it is storable at only small rates. For these reasons, power plants have to regulate any imbalances between supply and demand, and, in particular, need to cope with unpredictable changes in the customer load. For that purpose regulating power plants are used, which mostly run in part load and with reduced efficiency. Alternatively fast power plants such as open cycle gas turbines may be used, which can start up within short time. Beyond the cover of the fluctuating load of the customer side, these power plants must also adjust to the increasing share of time-varying power production on the supply side, mostly caused from fluctuating renewables, notably wind.

Germany is the country with the highest installed wind power capacities worldwide. In the year 2006, there was approximately 20 GW installed (about 16.6% of the total installed power in Germany) and with the planned offshore development it could be up to 50 GW in 2030. Thereby, the sometimes strong and rapid fluctuations of the wind energy fed into the electrical network as well as the regional concentration in the north of the country increasingly pose problems to the network operators and power suppliers [7, 17]. Conventional fuel consumption may be saved by downregulating conventional (back-up) power plants, but investments in the power plant park can hardly be saved.

In this context, electrical energy storages offer a possibility to decouple supply and demand and to achieve a better capacity utilization as well as a higher efficiency of existing power plants. The changing context has led to an increased interest in such possibilities over the last few years. Yet with the liberalization of the electricity markets, the economics of storages have to be valued against market prices as established at the energy exchanges. Also the operation of storages will mostly not follow local imbalances of demand and supply, but rather try to benefit from market price variations. In particular, the (partial) unpredictability of market prices as well as of wind energy supply have to be taken into account. Things are complicated further through daily, weekly, seasonal, and other cyclic patterns in demand, supply, and prices. This requires a valuation of storages (and other options) over periods as long as 1 year.

Cost optimal operation planning under uncertainty for such long time periods poses a huge challenge to conventional stochastic programming methods. In this paper we investigate a novel approach, reducing complexity by applying recombining scenario trees. The latter are used to analyze a regional energy system model that is described in Sect. 15.2. Sect. 15.3 presents the decomposition approach based on recombining trees, whereas Sects. 15.4 and 15.5 are devoted to the results obtained so far.

15.2 Model Description

To study the economics of storages, a fundamental model is used. Combining technical and economical aspects, the model describes the energy supply of a large city, the available technologies for electricity generation, and the demand. An optimal load dispatch has to consider the marginal generation costs as well as the impact of other system restrictions such as start up costs, etc. Most important restriction of the model is the covering of the demand according to a given profile. For this purpose, energy can be produced by conventional power plants, procured as wind energy, and purchased on the spot market, cf. Fig. 15.1.

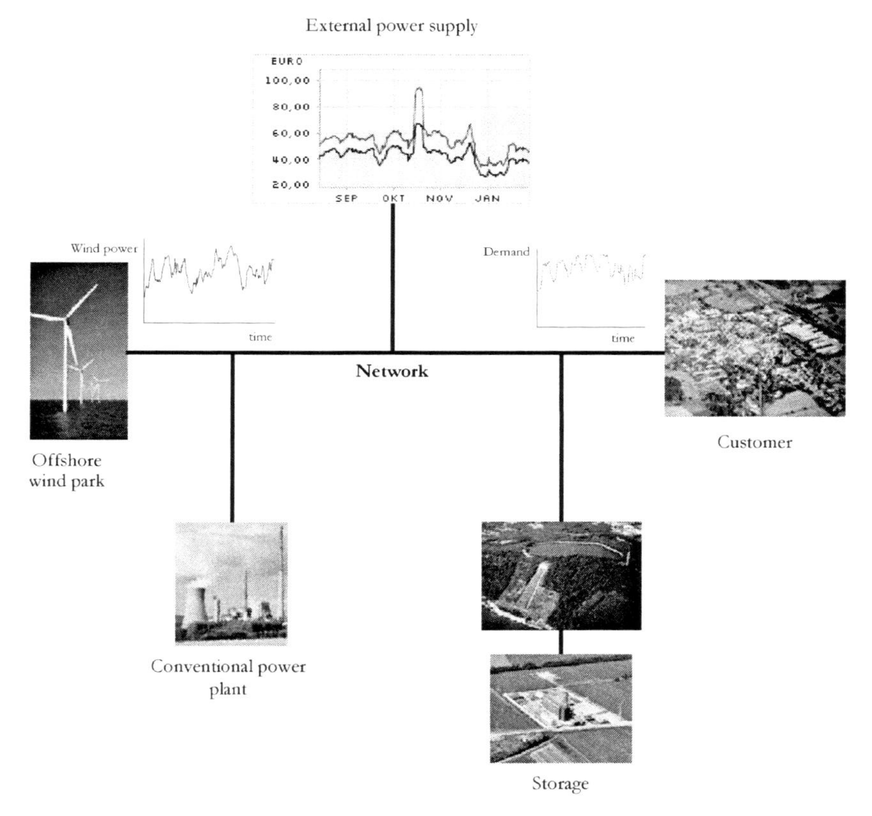

Fig. 15.1. Scheme of the fundamental model

Uncertainty in the amount of available wind energy and electricity prices is modeled by a multivariate stochastic process that can be represented by a recombining scenario tree. Thus, the proposed model combines many features of generation scheduling models (unit commitment and load dispatch) as found typically in energy system models [15,16]. In the following, the model is discussed in detail. Table 15.1 gives an overview of the notation used.

Under the assumption of power markets with efficient information treatment and without market power, the market results correspond to the outcomes of an optimization carried out by a fully informed central planner. If electricity demand is assumed to be price inelastic, welfare maximization is equivalent to cost minimization within the considered power network. Thereby, the total costs TC are given as the sum of import costs IC_t, operating costs $OC_{t,i}$, and startup costs $SC_{t,i}$ over all time steps t and unit types i:

$$TC = \sum_{t=1}^{T} \left(IC_t + \sum_{i} OC_{t,i} + SC_{t,i} \right) . \qquad (15.1)$$

Table 15.1. Notation used by the model

Variables			
Q	Production	IC	Import costs
H	Storage level	SC	Start-up costs
L	Capacity	OC	Operating costs
		TC	Total costs
Indices			
t	Time step	com	Compressing power
T	Final time	pum	Pumping power
i	Unit type	imp	Import power
stu	Start-up	wind	Wind power
Parameters			
D	Demand	c^{stu}	specific start-up costs
W	Wind power	c^{imp}	specific import costs
ℓ	Load factor	c^{oth}	other variable costs
η^0, η^m	Efficiency	c^{fuel}	fuel price

The costs for power import at time t are given by

$$IC_t = c_t^{\text{imp}} Q_t^{\text{imp}}. \tag{15.2}$$

For the operating costs $OC_{t,i}$, an affine function of the plant output $Q_{t,i}$ is assumed. An exact description of the plant operation costs requires a mixed-binary nonlinear formulation due to the dependency of the plant efficiency on the power output and the startup behavior. This is hardly feasible due to the high level of time detail. An appropriate linearization can be done by defining an additional decision variable for each plant type, the *capacity currently online* $L_{t,i}^{onl}$ [18]. The capacity online forms an upper bound on the actual output. Multiplied with the minimum load factor, it is also a lower bound on the output for each power plant. Hence, operating costs can be decomposed in fuel costs for operation at minimum load, fuel costs for incremental output, and other variable costs:

$$OC_{t,i} = \frac{c_{i,t}^{\text{fuel}}}{\eta_i^0} \ell_i L_{t,i}^{\text{onl}} + \frac{c_{i,t}^{\text{fuel}}}{\eta_i^m} (Q_{t,i} - \ell_i L_{t,i}^{\text{onl}}) + c_i^{\text{oth}} Q_{t,i}. \tag{15.3}$$

Here, η_i^m denotes the marginal efficiency for an operating plant and η_i^0 the efficiency at the minimum load factor ℓ_i. With $\eta_i^m > \eta_i^0$, the operators have an incentive to reduce the capacity online (for details see [18]).

Besides operating costs, start-up costs may influence the power scheduling decisions considerably. The start-up costs of unit i at time t are given by

$$SC_{t,i} = c_i^{\text{stu}} L_{t,i}^{\text{stu}}, \tag{15.4}$$

where $L^{\text{stu}}_{t,i}$ is the start-up capacity given by

$$L^{\text{stu}}_{t,i} = \max(0, L^{\text{onl}}_{t,i} - L^{\text{onl}}_{t-1,i}). \tag{15.5}$$

Covering the demand at time step t is ensured by

$$\sum_i Q_{t,i} + Q^{\text{wind}}_t + Q^{\text{imp}}_t \geq D_t + \sum_i Q^{\text{pum}}_{t,i} + \sum_i Q^{\text{com}}_{t,i}, \tag{15.6}$$

i.e., the supply at time t is given by the sum of the power production $Q_{t,i}$, the imported energy Q^{imp}_t, and the wind energy supply Q^{wind}_t. The total demand equals the sum of the exogenously given domestic demand D_t and the pumping and compressing energies $Q^{pum}_{t,i}$ and $Q^{com}_{t,i}$ used to fill the pumped hydro storage and compressed-air storage, respectively.

The operation levels of the units, pumps, and air compressors are constrained by the available capacity,

$$Q_{t,i} \leq L_{t,i}, \qquad Q^{\text{pum}}_{t,i} \leq L^{\text{pum}}_{t,i}, \qquad Q^{\text{com}}_{t,i} \leq L^{\text{com}}_{t,i}, \tag{15.7}$$

whereas the wind energy supply is bounded by the available wind energy at time t,

$$Q^{\text{wind}}_t \leq W_t. \tag{15.8}$$

For the storage plants, storage constraints need to be considered and the filling and discharging has to be described. This leads to the following storage level equation, linking the storage level $H_{t,i}$ at time t with the level $H_{t-1,i}$ at time $t-1$, both expressed in energy units. For the pumped hydro units, this reads as

$$\begin{aligned} H_{t,i} = H_{t-1,i} - \frac{1}{\eta^m_i} Q_{t,i} - \left(\frac{1}{\eta^0_i} - \frac{1}{\eta^m_i}\right) \ell_i L^{\text{onl}}_{t,i} \\ + \eta^{m,\text{pum}}_i Q^{\text{pum}}_{t,i} + (\eta^{0,\text{pum}}_i - \eta^{m,\text{pum}}_i) \ell_i L^{\text{onl,pum}}_{t,i} \end{aligned} \tag{15.9}$$

for $t = 1, \ldots, T$, where $H_{0,i}$ denotes the initial fill level. Additionally, as an adequate terminal condition we require the initial and terminal fill levels of the reservoirs to be fixed at the minimum fill level $H^{\min}_i$. Further, the storage level at time step t is also limited by the minimum and maximum storage levels,

$$H^{\min}_i \leq H_{t,i} \leq H^{\max}_i. \tag{15.10}$$

Similar capacity constraints are formulated for the compressed-air units. Additionally, all variables have to fulfill nonnegativity conditions.

The objective of the optimization is to find a decision process satisfying the constraints (15.5)–(15.10), being nonanticipative with respect to the stochastic process $(W_t, c^{\text{imp}}_t)_t$, and minimizing the expected total costs $\mathbb{E}[TC]$.

15.3 Decomposition Using Recombining Scenario Trees

In this section, we present the solution method based on recombining scenario trees that has been developed in [11], and sketch a method for generating recombining scenario trees.

15.3.1 Problem Formulation

The optimization problem presented in Sect. 15.2 can be written as a linear multistage stochastic program:

$$\min \mathbb{E}\left[\sum_{t=1}^{T} \langle b_t(\boldsymbol{\xi}_t), x_t\rangle\right] \tag{15.11}$$

$$\text{s.t. } x_t \in X_t, \quad x_t \in \sigma(\boldsymbol{\xi}^t), \qquad t = 1, \ldots, T,$$

$$A_{t,0}x_t + A_{t,1}x_{t-1} = h_t(\boldsymbol{\xi}_t), \qquad t = 2, \ldots, T. \tag{15.12}$$

Thereby, the vector x_t contains all decision variables at time stage t. The sets X_t are closed and polyhedral and model deterministic, static linear constraints at time t, i.e., the conditions (15.6), (15.7), and (15.10). The identities (15.12) describe the random and time-coupling constraints (15.5), (15.8), and (15.9). The uncertainty concerning the future wind energy input and spot prices is modeled by the bivariate discrete time stochastic process $\boldsymbol{\xi} = (\boldsymbol{\xi}_t)_{t=1,\ldots,T}$, that enters into the optimization model through the costs $b_t(\cdot)$ and the right-hand sides $h_t(\cdot)$, which are assumed to depend affinely linear on $\boldsymbol{\xi}_t$ for $t = 1, \ldots, T$. Furthermore, $\boldsymbol{\xi}$ defines the nonanticipativity constraints, i.e., a decision x_t at time t must depend exclusively on observations made until t. This is formalized by the condition $x_t \in \sigma(\boldsymbol{\xi}^t)$, where $\boldsymbol{\xi}^t$ denotes the vector $(\boldsymbol{\xi}_1, \ldots, \boldsymbol{\xi}_t)$.

To render possible a numerical solution of (15.11), every $\boldsymbol{\xi}^t$ is assumed to take values in a finite set $\Xi^t = \{\xi^t_{(1)}, \ldots, \xi^t_{(n_t)}\}$. Consequently, the process $\boldsymbol{\xi}$ can be represented by a scenario tree, cf., e.g., [2]. Then, (15.11) can be formulated as a (large-scale) deterministic linear optimization problem that can be solved, in principle, by means of available solvers. However, with growing time horizon T, problem (15.11) becomes too large to be solved as a whole and one has to resort to decomposition techniques, e.g., temporal decomposition. To this end, one considers certain time stages $0 = R_0 < R_1 < \ldots < R_n < R_{n+1} = T$, and defines the cost-to-go function at time R_j and state $(x_{R_j}, \xi^{R_j}_{(i)}) \in X_{R_j} \times \Xi^{R_j}$ recursively by $\mathcal{Q}_{R_{n+1}}(\cdot,\cdot) := 0$ and the Bellman Equation

$$\mathcal{Q}_{R_j}(x_{R_j}, \xi_{(i)}^{R_j}) := \qquad (\mathcal{Q}_{R_j})$$

$$\min \mathbb{E}\left[\sum_{t=R_j+1}^{R_{j+1}} \langle b_t(\boldsymbol{\xi}_t), x_t \rangle + \mathcal{Q}_{R_{j+1}}(x_{R_{j+1}}, \boldsymbol{\xi}^{R_{j+1}}) \,\middle|\, \boldsymbol{\xi}^{R_j} = \xi_{(i)}^{R_j} \right]$$

$$\text{s.t. } x_t \in X_t, \quad x_t \in \sigma(\boldsymbol{\xi}^t), \qquad t = R_j + 1, \ldots, R_{j+1},$$

$$A_{t,0}x_t + A_{t,1}x_{t-1} = h_t(\boldsymbol{\xi}_t), \qquad t = R_j + 1, \ldots, R_{j+1},$$

for $j = 1, \ldots, n$. Using this notation, problem (15.11) can be reformulated in terms of Dynamic Programming:

$$\min \quad \mathbb{E}\left[\sum_{t=1}^{R_1} \langle b_t(\boldsymbol{\xi}_t), x_t \rangle + \mathcal{Q}_{R_1}(x_{R_1}, \boldsymbol{\xi}^{R_1}) \right] \qquad (\mathcal{Q}_0)$$

$$\text{s.t.} \quad x_t \in X_t, \quad x_t \in \sigma(\boldsymbol{\xi}^t), \qquad t = 1, \ldots, R_1,$$

$$A_{t,0}x_t + A_{t,1}x_{t-1} = h_t(\boldsymbol{\xi}_t), \qquad t = 2, \ldots, R_1,$$

and solved by, e.g., the *Nested Benders Decomposition method* [1,12,14]. Furthermore, a modification of this algorithm as proposed in [11] allows to exploit the structure of *recombining scenario trees* for simultaneous cutting plane approximations. This approach indeed enables to solve problem $(\mathcal{Q}_0)$ for longer time horizons T and large number of scenarios.

15.3.2 Recombining Scenario Trees

At time t, the scenario tree representing $\boldsymbol{\xi}$ has $n_t = |\Xi^t|$ nodes, that are denoted by $u = 1, \ldots, n_t$. The node u corresponds to the event $\{\boldsymbol{\xi}^t = \xi_{(u)}^t\}$. A special situation is given whenever the subtrees associated at some nodes u and k at time R_j coincide, i.e., the corresponding conditional distributions of $(\boldsymbol{\xi}_t)_{t=R_j+1,\ldots,T}$ are equal:

$$\mathbb{P}\left[(\boldsymbol{\xi}_t)_{t=R_j+1,\ldots,T} \in \cdot \;|\boldsymbol{\xi}^{R_j} = \xi_{(u)}^{R_j}\right] = \mathbb{P}\left[(\boldsymbol{\xi}_t)_{t=R_j+1,\ldots,T} \in \cdot \;|\boldsymbol{\xi}^{R_j} = \xi_{(k)}^{R_j}\right]. \tag{15.13}$$

As far as it concerns the tree representation of the process $\boldsymbol{\xi}$, property (15.13) would allow to recombine the nodes u and k, and recombining at several time stages R_j may prevent the node number to grow exponentially with the number of time stages. Unfortunately, recombining is not allowed under time coupling constraints (15.12) since the scenario-dependent control $x_{R_j}(\boldsymbol{\xi}^{R_j})$ will not be equal on $\{\boldsymbol{\xi}^{R_j} = \xi_{(u)}^{R_j}\}$ and $\{\boldsymbol{\xi}^{R_j} = \xi_{(k)}^{R_j}\}$, in general. However, (15.13) can be useful since it entails equality of the cost-to-go functions $\mathcal{Q}_{R_j}(\cdot, \xi_{(u)}^{R_j})$ and $\mathcal{Q}_{R_j}(\cdot, \xi_{(k)}^{R_j})$. This is exploited by the solution algorithm presented in Sect. 15.3.3.

In the remaining part of this section we sketch a method for generating scenario trees with property (15.13) for some nodes and several time stages

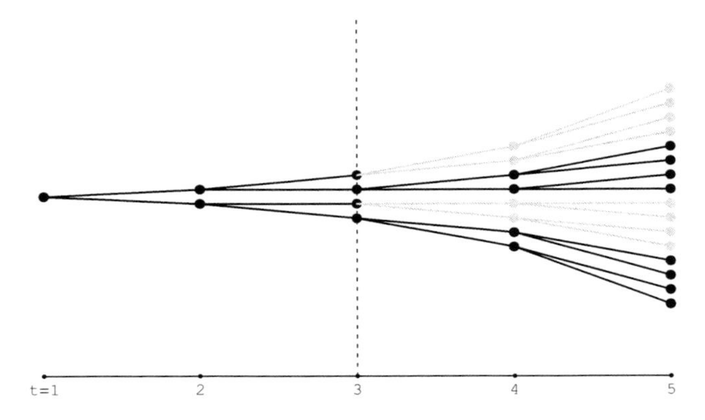

Fig. 15.2. Scenario tree with property (15.13), $R_1 = 3$, and $m_{R_1} = 2$, i.e., two different subtrees are associated at time stage 3. (The black and the gray subtrees coincide, respectively.)

Table 15.2. Notation used by Algorithm 1

ζ_t^i	value of trajectory i at time t
$\xi_{t,(u)}$	value of the random variable $\boldsymbol{\xi}_t$ in node u
m_{R_j}	number of subtrees with root node at time R_j ($m_{R_0} := 1$)
(h, u)	node u of some subtree h
$n_{R_j}(h)$	number of nodes at time R_j of some subtree h
$s_{t+1}(u)$	number of nodes at time $t + 1$ descending from some node u at time t
$\underline{t}$	time parameter for short-term history clustering
$C_t^{(h,u)}$	subset of $\{1, \ldots, N\}$, indicating trajectories ζ^i going at time t through node (h, u)
$\underline{C}_{R_j}^{(h)}$	subset of $\{1, \ldots, N\}$, indicating trajectories ζ^i lying in subtree h with root node at time R_j

R_j, $j = 1, \ldots, n$. It is a modification of the *forward tree construction* [9], also based on successive stagewise clustering of a set of sampled trajectories $\zeta^i = (\zeta_1^i, \ldots, \zeta_T^i), i = 1, \ldots, N$, that coincide in $t = 1$. Basically, it consists of constructing nonrecombining subtrees for every time period $[R_j + 1, R_{j+1}]$ (Step 2), and assigning to several nodes at time R_{j+1} the same subtree for the subsequent time period (Step 1). Thereby, two nodes at time R_{j+1} obtain the same subtree h, whenever the values of $\boldsymbol{\xi}$ in these nodes are close for some time $\underline{t}$ before R_{j+1}. Table 15.2 explains the notation used. The values of $m_{R_j}, n_{R_j}(h)$, and $s_{t+1}(u)$, determining the structure of the scenario tree, may be predefined or, as proposed in [9], determined within the algorithm to not exceed certain local error levels. Whenever the sampled trajectories come from a time series model, the parameter $\underline{t}$ may be chosen according to the latter.

Algorithm 1 (Generation of a recombining scenario tree).
Initialization: Set $C_1^{(1,1)} := \underline{C}_{R_0}^{(1)} := \{1, \ldots, N\}, \xi_1^{(1)} := \zeta_1^1$ and $\underline{C}_{R_j}^{(h)} := \varnothing$ for all h and $j \geq 1$.
For $j = 0, \ldots, n$ (*j-th recombination time stage*):

1. If $j > 0$: *Short-term history clustering for subtree assignment.*
   Find an index set $A = \{a_1, \ldots, a_{m_{R_j}}\} \subset \{1, \ldots, N\}$ with minimal

   $$\sum_{h=1,\ldots,m_{R_{j-1}}} \sum_{u=1,\ldots,n_{R_j}(h)} \min_{a_l \in A} \sum_{i \in C_{R_j}^{(h,u)}} \|(\zeta_{R_j-\underline{t}}^i, \ldots, \zeta_{R_j}^i) - (\zeta_{R_j-\underline{t}}^{a_l}, \ldots, \zeta_{R_j}^{a_l})\|.$$

   Pass through all nodes (h, u) at time R_j:
   (a) Consider an $a_{h'} \in A$ that is close to node (h, u) in the sense that

   $$a_{h'} \in \arg\min_{a_l \in A} \sum_{i \in C_{R_j}^{(h,u)}} \|(\zeta_{R_j-\underline{t}}^i, \ldots, \zeta_{R_j}^i) - (\zeta_{R_j-\underline{t}}^{a_l}, \ldots, \zeta_{R_j}^{a_l})\|.$$

   (b) Node (h, u) obtains subtree h', i.e., one has to update

   $$\underline{C}_{R_j}^{(h')} := \underline{C}_{R_j}^{(h')} \cup C_{R_j}^{(h,u)}.$$

   In particular, all trajectories ζ^i belonging to node (h, u) will be used for the construction of subtree h' of the subsequent timeperiod.
2. *Subtree generation.*
   For every subtree $h = 1, \ldots, m_{R_j}$ of period $[R_j + 1, R_{j+1}]$:
   (a) Find an index set $A = \{a_1, \ldots, a_{n_{R_j}(h)}\} \subset \underline{C}_{R_j}^{(h)}$ with minimal

   $$\sum_{i \in \underline{C}_{R_j}^{(h)}} \min_{a_l \in A} \|\zeta_{R_j+1}^i - \zeta_{R_j+1}^{a_l}\|.$$

   Find a partition $C_{R_j+1}^{(h,u)}$, $u = 1, \ldots, n_{R_j+1}(h)$, of $\underline{C}_{R_j}^{(h)}$ with

   $$C_{R_j+1}^{(h,u)} \subset \{i \in \underline{C}_{R_j}^{(h)} : a_u \in \arg\min_{a_l \in A} \|\zeta_{R_j+1}^i - \zeta_{R_j+1}^{a_l}\|\}.$$

   Define the value of $\boldsymbol{\xi}_{R_j+1}$ on node (h, u) by $\xi_{R_j+1,(h,u)} := \zeta_{t+1}^{a_u}$.
   For every subtree $\tilde{h}$ of the preceding period $[R_{j-1} + 1, R_j]$ and every node $(\tilde{h}, \tilde{u})$ at time R_j: Define the *transition probability* from node $(\tilde{h}, \tilde{u})$ to node (h, u) by

   $$\mathbb{P}_{R_j+1|R_j}[(h, u)|(\tilde{h}, \tilde{u})] := \begin{cases} \frac{|C_{R_j+1}^{(h,u)}|}{|C_{R_j}^{(h)}|} & \text{if } C_{R_j}^{(\tilde{h},u)} \subset \underline{C}_{R_j}^{(h)}, \\ 0 & \text{else.} \end{cases}$$

(b) For $t = R_j + 1, \ldots, R_j - 1$:
For every node $\tilde{u}$ of subtree h at time t:
Find an index set $A = \{a_1, \ldots, a_{s_{t+1}(\tilde{u})}\} \subset C_t^{(h,\tilde{u})}$ with minimal

$$\sum_{i \in C_t^{(h,\tilde{u})}} \min_{a_l \in A} \|\zeta_{t+1}^i - \zeta_{t+1}^{a_l}\|.$$

Find a partition $C_{t+1}^{(h,u)}, u = 1, \ldots, s_{t+1}(\tilde{u})$, of $C_t^{(h,\tilde{u})}$ with

$$C_{t+1}^{(h,u)} \subset \{i \in C_t^{(h,\tilde{u})} : a_u \in \arg\min_{a_l \in A} \|\zeta_{t+1}^i - \zeta_{t+1}^{a_l}\|\}.$$

Set $\xi_{t+1,(h,u)} := \zeta_{t+1}^{a_u}$ and $\mathbb{P}_{t+1|t}[(h,u)|(h,\tilde{u})] := \frac{|C_{t+1}^{(h,u)}|}{|C_t^{(h,\tilde{u})}|}$.

The determination of the index sets A is a $k-$mean problem and, thus, an NP-hard combinatorial optimization problem [5]. While it is possible for small values $m_{R_j}, n_{R_j}(h)$, and $s_{t+1}(u)$ to find optimal sets A by enumeration, larger values demand for heuristics, e.g., the *forward selection* proposed in [3, 8].

In Step 1 of Algorithm 1, several nodes $(\tilde{h}, \tilde{u})$ at time R_j obtain the same subtree h. For notational convenience, we pick out a representative amongst them and denote the associated value of $\boldsymbol{\xi}^{R_j}$ by $\lambda_h^{R_j}$. The following function will be used in Sect. 15.3.3 and maps a node $(\tilde{h}, \tilde{u})$ with subtree h to the corresponding representative node:

$$\boldsymbol{\lambda}^{R_j} : \Xi^{R_j} \to \{\lambda_1^{R_j}, \ldots, \lambda_{m_{R_j}}^{R_j}\} =: \Lambda^{R_j},$$

$$\boldsymbol{\lambda}^{R_j}(\xi_{(\tilde{h},\tilde{u})}^{R_j}) := \lambda_h^{R_j} \text{ whenever } C_{R_j}^{(\tilde{h},\tilde{u})} \subset \underline{C}_{R_j}^{(h)}.$$

15.3.3 Solution Algorithm

In [11], it was shown how to modify a Nested Benders Decomposition [1, 12, 14] of problem (15.11) to exploit the recombining property (15.13) of the process $\boldsymbol{\xi}$. In the following, we sketch this modified algorithm.

Let us consider the formulation $(\mathcal{Q}_0)$ of problem (15.11). A Nested Benders Decomposition successively approximates the piecewise-linear convex functions $x_{R_j} \mapsto \mathcal{Q}_{R_j}(x_{R_j}, \xi_{(u)}^{R_j})$ by a set of supporting hyperplanes and evaluates them in an adaptively chosen sequence of points x_{R_j}. Whenever two nodes u and k at time R_j fulfill (15.13), the functions $\mathcal{Q}_{R_j}(\cdot, \xi_{(u)}^{R_j})$ and $\mathcal{Q}_{R_j}(\cdot, \xi_{(k)}^{R_j})$ coincide, and, thus, they may be approximated simultaneously.

To this end, we define the following underestimating functions: We set $\mathcal{Q}_{R_{n+1}}^{LC}(\cdot, \cdot) := 0$ and for $j = n, \ldots, 0$, $\bar{x}_{R_j} \in X_{R_j}$, and $\lambda_i^{R_j} \in \Lambda^{R_j}$ let

$$\mathcal{Q}^L_{R_j}(\bar{x}_{R_j}, \lambda_i^{R_j}) := \qquad (\mathcal{Q}^L_{R_j})$$

$$\min \mathbb{E}\left[\sum_{t=R_j+1}^{R_{j+1}} \langle b_t(\boldsymbol{\xi}_t), x_t\rangle + \mathcal{Q}^{LC}_{R_{j+1}}(x_{R_{j+1}}, \boldsymbol{\lambda}^{R_{j+1}}(\boldsymbol{\xi}^{R_{j+1}})) \middle| \boldsymbol{\xi}^{R_j} = \lambda_i^{R_j}\right]$$

$$\text{s.t.}\; x_t \in X_t, \quad x_t \in \sigma(\boldsymbol{\xi}^t), \qquad t = R_j+1,\dots,R_{j+1},$$
$$A_{t,0}x_t + A_{t,1}x_{t-1} = h_t(\boldsymbol{\xi}_t), \quad t = R_j+1,\dots,R_{j+1},$$
$$x_{R_j} = \bar{x}_{R_j}. \tag{15.14}$$

Thereby, $\mathcal{Q}^{LC}_{R_{j+1}}(\cdot, \lambda_i^{R_{j+1}})$ is an approximation of $\mathcal{Q}^L_{R_{j+1}}(\cdot, \lambda_i^{R_{j+1}})$ by supporting hyperplanes that is easy to evaluate and that will be properly defined in (15.15) below. Problem $(\mathcal{Q}^L_{R_j})$ is often referred to as the *master problem.* Note, that in contrast to the classical Nested Benders Decomposition, the *same* approximation $\mathcal{Q}^{LC}_{R_{j+1}}(\cdot, \lambda_i^{R_{j+1}})$ can be used in the objective function of $(\mathcal{Q}^L_{R_j})$ for all nodes with the same subtree i, i.e., whenever $\boldsymbol{\lambda}^{R_{j+1}}(\boldsymbol{\xi}^{R_{j+1}}) = \lambda_i^{R_{j+1}}$. Thus, this decomposition into subproblems for each timeperiod allows one to exploit the recombining nature of the process ξ. This had not been possible with other decomposition algorithms like scenario decomposition [12].

The function $\mathcal{Q}^{LC}_{R_j}(\cdot, \lambda_i^{R_j})$ is used to induce a feasible solution at stage R_j and to approximate the value of $\mathcal{Q}^L_{R_j}(\cdot, \lambda_i^{R_j})$ on its domain. For the latter purpose, given a point $\bar{x} \in X_{R_j}$ with $\mathcal{Q}^L_{R_j}(\bar{x}, \lambda_i^{R_j}) < \infty$, an *optimality cut* supporting $\mathcal{Q}^L_{R_j}(\cdot, \lambda_i^{R_j})$ is given by $\mathcal{Q}^L_{R_j}(\bar{x}, \lambda_i^{R_j}) + \langle \pi, x_{R_j} - \bar{x}\rangle \le 0$, where π denotes the dual variables corresponding to the constraint (15.14) in an optimal solution of problem $(\mathcal{Q}^L_{R_j})$. To induce feasibility at time stage R_j, a point $\bar{x} \in X_{R_j}$ that is infeasible for $(\mathcal{Q}^L_{R_j})$ is cut off using a *feasibility cut* $\langle d, x\rangle + e \le 0$. This cut is computed by solving an auxiliary problem, cf. [11], and has the property $\langle d, \bar{x}\rangle + e > 0$ and $\langle d, x\rangle + e \le 0$ for all $x \in X_{R_j}$ with $\mathcal{Q}^L_{R_j}(x, \lambda_i^{R_j}) < \infty$.

Hence, an approximation of $\mathcal{Q}^L_{R_j}(\cdot, \lambda_i^{R_j})$ by means of optimality cuts $C_{\text{opt}}(\lambda_i^{R_j})$ and feasibility cuts $C_{\text{feas}}(\lambda_i^{R_j})$ is given by

$$\mathcal{Q}^{LC}_{R_j}(x_{R_j}, \lambda_i^{R_j}) := \max_{(\bar{x},\bar{\pi}) \in C_{\text{opt}}(\lambda_i^{R_j})} \mathcal{Q}^L_{R_j}(\bar{x}, \lambda_i^{R_j}) + \langle \bar{\pi}, x_{R_j} - \bar{x}\rangle \tag{15.15}$$
$$\text{s.t.} \quad \langle d, x_{R_j}\rangle + e \le 0, \quad (d,e) \in C_{\text{feas}}(\lambda_i^{R_j}).$$

The solution algorithm processes the master problems $(\mathcal{Q}^L_{R_j})$, $j = 0,\dots,n$, of the decomposed scenario tree in a forward or backward manner. At each time stage R_j, each master problem $\mathcal{Q}^L_{R_j}(\cdot, \lambda_i^{R_j})$, $\lambda_i^{R_j} \in \Lambda^{R_j}$, is evaluated for a set $Z_j(\lambda_i^{R_j})$ of controls x_{R_j}. If $\mathcal{Q}^{LC}_{R_j}(x_{R_j}, \lambda_i^{R_j}) < \mathcal{Q}^L_{R_j}(x_{R_j}, \lambda_i^{R_j})$, the approximation $\mathcal{Q}^{LC}_{R_j}(\cdot, \lambda_i^{R_j})$ (and all master problems that use $\mathcal{Q}^{LC}_{R_j}(\cdot, \lambda_i^{R_j})$) is updated by generating new optimality or feasibility cuts. Further, in the forward mode, new control points $x_{R_{j+1}}$ are generated from the solution of the master problem $(\mathcal{Q}^L_{R_j})$ to form the sets $Z_{j+1}(\lambda_i^{R_{j+1}})$ for $\lambda_i^{R_{j+1}} \in \Lambda^{R_{j+1}}$. Since each such evaluation contributes several new controls $x_{R_{j+1}}$ to the sets $Z_{j+1}(\lambda_i^{R_{j+1}})$, the latter can grow exponentially with increasing j. It was shown in [11] how the problem structure allows to deal with this difficulty.

The algorithm stops when either the first timeperiod master problem $(\mathcal{Q}^L_0)$ is infeasible, or all master problems could be solved to optimality and the generation of cuts has stopped. In the former case, also problem $(\mathcal{Q}_0)$ is infeasible, in the latter, the problem has been solved to optimality. Another stopping criteria which allows to stop the algorithm when the error falls below a given tolerance, is also discussed in [11]. A more detailed description of the Nested Benders Decomposition Algorithm can be found in [1, 6, 12].

15.4 Case Study

We study a power generating system, consisting of a hard coal power plant to cover the minimum and medium load, and two fast gas turbines on different power levels to cover the peaks. The operating parameters of these units rely on real data. Furthermore, the model contains an offshore wind park, a pump storage power plant (PSW) with the basic data of the PSW Geesthacht, Germany, and a compressed-air energy storage (CAES) with the operating parameters of the CAES Huntorf, Germany. Further source of power supply is the EEX spot market.

The time horizon considered for the optimization is 1 year and a hourly discretization is used, i.e., the model contains $T = 8,760$ time stages.

The stochastic wind power process is represented by a time series model fitted to historical data and scaled to the size of the offshore wind park regarded. To take into account the interdependency between wind power and spot price behavior, the *expected* spot market prices are calculated from a fundamental model that is based on the existing power plants in Germany and their reliability, prices for fuels and CO_2, the German load, and the wind power process above. Fluctuation of the spot prices around their expected value are modeled by a further time series model. This hybrid approach was used to generate 1,000 trajectories, containing hourly values of wind power and spot prices in the course of 1 year. These trajectories were used to generate a recombining tree by Algorithm 1 of Sect. 15.3.2. The resulting scenario tree branches three times per day in a binary way. Recombination into $m_{R_j} = 3$ different subtrees took place once a day, i.e., $R_j = j \cdot 24$, $j = 1, \ldots, 364$.

15.5 Numerical Results

The optimization problem was solved with varying model parameters. To this end, a *base setting* was defined, with wind power of approximately 50% of the totally installed plant power and storage sizes corresponding to the

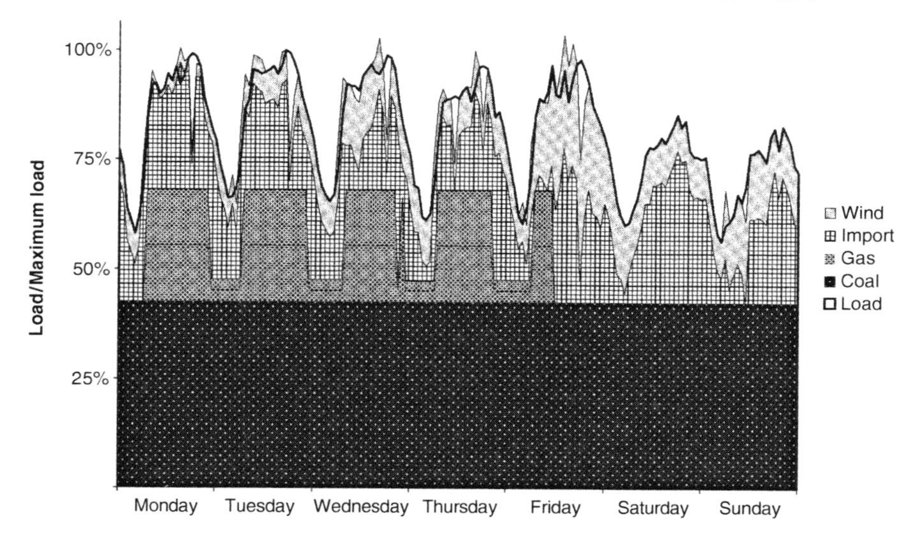

Fig. 15.3. Optimal power scheduling in a winter week

aforementioned CAES and PSW units. Coming from this setting, variations with higher and lower levels of installed wind power and different storage dimensions were calculated. In the following some results are presented.

The optimal operation levels along a randomly chosen scenario from the base setting during a winter week are depicted in Fig. 15.3. Whenever the power production exceeds the demand curve, energy is put into the storages, whereas the white spaces under the demand curve represent the output of the storage plants. The operation levels of the thermal units show the usual characteristics and availability of wind power obviously reduces imports from the spot market. The storage units are mainly used to cover the peaks and are only marginally used during the weekend. In this model, the contribution of the operating costs to the power supply costs amount to 2.08 Eurocents kWh^{-1} with using storage plants and 2.10 Eurocents kWh^{-1} without using storage plants. Fig. 15.4 shows the optimal output and fill level of the CAES (as a fraction of maximum discharge power and maximum fill level, respectively) in comparison to the actual power price. The minimum fill level of the CAES is 60%. Obviously, the storage plant discharges in times of high spot prices on weekdays. The aforementioned marginal usage of storage plants during the weekend coincides with lower power prices over this period.

To study the impact of the share of wind power on the system, the optimization problem was solved again with doubled wind power capacity. The results along the same scenario and for the same winter week are depicted in Fig. 15.5. While this extension does not lead to significant changes of the thermal units, it enables to largely reduce the amount of energy bought at the spot market. Fig. 15.6 shows the operation of the CAES in the course of

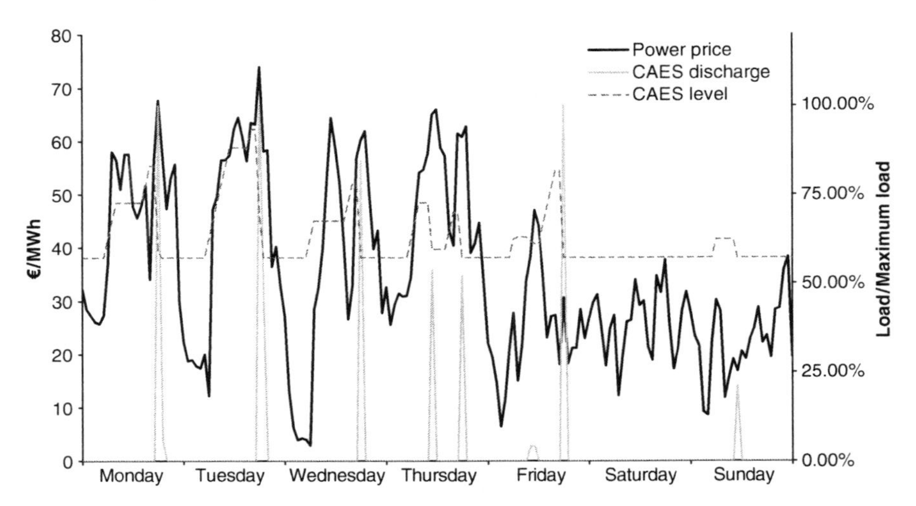

Fig. 15.4. Spot market price and CAES output in a winter week

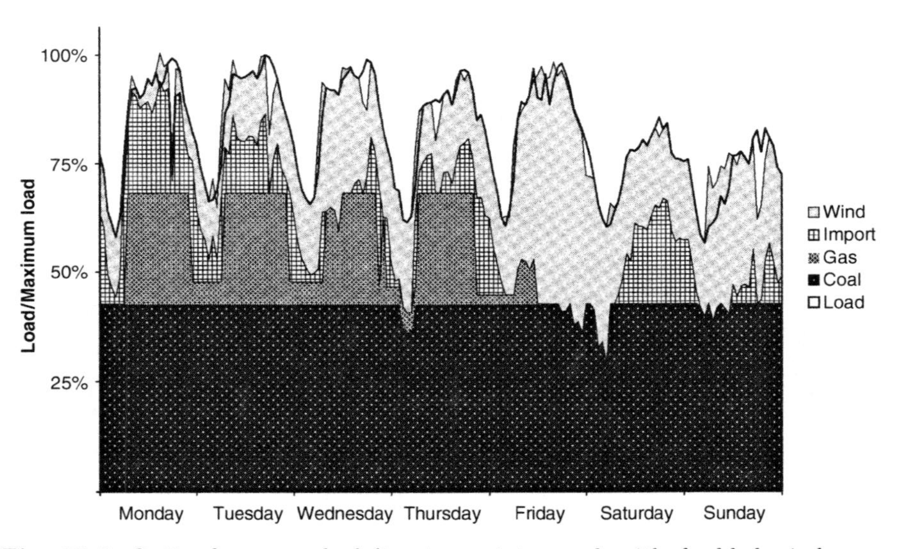

Fig. 15.5. Optimal power scheduling in a winter week with doubled wind power capacity

the week. Again, the CAES is mainly used at peak times to avoid expensive imports from the spot market. It can be seen that the availability of more wind power in the system can lead both to more and to less extraction of stored energy. This is due to the fact that, on the one hand, with more wind power more energy may be stored and therefore extracted (Sunday). On the other hand, less power has to be generated in times with high wind power (from Wednesday to Friday).

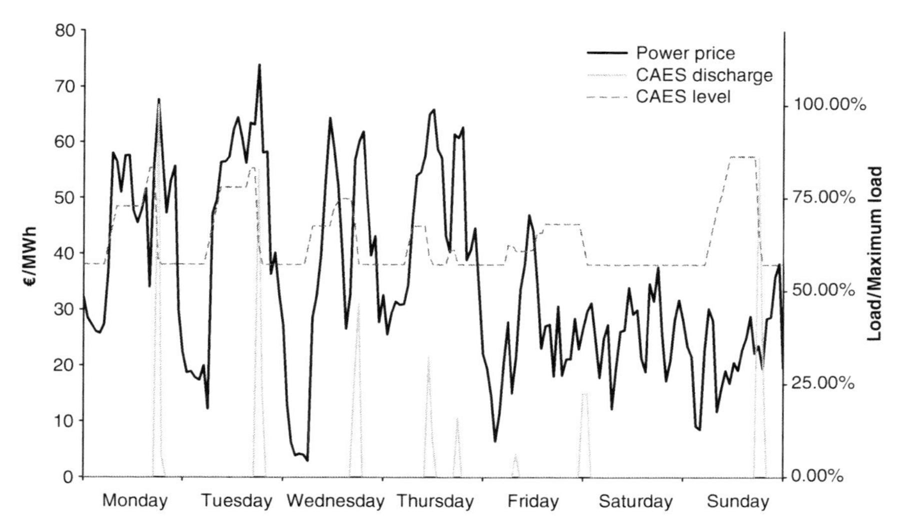

Fig. 15.6. Spot market price and CAES output in a winter week with doubled wind power capacity

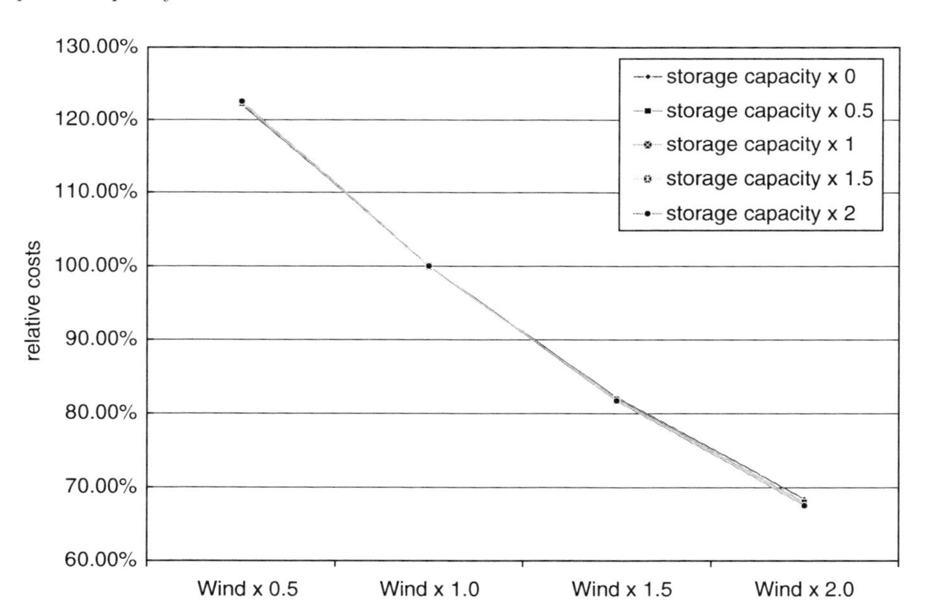

Fig. 15.7. Minimal expected costs depending on the wind power capacity installed for different storage capacities installed

The optimization problem was solved further times with varying quantities of installed wind power and storage capacities. Fig. 15.7 shows the minimal expected costs depending on the wind power capacity for different storage

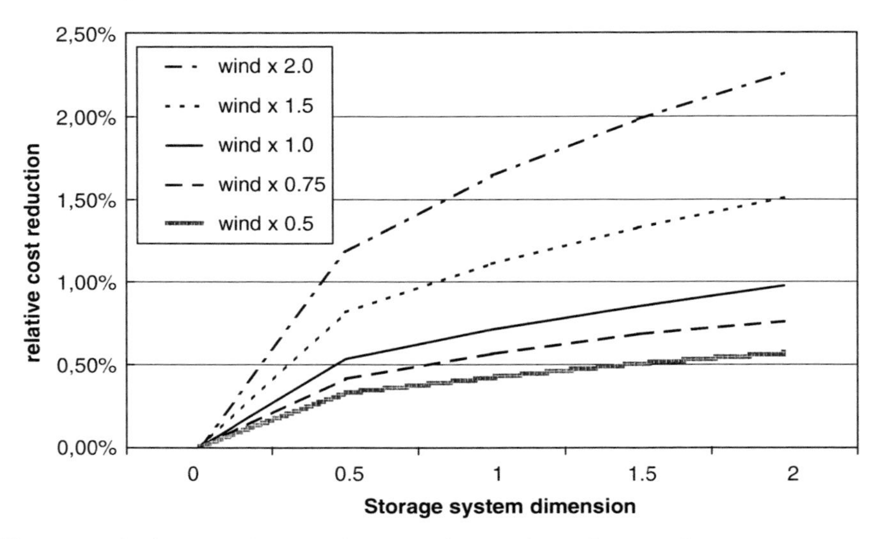

Fig. 15.8. Reduction of minimal expected costs depending on the storage capacity installed for different wind power capacities installed

capacities. Thereby, a wind factor of y stands for an amount of wind power being y times the wind power of the *base setting*. In relation to the wind power capacity, the impact of an extension of the storage system on the costs appears to be rather marginal and the individual curves are almost superposed. Thus, to analyze the latter, Fig. 15.8 shows the relative reduction of costs that can be achieved by the use of storage systems of different dimensions, where a model without storages generates operating costs of 100%. Again, a storage system dimension of y corresponds to y times the dimension of the base setting. The results clearly show, that the relative cost reduction due to storage use is the highest in the twice-wind-setting, and in all settings the most prevailing gradient is between no use and the use of the half dimension of storage sizes. Hence, it seems promising to study expansion models for cost-optimal storage sizes, taking into account operational as well as investment costs.

The optimization algorithm was implemented in C++ and the master problems were solved with CPLEX 10.0 [10]. Running time on a PC with 2.4 GHz CPU and 2 GB RAM was 10 min, approximatively.

15.6 Conclusions and Outlook

We applied a decomposition method for linear multistage stochastic optimization problems proposed by [11] to optimal scheduling within a regional energy system including wind power and energy storages. It has been shown that this approach relying on recombining scenario trees allows to handle multistage problems with large numbers of scenarios and including time-coupling constraints, and, therefore, it is suitable for optimizing and analyzing energy systems.

In principle, the recombining tree decomposition approach allows for discrete decision variables in the first time stage and, hence, this method seems to be also appropriate to find optimal first-stage investment decisions within expansion models. This could be one aspect of future studies. However, further research is needed to extend the decomposition approach to more general optimization models, in particular those including discrete variables in later time stages. The latter would allow to adapt numerous aspects of the energy system model to achieve a more detailed picture of the system and its constraints.

Another aspect of future research could be the extension of the decomposition approach to optimization problems including multiperiod risk measures. For this purpose, especially the class of polyhedral risk measures [4] seems to be suitable, since the linear structure of the optimization problem is maintained.

Acknowledgement. This work was supported by the German Ministry of Education and Research (BMBF) within the topic "Dezentrale regenerative Energieversorgung: Innovative Modellierung und Optimierung" and the "Wiener Wissenschafts-, Forschungs- und Technologiefonds" in Vienna (`http://www.univie.ac.at/crm/simopt`).

References

1. J. R. Birge. Decomposition and partitioning methods for multistage stochastic programming. *Operations Research*, 33(5):989–1007, 1985.
2. J. Dupačová, G. Consigli, and S. W. Wallace. Scenarios for multistage stochastic programming. *Annals of Operations Research*, 100:25–53, 2000.
3. J. Dupačová, N. Gröwe-Kuska, and W. Römisch. Scenarios reduction in stochastic programming: An approach using probability metrics. *Mathematical Programming*, 95(A):493–511, 2003.
4. A. Eichhorn and W. Römisch. Polyhedral risk measures in stochastic programming. *SIAM Journal on Optimization*, 16:69–95, 2005.
5. M. R. Garey and D. S. Johnson. *Computers and Intractability - A Guide to the Theory of NP-Completeness.* Freeman, New York, 1979.
6. H. I. Gassmann. MSLiP: a computer code for the multistage stochastic linear programming problem. *Mathematical Programming*, 47:407–423, 1990.
7. E. Handschin, F. Neise, H. Neumann, and R. Schultz. Optimal operation of dispersed generation under uncertainty using mathematical programming. *International Journal of Electrical Power and Energy Systems*, 28:618–626, 2006.
8. H. Heitsch and W. Römisch. Scenario reduction algorithms in stochastic programming. *Computational Optimization and Applications*, 24:187–206, 2003.
9. H. Heitsch and W. Römisch. Scenario tree modeling for multistage stochastic programs. *Mathematical Programming* to appear, 2008.

10. ILOG, Inc. CPLEX 10.0. http://www.ilog.com/products/cplex.
11. C. Küchler and S. Vigerske. Decomposition of multistage stochastic programs with recombining scenario trees. *Stochastic Programming E-Print Series*, 9, 2007. http://www.speps.org.
12. A. Ruszczyński. *Stochastic Programming, A. Ruszczyński and A. Shapiro (Editors)*, chapter Decomposition Methods, chapter 3, pages 141–221. Elsevier, Amsterdam, 2003.
13. A. Ruszczyński and A. Shapiro, editors. *Stochastic Programming*. Handbooks in Operations Research and Management Science. Elsevier, Amsterdam, 2003.
14. R. M. Van Slyke and R. Wets. L-shaped linear programs with applications to optimal control and stochastic programming. *SIAM Journal of Applied Mathematics*, 17(4):638–663, 1969.
15. D. Swider, P. Vogel, and C. Weber. Stochastic model for the european electricity market and the integration costs for wind power. Technical report, GreenNet Report on WP 6, 2004.
16. D. Swider and C. Weber. The costs of wind's intermittency in Germany: Application of a stochastic electricity market model. *European Transactions on Electrical Power*, 17(2):151–172, 2007.
17. H.-J. Wagner. *Wind Energy Utilization, N. Bansal and J. Mathur (Editors)*, chapter Wind Energy and Present Status in Germany. Anamaya, New Delhi, 2002.
18. C. Weber. *Uncertainty in the Electric Power Industry: Methods and Models for Decision Support*. Springer, New York, 2005.

16

Stochastic Model of the German Electricity System

Nina Heitmann and Thomas Hamacher

Summary. We present a model of the German electricity system which is able to describe the system in arbitrary spatial and temporal resolution. Due to the high temporal resolution the model generator is particularly suitable to analyse the influence of uncertain and fluctuant parameters like the wind supply to the existing electricity system. Germany is represented by 29 knots within Germany and 13 knots of neighbouring countries. Major transmission lines between these knots are modelled in a stylized manner. The model calculates the optimal capacities as well as the energy flows. We will discuss stochastic programming options as latest extension of the model generator. The stochastic parameters are the fuel costs respectively the CO_2 prices. It is also attempted to handle the supply of wind energy in a stochastic way.

16.1 Introduction

The German electricity system is expected to undergo major transitions in the future. Drivers of the expected change are the political decision to phase out nuclear energy and the confession to fulfil the Kyoto Protocol. But also uncertain factors like weather conditions or price developments have influence on the decision, therefore it is important to make risk analysis. With the help of stochastic programming techniques such uncertainties can be trapped to a certain point. Especially variation in prices can be recognized with the integration of stochastic elements. One alternative to stochastic models is the use of numerous scenarios for each uncertain parameter. The advantage of stochastic models is that the model proposes one unique solution to the decision maker, while the scenario techniques leave it to the judgment of the analyst to choose the most "likely" scenarios.

16.2 Model

16.2.1 Two-Stage Stochastic Linear Program

A linear program where the coefficients are characterized by uncertainty, that means that instead of a constant parameter a random variable is given, is called stochastic linear program. In a two-stage stochastic linear program the variables can be grouped into variables which are independent of the random variable, they have to be decided at a first stage under uncertainty. These variables are called first-stage variables. The other variables are influenced by the realization of the random event, so they can be decided at a second stage with full information when the specific development of the random variable is known. These variables are called second-stage variables. After J.R. BIRGE a basic two-stage stochastic linear program can be written as follows:

$$min \left\{ c^T x + \Theta(x) | Ax = b, x \geq 0 \right\} \tag{16.1}$$

where

$$\Theta(x) = E_\xi Q(x, \xi(\omega)) \tag{16.2}$$

and

$$Q(x, \xi(\omega)) = min_y \left\{ q(\omega)^T y(\omega) | Wy(\omega) = h(\omega) - T(\omega)x, y(\omega) \geq 0 \right\} \tag{16.3}$$

The first-stage decisions are represented by the vector $x \in \mathbb{R}^{n_1}$. The vectors $c \in \mathbb{R}^{n_1}$ and $b \in \mathbb{R}^{m_1}$ and the matrix $A \in \mathbb{R}^{m_1 \times n_1}$ correspond to x, whereas the vectors $q \in \mathbb{R}^{n_2}$ and $h \in \mathbb{R}^{m_2}$ and the matrices $T \in \mathbb{R}^{m_2 \times n_1}$ and $W \in \mathbb{R}^{m_2 \times n_2}$ correspond to the second-stage decision vector $y \in \mathbb{R}^{n_2}$. The stochastic components of the second-stage data, can be written in one vector $\xi(\omega) = (q^T(\omega), h^T(\omega), T_{1.}(\omega), .., T_{m_2.}(\omega))$, where the i-th row of T is written as $T_{i.}(\omega)$. Each component of q, h and T is a possible random variable and is influenced by each realization of the random event $\omega \in \Omega$. That means for a given realization ω the second-stage data $q(\omega), h(\omega), T(\omega)$ are known and so the second-stage variables $y(\omega)$ can be calculated [2].

In the model of the German energy production and transmission system the random event ω has influence on q, the costs concerning the energy production (fuel prices and CO_2 prices) and on one row of T, the equation describing the energy production with wind. The right-hand-side vector h of the second stage problem is constant for all random events ω, therefore we write h instead of $h(\omega)$. Furthermore there exist no equations that are related only to the first stage problem. It is assumed that the random variable ξ has a discrete probability distribution. Taking all these facts into consideration the problem 16.1–16.3 can be reformulated as follows:

$$min\left\{c^Tx+\sum_{\omega}p(\omega)q^T(\omega)y(\omega)|Wy(\omega)=h-T(\omega)x,x\geq 0,y(\omega)\geq 0\right\}$$

where $p(\omega) \in [0,1]$ is the probability of the random event ω. The assumption of discrete probabilities allows using an algorithm for linear optimization problems to solve the stochastic optimization problem. The most efficient solving algorithm provided by CPLEX is the barrier method. The barrier method is one of the interior point methods, being used to solve linear and nonlinear convex optimization problems. A detailed description of this algorithm can be found in [5].

16.2.2 Model Generator

The model of the German electricity system was developed with the model generator URBS. URBS is a model generator for general linear optimization problems which is written in GAMS ("General Algebraic Modelling System"). Data in- and output can be done with MS Excel and Visual Basic to simplify the data evaluation. URBS is able to consider the problem in a high temporal and spatial resolution over a time frame of 1 year, therefore it's particularly suitable to analyze the influence of uncertain and fluctuant parameters to the system.

The model is formulated in terms of a two stage stochastic linear program. The fuel prices, the CO_2 prices and the wind supply can be managed as stochastic features. The first-stage variables contain the capacities for power plants, transmission lines and storage plants, they have to be calculated under uncertainty, e.g., before the development of the gas price is known. All energy flows belong to the second-stage variables; they can be decided under certainty for a specific realization of the stochastic parameter. The objective is to plan and generate a mix of power plants which causes the minimal total costs. The objective function contains the investment costs, fixed costs and variable costs as well as the fuel and CO_2 emission prices. General terms and conditions (like the satisfaction of demand) and specifications of the power plants (efficiency, power-change coefficient, idle time is managed with the limitation of full load hours etc.) are implemented with the help of parameters and equations. In general the model contains equations for the commodities, the process, the transmission and the storage.

16.2.3 The Electricity Model of Germany

Germany is represented by 29 knots within Germany and 13 knots of the neighboring countries. Each node has several types of power plants: coal, lignite and gas fired plants, nuclear plants, wind turbines for off- and onshore regions and two types of hydro plants (run-of-river power plant, pump storage plants). Each neighbored country is mapped to one node having an energy supply and an energy demand (hourly resolution). The different sites (nodes) are

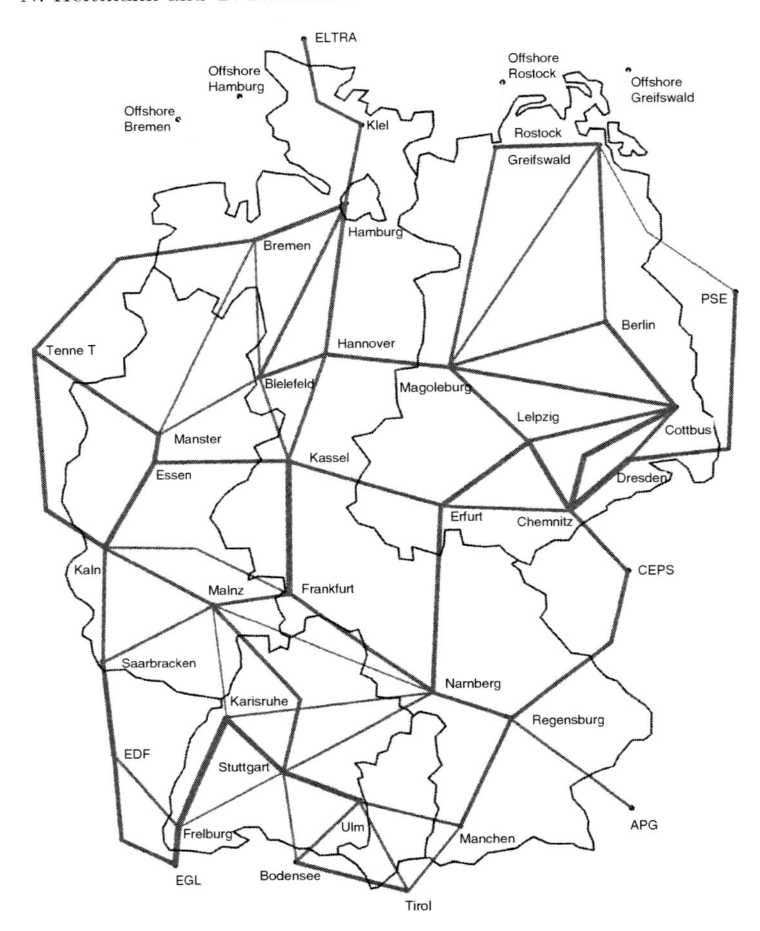

Fig. 16.1. Model of Germany

connected by transmission lines. The 220/380 kV-lines of the German transmission grid were merged to few transmission lines representing the whole network. The investigations are based on a very complex model of the complete UCTE-net, which was implemented in DIgSILENT PowerFactory [4]. In Fig. 16.1 the model of Germany is shown. The software tool DIgSILENT PowerFactory [1] is an engineering model of the power system including power plants, transmission lines, transformers and end-users. Modelling of the dynamic behaviour of all the components – including the mechanical and heat components – down to very short time scales is possible. The dispatching and control philosophy of various networks can also adequately be simulated.

16.2.4 Input Data

Stochastic Parameters and Costs

The specific costs for the power plants derive from the *Bremer Energie Institut* [7] and from GEMIS [6]. The fuel prices were taken from the statistical pocketbook for energy and transport figures [3].

Depending on the scenario either the fuel prices, the CO_2 price or the wind year were handled stochastically. The stochastic parameters are given discrete probabilities with three different specifications. In the following scenarios the realization of the stochastic parameters are denoted with p1 (low gas price, CO_2 prices respectively wind supply), p2 (middle gas price, CO_2 prices respectively wind supply) and p3 (high gas price, CO_2 prices respectively wind supply). For the following scenarios an equipartition was used.

For the German model different power plants are used: Nuclear plants, lignite and coal fired plants, combined cycled gas plants (CC), gas turbines (GT), wind on- and offshore turbines and hydro plants. To store the energy pump storage plants were used. The specific costs used for the scenarios are listed in Table 16.1.

Time Steps

Based on the wind data of the world wind atlas [8] the wind velocities were transformed into wind power for each modelled region so that finally hourly values for each site were available. The consumer load is given as time series. It was taken from a detailed UCTE-dataset of the university of Rostock.

The distribution of hours according to their energy demand and wind supply are outlined in Fig. 16.2. Each of the 8,760 h in 1 year is represented in one point. To simplify matters instead of looking at the distribution of the demand and supply data of all sites separately, the sum of demand and supply data of all sites is regarded. The rectangle which shows the distribution is divided into nine categories, where as well the energy demand as the wind supply data

Table 16.1. Costs

	InvCost [€/kW]	FixCost [€/kW]	Variable cost [€ct/kWh]	Fuel Prices [€ct/kWh$_{th}$]
Nuclear	2,050	62	0.51	0.41
Lignite	1,210	20	0.5	0.59
Coal	1,018	22	1.3	0.87
GT	256	10	0.05	1.4, 2.4, 3.5
CC	523	13	0.12	1.4, 2.4, 3.5
Wind-onshore	900	38	0.1	
Wind-offshore	1,700	50	0.1	
Hydro	4,500	20.5	0.01	

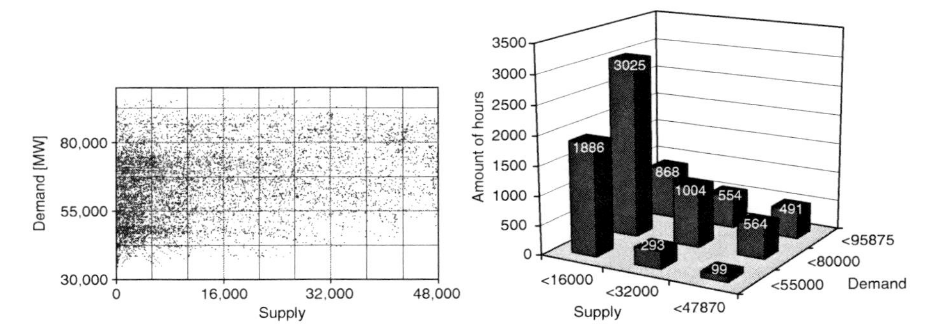

Fig. 16.2. Distribution of energy demand and wind supply of 1 year

are split into three levels (low, medium, high). The scatter plot shows each hour whereas the bar chart at the right hand side illustrates the amount of hours in each category. Most of the hours have a low wind supply. From the demand side a medium energy demand is represented most frequently. For example 3,025 of the 8,760 h are accumulated at a medium demand between 55,144 and 75,500 MW and low wind supply. But especially the hours with low wind supply and a high energy demand are responsible for the installation of power plant capacities.

Due to limitations of computational power in ordinary PCs it is necessary to decide on a number of time steps and their weighting, which stands for the actual year. To represent one realistic year the imbalance of the distribution has to be considered in the choice of hours and their weights. One possibility to weight the chosen hours adequately is to give them the weight of the amount of hours of the category, where the hour was taken from. We need at least 1 h from each category. Furthermore the sum of all weights has to be the amount of all hours in 1 year, that means 8,760. For the following scenarios 174 h were chosen: 4 days of a summer week (Thursday–Sunday) and 3 days of a winter week (Monday–Wednesday) plus six additional hours to have a representative of all categories. Their weighting results in an energy demand of 645 TWh per year, the energy which is exported to the surrounding countries is included in that number.

16.3 Scenarios

Several scenarios were developed which are represented in the following section. Starting from a basis scenario the influence of wind integration in the existing system as well as the limitation of CO_2 emission were studied. In addition to that a scenario describing the phase out of nuclear energy was investigated.

For all scenarios and all sites the existing capacities for lignite fired plants and nuclear plants were set as upper capacity boundaries. Moreover it was assumed that the installed capacities for hydro plants are not expandable and the wind turbines have an upper capacity limit of about 45 GW (20 GW for onshore and 25 GW for offshore turbines).

According to the stochastic parameters the scenarios can be classified into several groups. In the first group the fuel prices for gas were regarded as stochastic parameter. Subject to that condition the installation of wind turbines were varied and so changes to the power plant mix, to the energy production and to changes in costs (marginal costs for demand and capacities) were analyzed. In addition to that the limitation of CO_2 emission was studied by restricting the boundary to its minimum. These scenarios are called $250co2$, $200co2$, $150co2$ and $112co2$ where the number at the beginning of name denotes the upper limit of the CO_2 emission in Mt.

At the other scenarios the fuel prices were not regarded as stochastic parameters anymore. The prices were set to a constant level. The fuel price for gas was set to a medium level of €2.4 ct/kWh. Instead of that the boundary of the CO_2 price was treated as stochastic parameter in the scenario CO_2 *Price* respectively the wind supply in the scenario *WindSup*.

16.3.1 List of Scenarios

NAME	DESCRIPTION
*Basis*1	Basis scenario no limitation and no installations
*Wind*50	Installation of wind turbines is forced to 50% of upper capacity limit
*Wind*100	Installation of wind turbines is forced to 100% of upper capacity limit
250*co*2	Limitation of CO_2 emission to 250 Mt
200*co*2	Limitation of CO_2 emission to 200 Mt
150*co*2	Limitation of CO_2 emission to 150 Mt
112*co*2	Limitation of CO_2 emission to 112 Mt
CO_2*Price*	Stochastic Parameter: CO_2 price (10, 25, 50 €/kg)
WindSup	Stochastic Parameter:Wind Supply
ZeroUran	Scenario without nuclear energy

16.3.2 Wind

The aim of the wind scenario was to analyze the integration of wind energy into the existing system, to see whether changes in manner of plant installation and the production occur. Compared to conventional power plants wind turbines are very expensive, that means in an optimization model that minimizes the overall system cost wind turbines are not installed (see scenario *Basis*1). Nevertheless the installation of wind turbines can be forced by setting lower capacities limits which was done here. The components of the second-stage data that concern the fuel prices for gas are influenced by uncertainty, all other components are fixed.

Capacity

An increase of wind turbines (from 0 to 45 GW) leads to a decrease of coal fired power plants (from 22 to 16 GW) and combined cycle gas fired plants (from 39 to 28 GW), gas turbines are needed as back-up to balance the wind fluctuation, so they rise from only 6 to 22 GW (see Fig. 16.3). Nuclear plants and lignite fired plants are installed up to their maximum. The hydro plants are fixed manually. The total installed capacities rise from 111 GW in the *Basis*1 scenario to 133 GW in scenario *Wind*50 to 155 GW in scenario *Wind*100. That means for 1 GW of wind power 1 GW of back-up capacity is needed.

Wind power provokes additional transmission flow, mainly in the north western part of Germany. In hours with a high wind supply in wind regions depending on the amount of installed wind turbines, there is a surplus of energy which has to be transmitted to the adjacent sites. In scenario *Wind*50 new transmission lines become necessary between Hamburg and Kiel, to transport the offshore wind power from Kiel to Hamburg . In scenario *Wind*100 additional wind power has to be transmitted from Kiel to Hamburg and from Bremen to Münster respectively Bielefeld (see Fig. 16.4).

Production

The electricity production for the wind scenarios is shown in Fig. 16.5. Due to the stochastic implementation of the fuel prices there exist three sets of production for each scenario, one for each realization of the gas price; p1

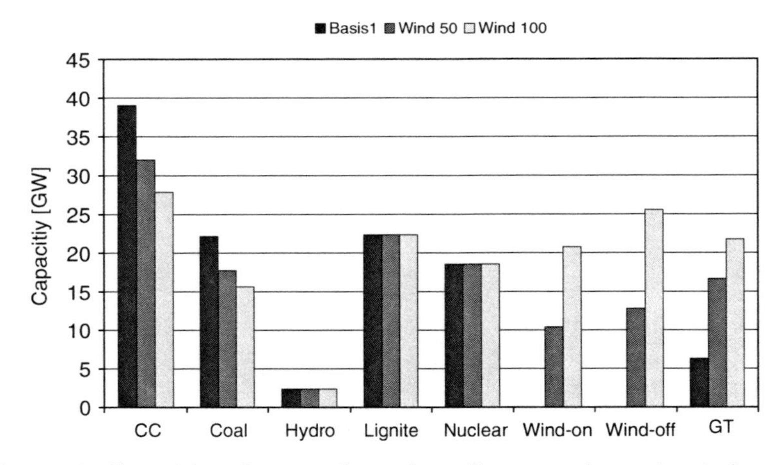

Fig. 16.3. Capacities of power plants depending on a change in wind power

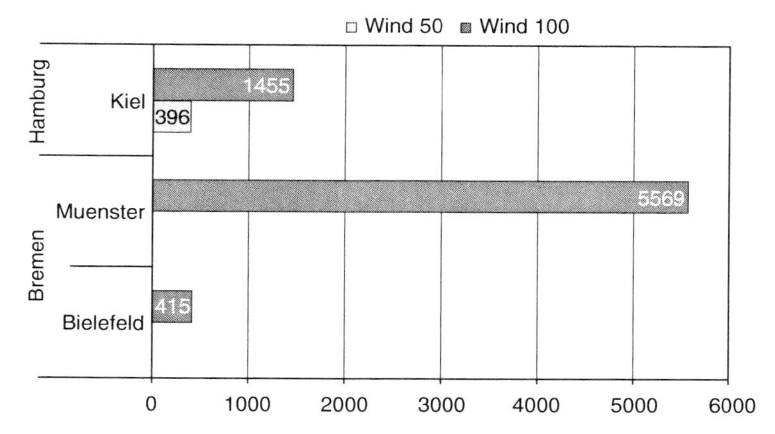

Fig. 16.4. Installation of wind power provokes an upgrade of the transmission system [MW]

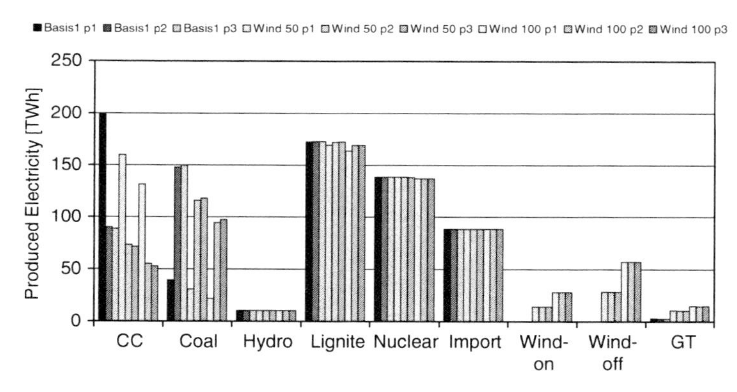

Fig. 16.5. Produced electricity by power plants depending on a change in wind power

denotes the case with low gas prices, p2 the case with medium gas prices and p3 the case with high gas prices. The installed capacities are for each realization the same (see 16.3).

The variation of fuel prices is mainly compensated by the combined cycle gas plants and the coal fired plants. For a high gas price the coal fired plants triple their production, the full load hours rise from 1,738 to 6,630 in scenario *Wind*50 and from 1,391 to 6,199 in scenario *Wind*100 respectively, whereas the combined cycle gas plants cut their production by approximately 50% in comparison to the case of low gas prices. The full load hours of the combined cycle plants decrease from 4,989 to 2,219 in scenario *Wind*50 and from 4,719 to 1,887 in scenario *Wind*100 respectively. The production with lignite fired plants show only small differences. The base load is provided by the lignite fired (163–172 TWh) and nuclear plants (137–138 TWh), coal fired (22–150 TWh) and combined cycle gas plants (53–198 TWh) meet the middle load and the wind and gas turbines provide the peak load. With the installation of 25 GW offshore wind power (scenario *Wind*100) an energy production of nearly 60 TWh can be achieved.

The CO_2 emission from scenario *Basis*1 to scenario *Wind*100 can be reduced by about 43 Mt from 283 to 240 Mt.

Marginal costs of demand

The marginal costs of demand denote the increase of the total costs (objective function) for an additional unit of demand. The marginal costs are an important indication for the price development of electricity. In perfect markets the price of a commodity is given by the marginal costs. Figure 16.6, for example, shows the differences of marginal costs between the scenarios *Basis*, *Wind*50 and *Wind*100 for two significant hours. Whereas Fig. 16.7 exemplifies the development and range of marginal costs of scenario *Wind*100. The costs range from 0.027 up to 0.140 €/kWh, depending on the demand and therefore what kind of power plants are used. The power plant with the highest operational cost which is used to satisfy the demand is responsible for a build up of marginal costs. That means an increase of demand leads as well to an increase of marginal costs, because additional power plants are used with presumably higher operational costs. Figure 16.6 illustrates that the integration of wind power does not lead to lower marginal costs in general; in fact it depends on the wind supply at the specific hour. Here 2 h were selected 1 h with high wind supply and 1 h with nearly no wind supply to see the influence of installation of wind turbines on the electricity price. In hours

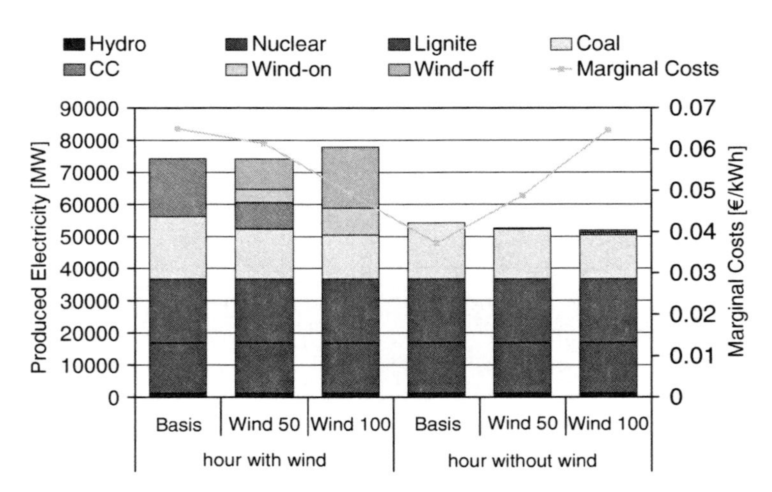

Fig. 16.6. Comparison of marginal costs and the electricity generation between the scenarios *Wind*100, *Wind*50 and the basis scenario

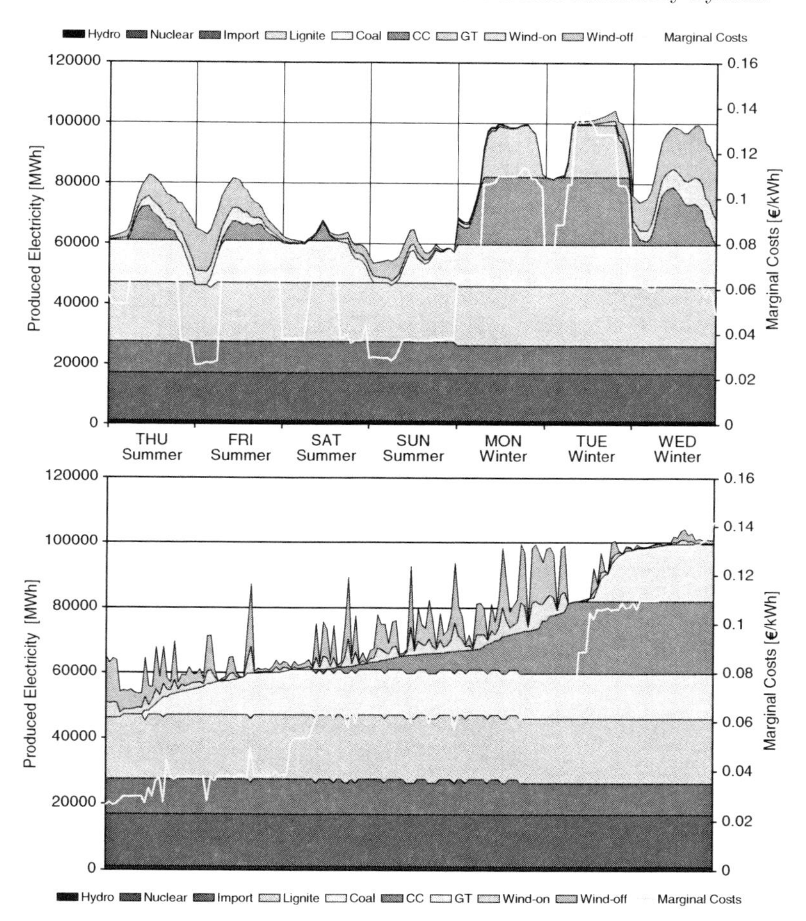

Fig. 16.7. Development of marginal costs against the energy production for the scenario with 100% of wind power ($Wind100$)

with a high wind supply the marginal costs are lower the more wind turbines are installed, because the wind turbines replace the operation of conventional plants, like combined cycle gas plants and coal fired plants, having high operational costs (see Fig. 16.6, compare also Fig. 16.7). But in hours with no wind the marginal costs rise when more wind turbines are installed, because an installation of wind turbines leads to a replacement of coal fired plants by combined cycle plants which are needed as back-up plants and which are more expansive in their operation. In Fig. 16.7 the energy production for each single hour of the $Wind100$ scenario for a high gas price and the corresponding marginal costs are illustrated. The marginal costs show high seasonal and weekly variations. In winter times and on weekdays when the demand is high the price for electricity is higher. In addition to that the daily differences become clear. During night time one can notice falling prices for electricity. The base load is provided by nuclear and lignite fired plants, but also hydro plants and energy import contribute to the hourly power supply. When the wind blows first the gas turbines go off line, because they are only used in the demand peaks. But also the energy production with middle load plants like combined cycle power plants and coal fired plants is influenced. In few hours even the lignite fired plants go off line. When the wind blows the marginal costs drop down, that means the marginal costs are not dependent on the energy demand itself, they are dependent on the energy demand reduced by the wind supply. For example, a day in winter with a high demand and with a high wind supply has the same marginal costs like a day in summer (see WED, winter) with a much lower energy demand. To outline the correlation between the residual demand and the marginal costs the data of the upper picture was sorted according to electricity production without wind power which is shown in the lower graph of Fig. 16.7. Here you can see that there is a high correlation between the marginal costs and the residual demand. If a new kind of power plant is needed to produce the required demand, there is a clear increase of marginal costs, which equal the variable costs (fuel and emission prices included) of that specific plant. If a restructuring gets necessary due to an increase in demand also these costs are included in marginal costs, for example costs for transmission and storage or installation costs for a new power plant. The wind fluctuation is reflected in the marginal costs, i.e., the variation of wind supply leads to instability of the electricity price. Table 16.2 resumes the basic figures of energy demand (wind energy excluded), marginal costs, the order of power plants and the highest variable costs of those power plants.

Marginal Costs of Capacities

Marginal costs of capacities which are shown in Fig. 16.8 for wind onshore turbines and in Fig. 16.9 for wind offshore turbines are the increase of the total costs (objective function) if one additional unit of wind power is forced into the system. For power plants, which are anyhow chosen by the optimization, these values are zero until the optimal capacity is reached. It is obvious that there exist variations of the marginal costs between the different sites as well as

Table 16.2. Energy demand [GWh], marginal cost [€/kWh], order of power plants and their variable costs [€/kWh]

Demand	Marginal costs	Variable costs	Power plants
27.4-47.1	0.026-0.029	0.027	Hydro, Nuclear, Lignite
47.1-60.9	0.027-0.056	0.037	Hydro, Nuclear, Lignite, Coal
60.9-82.1	0.057-0.089	0.065	Hydro, Nuclear, Lignite, Coal, CC
82.1-99.5	0.102-0.142	0.106	Hydro, Nuclear, Lignite, Coal, CC, GT

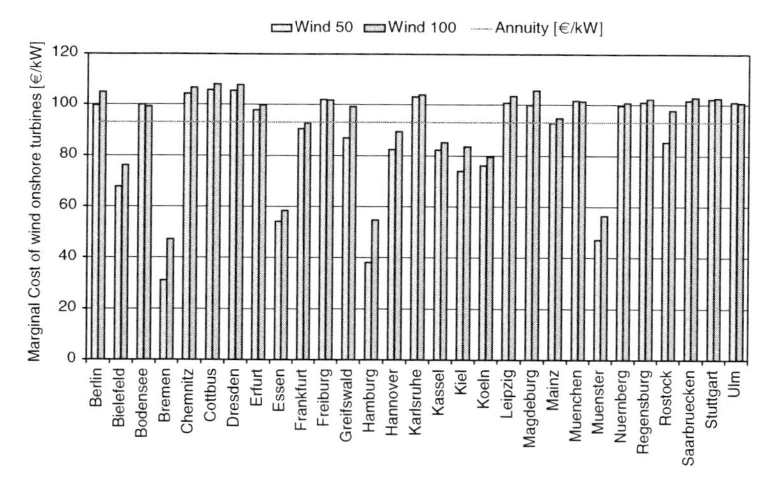

Fig. 16.8. Marginal costs of capacities for onshore wind turbines for the scenario with 100% and 50% of wind power ($Wind100$ and $Wind50$)

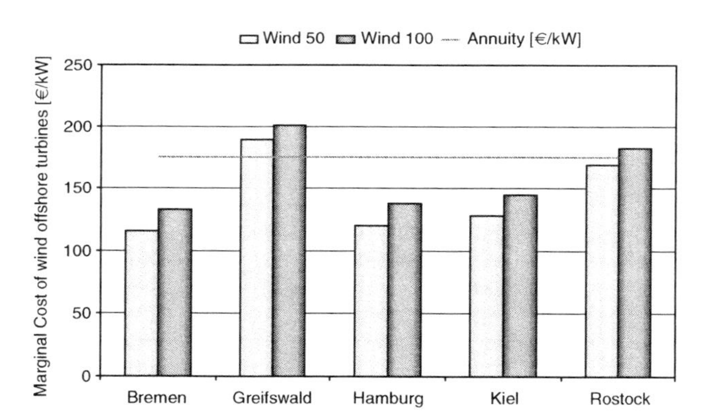

Fig. 16.9. Marginal costs of capacities for offshore wind turbines for the scenario with 100% and 50% of wind power ($Wind100$ and $Wind50$)

between the scenarios. To install 1 MW more (3,920 MW instead of 3,919 MW) of wind turbines in Bremen would provoke an increase of 48 k€ (scenario $Wind100$), at other sites like Dresden an increase of one MW would cause additional costs of 108 k€.

The height of the marginal costs is a composition of the annuity and variable costs of the specific power plant (here wind turbines) minus the costs of capacities and their operating costs getting redundant due to the new installations plus additional costs for transmission, storage or relevant changes to the existing system. The horizontal line in Figs. 16.8 and 16.9 displays the yearly amount of investment (annuity) with a discount rate of 6% being used for the optimization. For onshore wind turbines the annuity is 93 €/kW and for offshore wind turbines 175 €/kW.

How much the yearly investment can be reduced depends on the location of the wind turbine. At a site with a huge wind supply, like in Bremen, Essen, Hamburg or Münster, one MW of wind turbine replaces the energy production with conventional plants in many hours, in which these operating costs can be saved. At some sites, like Berlin, Dresden, Karslruhe, Leipzig etc., in contrast, the additional cost of a wind turbine can not be reduced by the reason of a lack of wind. The installation of a wind turbine is redundant and would lead only to dispensable costs. The back-up plants with the highest operating costs are the first plants being replaced by the wind turbines. If more wind turbines are installed also the energy production with plants, having lower operational costs, is reduced, this explains the increase of marginal costs from the scenario $Wind50$ to $Wind100$.

16.3.3 CO_2 Emission

With the CO_2 emission scenarios the minimal possible CO_2 emission per year should be determined and what kind of power plant structure is necessary to achieve that objective. In these scenarios the yearly emission was limited to 250, 200, 150 and 112 Mt, where 112 Mt is the minimal possible CO_2 emission.

Capacities

Figure 16.10 shows the optimal capacities for the CO_2 emission scenarios. With the reduction of CO_2 the installation of coal and lignite plants has to be

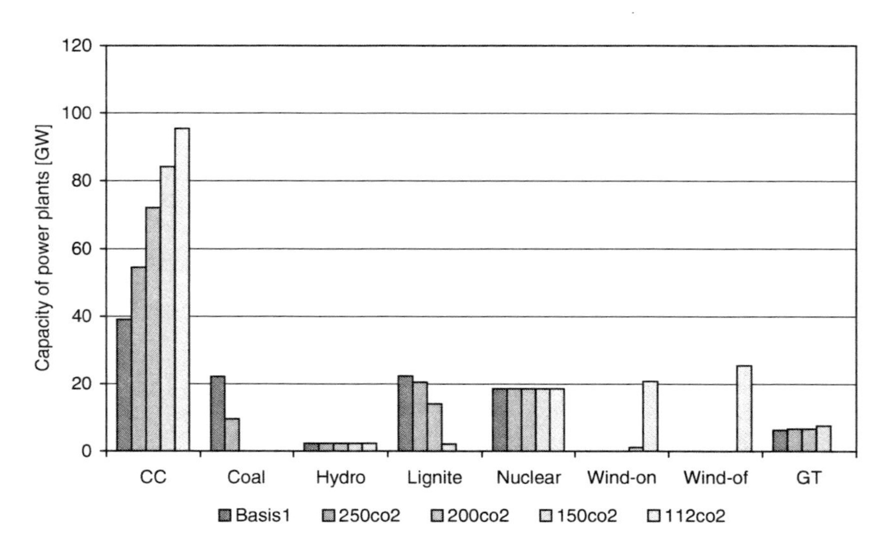

Fig. 16.10. Capacities of power plants depending on a change CO_2 restriction

diminished due to their high emission factor. Coal is not used any more from an emission of 200 Mt downwards, whereas lignite is used until an emission of 150 Mt. The coal and lignite plants are replaced by the gas plants. Combined cycle gas plants grow from 40 to 95 GW, that's a growth of 240%. The gas turbines denote only a small increase. Nuclear power is needed in all cases: the installation is constant with 18.6 GW between all runs. Wind onshore is already profitable for an reduction of emission of 150 Mt, but only with an installation of 1 GW. For the scenario with minimal emission (112 Mt) the installation of wind onshore turbines grows to 20.8 GW and wind offshore turbines to 25.6 GW, which is the upper boundary of the capacities. The gas turbines as back-up plants are replaced by the combined cycle power plants at the last of those scenarios. The total capacities increase from 111 GW (Basis1) over 112 GW (250co2), 114 GW (200co2), 116 GW (150co2) up to 163 GW (112co2). That shows again the necessary back-up capacities, which are effected by the installation of wind power.

A limitation of emission to 112 Mt requires additional capacities of transmission lines which are shown in Fig. 16.11. The flows are mainly located in the north western part of Germany, directed from the North with wind power to the South. Bremen for example has only wind turbines and combined cycle gas plants, therefore in hours with high wind supply there is a surplus of energy in Bremen which is transmitted to Münster and even further to Essen. In Münster and Essen the wind power is then regulated by the combined cycle power plants which go off-line when the wind energy comes from Bremen. Even the nuclear plants go off-line in Münster when the supply is very high. About 2,000 MWh of the energy which comes from Bremen is needed in Münster, the rest (maximum 10,170 MWh) is delivered to Essen. Another area where new capacities have to be installed is the region around Hamburg, Kiel, Bielefeld, Hannover and Kassel.

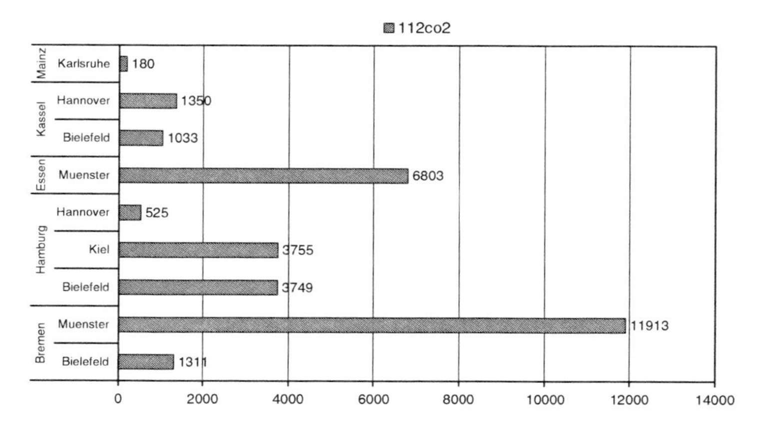

Fig. 16.11. New transmission lines [MW]

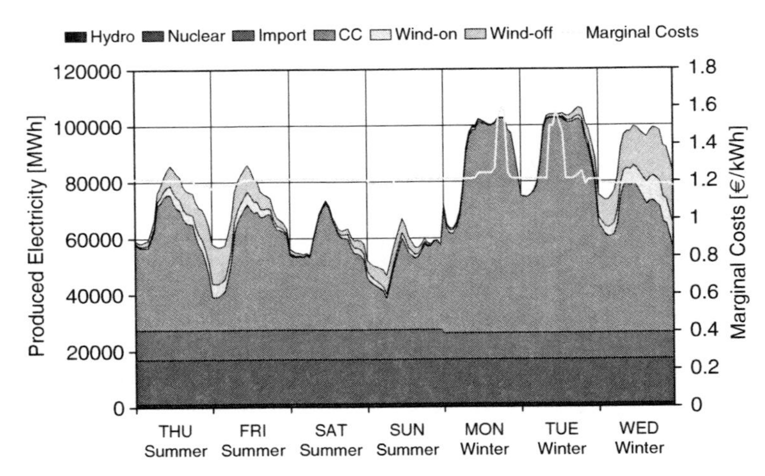

Fig. 16.12. Development of marginal costs against the energy production for the scenario with 112 Mt CO_2 emission

Marginal Cost and Production

Figure 16.12 shows the energy production for each hour (day) and the corresponding marginal cost for the scenario with minimal CO_2 emission (112*co*2). For that reason only power plants with no or only little emissions are used. Most of the electricity is provided by the combined cycle gas plants having the highest variable cost. These plants are used in every single hour, that's why the marginal costs are on a constant level. The base load is provided by hydro and nuclear plants. The rest is imported. The system is working at its limit because of the high limitation of CO_2 emission, which can also be seen in the marginal costs. The height of the marginal costs shows that such a power plant mix is uneconomically, the price for electricity is too high (1.2 €/kWh). The influence of wind on the marginal costs is very small, there can be seen only a little decrease of marginal costs between THU and FRI in summer time and SUN morning and TUE evening and WED in winter time. The seasonal, weekly and daily variation of the marginal costs has vanished. The two peaks of marginal costs show the hours with maximal demand.

16.3.4 Uranium

One scenario was calculated to simulate the phasing out of nuclear energy (see Figure 16.13). The optimization shows that instead of nuclear plants combined cycle gas plants and especially coal fired plants are built. Instead of 18.6 GW of nuclear plants the combined cycle plants rise from 39.04 GW to 43.36 GW and the coal fired plants from 22.16 to 37.71 GW. The installation of gas turbines is reduced from 6.32 GW in the Basis1 scenario to 4.65 GW in the scenario without nuclear power. Wind becomes efficient when a restriction to the CO_2 emission of 200 Mt is stipulated.

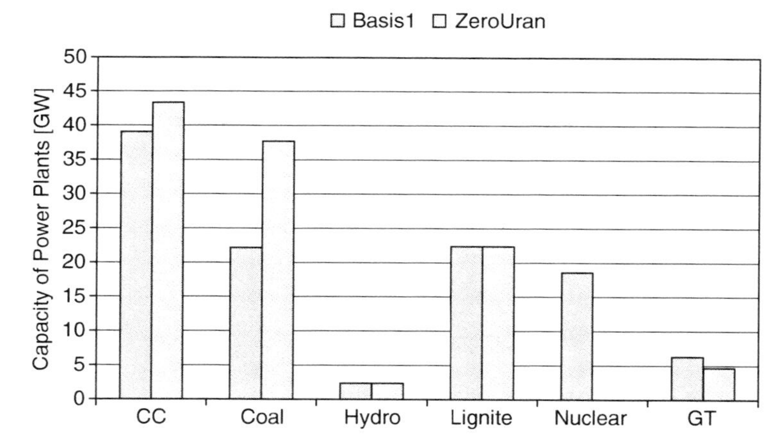

Fig. 16.13. Installation of power plants with and without nuclear energy

Without nuclear energy the CO_2 emission can be reduced only to 160 Mt. Depending on the power plants which replace the nuclear plants the increase of CO_2 emission varies notably. If 138.47 TWh of nuclear energy is substituted by the combined cycle gas plants, the increase of CO_2 emission is 48 Mt, but if it is replaced by the coal fired plants, the rise is already 92 Mt.

16.3.5 Stochastic Parameter: CO_2 price

The CO_2 price is affected by changing values. It is useful to take such variation into account for the power plant planning. This can be realized in the model with the help of stochastic parameters. In the scenario CO_2*Price* the CO_2 price varies from 10 €/t (case p1), 25 €/t (case p2) to 50 €/t (case p3).

The total installed capacity of 113 GW in the CO_2*Price* scenario is only 2% higher than the capacity of the *Basis*1 scenario with 111 GW. Hydro and nuclear plants are installed up to their limit to 2.4 GW respectively 18.6 GW for the base load. Lignite plants are installed up to 17.8 GW. In case p1 and p2 for a CO_2 price of 10 €/t respectively 25 €/t they are used as base load plants, in case p3 with a CO_2 price of 50 €/t they provide only the peak load, with 2,163 instead of 7,709 full load hours (compare Fig. 16.14). Combined cycle gas plants build the biggest part; they have a capacity of 68 GW, which are 60%. The missing electricity in case p3, which is produced in the other cases p1 and p2 by the lignite fired plants, is provided by the combined cycle power plants. Instead of 270 TWh in case p1 and p2 they produce 369 TWh. The peak load (2.8 TWh) is produced with gas turbines with a capacity of 6.7 GW. This scenario does not install wind turbines and coal fired plants.

Figure 16.14 compares the hourly energy production for two different CO_2 prices. For each CO_2 price the marginal costs are on a constant level, only in hours when gas turbines are used an appreciation of marginal costs becomes

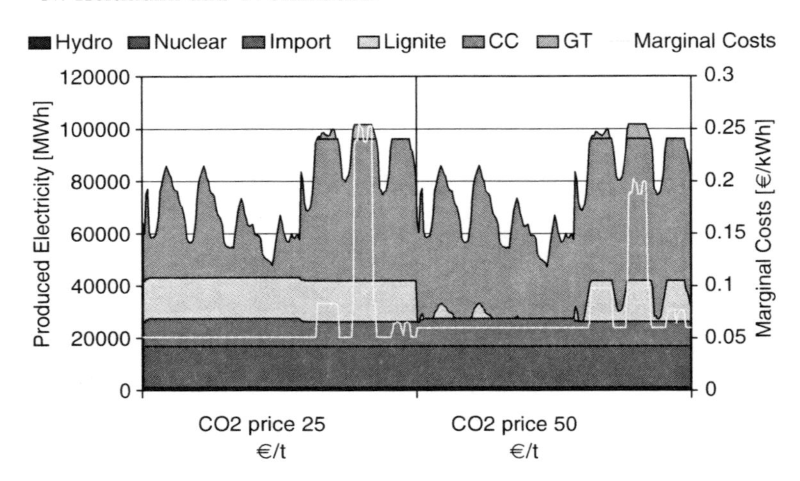

Fig. 16.14. Energy production for changing CO_2 prices

apparent. The rise in CO_2 price badges clearly in the marginal costs which are higher on the right hand side for a CO_2 price of 50 €/t. The CO_2 emission can be reduced by 53 Mt from 216 to 163 Mt.

16.3.6 Stochastic Parameter: Wind Supply

Another possibility for the stochastic program is to regard the wind supply in a stochastic way. On the one hand it is possible to implement each hour in a stochastic manner, but that would cause immense computing time and the high time resolution, the manner of operation of the power plant which is one of the favourite features of this model generator would become less important. On the other hand it is feasible to provide the wind supply data as time series, but instead of only one possible time series there can be used different ones to reproduce the change of wind supply between the years. This is done in scenario *WindSup*. For that scenario different time steps were chosen as in the scenarios before. Here one complete week was taken to look mainly at the manner of operation and the corresponding electricity price. For that scenario the power plant capacities were predetermined. Amongst others wind turbines were installed manually to see the influence of wind power on the marginal costs.

Capacity and Production

The optimal power generation and the marginal cost for the case of high wind supply (MC high), medium wind supply (MC medium) and the case of a low wind year (MC low) are shown in Fig. 16.15. Differences in the marginal costs become apparent especially when the wind blows (see FRI). The marginal

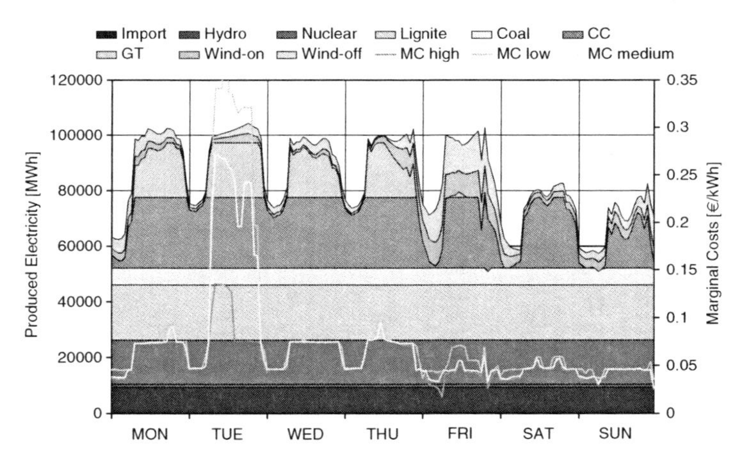

Fig. 16.15. Energy production and marginal costs depending on the wind supply

costs for the year with high wind supply are always the lowest ones, this shows that wind power beats down the price of electricity.

For the case of low wind supply, the decisive factor that is responsible for the building of power plant capacities is on TUE. Here the maximal demand is reached and so one additional unit of power plant has to be installed for a rise in demand, therefore the costs increase steeply and obtain the height of the investment cost of the cheapest power plant that could be installed. The crucial hour is in case of a low wind supply, but also in the medium and high wind year the top of demand can be noticed in the marginal costs.

16.3.7 Emission and Costs

In the following section the CO_2 emissions are opposed to the total costs of each scenario, which is displayed in Fig. 16.16. Without further restriction 282 Mt of CO_2 are emitted by the electricity production with the underlying power plant mix. This emission is relatively low because a new generation of coal and lignite plants with high efficiencies were used for the optimization.

In general an increase in fuel prices for gas provokes an increase in CO_2 emission because gas plants are reduced by coal or lignite fired plants having higher emissions, whereas an increase of CO_2 price diminishes the emission. The wind scenarios ($Wind50$ and $Wind100$) show a steady decrease of emissions but an increase of total costs. The emission that can be saved is about 44 Mt and the cost increase is 4,863.54 M€, that means saving one tonne of CO_2 costs about 111 €.

With the help of the CO_2 restriction the emission can be reduced to 112 Mt, that's a reduction of 171 Mt compared to the basis scenario and the cost increase is 7,396.27 M€; saving one t of CO_2 therefore costs 44 €. In contrast a restriction to 150 Mt has a cost increase of only 2,230.40 M€ and the

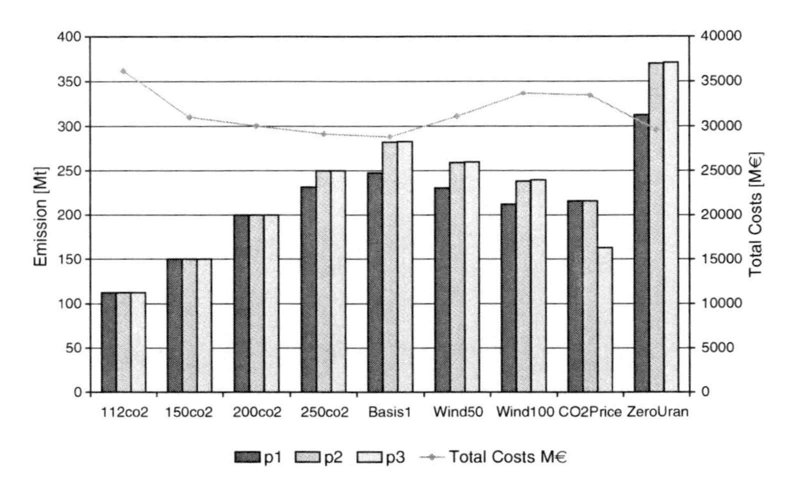

Fig. 16.16. Comparison of total costs and CO_2 emission for all scenarios

CO_2 emission is reduced by 132 Mt, so saving one tonne of CO_2 costs 11 €. That means ecologically and economically it is more reasonable to support the idea of CO_2 emission reduction than to integrate wind power. The integration of wind power is cost intensive and contributes proportionally little to a reduction of emission.

The problem with phasing out nuclear energy is not the cost factor; rather it entails a growth of the greenhouse gases. To avoid the installation of nuclear plants causes an enormous increase of CO_2. Even if nuclear power is replaced by high efficient coal plants, the increase is huge. The minimal emission that can be achieved without nuclear power is 160 Mt, but in that case all nuclear power has to be replaced with combined cycle gas turbines.

16.4 Conclusion and Outlook

A stochastic model of the German electricity system was developed, which is able to describe wind energy in details. Stochastic programming offers a much better way to deal with uncertainties like gas and CO_2 prices. The model is capable to handle various input parameters in a stochastic fashion, namely the gas prices, the CO_2 prices and the wind supply. Scenarios were developed with different constraints on the installation of wind energy and nuclear. Gas price variations are counterbalanced by a stronger power generation with coal and CO_2 price variations by the use of lignite and gas. Integration of wind energy into the system in the magnitude of 45 GW requires strong back-up and transport capacities and is not the cheapest option for CO_2 reductions. The variation of wind supply has strong impact on the electricity price. In the future we will investigate the implementation of other renewable sources like solar and the possible impact of novel storage technologies.

References

1. *DIgSILENT GmbH.* Gemeringen, Germany.
2. J.R. Birge and F. Louveaux. *Introduction to Stochastic Programming.* Springer Series in Operations Research. Springer, New York, USA, 1997.
3. European Comission, Directorate-General of Energy, and Transport in co-operation with Eurostat. Energy and transport in figures 2006, 2006.
4. T. Haase. *Anforderungen an eine durch erneuerbare Energien geprägte Energieversorgung – Untersuchung des Regelverhaltens von Kraftwerken und Verbundnetzen.* PhD thesis, Universtität Rostock, Rostock, Germany, 2005.
5. N. Heitmann. Lösung energiewirtschaftlicher Probleme mit Hilfe linearer Programmierung. Master's thesis, Universität Augsburg, 2005.
6. Öko-Institut, Institut für angewandte Ökologie e.V. *Global Emission Model for Integrated System (GEMIS).* Version 4.3.
7. W. Pfaffenberger and M. Hille. Investitionen im liberalisierten Energiemarkt: Optionen, Marktmechanismen, Rahmenbedingungen. Bremer-Energie-Institut, Bremen, Germany, 2004. Abschlussbericht.
8. Sander und Partner GmbH, Schweiz. *World-Wind-Atlas.*

17

Optimization of Risk Management Problems in Generation and Trading Planning

Boris Blaesig and Hans-Jürgen Haubrich

Summary. Due to increased cost pressure on power generation and trading companies, caused by operation under market conditions, a cost efficient management of the risks becomes more important. As a result of the liberalization of the markets for electrical energy companies are exposed to higher uncertainties in power generation and trading planning, e.g., the volatility of the prices for electrical energy and for primary energies, especially natural gas. Additionally, bankruptcies of companies in the energy sector, e.g., ENRON or TXU Europe, have demonstrated that the loss of trading partners may cause a major disprofit, if not hedged appropriately. Together with risk management regulations, the need for risk management in generation and trading planning is increasing.

The objective of this work is the development of adequate methods for generation and trading planning, i.e., maximization of the contribution margin, taking the risks into account. The risk management process comprises identification and analysis of both risks and their impacts as well as the control of the occurring risks.

In this work two approaches, a separate expost and an integrated risk management method, have been developed using appropriate algorithms [2]. The expost approach uses the schedule of the power plants from the generation planning as given input data and optimizes the trading decisions by means of risk management concepts. The integrated approach yields the optimal generation and trading decision in terms of maximal contribution margin as well as minimal risk in one step.

The multicriterial optimization of the maximal contribution margin as well as the minimal risk is implemented either by risk constraints which limit the risk to a maximum or by utility functions which map the combination of contribution margin and risk to a single criterion.

The investigations of different systems demonstrate the results of the different risk management methods, whereas in this paper the results of a thermal dominated typical German generation and trading company are discussed.

Investigation of the effectiveness of the risk management methods using different power markets show improvement of the risk control participating in these markets compared to the negligence of these opportunities. Entering markets for weather and primary energy derivatives can reduce the risk of the portfolio.

The investigations show the tradeoff between contribution margin and risk. Depending on the risk aversion of the company the risk can be reduced for the tradeoff of a lower contribution margin. Comparing the results of the expost and the integrated risk management, it can be summarized that the integrated approach is more effective. This is due to the advantage of the integrated risk management method using both redispatch of the power plants for risk management purposes and even more important for adaptation to changed trading decisions.

17.1 Introduction and Motivation

Market-based prices for electrical energy in new deregulated markets increased the cost pressure on power generation and trading companies. Before the deregulation energy companies had the possibility to raise their energy price in the long-term time horizon to pass on the costs to the end-user. Due to the improved maturity of the electricity markets, the electricity price has become the factor determining the operation of the power plants. As a result of this process electricity companies are directly exposed to these markets and are therefore directly responsible for their profits and losses. This necessitates a cost-efficient management of the risks.

As a result of the deregulation of the power markets the companies are additionally exposed to greater uncertainties within power generation and trading planning. The volatility of the electrical energy prices has increased compared to long-term energy contracts in times of the monopoly. A comparable development can be observed in the market for primary energies. Especially the deregulation of the gas market headed to more volatile prices of gas [19]. The option for consumers to change their power supplier added a new uncertainty in the planning process for distributing electrical energy.

Next to increased risks, bankruptcies of companies in the energy sector demonstrated that the loss of trading partners can cause great disprofits, if not hedged accordingly (e.g., ENRON, TXU-Europe). As a result risk management in the energy sector is essential, as already demanded by certain regulation like the KonTraG in Germany [5]. Therefore risk management methods have to be considered in all phases of the planning process and are thus elementary in the generation and trading planning.

Despite of the numerous research approaches in the field of risk management in the generation and trading planning there is a lack of broad investigations including all relevant uncertainties for the planning process. Therefore, this work analyzes the system for generation and trading planning and shows how risk management methods can be applied. Beyond the analysis of the risks an approach of how the risk can be modeled and used in a planning tool will be presented in Sect. 17.2. The optimization tool is explained in detail in Sect. 17.3 and in exemplary investigations in Sect. 17.4 it will be demonstrated how the risk can be controlled. Section 17.5 exposes the conclusions.

17.2 Analysis and Modeling

17.2.1 Risk Management in Generation and Trading Planning

In generation and trading planning the portfolio of assets for electrical energy for a certain company is optimized. Looking at risk management in generation and trading it is important to specify a time horizon. In the short-term time horizon other uncertainties have to be considered compared to mid- or long-term planning. In general, the longer the planning horizon, the more uncertainties have to be considered and the risk management instruments become rare and are also exposed to uncertainties. Therefore companies use generation and trading planning in general with a time midterm horizon from 1 month to 1 year to receive trading recommendations for electrical energy and fuel markets. Market alternatives between the wholesale market and the market for system reserve need to be traded off. Regarding such a time horizon, as it is used in this work, investigations have shown that a time pattern of 1 h is sufficient [11].

In Fig. 17.1 the system of a European generation and/or trading company is depicted. Looking at the value chain of such a company consisting of a number of optional system components the planning uncertainties displayed in this figure have to be considered.

Electrical energy is generated by hydraulic and thermal power plants. Hydraulic power plants depend on the amount of water flowing into the reservoirs. Thermal power plants have a certain reliability meaning that outages may occur. Primary energies (fuels) need to be procured at fuel markets or bilaterally from trading partners. Prices at fuel markets can be uncertain resulting in a price risk. The operation of thermal power plants produces greenhouse gas emissions and therefore requires emission certificates. The

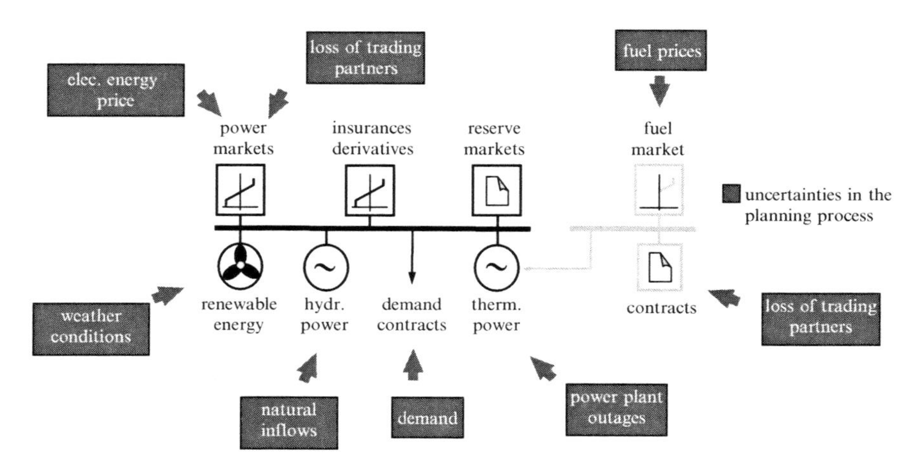

Fig. 17.1. System overview

reduction mechanism is based on a cap-and-trade approach, resulting in uncertain prices for emission certificates, which have to be considered in the planing process. The counterpart risk that is based on bankruptcy of trading partners can lead to a loss of energy delivery or payment.

Besides generating electrical energy, companies can buy or sell energy at the different electrical energy markets with different uncertain prices. Markets for system services, mainly the system reserve, are market alternatives for generation units. For risk management reasons insurances or derivatives, i.e., for fuels or on weather gain in importance.

With a high relevance in Germany, the wind fed into the system is another uncertainty especially for wind farm operators. Other renewable energies also depend on weather conditions, resulting in an uncertainty for the operators of these assets. The demand contracts including the load of end-users depending on the demand of all customers need to be fulfilled.

Using the scenario analysis as an appropriate method to model the planning uncertainties in generation and trading planning [2] the uncertainties are modeled by a scenario tree that starts with a deterministic root and then branches with increasing time to represent possible future developments [16]. The scenario analysis allows the suggestion of definite decisions in the deterministic root under consideration of the uncertainty in the far future. Therefore the results leave the flexibility to react on the uncertain future development of the input parameters but calculates a robust contribution margin.

In [2] the system components with its risks and hedging instruments are further analyzed and the resulting model of the components are derived. In this paper, it is first demonstrated how risk management concepts are included in the optimization problem (see Sect. 17.2.2). The resulting model of the different components and the risk management concept are presented in Sect. 17.2.3.

17.2.2 Concepts of Risk Management

In literature different approaches of risk modeling can be found [2]:

1. The risks ρ can be formulated in a constraint meaning that the resulting risk is limited to a certain upper level.
2. The risks can be implemented in the objective function so that the risks are minimized together with the maximization of the contribution margin.

The portfolio theory [9] provides the basis for modeling risks and is applied to generation and trading planning: Fig. 17.2 shows two assets from which the first has a high risk as well as a high expected contribution margin, consisting of the difference between variable revenues and variable costs. Whereas the second has a low expected contribution margin and a low risk. By mixing these two positions, different portfolios can be realized, lying on the thick solid curve between the two extreme portfolios of just one asset. The increasing part of the curve consists of the portfolios with the maximal

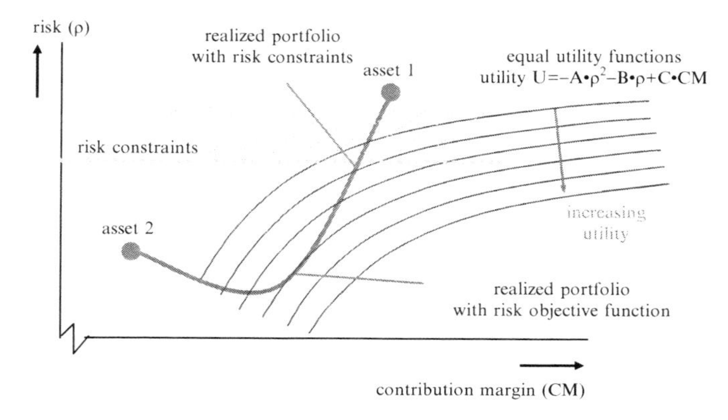

Fig. 17.2. Risk management in constraints and objective functions

contribution margin for a given risk and therefore dominates the decreasing part. Hence these portfolios are called efficient frontier of the portfolio function. The shape of this curve is determined by the correlation between the two assets. Return and risk of the realized portfolio depends on the composition of the portfolio. The resulting risk of the portfolio can be lower compared to the extreme positions due to the markowitz diversification effect. Totally (negative) correlated assets allow a construction of a risk free portfolio [1].

Using an objective function maximizing the contribution margin, which represents a risk neutral decision maker, the portfolio at the right end, consisting exclusively out of asset 1, is realized, neglecting the influence of the risk. Risk management can restrict the resulting risk to a maximum limit (dotted line in Fig. 17.2). In this case the portfolio, which results in the maximum contribution margin below the risk constraint, will be selected:

$$\rho \leq \rho^{\max} \tag{17.1}$$

It is possible to define equal-utility functions (see light solid curves in Fig. 17.2) where each portfolio on the curve yields an equal utility U (e.g., with a quadratic dependency on risk and expected return or contribution margin $\overline{CM}$) to the user as all the other portfolios on the curve. The user is therefore indifferent about selecting a portfolio on such an equal-utility function.

$$U = -A \cdot \rho^2 - B \cdot \rho + C \cdot \overline{CM} \tag{17.2}$$

As depicted in Fig. 17.2, for each utility an equal-utility function can be defined. Therefore an infinite number of equal-utility functions exists, which do not cross each other. The arrow indicates increasing utility. The shape of the equal-utility functions depends on the risk preference of user, which makes the parameterization (A, B, and C) of these functions difficult. For this reason the method of using risk constraints is prefered in practice. But in this paper a practical way of determining the whole portfolio function using equal utility functions is discussed, which has the advantage of leaving a high degree of freedom to the decision maker (see Sect. 17.4.3).

The user realizes preferably a portfolio on the efficient frontier spending the highest utility. Therefore, the indicated portfolio on the intersection of the efficient frontier and the equal-utility function with the highest utility is selected for the case when risk management is implemented in the objective function.

This reflection is independent from the used risk measure. In order to implement the risk in an optimization method, an appropriate risk measure needs to be selected. In literature many risk measures are discussed for different applications. To introduce a risk measure in power generation and trading planning a set of requirements, like coherence, statistical dominance and the practical relevance, can be introduced. A comprehensive discussion of the different requirements and the risk measures can be found in [2].

While analyzing different risk measures as the variance, the semivariance, the Greeks, the lower-partial-moments, and different Value-at-Risk approaches including the Conditional Value-at-Risk, it turned out that the Conditional Value-at-Risk ($CVaR$) is the most suitable risk measure under the selected requirements, especially under the aspect of practical relevance [2,14]. Therefore the $CVaR$ is selected for the further work.

The $CVaR$ represents the weighted average of the losses (negative contribution margin CM) within a certain probability α. Figure 17.3 shows a probability density function p for the contribution margin of an asset. The area covering the worst α realizations is marked in the figure. The $CVaR_\alpha$ is the weighted average of returns within the probability α.

Considering discrete probability functions resulting in case of the scenario analysis, the calculation of the $CVaR$ can be formulated as an optimization problem itself while summing up the contribution margin for all scenarios $s \forall Sz$, each with the probability pr_s:

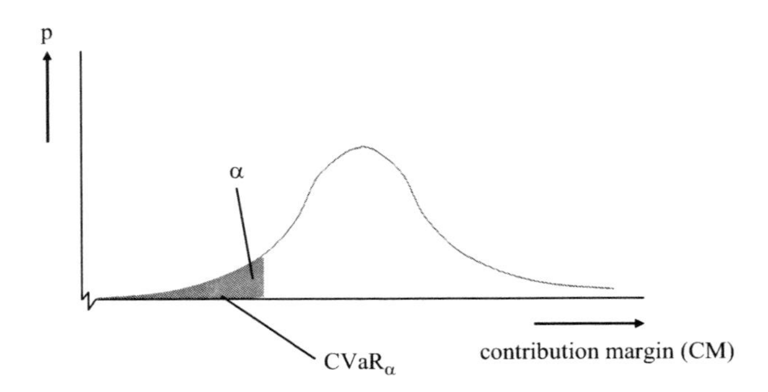

Fig. 17.3. Conditional value-at-risk

$$CVaR_{\alpha} = \min_{r}\left(r + \frac{1}{\alpha}\cdot\sum_{s\forall Sz}\left[\max\left(0, -CM_{s} - r\right)\cdot pr_{s}\right] : r \in \mathbf{R}\right)$$

$$= \min_{y_0,\mathbf{y}_1}\left(y_0 + \frac{1}{\alpha}\cdot\sum_{s\forall Sz}\left[y_{1,s}^{(1)}\cdot pr_{s}\right] : y_0 \in \mathbf{R}, \mathbf{y}_1 \in \mathbf{R}_{+}\times\mathbf{R}_{+}\right) \quad (17.3)$$

with $y_0 + y_{1,s}^{(1)} - y_{1,s}^{(2)} = -CM_{s} \quad s\forall Sz$

Considering a multidimensional *CVaR*, the properties concerning the coherence and the statistical dominance can be preserved [7]. Therefore multitime horizon and multidimension in the quantile α shall be included in the risk measure so that (17.3) is extended to multitime sections RA and multiquantiles Qu [2].

17.2.3 Resulting Model

In the process of generation and trading planning, generation companies maximize their contribution margin, i.e., the difference of the revenues of energy trades and the costs for generating and purchasing electrical energy by means of risk management concepts.

As shown in Sect. 17.2.2 the objective function of the resulting model consists of maximizing the utility:

$$\max\left(U\right) = \max(-A\cdot\rho^{2} - B\cdot\rho + C\cdot\overline{CM}) \quad (17.4)$$

Using the *CVaR* the risk is specified by (17.3) whereas the contribution margin consists of the contribution margin of the different system components.

Thermal power plants *TP* generate electrical energy using different primary energies in their thermomechanical cycle. The hourly power output of each unit is used as the decision variable. Each unit has to comply with technical restrictions like the minimum and maximum power output level, minimum up- and down-times as well as nonlinear efficiency curves. Furthermore the accruing costs during start-up and normal operation have to be considered accordingly.

To procure fuels for thermal power plants, the trade of fuels, i.e., hard-coal and natural gas, gains in importance in the planning process of generation companies. Markets for hard-coal are already well established whereas the markets for natural gas emerge due to the recent deregulation of the gas sector. In analogy to the electricity markets spot and futures markets for gas develop.

Hydro power plants *HP* can be separated into storage and pump storage as well as run-of-river power plants. Especially in mountainous regions hydro power plants are usually interconnected and characterized by interdependencies for the operation. The generation of electricity in hydro power plants is a nonlinear function of the water flow and the hydraulic head. Hydro power plants are characterized by neglectable variable costs. In contrast to the thermal power plants the rate of flow of each machine for every hour is used as the decision variable to represent the hydraulic problem using a network flow model.

Due to the deregulation of the electricity markets in Europe new power markets have been established. Electrical energy can be traded bilaterally (over-the-counter; OTC) or at power exchanges. At the spot market Sp electrical energy can be sold and purchased day ahead, whereas futures markets Fu allow trading long-term contracts at power exchanges and forward markets Fo refer to over-the-counter trading. The most liquid long-term products are base and peak which can be traded up to several years in advance [8]. In addition, options on electrical energy provide opportunities of risk management in combination with futures. For all markets the traded and executed amounts define the decision variables.

As a risk management instrument the market for options Op on electrical energy gained in importance over the last couple years.

Transmission system operators have to procure reserve power to facilitate a secure network operation. For generation companies these newly established markets for system reserve offer an alternative to wholesale markets.

In addition to the markets for electrical energy and reserve, it is possible to trade derivatives De and insurances In on correlated values, e.g., fuel or weather derivatives.

The consideration of the components described above leads to a complex optimization problem with the objective function shown in (17.4) separating the contribution margin for the relevant components with the multidimensional $CVaR$ as a risk measure:

$$\begin{aligned} \max(U) = \min \Bigg(& A \cdot \left(\frac{1}{n_{Qu}} \cdot \sum_{a \forall Qu} y_{0,a} + \frac{1}{\alpha_a} \cdot \sum_{r \forall RA} \sum_{s \forall Sz_r} \left[y^{(1)}_{r,s,a} \cdot pr_s \right] \right)^2 \\ & + B \cdot \left(\frac{1}{n_{Qu}} \cdot \sum_{a \forall Qu} y_{0,a} + \frac{1}{\alpha_a} \cdot \sum_{r \forall RA} \sum_{s \forall Sz_r} \left[y^{(1)}_{r,s,a} \cdot pr_s \right] \right) \\ & - C \cdot \Bigg(\sum_{t=1}^{T} \sum_{s \forall Sz_t} \Bigg(\sum_{l \forall TP} CM^{TP_l}_{t,s} + CM^{Sp}_{t,s} + CM^{Fu}_{t,s} + \\ & \qquad CM^{Fo}_{t,s} + CM^{Op}_{t,s} + CM^{In}_{t,s} + CM^{De}_{t,s} \Bigg) \cdot pr_s \Bigg) \Bigg) \end{aligned} \tag{17.5}$$

The solution space is constrained by technical and economic properties of the different components, the load and reserve balance of the system and the constraints for the $CVaR$, which are discussed in detail in [2].

Considering the technical and economical properties of the system the resulting optimization model is characterized by nonlinearities, system and time spanning constraints as well as integer decisions [2].

17.2.4 Methodology

As shown in earlier works [3], there are two main approaches of risk management in generation and trading planning.

1. The first concept separates the generation and trading planning (GTP) from the risk management. In this case the generation and trading planning in done first. The results of this step are the schedule for the generation assets as well as the trading recommendations. These decisions are processed in a second step the risk management. The schedule for the generation assets is not changed and hence kept constant in this step. The trading decisions are revised by risk management concepts in this stage. Therefore these decisions are made on the basis of maximising the contribution margin and considering risks whereas the generation assets are dispatched disregarding the risk. Up to now generation companies use – if at all– a risk management concept like this expost management.
2. The second approach which is derived in Figs. 17.4 and 17.5 combines the two planning steps of generation and trading planning and risk management and is therefore called integrated risk management. This concept allows to determine decision support for the generation assets as well as for the markets on the basis of maximising the contribution margin and considering risks. Therefore synergies between the two steps can be achieved.

In this work the results of the two different methods will be compared in order to quantify the synergies between generation and trading planning and risk management (see Sect. 17.4.2).

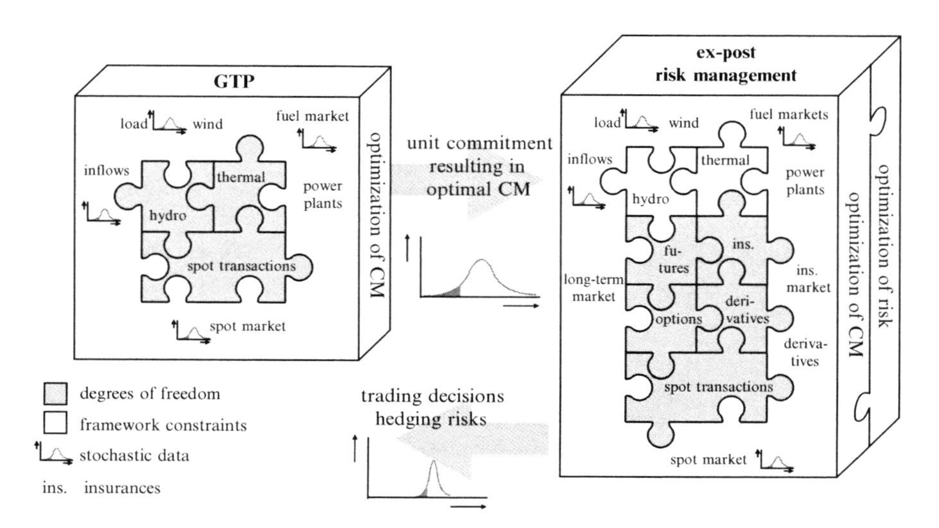

Fig. 17.4. Expost risk management

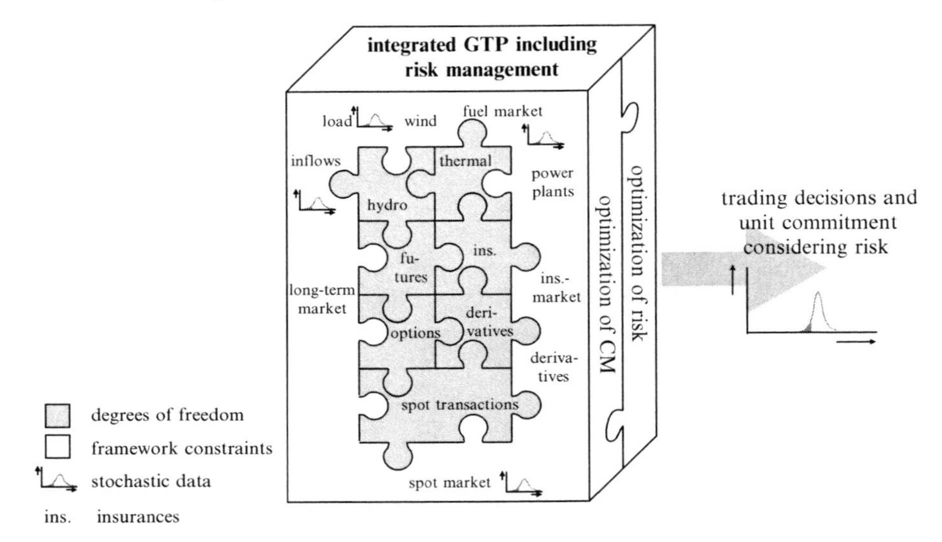

Fig. 17.5. Integrated risk management

17.3 Optimization Method

As discussed in Sect. 17.2.4 two different risk management concepts are investigated in this work. This necessitates in the implementation of optimization methods for these two concepts. The complete optimization problem is derived in Sect. 17.2.3 whereas the degrees of freedom in the optimization problem depend on the risk management concept. Therefore the two concepts of expost risk management and integrated risk management are discussed in the following sections separately.

17.3.1 Expost Risk Management

Starting from a generation and trading planning the decisions for the operation of the power plants are taken on to the expost risk management. Therefore the optimization task of planning the generation decisions needs to be solved. This can be done by different tools [4, 6, 15, 17]. In this work the planning tool based on [11, 13, 18] is used for comparability reason with respect to the algorithms implemented in the integrated risk management.

The expost risk management solves the remaining optimization problem which consists of the trading decision based on risk management concepts resulting in the following objective function:

$$\max(U) = \min\left(A \cdot \left(\frac{1}{n_{Qu}} \cdot \sum_{a \forall Qu} y_{0,a} + \frac{1}{\alpha_a} \cdot \sum_{r \forall RA} \sum_{s \forall Sz_r} \left[y^{(1)}_{r,s,a} \cdot pr_s\right]\right)^2 \right.$$
$$+ B \cdot \left(\frac{1}{n_{Qu}} \cdot \sum_{a \forall Qu} y_{0,a} + \frac{1}{\alpha_a} \cdot \sum_{r \forall RA} \sum_{s \forall Sz_r} \left[y^{(1)}_{r,s,a} \cdot pr_s\right]\right)$$
$$- C \cdot \left(\sum_{t=1}^{T} \sum_{s \forall Sz_t} \left(CM^{Sp}_{t,s}\left(P^{Sp}_{t,s}\right) + \right.\right.$$
$$CM^{Fu}_{t,s}\left(P^{Fu}_{c,t,s}, \left(P^{Fu}_{c,t,s}\right)^2\right) +$$
$$CM^{Fo}_{t,s}\left(P^{Fo}_{c,t,s}, \left(P^{Fo}_{c,t,s}\right)^2\right) +$$
$$CM^{Op}_{t,s}\left(P^{Op}_{c,t,s}, \left(P^{Op}_{c,t,s}\right)^2, P^{Op}_{ex,t,s}\right) +$$
$$CM^{In}_{t,s}\left(P^{In}_{c,t,s}, \left(P^{In}_{c,t,s}\right)^2, P^{In}_{t,s}\right) +$$
$$\left.\left.\left. CM^{De}_{t,s}\left(P^{De}_{c,t,s}, \left(P^{De}_{c,t,s}\right)^2, P^{De}_{t,s}\right)\right) \cdot pr_s\right)\right) \tag{17.6}$$

The objective function consists of different components to control the risk as well as the contribution margin. Equation 17.6 shows the dependencies of the contribution margin from the trading decisions P of the different markets. For example, the contribution margin for the spot market Sp depends linear on traded energy at the spot market $P^{Sp}_{t,s}$ in each hour t and each scenario s. For the future and forward markets the dependency is quadratic with respect to the contracted energy. For the options market the model results in a quadratic dependency from the contracts traded and a linear dependency from the executed amount $P^{Op}_{ex,t,s}$. Hence, the resulting objective function contains linear as well as quadratic terms [2].

The constraints of the optimization problem can be separated into constraints for the $CVaR$ (see (17.3)), the system spanning constraints and the constraints for the different components [2]. For the expost risk management the system consists of the different markets which are modeled using linear constraints. Because of the quadratic dependency of the contribution margin from the decision variables the constraints for the $CVaR$ have quadratic terms. This results in quadratic and linear constraints.

Due to the quadratic objective function, a quadratic programming approach is adequate to used. The quadratic constraints are linearized and therefore a Successive Quadratic Programming (SQP) algorithm is used to solve the introduced optimization problem in this work [2]. This algorithm linearizes the constraints in a working point and solves the resulting quadratic problem with linear constraints. This result will then linearized repeatedly in the new working point until a sufficient convergence is achieved. This allows the consideration of the nonconvex quadratic constraints in the optimization problem.

17.3.2 Integrated Risk Management

In contrast to the expost risk management, the integrated risk management concept solves the planning task of power generation and trading planning as well as risk management in one integrated step. Therefore the optimization problem is more complex compared to the expost risk management. All the components shown in Fig. 17.1 together with the risk management need to be included in the problem so that the following objective function needs to be solved:

$$
\begin{aligned}
\max(U) = \min \Bigg(& A \cdot \left(\frac{1}{n_{Qu}} \cdot \sum_{a \forall Qu} y_{0,a} + \frac{1}{\alpha_a} \cdot \sum_{r \forall RA} \sum_{s \forall Sz_r} \left[y_{r,s,a}^{(1)} \cdot pr_s \right] \right)^2 \\
& + B \cdot \left(\frac{1}{n_{Qu}} \cdot \sum_{a \forall Qu} y_{0,a} + \frac{1}{\alpha_a} \cdot \sum_{r \forall RA} \sum_{s \forall Sz_r} \left[y_{r,s,a}^{(1)} \cdot pr_s \right] \right) \\
& - C \cdot \Bigg(\sum_{t=1}^{T} \sum_{s \forall Sz_t} \Bigg(\sum_{l \forall TP} CM_{t,s}^{TP_l} \left(P_{t,s}^{TP_l}, \left(P_{t,s}^{TP_l} \right)^2, CM_{t,s}^{st,TP_l} \right) + \\
& \qquad CM_{t,s}^{Sp} \left(P_{t,s}^{Sp} \right) + \\
& \qquad CM_{t,s}^{Fu} \left(P_{c,t,s}^{Fu}, \left(P_{c,t,s}^{Fu} \right)^2 \right) + \\
& \qquad CM_{t,s}^{Fo} \left(P_{c,t,s}^{Fo}, \left(P_{c,t,s}^{Fo} \right)^2 \right) + \\
& \qquad CM_{t,s}^{Op} \left(P_{c,t,s}^{Op}, \left(P_{c,t,s}^{Op} \right)^2, P_{ex,t,s}^{Op} \right) + \\
& \qquad CM_{t,s}^{In} \left(P_{c,t,s}^{In}, \left(P_{c,t,s}^{In} \right)^2, P_{t,s}^{In} \right) + \\
& \qquad CM_{t,s}^{De} \left(P_{c,t,s}^{De}, \left(P_{c,t,s}^{De} \right)^2, P_{t,s}^{De} \right) \Bigg) \cdot pr_s \Bigg) \Bigg)
\end{aligned}
\tag{17.7}
$$

In analogy to the objective function of the expost risk management (17.7) presents the dependency of the objective function from the decision variables. The equation structure for the market components is set up in analogy to (17.6). Additionally the thermal power plants TP need to be included in the objective function, resulting in a quadratic dependency from the energy output of the plants and a dependency on the starting decisions st, which are integer decisions. The contribution margin of the starting decisions consists of the difference of energy contribution during the starting process and the costs for starting the unit. This results in a nonlinear, nonquadratic objective function with integer decisions [2].

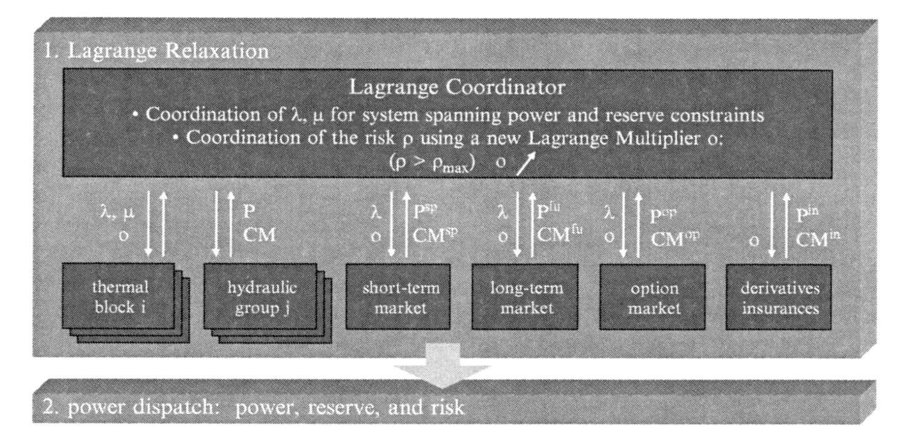

Fig. 17.6. Integrated risk management

The constraints of the problem are component related, system related, and for the *CVaR*, resulting in linear and nonlinear constraints which are system as well as time spanning [2]. Therefore it is not feasible to solve this optimization problem with the desired modeling accuracy in a closed formulation.

In this work a two step approach is used to solve the resulting problem (see Fig. 17.6). In the first stage the optimal generation schedule as well as the trading decisions are determined using a decomposition approach which divides the problem into subproblems on a system level. These subproblems are solved separately by means of the most appropriate algorithms. Due to this iterative Lagrange Relaxation used in stage one, the system constraints, i.e., the energy and reserve balance as well as the risk, may be violated. Therefore a second stage performs a power dispatch on the basis of the integer decisions determined in the first stage, optimizing the remaining continuous nonlinear problem [10].

Lagrange Relaxation

The Lagrange Relaxation optimizes the generation portfolio together with the presented markets. The overall problem is divided into its subproblems which are solved as independent problems using appropriate algorithms. In order to coordinate the system spanning constraints for electrical energy, reserve, and risk Lagrange multipliers, that can be interpreted as price incentives for electrical energy λ as well as reserve power μ, are passed on to the submodules. The Lagrange multipliers for risk o can be interpreted as incentives for costs avoidance in the scenarios responsible for the high risk.

The submodules optimize the specific problems for the system components and return the resulting generated energy as well as the reserve power to the Lagrange Coordinator. The Lagrange multipliers are updated depending on the convergence of each system spanning constraint and the subproblems are reoptimized. Considering the willingness to take risks in the objective function, the Lagrange multipliers are updated in every iteration of the Lagrange Relaxation to control the maximum allowed risk and the relationship between risk and contribution margin. Due to this iterative process, the system constraints converge until a sufficient level of accuracy or a until maximum number of iterations is reached.

The thermal power plants are optimized individually against the price and risk avoidance incentives. Due to the integer decisions of the thermal blocks, stochastic dynamic programming is applied to determine the optimal schedule [2]. As shown in Fig. 1, different fuel markets and the emission certificates market can be considered [12]. The prices for these markets as well as the outage behavior of the power plants are modeled with stochastic data.

A hydraulic network model is developed for each interconnected group of hydro power plants, so that the interconnections between the reservoirs as well as the distribution of the limited available water in the planning horizon are considered. Since hydro power plants have negligible variable costs, the optimization model determines the optimal schedule considering the price incentives for electrical energy and reserve. Successive Linear Programming is used to incorporate the nonlinearities in the model [11]. The natural inflow to the reservoirs is modeled as a stochastic process.

Electrical energy is traded at different markets which are modeled as separate components. The hourly spot market can be solved analytically since the decisions for the hourly products can be determined independently. However, the futures and forward markets consider the price behavior for the time period in which the products can be traded. Due to the time spanning dependencies, a Quadratic Programming approach is selected to determine the optimal decisions for trading future and forward contracts. With the consideration of the executed amount the options market can be implemented in analogy to the futures market using Quadratic Programming [2]. For the derivative and the insurance market Quadratic Programming is applied as well.

Compared to earlier works considering generation and trading planning tools [11, 13, 18] this optimization method needs to control the system spanning constraints for electrical energy and reserve as well as the resulting risk. The coordination of the resulting risk is done by new Lagrange multipliers for risks, which are incentives to avoid costs in the scenarios responsible for a high *CVaR*. Figure 17.7 illustrates the developed update strategy of these Lagrange multipliers and the coordination of the risk.

In the first iteration of the Lagrange Relaxation the Lagrange multipliers for risk are initialized with zero, which results in the negligence of the risk. This first result is a generation and trading planning disregarding risk management. Figure 17.7 shows the distribution function of the contribution margin in the different scenarios. The worst scenarios considered in the calculation of the *CVaR* and are marked in gray. The Lagrange multipliers of these scenarios are increased depending on the violation of the risk con-

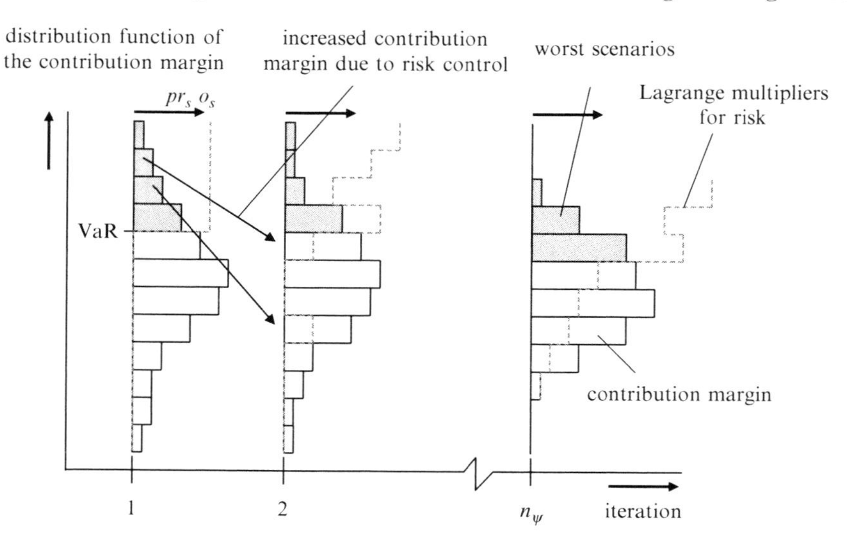

Fig. 17.7. Update of the Lagrange multipliers for risk

straint or depending on the relationship of risk and contribution margin [2]. The multipliers in the worst scenarios act as incentives for cost avoidance in the submodules in the following iterations, so that the costs in the relevant scenarios are increased in all submodules.

The results of the following iterations are based on this procedure. Due to the cost avoidance incentives the worst scenarios from iteration one are increased in contribution margin, as shown in Fig. 17.7. But other scenarios are decreased in the resulting contribution margin, so that – given the case that the update parameters are set up correctly – the distribution function of the contribution margin becomes narrower and the risk can be reduced.

Due to this iterative process the optimization algorithm tries to increase the contribution margin of the worst scenarios at the expense of the better ones. This is the general concept of risk control. The result of this procedure is the distribution function of contribution margin, which satisfies the risk management concepts. The resulting Lagrange multipliers for the different iterations are also shown in Fig. 17.7. At the end the value of these multipliers can be unequal to zero for not just the worst scenarios, but for (nearly) all because the update process of each iteration is based on the values of the iteration before. Therefore the multipliers represent incentives to avoid costs in the respective scenarios.

By this method the algorithm convergence so that the system spanning constraints are fulfilled up to a certain level. The Lagrange Relaxation stops when a certain level of convergence or a maximum number of iterations is reached.

Power Dispatch

Since it cannot be guaranteed that the system spanning constraints are fulfilled precisely in the iterative Lagrange Relaxation, the integer decisions derived in the second stage are used to set up a continuous problem which is solved in the power dispatch. Because of the nonlinear problem structure a Successive Quadratic Programming (SQP) approach is selected.

The results of the two stage optimization method are the optimal generation and trading decisions based on the maximal contribution margin considering risk management.

17.4 Exemplary Results

17.4.1 Data Models

The used data model is constructed to represent approximately 5% of the installed capacity of the german generation system. It consists of one hydraulic power plant and ten thermal generation units, all with typical technical parameters, using the respective primary energies in the proportion of the german system. A detailed description of the generation system and the data model can be found in [2].

As shown in Fig. 17.1 the system consists of different markets. The market for electrical energy is modeled as a short-term spot market, a future and an option market. Additionally a reserve contract is considered and the markets for fuel and weather derivatives allow hedging against price and quantity risks.

In the investigations the price risk of the spot market as well as the price risk for gas and hard coal are considered using a scenario tree with 50 scenarios constructed by the reduction of 1,000 multivariate scenarios [16].

17.4.2 Convergence

In this section the convergence of the expost risk management is investigated. Therefore the expost risk management with different utility functions is applied following a generation and trading planning. Figure 17.8 shows the expected contribution margin after the expost risk management depending on the number of iterations of the SQP-loop (see Sect. 17.3.1) and the parameters of the utility function (see Sect. 17.2.2). The parameters $A = 0$ and $C = 1$ of the utility function are kept constant, so that the parameter B represents the risk aversion of the user. $B = 0$ represents a risk-neutral user whereas the risk aversion increases with higher B.

In analogy to Fig. 17.8, Fig. 17.9 presented the corresponding $CVaR$ of the results in dependency of the number of iterations and the risk aversion.

The results of expected contribution margin and $CVaR$ are depicted in relative values based on one selected reference value. Looking at the results

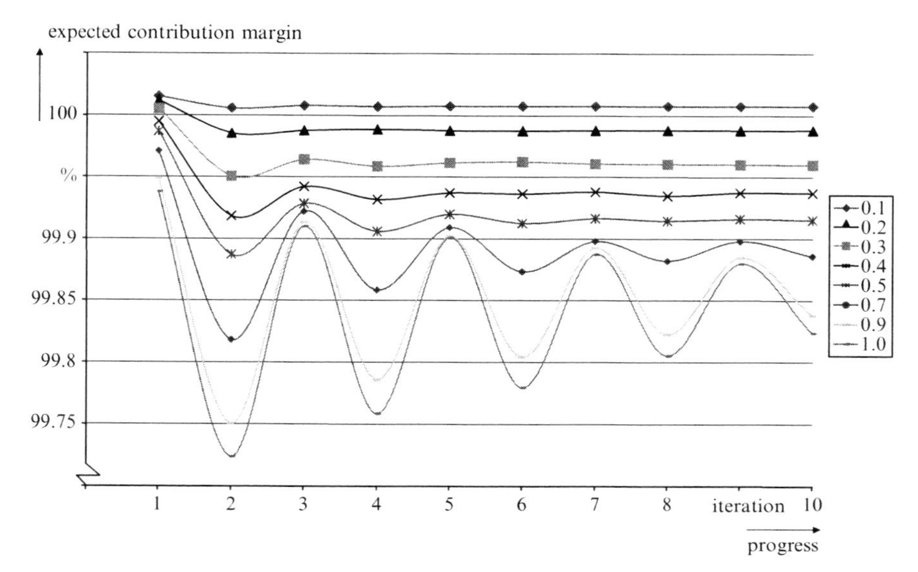

Fig. 17.8. Convergence of the expected contribution margin

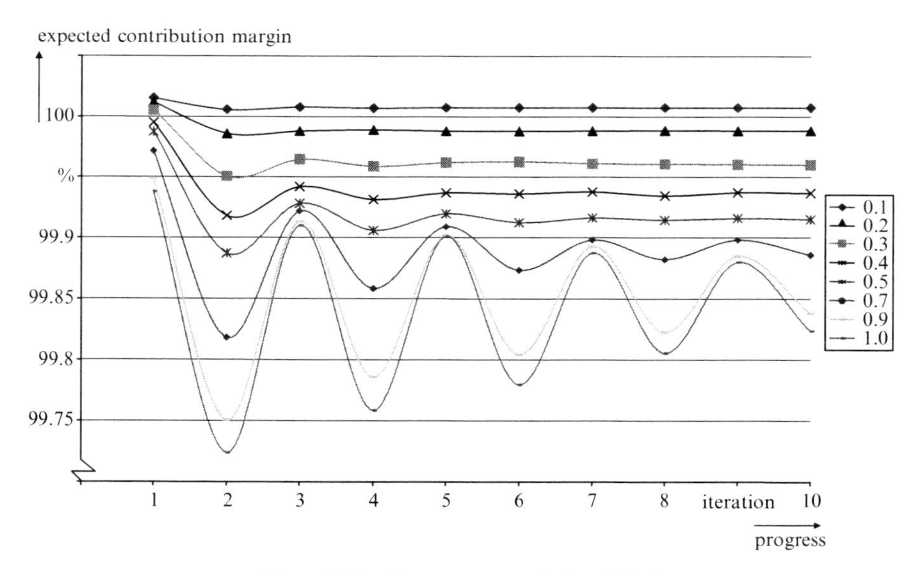

Fig. 17.9. Convergence of the *CVaR*

after one exemplary iteration, e.g., after the first iteration, it can be observed that the expected contribution margin and the *CVaR* decrease with higher risk aversion. This shows the tradeoff between expected contribution margin and risk.

Depending on the risk aversion the results of expected contribution margin and *CVaR* are converging with the increasing number of iterations. For a low risk aversion an adequate convergence can be reached directly after a few iterations, whereas the necessary number of iterations for convergence is increasing with higher risk aversion. After 10 iterations the results are fluctuating by less then 1‰ even for higher risk aversions. The fluctuations in the expected contribution margin are slightly higher compared to the fluctuations in risk. Concluding it can be stated that up to a risk aversion of $B = 0.5$ five iterations are sufficient for a good convergence whereas for a higher risk aversion a higher number of iterations is necessary.

17.4.3 Portfolio Function

The evaluation of generation assets under consideration of different risk preferences of the company demonstrates that, the expected contribution margin can be traded off with the exposed risk. The following investigations are carried out in three stages:

1. In the first stage the generation assets and the trading portfolio are optimized against the short term market for electrical energy.
2. The trading portfolio is extended to all markets for electrical energy within the second stage, i.e., the future and option market can be used as well as the spot market.
3. In the third stage the markets for derivates and insurances can also be used as hedging instruments. In this stage the company has the highest degree of freedom by trading at all markets.

The portfolio functions for this thermal dominated system are depicted in Fig. 17.10 for one exemplary year. For risk neutral operation the objective function consists of the expected contribution margin neglecting the deviation of the contribution margin marked by the solid point or square at the beginning of the portfolio function.

Depending on the stage the results show different effects. In the first stage the portfolio function shrinks to one single point meaning that only the redispatchment of the thermal generation asset cannot be used to control the risk. However in the second stage a risk reduction can be seen. The option and future market is used to substitute the spot market which results in a lower risk [2]. It can be observed that the risk can be lowered for investors with increasing risk aversion. However, the expected contribution margin reduces in combination with a lower risk, resulting in the respective curve. Initially, a reduction of the risk can be achieved for a relatively small price in expected contribution margin. For a stronger risk reduction the price in expected contribution margin is relatively high. The third stage even improves this effect by means of derivatives and insurances.

Comparing the expost risk management with the integrated approach it can be noticed that the integrated approach has an advantage for the purpose

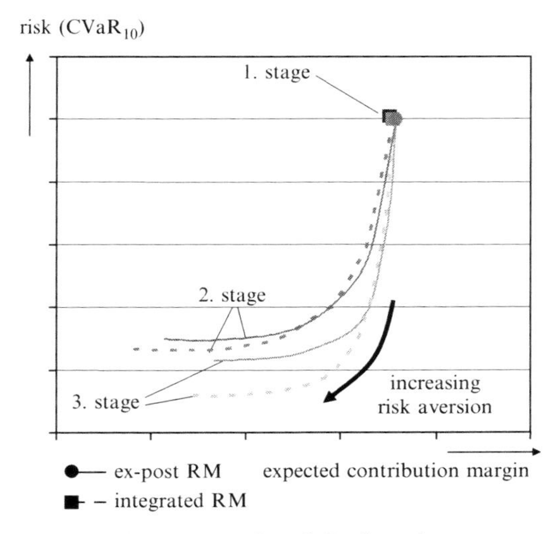

Fig. 17.10. Portfolio function

of risk control. This is due to the effect that the generation assets can be optimized together with the trading decisions and can therefore be adjusted to these decisions. This effect can be seen in both the second and the third stage.

These effects demonstrate, that the investor needs to select a commitment strategy with the help of the portfolio curve depending on his risk aversion using all possible hedging options. The integrated risk management methods is thereby the preferred method. The here presented method of generating a portfolio curve leaves the decision maker the freedom the choose his optimal portfolio depending on his risk aversion.

Taking a look at the size of the optimization problem and the practical requirements in needed memory (RAM) and processing time (CPU time), it is remarkable that the two risk management approaches have similar requirements. In the two-stage expost risk management approach the generation and trading planning uses the largest amount of CPU time so that the CPU time of the risk management can almost be neglected. For the integrated risk management approach the required CPU time and memory are virtually independent from the risk aversion with a slightly higher demand of resources compared to the generation and trading planning. In conclusion it can be stated that the integrated risk management approach uses about 10% more CPU time and neglectable more memory compared to the two-step approach.

The size of the optimization problem of the integrated risk management depends on many parameters. To approximate the problem size, the power dispatch which solves the remaining problem will be used to give an impression of the total problem dimension. The problem for the presented thermal dominated system consists of 3.2 Mio. variables and 500 Tsd. constraints with 13.2 Mio. nonzero entries. 2.2 Mio. variables are integer variables which are decided on in the Lagrange Relaxation, so that they are modeled continuous in the power dispatch or neglected if the value is zero. The total computation time of the integrated risk management is 13.5 h for a risk aversion of $B = 1$ using a Sun Fire E6900 Ultra Sparc IV with 1.2 GHz.

17.5 Conclusions

Increasing competition in the energy sector and grown uncertainties in the generation and trading planning led to higher risks for generation and trading companies. Furthermore negative examples of failed risk management (e.g., Enron) and laws demanding to include risk management methods in the business process, increased the sensibility for risk management in the energy sector. This results in a high demand for risk management methods in all planning phases including the generation and trading planning. Therefore risk management methods for generation and trading planning have been developed in this work.

The risk management methods include a limitation of the risk either by a so called risk constraints or by the introduction of utility functions to trade off expected contribution margin and risk. These two concepts are used to generate a portfolio function, so that the user (the company) can select his optimal portfolio on the basis of his willingness to take risk.

In the scope of this work two methods of risk management in the generation and trading planning were developed: An expost risk management that takes the power plant decisions from a previous generation and trading planning and optimizes the trading decisions by means of risk management. An integrated approach optimizes the power plant and trading decisions in one combined step, so that the power plant decisions are based on risk management concepts and the synergies between generation on the one hand and trading planning on the other hand can be realized.

The presented optimization problem depends on the risk management concept and is characterized in the case of the integrated method by nonlinearities, integer decisions, as well as time and system spanning constraints, so that Lagrange Relaxation is used to solve this problem. The expost risk management approach has – due to the adaption of the power plant decisions – a quadratic structure, so that Quadratic Programming is used.

Both risk management concepts demonstrate that the risk can be controlled, but a decrease of expected contribution margin has to be accepted for the risk reduction. These costs are relative small for an initial reduction in risk but increase due to saturation effects for higher reductions.

For the considered system dominated by thermal power plants a risk control by changing the schedule of the power plants is not possible. For the case that the plants are marketed against different markets for electrical energy the risk can be controlled depending on the risk preference of the user.

Including derivatives on primary energy increases the effectiveness of the risk management.

In the considered hydraulic system the redispatch of hydraulic power plants allows the reduction of the exposed risk. But the risk control by different markets for electrical energy is more effective compared to the redispatch. Due to the missing price spikes in future markets, it is – compared to thermal power plants – not so advantageous to market hydraulic power plants against future markets. However the market for weather derivatives offers a high potential to reduce the risk for hydraulic power plants.

Comparing the two risk management methods, the integrated risk management process requires a neglectable longer computation time compared to the two step approach. Contrariwise the results of the integrated method demonstrate a more effective risk reduction since the power plant decisions are made by means of risk management and the synergies between the power plant and the trading decisions can be achieved.

References

1. P. Albrecht and R. Maurer. *Investment- und Risikomanagement*. Schäffer-Poeschel Verlag, Stuttgart, 2002.
2. B. Blaesig. *Risikomanagement in der Stromerzeugungs- und Handelsplanung*, volume 113. Aachener Beiträge zur Energieversorgung, Klinkenberg, Aachen, 2007.
3. B. Blaesig and H.-J. Haubrich. Methods of Risk Management in the Generation and Trading Planning. In *IEEE PowerTech Proceedings*, St. Petersburg, 2005.
4. H. Brand and Chr. Weber. Stochastische Optimierung im liberalisierten Energiemarkt: Wechselwirkungen zwischen Kraftwerkseinsatz, Stromhandel und Vertragsbewirtschaftung. In *Optimierung in der Energieversorgung, VDI-Berichte 1627*, pages 173–183, Düsseldorf, 2001.
5. Deutscher Bundestag. Gesetz zur Kontrolle und Transparenz im Unternehmensbereich (KonTraG). Bundesgesetzblatt year 1998, Part I, No. 24, 1998.
6. C.-P. Cheng, C.-W. Liu, and C.-C. Liu. Unit commitment by Lagrangian relaxation and genetic algorithms. In *IEEE Transactions on Power Systems*, pages 707–714, May 2000.
7. A. Eichhorn and W. Römisch. Polyhedral risk measures in stochastic programming. *SIAM Journal on Optimization*, 2004.
8. European Energy Exchange AG. `www.eex.de`, 01.07.2007, 2007.
9. H. Garz, S. Günther, and C. Moriabadi. *Portfolio-Management*. Banka-kademie Verlag GmbH, 2002.
10. Th. Hartmann, B. Blaesig, G. Hinüber, and H.-J. Haubrich. Stochastic Optimization in Generation and Trading Planning. In *Waldmann K.-H., Stocker U. M.: Operations Research Proceedings 2006*. Springer, Berlin, 2007.
11. B. Krasenbrink. *Integriete Jahresplanung von Elektrizitätserzeugung und –handel*, volume 81. Aachener Beiträge zur Energieversorgung, Klinkenberg, Aachen, 2002.

12. H. Neus. *Integrierte Planung von Brennstoffbeschaffung und Energieeinsatz zur Stromerzeugung*, volume 95. Aachener Beiträge zur Energieversorgung, Klinkenberg, Aachen, 2003.
13. H. Neus, H. K. Schmöller, B. Pribićević, and H. P. Flicke. *Integrated Optimisation of Power Generation and Trading – Requirements and Practical Experience*, volume 87. Annual Report 2002 of the Institute of Power Systems and Power Economics of RWTH Aachen University, Aachener Beiträge zur Energieversorgung, Klinkenberg, Aachen, 2002.
14. R. T. Rockafellar and S. Uryasev. Conditional Value-at-Risk for General Loss Distributions. *Journal of Banking and Finance*, 26:1443–1471, 2002.
15. W. Römisch. Optimierungsmethoden für die Energiewirtschaft: Stand und Entwicklungstendenzen. In *Optimierung in der Energieversorgung, VDI-Berichte 1627*, pages 23–36, Düsseldorf, 2001.
16. H. K. Schmöller, Th. Hartmann, I. Kruck, and H.-J. Haubrich. Modeling Power Price Uncertainty for Midterm Generation Planning. In *IEEE PowerTech Proceedings*, Bologna, 2003.
17. S. Sen, L. Yu, and T. Genc. A Stochastic Programming Approach to Power Portfolio Optimization. *Operations Research*, pages 55–72, 2006.
18. B. Stern. *Kraftwerkseinsatz und Stromhandel unter Berücksichtigung von Planungsunsicherheiten*, volume 78. Aachener Beiträge zur Energieversorgung, Aachen, 2001.
19. The European Parliament and the Council of the European Union. Directive 2003/55/EC of 26 June 2003 concerning common rules for the internal market in natural gas and repealing Directive 98/30/EC. Official Journal of the European Union, No. L 176/57, 15.07.2003.

18

Optimization Methods Application to Optimal Power Flow in Electric Power Systems

Virginijus Radziukynas and Ingrida Radziukyniene

Summary. Optimal power flow is an optimizing tool for power system operation analysis, scheduling and energy management. Use of the optimal power flow is becoming more important because of its capabilities to deal with various situations. This problem involves the optimization of an objective functions that can take various forms while satisfying a set of operational and physical constraints. The OPF formulation is presented and various objectives and constraints are discussed. This paper is mainly focussed on review of the stochastic optimization methods which have been used in literature to solve the optimal power flow problem. Three real applications are presented as well.

18.1 Introduction

As power industry is moving to a competitive market, its operation is strongly influenced. In the deregulated environment, the security and economical issues of power systems are coordinated tightly than before. Thus, the need for fast and robust optimization tools that consider both security and economy is more demanding than before to support the system operation and control [34]. Optimization methods have been widely used in power system operation, analysis and planning. One of the most significant applications is optimal power flow (OPF). Since its introduction by Carpentier in the 1962 as a network constrained economic dispatch problem [12], OPF has been studied and widely used in power system operation and planning, due to its capability of integrating the economic and security aspects of a power system into one mathematical formulation.

The OPF issue is one of the most important problems faced by dispatching engineers to handle large-scale power systems in an effective and efficient manner. It is a particular mathematical approach of the global power system optimization problem that aims at determining the least control movements to keep power system at the most desired state [23]. The OPF provides a useful support to the operator to overcome many difficulties in the planning, operation and control of power systems. Thus, it represents a flexible and powerful tool, which is widely used in many applications, such as constrained economic dispatch and voltage control problems [32]. However, the complexity of optimal power flow increases dramatically with large-scale networks, which often discourages the utilization of this powerful tool in many applications [23]. The OPF problem aims to achieve an optimal solution of a specific power system objective function, such as fuel cost, by adjusting the power system control variables, while satisfying a set of operational and physical constraints [24]. The control variables include the generator real powers, the generator bus voltages, and the tap ratios of transformer and the reactive power generations of reactive power (VAR) sources, such as capacitor banks, static VAR compensators (SVC), and static synchronous compensators (STATCOM). State variables are slack bus power, load bus voltages, generator reactive power outputs, and network power flows [12]. The constraints include inequality ones which are the limits of control variables and state variables; and equality ones which are the power flow equations. In its most general formulation, OPF is a nonlinear, non-convex, large-scale, static optimization problem, with both continuous and discrete control variables [6].

In the past few decades, many stochastic optimization methods have been developed and their applications to global optimization problems become attractive because they have better global search abilities over conventional optimization algorithms [12]. This review will address the optimal power flow problem and stochastic optimization methods used to solve it.

18.2 Overview of Optimal Power Flow

OPF can help in solving many problems. There are some scenarios of OPF contribution to the analysis of power systems [23]:

- In the standard description of the OPF problem, if an empty set is specified for the controls, the algorithm reduces directly to a typical power flow problem. The procedures in this case depend on the bus mismatch equations and provide the same state solution like the classic power flow, including bus voltages and branch flows.
- OPF may be associated with the constrained economic dispatch to define the optimal allocation of loads among the generators by specifying the generation cost characteristics, the network model and the load profile [50].
- OPF can also be used to minimize the total real power loss through reactive power dispatch [32]. In this case, only reactive controls such as transformer tap positions, shunt capacitors and reactors, and excitation systems are used to minimize the total losses in the entire network, or in a subset of the network.
- OPF can be used to define feasible solutions or indicates if one exists using the so-called minimum of control movements strategy [47]. According to this strategy, the objective of the optimization process is to minimize the cost function based on control deviations from the base case [23].

Researchers proposed different mathematical formulations of the OPF problem that can be classified into [3]:

- Linear problem in which objectives and constraints are given in linear forms with continuous control variables.
- Nonlinear problem where either objectives or constraints or both combined are linear with continuous control variables.
- Mixed-integer linear and linear problems when control variables are both discrete and continuous.

The mathematical formulation of the OPF problem is a well known optimization problem. In general this can be formulated as follows [46]:

$$\begin{aligned} \text{Minimize} \quad & f(u,x) \\ \text{Subject to} \quad & g(u,x) = 0 \\ & h(u,x) \leq 0 \end{aligned} \tag{18.1}$$

where u is the set of controllable variables in the system; x is the set of dependent variables called state variables; objective function $f(u,x)$ is a scalar function that represents the power system's operation optimization goal, which could be the total generation cost, total network loss, corridor transfer power, total cost of compensation and so on [24]; $g(u,x)$ is a vector function with conventional power flow equations and other special equality constraints such as the limit of the number of potential VAR compensators; $h(u,x)$ is a vector of inequality constraints that are physical and operational limits of the power system.

The control variables may include generator active power output, regulated bus voltage magnitude, variable transformer tap settings, phase shifters, switched shunt reactive devices, and load to shed in special conditions. The state variables may include voltage magnitudes at load buses, voltage phase angle at every bus, and line flows.

The conventional OPF constraints may include the normal state (base case) power-flow limits and the contingency state power flow limits, but there are proposals to include the voltage stability limits, under both normal state and contingency state, due to the increased pressure of voltage stability and stressed transmission systems. These different constraints are the key of the classification of various optimization models, identified as OPF model, security-constrained OPF (SCOPF) model, and SCOPF with voltage stability constraints (SCOPF-VS), the present state-of the-art in reactive power planning (RPP) [46]. In some works, the consideration of contingency analysis and voltage stability may be included in the objective functions. The relationship in terms of feasible region among the three formulations is shown in Fig. 18.1. However one of the objective models may be combined with one of the constraint models to formulate the problem.

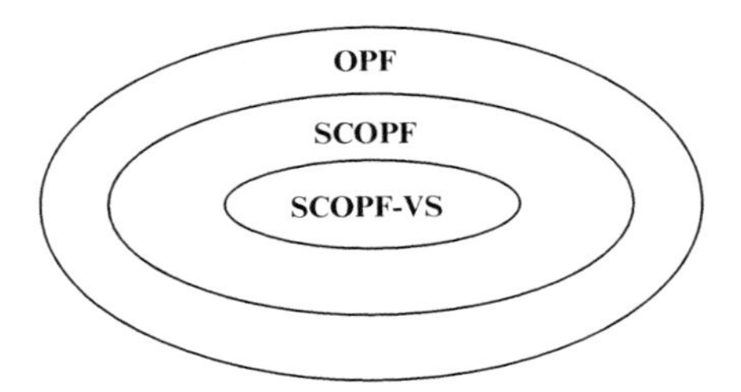

Fig. 18.1. Relationship of different OPF models

18.2.1 Objectives of Optimal Power Flow

The objective function may be cost-based, which means to minimize the possible cost of new reactive power supplies, the cost of real power losses or the fuel cost. Other possible objectives may be to minimize the deviation from a given schedule of a control variable (such as voltage) or to maximize voltage stability margin. The detailed discussions are presented as follows.

Minimize VAR Cost

Generally, there are two VAR source cost models for minimization. The first formulation is to model VAR source costs with $C_1 \cdot Q_c$ that represents a linear function with no fixed cost. This model considers only the variable cost relevant to the rating of the newly installed VAR source Q_c and ignores the fixed installation cost in \$/(MVar·hour) [46].

The second formulation with the format $(C_0 + C_1 \cdot Q_c) \cdot C_0$ considers the fixed cost, C_0 (\$/hour), which is the lifetime fixed cost prorated to per hour, in addition to the incremental/variable cost, C_1 (\$/hour). This is a more realistic model of VAR cost, but this would complicate the problem from NLP to mixed-integer NLP (MINLP), because there is a binary variable x indicating whether the VAR source will be actually installed or not. This is a slight difference in the cost model, however, it leads to dramatic difference in the optimization model and an adequate method is needed for it.

Minimize VAR Cost and Real Power Losses

This objective may be divided into two groups: to minimize $C_1(Q_c)+C_2(P_{loss})$ and to minimize $(C_0 + C_1 \cdot Q_c) \cdot x + C_2(P_{loss})$. Here $C_2(P_{loss})$ expresses the cost of real power loss. In [14] it is considered the real power losses consumed not only in the base case but in all contingency cases. So the objective can be written as follows:

$$\min F = C_1(Q_c) + \sum_{k=0}^{N_C} C_2(P_{loss})_k \tag{18.2}$$

where $k(=0, 1, \ldots, L, \ldots, N_c)$ represents the kth operating case. Here, considered are the base case ($k = 0$), the contingency cases under preventive mode ($k = 1, \ldots, L$), and the contingency cases under corrective mode ($k = L + 1, \ldots, Nc$).

Minimize VAR Cost and Generator Fuel Cost

This objective comprises of the sum of the costs of the individual generating units

$$C_T = \sum_{i=1}^{n} f_i(P_{gi}) \tag{18.3}$$

where $f_i(P_{gi} = a_{0i} + a_{1i}P_{gi} + a_{2i}P_{gi}^2)$ is the common generator cost-versus-MW curves approximately modelled as a quadratic function, and a_{0i}, a_{1i}, a_{2i} are cost coefficients [46].

Minimize Deviation from a Specified Point

This objective differs from the previous objective functions, because it is not cost-oriented but it is usually defined as the weighted sum of the deviations of the control variables, such as bus voltages, from their given target values. In minimization of voltage deviation, i.e. $\sum_i (V_{imax} - V_i)$, the subscript i represents different buses for voltage regulation.

Voltage Stability Related Objectives

As power systems are more stressed, voltage becomes a poor indicator of system collapse conditions and involvement of voltage stability becomes more and more important. The voltage stability is usually represented by a P–V (or S–V) curve (Fig. 18.2). The nose point of the P–V curve is called the point of collapse (PoC), where the voltage drops rapidly with an increase of load. Hence, power-flow solution fails to converge beyond this limit, which indicates voltage instability caused due to the lack of reactive power.

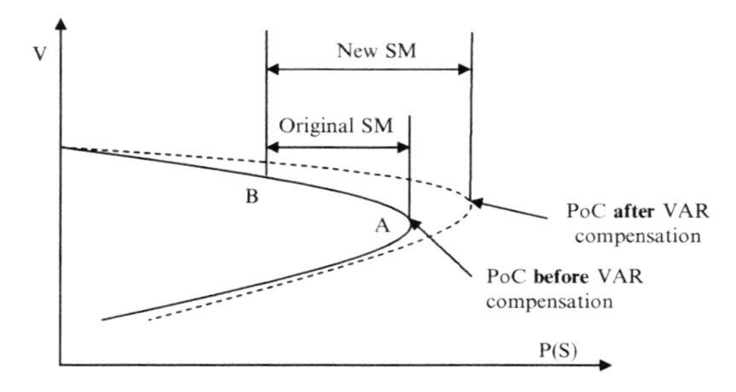

Fig. 18.2. Voltage stability curve

The objective can be to increase the static voltage stability margin (SM) expressed as follows:

$$SM = \frac{\sum_i S_i^{critical} - \sum_i S_i^{normal}}{\sum_i S_i^{critical}} \tag{18.4}$$

where S_i^{normal} and $S_i^{critical}$ are the MVA loads of load bus at normal operating state **B** and the voltage collapse critical state (PoC) **A** as shown in Fig. 18.2.

Multi-Objective (MO)

Chen and Liu [8] proposed the MO model of three objective functions with directly incorporated voltage stability margins (SM). In (18.5) the first objective function is minimization of operation cost caused by real power losses and VAR source investment cost; the second is maximization of the voltage stability margin (SM); the third is minimization of the voltage magnitude deviation, where $\Phi(x) = 0$ if $x < 0$; and $\Phi(x) = x$ otherwise; V_i^{ideal} is the specifically desired voltage at load bus i and is usually set to be 1 p.u.; and Δv_i is the tolerance of maximum deviation in the voltage.

$$\begin{aligned} \min F_1 &= (C_0 + C_1 Q_c) \cdot x + C_2(P_{loss}) \\ \max F_2 &= 1 - SM \\ \min F_3 &= \sum_i \frac{\Phi(|V_i - V_i^{ideal}| - \Delta v_i)}{V_i} \end{aligned} \tag{18.5}$$

Reference [27] presents another MO problem with the three objectives: $\min F_1 = C_1 Q_c$, $\min F_2 = C_2(P_{loss})$, and $\max F_3 =$ the maximum loadability associated with the critical state.

18.2.2 Constraints in OPF

Conventional OPF Constraints

OPF constraints may be classified as the power flow equality constraints, physical limits of the control variables, physical limits of the state variables, and other limits such as power factor limits [24].

- Power flow constraints
  $P_{gi} - P_{li} - P(V, \theta) = 0$ (active power balance)
  $Q_{gi} + Q_{ci} - Q_{li} - Q(V, \theta) = 0$ (reactive power balance)
- Control variables limits
  $P_{gi}^{min} \le P_{gi} \le P_{gi}^{max}$ (active power generation limits)
  $V_{gi}^{min} \le V_{gi} \le V_{gi}^{max}$ (PV bus voltage limits)
  $V_l^{min} \le V_l \le V_l^{max}$ (transformer tap change limits)
  $Q_{ci}^{min} \le Q_{ci} \le Q_{ci}^{max}$ (VAR source size limits)

- State variables limits
  $Q_{gi}^{min} \le Q_{gi} \le Q_{gi}^{max}$ (reactive power generation limits)
  $V_i^{min} \le V_i \le V_i^{max}$ (PQ bus voltage limits)
  $|LF_l| \le LF_l^{max}$ (line flow limit)
- Other limits

where P_{gi} - generator active power output; P_{li} - load active power; Q_{gi} - generator reactive power output; Q_{ci} - VAR source installed at bus; Q_{li} - load reactive power; V_{gi} - PV bus voltage; T_l - transmission tap change; V_i - bus voltage; LF_l - transmission line flow.

Additional Constraints

In the conventional OPF formulation, transient stability constraints are usually excluded, however, the systems operated at the point suggested by the OPF, may fail to maintain transient stability when subject to a credible contingency. As systems are operated closer to their limits, it is critical that the system is modelled appropriately and that control actions take into account stability margins [28]. Mathematically, OPF with transient stability is an extended OPF with additional equality and inequality constraints [13]. The additional equality constraints consist of the dynamic equation of the system. For each contingency, e.g. the mth contingency, the dynamics of the disturbed system can generally be divided into three stages, i.e. pre-fault stage, fault stage, post fault stage [26], and they can be described as:

$$F(x_0, y_0, u) = 0 \quad G(x_0, y_0, u) = 0 \quad t = 0 \tag{18.6a}$$

$$\dot{x}^{(m)}(t) = F_1^{(m)}(x^{(m)}(t), y^{(m)}, u) \quad G_1^{(m)}(x^{(m)}(t), y^{(m)}(t), u) = 0 \quad t \in (0, t_{cl}^{(m)}] \tag{18.6b}$$

$$\dot{x}^{(m)}(t) = F_2^{(m)}(x^{(m)}(t), y^{(m)}, u) \quad G_2^{(m)}(x^{(m)}(t), y^{(m)}(t), u) = 0 \quad t \in (t_{cl}^{(m)}, T] \tag{18.6c}$$

where $x(t)$ and $y(t)$ are the state variables and the algebraic variables of power system. x_0 and y_0 are the initial value of x and y respectively. t_{cl} is the fault clear time, T is the study period. Superscript 'm' denotes conditions associated to the m-th contingency.

The additional inequality constraints define the limits that the system behaviour should respect during the transient. In order to maintain the rotor-angle stability of the system and to ensure its acceptable dynamic behaviour, for a system with ng machines, nb buses and nl transmission lines, the additional inequality constraints can be in three categories:

- Rotor-angle stability constraints:

$$d(t) = \sum_{i=1}^{ng} P_{COI}^i(\theta(t)) \times (\theta^i(t) - \theta_{sep}^i) \le 0, \; i = 1, \ldots, ng \; t \in \left(t_{cl}^{(m)}, T\right] \tag{18.7}$$

where p_{COI}^{i} is the accelerating power, θ^{i} the rotor angle with respect to the centre of inertia (COI) reference frame and subscript 'sep' denotes the stable equilibrium point of the post-fault system.

- Limit on transient voltage for each bus:

$$V_j^{min} \leq V_j^{(m)} \leq V_j^{max} \; j = 1, \ldots, nb \; t \in \left(t_{cl}^{(m)}, T\right] \tag{18.8}$$

where V_j^{min} and V_j^{max} are respectively low and upper limits on transient voltage of the jth bus.

- Limit on the power oscillations for each transmission line:

$$S_l^{(m)} \leq S_l^{max} \; l = 1, \ldots, nl \; t \in \left(t_{cl}^{(m)}, T\right] \tag{18.9}$$

where S_l^{max} is the maximum limit on apparent power oscillations of the lth transmission line.

18.3 Stochastic Methods for OPF

This section will discuss the stochastic algorithms to solve the OPF problem [27]. The solution techniques for OPF have been developed over many years. Numerous numerical techniques had been used for this problem, such as an interior point method, a successive quadratic programming method, Lagrange Newton method, and others. However, these methods are designed for purely continuous-variable OPF. In reality, the power systems consist of several discrete control variables such as the switching shunt capacitor banks and transformer taps [21].

In the last decade, many new stochastic search methods have been employed to overcome the drawbacks of conventional techniques, such as genetic algorithms [6, 33–44] differential evolution [2, 4], chaos optimization algorithm [29, 30, 50] and ant colony [35, 38]. The results were promising and encouraging for further research in this direction. Several objectives can be defined in OPF problem. Recently, the evolutionary computation techniques have found many applications in power systems [1, 37], especially in the economic operation area. So for several different reasons, a reliable global optimization approach to reactive power dispatch problem would be of considerable value to both secure and economical operation of power systems. A number of approaches for solving optimal reactive power dispatch and voltage control problem of power systems has been developed, based on the recently introduced particle swarm optimization (PSO) algorithms [36–49].

18.3.1 Genetic Algorithm (GA)

This section engages into the concept of genetic algorithms that reflects the nature of chromosomes in genetic engineering. Genetic Algorithms (GAs) are a class of stochastic search algorithms that start with the generation of an initial population or set of random solutions for the problem at hand. Each individual solution in the population called a chromosome or string represents a feasible solution. The objective function is then evaluated for these individuals. If the best string (or strings) satisfies the termination criteria, the process terminates, assuming that this best string is the solution of the problem. If the termination criteria are not met, the creation of new generation starts, pairs, or individuals are selected randomly and subjected to crossover and mutation operations. The resulting individuals are selected according to their fitness for the production of the new offspring. Genetic algorithms combine the elements of directed and stochastic search while exploiting and exploring the search space [18].

The advantages of GA over other traditional optimization techniques can be summarized as follows:

- GA searches from a population of points, not a single point. The population can move over hills and across valleys. GA can therefore discover a globally optimal point, because the computation for each individual in the population is independent of others. GA has inherent parallel computation ability.
- GA uses payoff (fitness or objective functions) information directly for the search direction, not derivatives or other auxiliary knowledge. GA therefore can deal with non-smooth, non-continuous and non-differentiable functions that are the real-life optimization problems. OPF in FACTS is one of such problems. This property also relieves GA of the approximate assumptions for a lot of practical optimization problems, which are quite often required in traditional optimization methods.
- GA uses probabilistic transition rules to select generations, not deterministic rules. They can search a complicated and uncertain area to find the global optimum. GA is more flexible and robust than the conventional methods [19].

The first attempt of the application of genetic algorithms in power systems is in the load flow problem [49]. It has been found that the simple genetic algorithm (SGA) quickly finds the normal load flow solution for small-size networks by specifying an additional term in the objective function. The evolutionary programming has also been applied to the problem of reactive power dispatch [40]. A number of approaches to improving convergence and global performance of GAs have been investigated [49].

Authors [19] proposed a new GA approach to solve the optimal power flow control problem with FACTS, where UPFC is used as power flow controllers. UPFC can provide the necessary functional flexibility for optimal power flow control. This approach allows the combined application of phase angle control with controlled series and shunt reactive compensation. The total generation fuel cost is used as the objective function and the operation and security limits are considered. In most of GAs used in power system reactive power optimization, either all variables are regarded as continuous variables, or all variables are regarded as discrete variables. But in reality, the generator voltage magnitude is continuous, while reactive power installation and the transformer taps are discrete. The different characteristics of different variables are not considered in those methods. In [44] an integer/float mixed coding genetic algorithm is proposed. In [47] the application of Improved Genetic Algorithms (IGA) for optimal reactive power planning in loss minimisation scheme is presented. In this study, IGA engine was developed to implement the optimization of reactive power planning. The selection and steady state elitism combined with the conventional anchor spin techniques are incorporated into the traditional Genetic Algorithms (GA) for the development of the IGA. In each probing, identical initial population is supplied to the mechanism of IGA and traditional GA in order to have consistency during the initial population. Comparative studies on the results obtained from the IGA with respect to the traditional GA, indicating that IGA outperformed the traditional GA in terms of accuracy and number of iteration. The authors [6] for OPF problem applied the Enhanced GA (EGA), the simple GA with an added set of advanced and problem-specific genetic operators in order to increase its convergence speed and improve the quality of solutions [5]. The advanced features included in EGA implementation are Fitness Scaling (a linear transformation), Elitism, multi-parameter Hill-climbing and Elite Self-fertilization. The problem-specific operators applied to all chromosomes are the Gene Swap Operator, the Gene Cross-Swap Operator, the Gene Copy Operator, the Gene Inverse Operator and the Gene Max–Min Operator. To solve the OPF with discrete control variables in a more exact manner, Bakirtzis et al. [5] proposed an enhanced genetic algorithm, which needs only the power flow solutions for fitness evaluation, however, sacrificing the hard restriction on branch flow limits.

18.3.2 Differential Evolution (DE)

Differential evolution is a stochastic direct search optimization method. It was initially presented by Storn and Price in 1995 [17] as heuristic optimization method which can be used to minimize nonlinear and non-differentiable continuous space functions with real-valued parameters. This has been extended to handle mixed integer discrete continuous optimization problems [17]. The most important characteristics of DE is that it uses the differences of randomly sampled pairs of object vectors to guide the mutation operation instead of using the probability distribution function as other evolutionary algorithms (EAs). The main advantages of differential evolution are [4]:

- Simple structure, ease of use and robustness.
- Operating on floating point format with high precision.
- Effective for integer, discrete and mixed parameter optimization.

- Handling non-differentiable, noisy and/or time dependent objective functions.
- Effective for nonlinear constraint optimization problems with penalty functions, etc.

Like evolutionary algorithm (EA) family, DE also depends on initial random population generation, which is then improved using selection, mutation, and crossover repeated through generations until the convergence criterion is met. The main difference in constructing better solutions is that genetic algorithms rely on crossover while DE relies on mutation operation. This main operation is based on the differences of randomly sampled pairs of solutions in the population.

An optimization task consisting of D parameters can be represented by a D-dimensional real value vector. In DE, a population of NP solution vectors is randomly created at the start. Each of the NP vectors undergoes mutation, recombination and selection. A mutant vector for parameter vector xi,G is produced by selecting three vectors with different indices according (18.10).

$$v_{i,G+1} = x_{r1,G} + F(x_{r2,G} - x_{r3,G}) \tag{18.10}$$

where G is generation number, F is the scaling factor for mutation. It controls the speed and robustness of the search; a lower value increases the rate of convergence but also the risk of being stuck at the local optimum [4]. Recombination incorporates successful solutions from the previous generation. The trial vector $u_{i,G+1}$ is developed from the elements of the target vector, $x_{i,G}$, and the elements of the mutant vector, $v_{i,G+1}$ according (18.11).

$$u_{j,i,G+1} = \begin{cases} v_{j,i,G+1} & \text{if } rand_{j,i \leq CR} \text{ or } j = I_{rand} \\ x_{j,i,G} & \text{if } rand_{j,i > CR} \text{ and } j \neq I_{rand} \end{cases} \tag{18.11}$$

where $rand_{j,i}$ $U[0,1]$, I_{rand} is a randomly chosen index from $[1, 2, \ldots, D]$, $CR(0 \leq CR \leq 1.0)$ is a crossover constant.

The target vector $x_{i,G}$ is compared with the trial vector $v_{i,G+1}$ and the one with the lowest function value is admitted to the next generation (18.12).

$$x_{i,G+1} = \begin{cases} u_{i,G+1} & \text{if } f(u_{i,G+1} \leq f(x_{i,G}) \\ x_{i,G} & \text{otherwise} \end{cases} \tag{18.12}$$

In [4] proposed DE tool for reactive power and voltage control has been developed and tested on the Nigerian transmission grid modelled on the power world simulator in detail. This provides a platform to preset a multitude of scenarios under operational realism. In [2], multi-agent system and DE are integrated to form a multi-agent-based DE approach (MADE) for solving reactive power optimization in power markets. In MADE, an agent represents not only a candidate solution to the optimization problem but also an agent to DE. Firstly, a lattice-like environment is constructed, with each agent fixed on a lattice-point. In order to obtain the optimal solution, each agent competes and cooperates with its neighbours, and it also uses its own knowledge by self-learning. Making use of the evolution mechanism of DE, it can speed up the transfer of information among agents. Making use of these agent–agent interactions and evolution mechanism of DE, the proposed method can find high-quality solutions reliably, with better convergence characteristics and in a reasonably good computation time.

18.3.3 Chaos Optimization Algorithm

The chaos optimization algorithm (COA) is a stochastic search algorithm that differs from any of the existing evolutionary algorithms. Chaos, apparently disordered behaviour that is nonetheless deterministic, is a universal phenomenon that occurs in many systems in all areas of science [30]. Although it appears to be stochastic, it contains exquisite inner structure [50]. A chaotic movement can go through every state in a certain area according to its own regularity, and every state is obtained only once. The ergodicity, regularity and intrinsic stochastic property of chaos make chaotic optimization to obtain the global optimal solution. The COA can more easily escape from local minima than can other stochastic optimization algorithms [29, 30].

In [50] authors introduced an improved chaotic optimization to solve economic dispatch and OPF problem. It utilized chaotic variable to search the global optimal solution. In order to improve the precision of solution, the search space of the second carrier wave was gradually reduced. In addition, the descending information of the object was used to accelerate the convergence and one parameter was used to avoid search at on direction. The numerical results showed that the algorithm is high efficient and effective in meeting precision and constraints. In [29, 30] authors proposed a hybrid optimization algorithm, which combines COA and a linear interior point method, to deal with the OPF problem. The hybrid algorithm is structured with two stages. The COA takes the place of the first stage of the search, providing the potential for non-convex OPF problem, while the linear primal–dual interior point method is employed in the second stage. The OPF problem was solved by the interior point method as a sequence of linearized sub-problems in the neighbourhood of the global minima. The algorithm is capable of determining the global optimum solution to the OPF problem, and is not sensitive to starting points. To show the feasibility and potential of the proposed algorithm over the COA and SLP alone, IEEE-14, −30 and −57 bus test systems with multimodal objective functions was used in [29]. The results of the studies of the hybrid proposed algorithm were compared with those obtained from the COA and SLP alone. The results confirm the superiority of the algorithm.

18.3.4 Ant Colony

The Ant Colony Optimization (ACO) algorithm was introduced by Dorigo in [35]. It is a probabilistic technique for solving computational problems, which can be reduced to finding good paths through graphs. It mimics the behaviour of real ants establishing shortest route paths from their colony to feeding sources and back. In the algorithm, artificial ant colonies are in use, which have some memory, to find the shortest path via communicating information and cooperating with each other among individuals.

The optimum paths followed by ants are determined by their movements in a discrete time domain. While moving, ants lay some pheromone along their paths. The ants' decision to move from the present state r to the next state s is based on two measures. These are the length of the path which connects the present state to the next one, and the desirability measure (pheromone level). Each agent generates a completed path by choosing the next states to move to according to a probabilistic state transition rule (18.13). This rule reflects the preference of agents to move on shorter paths that connect the current state to the next state.

$$p_{rs}(k) = \begin{cases} \frac{\tau_{rs}^{\alpha}(t)\eta_{rs}}{\sum \tau_{rl}^{\alpha}(t)\eta_{rl}} & s, l \notin tabu_{(k)} \\ 0 & \text{otherwise} \end{cases} \tag{18.13}$$

where $\tau_{rs}(t)$ is the pheromone on the path rs at time t, and set an initial value $\tau_{rs}(0) = C$ (C is a small positive constant) and η_{rs} is the inverse of the length of path rs. The physical mean of $\tau_{rs}(t)\eta_{rs}$ is the pheromone of per unit of length, α is the weight factor. $tabu_k (k = 1, 2, \ldots, m)$ is a list of states which ants k positioned on state r cannot visit. The more pheromone a path has, the larger the transition probability of the path is, and more ants will choose the path. Once all agents have reached the final state and have identified the best-tour-so-far based on the value of the objective function, they update the pheromone level on the paths that belong to the best tour by applying a global pheromone updating rule (18.14).

$$\tau_{rs}(t + n) = \rho \cdot \tau_{rs}(t) + \sum_{k=1}^{m} \Delta\tau_{rs}(k) \tag{18.14}$$

where $\Delta\tau_{rs}(k) = \begin{cases} \frac{Q}{L_k} & \text{if } k\text{th ant uses edge}(r, s) \\ 0 & \text{otherwise} \end{cases}$, $0 < \rho < 1$ is a coefficient which represents the residual pheromone of trail in the process ants search their closed tours; Q is a constant; L_k is the length of the tour performed by the kth ant; $\Delta\tau_{rs}(k)$ is the quantity per unit of length of trail substance laid on edge (r, s) by the kth ant in the process that ants search their closed tours, $\Delta\tau_{rs}(k) = 0$ at time 0.

Currently, most works have been done in the direction of applying ACO to the combinatorial optimization problems [38]. ACO algorithms have an advantage over simulated annealing and GA approaches when the graph may change dynamically, since the ant colony algorithm can be run continuously and adapt to changes in real time [35]. The ACS algorithm proposed in [38] formulates the constrained load flow problem (CLF) as a combinatorial optimization problem. As an example, the settings of control variables (tap-settings, VAR compensation blocks, etc.) are combined in order to achieve optimum voltage values at the nodes of a power system. The graph that describes the settings of control variables of the CLF problem is mapped on the ant system (AS)-graph, which is the space that the artificial ants will walk. In this paper, for computational simplicity, the transition function considers only the trail intensity for the transition probability [31], i.e. the trail that more ants choose will have more probability to be selected.

18.3.5 Particle Swarm Optimization

Recently, Particle Swarm Optimization (PSO), a population based stochastic optimization technique developed by Kennedy and Eberhart in 1995 [24, 25] as an alternative to GA, has received much attention regarding its potential as a global optimization technique. In PSO, each potential solution is assigned a randomized velocity, and the potential solutions, called particles, fly through the problem space by following the current best particles. Unlike other evolutionary algorithms, PSO is capable of evolving toward global optimum with a random velocity by its memory mechanism and has better global search performance with faster convergence. PSO has been successfully applied to multiobjective optimization, constraints optimization, artificial neural network training, parameter optimization, and feature selection [41].

Moreover, PSO has some advantages over other similar optimization techniques such as Genetic Algorithm (GA), namely:

- PSO is attractive from an implementation viewpoint and there are fewer parameters to adjust [24, 35].
- In PSO, every particle remembers its own previous best value as well as the neighbourhood best; therefore, it has a more effective memory capability than the GA.
- In PSO, every particle remembers its own previous best value as well as the neighbourhood best; therefore, it has a more effective memory capability than the GA.
- PSO is more efficient in maintaining the diversity of the swarm (more similar to the ideal social interaction in a community), since all the particles use the information related to the most successful particle in order to improve themselves, whereas in GA, the worse solutions are discarded and only the good ones are saved; therefore, in GA the population evolves around a subset of the best individuals [35].
- PSO has comparable or even superior search performance for some hard optimization problems with faster and stable convergence rates [24].
- Unlike mathematical programming methods, PSO is not sensitive to starting points and forms of objective function.

Because of advantages of PSO, some researchers and scholars begin to apply it to solve optimization problems in electric power systems and gain success in some specific applications [41].

To solve constrained optimization problems, some modifications had been made to basic PSO. They can be categorized into:

- Method based on preference of feasible solutions over infeasible ones. This is performed in two ways. The first is when a particle is outside of feasible space, it will be reset to the last best value found. The second way is the preserving feasibility strategy, when updating the memories, all the particles keep only feasible solutions in their memory; during the initialization process, all particles are started with feasible solutions. em Method based on penalty functions. PSO is applied to a non-constrained optimization problem, which was converted by a non-stationary multi-stage assignment penalty function from constrained optimization problem. However, in case of incorrect penalty coefficients, the algorithm can converge slowly or prematurely [7].
- Method based on multi-objective optimization concept. Generation of better performer list (BPL) is obtained by implementing Pareto ranking scheme based on the constraint matrix. An individual that is not in the BPL improves its performance by deriving information from its closest neighbour in the BPL [41].

A PSO algorithm consists of a population formed by individuals (particles) updating continuously the knowledge of the given search space; each one represents a possible solution. Unlike to evolutionary algorithms, each particle moves in the search space with a velocity. Let t be a time instant. The new particle position is computed by adding the velocity vector to the current position

$$x^p(t+1) = x^p(t) + v^p(t+1) \tag{18.15}$$

being $x^p(t)$ particle p position, $p = 1, \ldots, s$, at time instant $t, v^p(t+1)$ new velocity (at time $t+1$) and s is population size. The velocity update equation is given by

$$v_j^p(t+1) = v(t)v_j^p(t) + \mu\omega_{1j}((y_j^p(t) - x_j^p(t)) + v\omega_{2j}(t)(\hat{y}_j(t) - x_j^p(t))) \tag{18.16}$$

for $j = 1, \ldots, n$, where $v(t)$ is a weighting factor (inertial), μ is the *cognitive* parameter and v is the *social* parameter. $\omega_{1j}(t)$ and $\omega_{2j}(t)$ are random numbers drawn from the uniform distribution $U(0,1)$, used for each dimension $j = 1, \ldots, n$. $y_j^p(t)$ is particle p position with the best objective function value and $\hat{y}_j(t)$ is a particle position with the best function value and can be describe by

$$\hat{y}(t) = \arg\min_{a \in A} = y^1(t), \ldots, y^s(t) \tag{18.17}$$

In [24] authors applied the PSO technique in power system to solve the challenging transient-stability constrained optimal power flow (TSCOPF) problem. The feasibility and robustness of the proposed PSO-based method for TSCOPF are demonstrated on the IEEE 30-bus and New England 39-bus systems with promising results. In [25] authors used PSO technique with reconstruction operators (PSO-RO) to solve an optimal power flow with security constraints (OPF-SC) problem. By the defined weight w, varying through the course of the PSO-RO run, the global search at the beginning of the iterative process, and the local searching at the end, can be improved. The proposed methodology is able to find feasible and satisfactory solutions for both normal state and for a group of credible contingencies. This solution guarantees that the power system is capable to pass from a pre-contingency state to a post-contingencies states, satisfying generation and links operative constraints. In [20] authors proposed an adaptive PSO algorithm (APSO). In this algorithm, inertia weight is nonlinearly adjusted by using population diversity information. By adding the mutation and crossover operator to the algorithm in the later phase of convergence, the APSO algorithm can not only escape from the local minimum's basin of attraction of the later phase, but also maintain the characteristic of fast speed in the early convergence phase. APSO algorithm is applied to reactive power optimization of power system. The simulation results of the standard IEEE-30-bus power system have indicated that APSO is able to undertake global search with a fast convergence rate and a feature of robust computation. It is proved to be efficient and practical during the reactive power optimization. In [3] authors presented a hybrid particle swarm optimization algorithm (HPSO) as a modern optimization tool to solve the discrete optimal power flow problem that has both discrete and continuous optimization variables. The objective functions considered are the system real power losses, fuel cost, and the gaseous emissions of the generating units. The proposed algorithm makes use of the PSO to allocate the optimal control settings while Newton–Raphson algorithm minimizes the mismatch of the power flow equations. Reference [7] presents the solution of optimal power flow using particle swarm optimization with penalty function. A non-stationary multistage assignment penalty function is used to convert the constrained optimization problem into a non-constrained optimization problem. The proposed PSO method is tested on the standard IEEE 30-bus system and it can obtain higher quality solutions efficiently in OPF problems by comparing with linear programming and genetic algorithm. However, inappropriate penalty coefficients can make the algorithm slow convergence or premature convergence. Deciding an optimal value of penalty coefficients is a difficult optimization problem itself. Reference [12, 16] improves particle swam optimization algorithm by incorporating a biology concept "passive congregation" to solve optimal power flow problems. Reference [43] presents an efficient mixed-integer particle swarm optimization with mutation scheme for solving the constrained optimal power flow with a mixture of continuous and discrete control variables and discontinuous fuel cost functions. In the improved particle swarm optimization algorithm [39], particles not only studies from itself and the best one but also from other individuals. These modifications above to basic PSO can effectively avoid premature convergence and obtain satisfactory results in the solution of OPF problems. Reference [10] presents a particle swarm optimization as a tool for loss reduction study. The study is carried out in two steps. First, by using the tangent vector technique, the critical area of the power system is identified under the point of view of voltage instability. Second, once this area is identified, the PSO technique calculates the amount of shunt reactive power compensation that takes place in each bus. The proposed approach has been examined and tested with promising numerical results using the IEEE 118-bus system. In [6] a comparative analysis of two mathematical programming methods with two meta heuristics has been presented. Mathematical programming methods have been proved robust and reliable for medium-size systems (up to 708 buses), even with the presence of discrete variables, which give a theoretical advantage to meta heuristics. On the other hand, meta heuristics have shown satisfactory behaviour in small-size systems, but failed to provide robust solutions with the necessary reliability in medium-size systems. It has been shown that meta heuristics do not scale easily to larger problems, since the execution time and the quality of the provided solution deteriorate with the increase of the number of control variables. The potential of meta heuristics to enhance their performance and provide satisfactory solutions for large-scale power systems has yet to be demonstrated. In [49] an improved particle swarm optimization approach (IPSO) has been developed for determination of the global or near global optimum solution for optimal reactive power dispatch and voltage control of power systems. The improved particle swarm optimization approach uses more particles' information to control the mutation operation. A new adaptive strategy for choosing parameters is also proposed to assure convergence of IPSO method. The performance of the proposed algorithm demonstrated through its evaluation on the IEEE 30-bus power system and a practical 118-bus power system shows that the IPSO is able to undertake global search with a fast convergence rate and a feature of robust computation. In [11] the original PSO was expanded to hybrid particle swarm optimizer with mutation (HPSOM) algorithms by combining with arithmetic mutation. Furthermore, the notion of mutation in the hybrid model was introduced from the genetic algorithm field. The proposed approach utilizes the local and global capabilities to search for optimal loss reduction by installing the shunt compensator. The optima found by the hybrid were better than by the standard PSO model, and the convergence speed was faster. In [48] the PSO method for OPF are demonstrated on the IEEE 30-bus test system with promising results. The results are compared to those of LP and GA. The results confirm the potential of the PSO method and show its effectiveness and superiority over LP and GA. Reference [45] presents a particle swarm optimization (PSO) method to deal with reactive power optimization problem in a province power system in China. The successful application to the practical Heilongjiang power system indicates the possibility of PSO as a practical tool for various optimization problems in power system. Reference [42] presents a particle swarm optimization for reactive power and voltage control (VVC) considering voltage security assessment (VSA).

18.3.6 Other Methods

For OPF problem the author [22] introduced a hybrid method integrating immune genetic algorithm and interior point method. The continuous variables are solved by nonlinear interior point method, and the discrete variables are solved by immune genetic algorithm. In [9], a multi-objective hybrid evolutionary strategy (MOHES) is presented for the solution of the comprehensive model for OPF formulated above. The hybridization of GA with SA is expected to affect a beneficial synergism of both. MOHES concentrates on the 'better' areas of the search space with the incorporation of the concept of acceptance number to guide the search. The greater modelling power of the method enables representation of nonlinear and discontinuous functions and discrete variables easily without involving approximations, and its enhanced search capabilities lead to better solutions. A complete set of non-inferior solutions representing the trade-off between various objectives is provided in a single run. This gives a larger number of alternatives and more flexibility to the operator in taking dispatching decisions. MOHES has been designed to use the small perturbation analysis to avoid computing the complete load flow in every fitness evaluation. This results in considerable savings in computational expense. Authors in [21] proposed an ordinal optimization theory-based algorithm to solve OPF problem for a good enough solution with high probability. Aiming for hard optimization problems, the ordinal optimization theory, in contrast to heuristic methods, guarantee to provide a top % solution among all with probability more than 0.95. The proposed approach consists of three stages. First, select heuristically a large set of candidate solutions. Then, use a simplified model to select a subset of most promising solutions. Finally, evaluate the candidate promising-solutions of the reduced subset using the exact model. Reference [15] proposes an application of a two-phase hybrid evolutionary programming. EP takes the place of the phase-1 of the search, providing the potential near optimum solution, and the phase-2 of a search technique using optimization by direct search and systematic reduction of the size of search region. Phase-2 algorithm is applied to rapidly generate a precise solution under the assumption that the evolutionary search has generated a solution near the global optimum.

18.4 Numerical Application

In this section three practical approaches for OPF problem are presented and their numerical results are discussed.

18.4.1 Application to Nigerian Grid

In [4] the authors have presented the application of the DE technique for solving the reactive power/voltage control problem in Nigerian grid. Floating point numbers were used for the parameter variables encoding. An initial population of size NP, which could be in the range from $2D$ to $100D$ (D is the total number of control devices) depending on the problem and the available computing facilities, was randomly generated within the parameter space.

The OPF problem was formulated as mixed-integer nonlinear programming. Generating units' voltage set-points as continuous variables assumed to operate within the range $(0.9 \leq Vi \leq 1.1)$. On-load tap changer (OLTC) transformers considered to have 20 tap positions with a discrete step of 0.01 within the range $(0.9 \leq Ti \leq 1.1)$. Number of reactors/condensers assumed to vary between 0 and the step size (nc_i) at each bus. The reproduction operation of DE can extend the search outside the range of the parameters. Any parameter that violates the limits after reproduction was replaced with random values.

A penalty function approach proposed in [17] is adopted in this study to handle the voltage limits violations. The objective function is formulated as follows:

$$f_{obj} = (P_{loss} + a) \cdot \prod_{i=1}^{nd} c_i^{b_i} \tag{18.18}$$

where:

$$c_i = \begin{cases} 1 + s_i V_{Ld} & \text{if } V_L i > V_{Li}^{max} \text{ or } V_{Li} > V_{Li}^{min} \\ 1 & \text{otherwise} \end{cases}$$

$$V_{Ld} = \begin{cases} V_{Li} - V_{Li}^{max} & \text{if } V_{Li} > V_{Li}^{max} \\ V_{Li}^{min} - V_{Li} & \text{otherwise} \end{cases}$$

$s_i \geq 1$ and $b_i \geq 1$. The constant a is used to ensure that only a non-negative value is assigned to the objective function. Constant s is used for appropriate scaling of the constraint function value. The exponent b modifies the shape of the optimization surface.

Proposed procedure was implemented using MATLAB V 7.1 R14 whereby the power flow calculation using the power world simulator on which the Nigerian 330 kV, 31-bus transmission grid was replicated in operational detail. It consists of following steps [4]:

1. At the initialization stage, the relevant DE parameters as shown in Table 18.1 are defined.
2. Run the base case Newton Raphson load flow on the power world simulator to determine the initial load bus voltage and active power losses, respectively.

Table 18.1. Optimum parameter settings for DE based tool

Control parameters	Differential Evolution
Maximum generation, gen^{max}	200
Number of control devices, D	32
Population size, np	$3D$
Scaling factor for mutation, F	0.8
Crossover constant, CR	0.6
Objective function scaling constant, a	7.0
Constraint function scaling constant, s	1
Optimization surface shape modifiers, b	1

Table 18.2. Effect of population sizes on DE performance

Population Size	D	$2D$	$3D$
Initial power losses (MW)	40.07	40.07	40.07
Final power losses (MW)	35.74	36.40	35.59
Power loss reduction (%)	10.81	9.16	11.18
Total no. of function evaluation	4,400	8,800	13,200
Gen. at minimum loss reductions	95	88	87
No. of function evaluation at minimum loss reduction	2,090	3,872	5,742

3. The randomly generated initial population comprises the control device variables within the parameter space. The objective function for each vector of the population is computed using. The vector with the minimum objective function value (the best fit) is determined.
4. Update of the generation count.
5. Mutation, crossover, selection and evaluation of the objective function are performed. If parameter violation occurs, the parameter value is generated randomly. The elitist strategy is applied.
6. If the generation count is less than the preset maximum number of generations, go to step 4. Otherwise the parameters of the fittest vector are returned as the desired optimum settings. With the optimal settings of the control devices, run the final load flow to obtain the final voltage profiles and the corresponding system active power losses.

The power system consists of seven generating units (four thermal units and three hydro), seven machine transformers equipped with tap changers, and compensation reactors of different discrete values located at eight buses. Three case studies were performed. In the first case, two of the four 75 MVar reactors at bus 8 and bus 10 were wrongly switched on. The approach was able to keep the voltage at all buses within limits for all the three population sizes (Table 18.2). There were also load reductions at some load points. In the second case, the system was initially operating as in previous case; later one transmission line was disconnected, this resulted in voltage limits violations at

12 buses. The algorithm was able to solve the voltage problem connected with 14.03% power loss reduction (from 42.05 to 36.15 MW). In the third case load modifications were performed and two transmission lines were opened. As a result of this action, both under and over voltage problems occurred in the power system. The approach succeeded in keeping the voltage at all buses within the limits and achieving power loss reduction of 4.20% (from 46.45 to 44.58 MW).

The obtained results showed that the proposed DE based reactive power dispatch is an efficient tool in keeping the abnormal bus voltages within the prescribed limits and simultaneously reducing lower system transmission power losses. According to authors [4], it is pertinent to curtail the number of control devices employed to alleviate bus voltage problems. It is also feasible to integrate a pre-selection mechanism into the DE to select the control devices a priori. This will be an added advantage to the computational time of the DE since the population size depends on the number of control variables.

18.4.2 Application to Heilongjiang Power System

In [45] authors applied PSO based approach to the reactive power optimization problem in a practical power system in China. The control variables are self constrained and dependent variables are incorporated into the objective function as penalty terms. The objective function is:

$$\min f = P_{loss} + \lambda_V \sum_{\alpha} \Delta V_L^2 + \lambda_Q \sum_{\beta} \Delta Q_G^2 \tag{18.19}$$

where f is the generalized objective function: λ_v and λ_Q are penalty factors and both equal to 1 here; ΔV_L and ΔQ_G are the violations of load-bus voltages and generator reactive powers; α and β are sets of buses whose voltage and reactive power generation violate their constraints, respectively. In PSO continuous variables for generator voltages and discrete variables for transformer taps and shunt capacitors were used and represented as the vector of a particle's position shown in (18.20). So, the reactive power optimization is a mixed discrete continuous nonlinear optimization problem.

$$x = [V_{G_1}, \ldots V_{G_{NG}}, K_{T_1}, \ldots K_{T_{NT}}, Q_{C_1}, \ldots Q_{C_{NC}}]^T \tag{18.20}$$

where NG is the number of generator voltages, NT is the number transformer taps and NC is the number of shunt capacitors.

The PSO based algorithm can be described in the following steps [45]:

1. Read the original data including power system data and the PSO parameters.
2. Set the generation counter $t = 0$, place the particles in the searching space $x_i(0)$ randomly and uniformly and assign a random and uniform velocity $v_i(0)$ for each particle.

Table 18.3. The reduction of real power loss and voltage violation by PSO and simplified gradient methods

Method	Real power loss P_{loss}(p.u.)	Real power loss Reduced $P_{loss}s$ %	No. of voltage violation	Voltage violation $V_{violation}$ %
Base case	0.945475	–	24	15.89
PSO	0.848951	10.21	0	0
Simplified gradients	0.921715	2.51	10	6.6

3. Calculate the fitness value of the initial particles by power flow calculation and objective function (18.19). $x_i(0)$ is set to $p_i(0)$ for each initial particle. The initial best evaluated value among the particle swarm is set to $g_i(0)$.
4. Let $t = t + 1$.
5. Update $v_i(t+1)$ and $x_i(t+1)$.
6. Calculate the fitness values of the new particles by power flow calculation and objective function.
7. Update $p_i(t+1)$ with $x_i(t+1)$ if $f(x_i(t+1)) < f(p_i(t))$ and update $g_i(t+1)$ with best $p_i(t+1)$ in the population swarm.
8. Go to step 4 until a criterion is met, usually a sufficiently good fitness value or a maximum number of generations.

The PSO based optimization algorithm was applied to the Heilongjiang province power system in China. The practical power system comprises of 151 buses and 220 transmission lines. The limits of control variables: generator voltages, transformer taps, and Shunt capacitors were set to their practical lower and upper values. The permissible ranges of dependent variables are limited according to the practical operating constrains.

The number of particles N was set to 50 in order to get a high quality solution within acceptable computation time. The stop criterion for the PSO was 100 generations and the PSO had been run 50 times with random initial values. To demonstrate the performance of PSO algorithm, the results of PSO were compared with that of the conventional simplified gradient method with the same objective function. The results are presented in Table 18.3. It should be noted that the real power losses are significantly decreased and the PSO optimal solution produces no voltage violations.

The PSO converges quickly under all cases. The authors note that after a lot of iterations, the particles tend to become homogeneous. They propose to keep fresh particles in generations because the PSO lacks of global search ability even when the global search ability is required to jump out of the local minimum in this case.

18.4.3 Application to the Kansai Electric Power System

The authors [42] proposed a method based on PSO for reactive power and voltage control (VVC) formulated (18.21) as a MINLP considering voltage stability assessment (VSA). Voltage security assessment is considered using a continuation power flow (CPFLOW) technique and a fast voltage contingency selection method.

$$\min f_c(x, y) = \sum_{i=1}^{n} Loss_i \tag{18.21}$$

where n is the number of branches, x is continuous variables, y is discrete variables, $Loss_i$ is power loss (P_{loss}) at branch i.

In PSO automatic voltage regulator (AVR) operating values (continuous variable), OLTC tap position (discrete variable) and the number of reactive power compensation equipment (discrete variable)are initially generated randomly between upper and lower bounds. The values is also modified in the search procedure between the bounds. Then, the corresponding impedance of the transformer is calculated for the load flow calculation.

The procedure for VSA consists of evaluation of the control strategy and various contingencies. The proposed VVC algorithm using PSO includes following steps [43]:

1. Initial searching points and velocities of agents are generated randomly.
2. P_{loss} to the searching points for each agent is calculated using the load flow calculation.
3. If the constraints are violated, the penalty is added to the loss (evaluation value of agent).
4. P_{best} is set to each initial searching point. The initial best evaluated value (loss with penalty) among p_{bests} is set to g_{best}.
5. New velocities are calculated.
6. New searching points are calculated.
7. P_{loss} to the new searching points and the evaluation values are calculated.
8. If the evaluation value of each agent is better than the previous pbest, the value is set to p_{best}. If the best p_{best} is better than g_{best}, the value is set to g_{best}. All of g_{bests} are stored as candidates for the final control strategy.
9. If the iteration number reaches the maximum iteration number, then go to step 9. Otherwise, go to step 4.
10. P–V curves for the control candidates and various contingencies are generated using the best g_{best} among the stored g_{bests} (candidates). If the MW margin is larger than the predetermined value, the control is determined as the final solution. Otherwise, select the next g_{best} and repeat the VSA procedure mentioned above.

The authors used only loss minimization as the objective function and checked whether the control strategy has enough voltage stability margins or not after loss minimization. Moreover, evaluation for each state is extremely time-consuming considering VSA during optimization procedure, and it is difficult to realize on-line VVC. Considering the trade-off between the optimal control and the execution time, the proposed method selected the way to handle the contingencies after generation of the optimal control candidates.

The proposed method is applied to a EHV system of Kansai Electric practical system with 112 buses. The model system has 11 generators for AVR control, 47 OLTCs with 9–27 tap positions, and 13 static condenser (SC) installed buses with 33 SCs for VVC.

PSO and reactive tabu search (RTS) are compared in 100 searching iterations. The average loss value by the proposed method is smaller than the best result by RTS. The authors say that PSO generates better solution than RTS with 96% possibility. It is obtained that the solution by PSO is converged to high quality solutions at about 20 iterations. The average calculation time by PSO is about four times faster than that by RTS. All of the best solutions by both PSO and RTS within 100 searching iterations are feasible solutions without voltage and power flow constraints violation in the simulation.

The authors applied the proposed method to large-scale systems as well, the results indicated the applicability of PSO to large-scale problems.

This paper showed the practical applicability of PSO to a MINLP and suitability of PSO for application to large-scale VVC problems. PSO only requires less than 50 iterations for obtaining sub-optimal solutions even for large-scale systems. Since many power system problems can be formulated as a MINLP, proposed PSO can be used as a practical tool for various MINLPs of power system operation and planning.

18.5 Concluding Remarks

The OPF problem aims to achieve an optimal solution of a specific power system objective function that may be cost-based, which means to minimize the possible cost of new reactive power supplies, the cost of real power losses or the fuel cost, while satisfying a set of constraints. OPF constraints may be classified as the power flow equality constraints, physical limits of the control variables, physical limits of the state variables, and other limits such as power factor limits.

As the power industry moves into a competitive environment and systems are operated closer to their limits, increasingly more transmission systems have become stability-limited. Therefore, it is critical that the system is modelled appropriately and transient stability should be the main concern in the operations. Mathematically, OPF with transient stability is an extended OPF with additional equality and inequality constraints. The additional equality constraints are a set of differential-algebraic equations which describe the system dynamics; the additional inequality constraints consist of angle stability constraints, and also some practical requirements of system dynamic behaviour.

OPF problem may be solved with classic optimization methods that include LP, NLP, or MINLP. Due to the nonlinearity of power systems, LP loses accuracy due to linear assumptions. Consideration of nonlinear algorithms and integer variables will make the running time much longer and the algorithm possibly less robust. Newer algorithms based on heuristic and intelligent searches such as EA, PSO and ACO can handle the integer variable very well, but need further investigation regarding performance under different systems.

The simulating results demonstrate that proposed methods can be successfully applied to practical power system, but the potential of metaheuristics to provide satisfactory solutions for large-scale power systems has yet to be demonstrated.

References

1. Abido M A. Multiobjective optimal power flow using strength pareto evolutionary algorithm. In *39th International Universities Power Engineering Conference, 2004. UPEC 2004*, volume 1, pages 457–461, September 2004.
2. Hosseini S H, Abbasy A, Tabatabaii I. A multiagent-based differential evolution algorithm for optimal reactive power dispatch in electricity markets. In *International Conference on Power Engineering, Energy and Electrical Drives, 12-14 April 2007. POWERENG 2007*, April 2007.
3. El-Hawary M E, AlRashidi M R. Hybrid particle swarm optimization approach for solving the discrete opf problem considering the valve loading effects. *IEEE Transactions on Power Systems*, 22(4):2030–2038, 2007.
4. Venayagamoorthy G K, Aliyu U O, Bakare G A, Krost G. Differential evolution approach for reactive power optimization of nigerian grid system. In *IEEE Power Engineering Society General Meeting 24-28 June, 2007*, pages 1–6, 2007.
5. Zoumas C E, Petridis V, Bakirtzis A G, Biskas P N. Optimal power flow by enhanced genetic algorithm. *IEEE Transactions on Power Systems*, 17(2):229–236, 2002.
6. Tellidou A, Zoumas C E, Bakirtzis A G, Petridis V Tsakoumis, Biskas P N, Ziogos N P. Comparison of two metaheuristics with mathematical programming methods for the solution of opf. In *Proceedings of the 13th International Conference on Intelligent Systems Application to Power Systems, 6-10 November 2005*, page 6, 2005.
7. Cao Y J, Chao B, Guo C X. Improved particle swarm optimization algorithm for opf problems. In *Proceedings of 2004 IEEE PES Power Systems Conference and Exposition*, 1:233–238, 2004.
8. Liu C C, Chen Y L. Optimal multi-objective var planning using an interactive satisfying method. *IEEE Transactions on Power Systems*, 10(2):664–670, 1995.
9. Patvardhan C, Das D B. Useful multi-objective hybrid evolutionary approach to optimal power flow. In *IEE Proceedings – Generation, Transmission and Distribution, 13 May 2003*, volume 150, pages 275–282, May 2003.
10. Lambert-Torres G, Esmin A A A. Loss power minimization using particle swarm optimization. In *International Joint Conference on Neural Networks, 2006. IJCNN '06.*

11. Zambroni de Souza A C, Esmin A A A, Lambert-Torres G. A hybrid particle swarm optimization applied to loss power minimization. *IEEE Transactions on Power Systems*, 20(2):859–866, 2005.
12. He S et al. An improved particle swarm optimization for optimal power flow. In *International Conference on Power System Technology, 21–24 November 2004*, pages 1633–1637, 2004.
13. Zimmerman R D, Gan D, Thomas R J. Optimal multi-objective var planning using an interactive satisfying method. *IEEE Transactions on Power Systems*, 15(2):535–540, 2000.
14. Lumbreras J, Parra V M, Gomez T, Perez-Arriaga I J. A security-constrained decomposition approach to optimal reactive power planning. *IEEE Transactions on Power Systems*, 6(3):1069–1076, 1991.
15. Prasanna R, Gopalakrishnan V, Thirunavukkarasu P. Reactive power planning using hybrid evolutionary programming method. In *Power Systems Conference and Exposition*, 3:1319–1323, 2004.
16. Prempaint E, Wu Q H, Fitch J, Mann S, He S, Wen J Y. An improved particle swarm optimization for optimal power flow. In *International Conference on Power System Technology. PowerCon*, 2:1633–1637, 2004.
17. Lampinen J, Zelinka I. Mixed integer-discrete-continuous optimization by differential evolution, part i: the optimization method. In P. Ošmera, editor, *Proceedings of MENDEL'99, 5th International Mendel Conference on Soft Computing, Brno University of Technology, Faculty of Mechanical Engineering, Institute of Automation and Computer Science*, pages 71–76, Brno, Czech Republic, 1999.
18. Musirin I, Kamal M F M, Rahman T K A. Application of improved genetic algorithms for loss minimisation in power system. In *National Power and Energy Conference, 2004. PEC on 2004. Proceedings*, pages 258–262, November 2004.
19. Chung T S, Leung H C. Optimal power flow with a versatile facts controller by genetic algorithm approach. In *International Conference on Advances in Power System Control, Operation and Management 2000. APSCOM-00*, pages 178–183, 2000.
20. Lu S, Ma J, Li Y, Li D, Gao L. Adaptive particle swarm optimization algorithm for power system reactive power optimization. In *American Control Conference, 9-13 July 2007. ACC'07*, pages 4733–4737, 2007.
21. Lin Ch -H, Lin S -Y, Ho Y -Ch. An ordinal optimization theory-based algorithm for solving the optimal power flow problem with discrete control variables. *IEEE Transactions on Power Systems*, 19(1):276–286, 2004.
22. Wong K P, Yan W, Xu G, Liu F, Chung C Y. Hybrid immune genetic method for dynamic reactive power optimization. In *International Conference on Power System Technology, 2006. PowerCon 2006.*
23. Azmy A M. Optimal power flow to manage voltage profiles in interconnected networks using expert systems. *IEEE Transactions on Power Systems*, 22:1622–1628, 2004.
24. Chan K W, Pong T Y G, Mo N, Zou Z Y. Transient stability constrained optimal power flow using particle swarm optimisation. *IET Generation, Transmission and Distribution*, 1(3):476–483, 2007.
25. Ramirez J M, Onate P E. Optimal power flow solution with security constraints by a modified pso. In *IEEE Power Engineering Society General Meeting, 24–28 June 2007*, pages 1–6, 2007.
26. Kundur P. *Power System Stability and Control.* McGraw-Hill, New Jersey, 1994.

27. Baran B, Ramos R, Vallejos J. Multiobjective reactive power compensation with voltage security. In *Transmission and Distribution Conference and Exposition: Latin America, 2004 IEEE/PES*, pages 302–307, November 2004.
28. Roman C, Rosehart W, Schellenberg A. Multiobjective reactive power compensation with voltage security. In *IEEE Power Engineering Society General Meeting, 18-22 June 2006*, page 7, 2006.
29. Zhijian H, Shengsong L, Min W. A hybrid algorithm for optimal power flow using the chaos optimization and the linear interior point algorithm. In *International Conference on Power System Technology, 2002. Proceedings. PowerCon*, 2:793–797, 2002.
30. Zhijian H, Shengsong L, Min W. Hybrid algorithm of chaos optimisation and slp for optimal power flow problems with multimodal characteristic. In *IEE Proceedings - Generation, Transmission and Distribution*, 150:543–547, 2003.
31. Liu Y -H, Teng J -H. A novel acs-based optimum switch relocation method. *IEEE Transactions on Power Systems*, 18(1):113–120, 2003.
32. Yesuratnam G, Thukaram D. Fuzzy - expert approach for voltage-reactive power dispatch. In *IEEE Power India Conference, 10-12 April 2006*, page 8, 2006.
33. Rajicic D, Todorovski M. An initialization procedure in solving optimal power flow by genetic algorithm. *IEEE Transactions on Power Systems*, 21(2):480–487, 2006.
34. Wu F, Tong X, Zhang Y. A decoupled semismooth newton method for optimal power flow. In *IEEE Power Engineering Society General Meeting, 18–22 June 2006*, 2006.
35. Harley R G, Venayagamoorthy G K. Swarm intelligence for transmission system control. In *IEEE Power Engineering Society General Meeting, 2007*, pages 1–4, June 2007.
36. Lee K Y, Vlachogiannis J G. Fuzzy logic controlled particle swarm for reactive power optimization considering voltage stability. In *The 7th International Power Engineering Conference, 2005. IPEC 2005*, pages 1–555, 2005.
37. Lee K Y, Vlachogiannis J G, Hatziargyriou N D. Power system optimal reactive power dispatch using evolutionary programming. *IEEE Transactions on Power Systems*, 10(3):1243–1249, 1995.
38. Lee K Y, Vlachogiannis J G, Hatziargyriou N D. Ant colony system-based algorithm for constrained load flow problem. *IEEE Transactions on Power Systems*, 20(3):1241–1249, 2005.
39. Huang Z -Q, Zhang J -W, Sun C -J, Wang C -R, Yuan H -J. A modified particle swarm optimization algorithm and its application in optimal power flow problem. In *Proceedings of 2005 International Conference on Machine Learning and Cybernetics, 18–21 August 2005*, 5:2885–2889, 2005.
40. Ma J T, Wu Q H. Power system optimal reactive power dispatch using evolutionary programming. *IEEE Transactions on Power Systems*, 10(3):1243–1249, 1995.
41. Zhao Z, Yang B, Chen Y. Survey on applications of particle swarm optimization in electric power systems. In *IEEE International Conference on Control and Automation, May 30, 2007–June 1, 2007. ICCA 2007*, pages 481–486, 2007.
42. Fukuyama Y, Takayama S, Nakanishi Y, Yoshida H, Kawata K. A particle swarm optimization for reactive power and voltage control considering voltage security assessment. *IEEE Transactions on Power Systems*, 15(4):1232–1239, 2001.

43. Gaing Z -L. Constrained optimal power flow by mixed-integer particle swarm optimization. In *IEEE Power Engineering Society General Meeting, 12-16 June 2005*, volume 1, pages 243–250, June 2005.
44. Meng F, Zhang H, Zhang L. Reactive power optimization based on genetic algorithm. In *International Conference on Power System Technology, 1998. Proceedings. POWERCON apos; 98*, volume 2, pages 1448–1453, August 1998.
45. Liu Y, Zhang W. Reactive power optimization based on pso in a practical power system. In *IEEE Power Engineering Society General Meeting, 6-10 June 2004*, volume 1, pages 239–243, June 2004.
46. Tolbert L M, Zhang W, Li F. Review of reactive power planning: Objectives,constraints, and algorithms. *IEEE Transactions on Power Systems*, 22(4):2177–2186, 2007.
47. Ren Z, Zhang Y. Optimal reactive power dispatch considering costs of adjusting the control devices. *IEEE Transactions on Power Systems*, 20(3):1349–1356, 2005.
48. Cao Y J, Zhao B, Guo C X. Improved particle swam optimization algorithm for opf problems. In *IEEE PES Power Systems Conference and Exposition, 10-13 October 2004*, volume 1, pages 233–238, October 2004.
49. Cao Y J, Zhao B, Guo C X. An improved particle swarm optimization algorithm for optimal reactive power dispatch. In *IEEE Power Engineering Society General Meeting, 12-16 June 2005*, volume 1, pages 272–279, June 2005.
50. Chuanwen J, Zhiqiang Y, Zhijian H. Economic dispatch and optimal power flow based on chaotic optimization. In *International Conference on Power System Technology, 2005. Proceedings. PowerCon 2002.*, volume 4, pages 2313–2317, 2002.

19

WILMAR: A Stochastic Programming Tool to Analyze the Large-Scale Integration of Wind Energy

Christoph Weber, Peter Meibom, Rüdiger Barth, and Heike Brand

Summary. Wind power is highly variable and partly unpredictable and therefore energy systems of the future have to cope with increased variability and stochasticity. The paper describes the use of a novel stochastic programming model to assess the impact of increased wind power generation on electricity systems. This WILMAR model takes explicitly the stochastic behavior of wind generation and the forecast errors into account. Also a detailed modeling of power plant, grid and market characteristics is performed. WILMAR thus allows to assess the impact of increased wind generation on reserve needs and usage, power plant operation and system cost.

Key words: Wind power, Electricity system, Stochastic programming, Electricity markets, Reserve power

19.1 Introduction

The integration of substantial amounts of wind power in a liberalized electricity system will impact both the technical operation of the electricity system and the electricity market. In order to cope with the fluctuations and the partial unpredictability in the wind power production, other units in the power system have to be operated more flexibly to maintain the stability of the power system. Technically this means that larger amounts of wind power will require increased capacities of spinning and nonspinning power reserves and an increased use of these reserves. Moreover, if wind power is concentrated in certain regions, increased wind power generation may lead to bottlenecks in the transmission networks. Economically, these changes in system operation have certainly cost and consequently price implications. Moreover they may also impact the functioning and the efficiency of certain market designs. Even if the wind power production is not bid into the spot market, the feed-in of the wind power will affect the spot market prices, since it influences the balance of demand and supply. As substantial amounts of wind power will require increased reserves, the prices on the regulating power markets are furthermore expected to increase. Yet this is not primarily due to the fluctuations of wind power itself but rather due to the partial unpredictability of wind power. If wind power were fluctuating but perfectly predictable, the conventional power plants would have to operate also in a more variable way, but this operation could be scheduled on a day-ahead basis and settled on conventional day-ahead spot markets. It is the unpredictability of wind power which requires an increased use of reserves with corresponding price implications. In order to analyze adequately the market impacts of wind power it is therefore essential to model explicitly the stochastic behavior of wind generation and to take the forecast errors into account. In an ideal, efficient market setting, all power plant operators will take into account the prediction uncertainty when deciding on the unit commitment and dispatch. This will lead to changes in the power plant operation compared to an operation scheduling based on deterministic expectations, since the cost functions for power production are usually nonlinear and not separable in time. For example, even without fluctuating wind power, start-up costs and reduced part-load efficiency lead to a trade-off for power plant operation in low demand situations, i.e., notably during the night. Either the power plant operator chooses to shut down some power plants during the night to save fuel costs while operating the remaining plants at full output and hence optimal efficiency. Or he operates a larger number of power plants at part load in order to avoid start-up costs in the next morning. This trade-off is modified if the next increase in demand is not known with (almost) certainty. So in an ideal world, where information is gathered and processed at no cost, power plant operators will anticipate possible future wind developments and adjust their power plant operation accordingly. The model presented in the following describes such an ideal and efficient market operation by using a stochastic linear programming model, which depicts "real world optimization" on the power market on an hourly basis with rolling planning. With efficient markets, i.e., also without market power, the market results will correspond to the outcomes of a system-wide optimization as described in the following. The cost and price effects derived for the integration of wind energy in this model should then provide a lower bound to the magnitude of these effects in the real, imperfect world. The remainder of this paper is organized as follows. In the next section, related work on wind energy integration is reviewed. Then the general approach to model the interaction between various markets within the so-called WILMAR model is discussed. WILMAR stands thereby for Wind Integration in Liberalized MARkets. In the subsequent section, key equations of the model are presented. In a separate section, key concepts of the model are highlighted, including notably the approach of rolling planning, which is applied to simulate real world planning processes. Finally, an application example is presented, looking at wind integration in Germany and the Scandinavian countries.

19.2 Existing Modeling Approaches

Wind energy integration has been evaluated in the past years using various modeling approaches. Thereby three broad strands of literature may be distinguished. First, an important issue dealt with in literature are changes in load flow due to increased wind power production. A preeminent example of this kind of studies is the phase I report of EWIS – the European wind integration study undertaken by the transmission system operators [9]. However here the focus is primarily on the electric load flows and not much on economic aspects. Another strand of literature looks in detail at the stability problems, which may occur when increasing the wind power infeed. [6] is an example for a recent publication in this field. Yet again this literature almost exclusively focuses on technical aspects of wind energy integration. Correspondingly an application of market or optimization models hardly does not occur. This is different for the third strand of literature, the work devoted to the costs of wind integration. [12] provides a first discussion and quantitative estimates of the additional costs induced by the installation of fluctuating renewables. In the context of optimal policy design, [8] and [10] give results on the costs of increased part-load operation, start-ups and backup costs for wind energy, without much detail however on the calculation methodology. [14] discusses the additional costs related to the integration of large amounts of renewables in the British electricity system, following closely the approach developed by [12]. In the various presentations given at [13], different approaches to the quantification of integration costs and also corresponding numerical values are given. [1] provide an overview of relevant cost components and discuss cost estimates taken from studies in various European countries. [19] derive the value of wind energy from an electricity system model, which includes explicitly the stochasticity of wind as well of hydro sources. Yet a simplified approach using so-called recombining trees is applied there. Since all of those approaches are not able to assess in detail the impact of increased wind, notably on required reserves and the interplay between information arrival, market prices and system operation, a detailed stochastic unit commitment model is presented in the following.

19.3 Markets and Unit Commitment

In a liberalized market environment it is possible not only to change the unit commitment and dispatch, but even to trade electricity at different markets. The WILMAR Joint Market model analyzes power markets based on a hourly description of generation, transmission and demand, combining the technical and economical aspects, and it derives hourly electricity market prices from marginal system operation costs. This is done on the basis of an optimization of the unit commitment and dispatch taking into account the trading activities of the different actors on the considered energy markets. In this model four electricity markets and one market for heat are included:

1. A day-ahead market for physical delivery of electricity where the Nord Pool market is taken as the starting point. This market is cleared at noon for the following day and is called the day-ahead market. The nominal electricity demand is given exogenously.
2. An intra-day market for handling deviations between expected production agreed upon the day-ahead market and the realized values of production in the actual operation hour. Regulating power can be traded up to 1 h before delivery. In the presented version of the Joint Market model the demand for regulating power is only caused by the forecast errors connected to the wind power production.
3. A day-ahead market for automatically activated reserve power (frequency activated or load-flow activated). The demand for these ancillary services is determined exogenously to the model.
4. An intra-day market for positive secondary reserve power (minute reserve) mainly to meet the N–1 criterion and to cover the most extreme wind power forecast scenarios that are neglected by the scenario reduction process (cf. below). Hence, the demand for this market is given exogenously to the model.
5. Due to the interactions of CHP plants with the day-ahead and the intra-day market, intra-day markets for district heating and process heat are also included in model. Thereby the heat demand is given exogenously.

19.4 Key Model Equations

The model is defined as a stochastic linear programming model. The stochastic part is presented by a scenario tree for possible wind power generation forecasts for the individual hours (cf. below). The technical consequences of the consideration of the stochastic behavior of the wind power generation is the partitioning of the decision variables for power output, for the transmitted power etc.: one part describes the different quantities of power sold or bought at the day-ahead market. After determination of the optimal values for these variables, they are fixed and do not vary for different scenarios. The other part describes contributions at the intra-day-market both for up and down regulation. The latter consequently depends on the scenarios. So for the power output of the unit group i at time t in scenario s we find $P_{i,s,t} = P_{i,t}^{DAY\_AHEAD} + P_{i,s,t}^{+} - P_{i,s,t}^{-}$. The variable $P_{i,t}^{DAY\_AHEAD}$ denotes the energy sold at the day-ahead market and has to be fixed the day before. $P_{i,s,t}^{+}$ and $P_{i,s,t}^{-}$ denote the positive and negative contributions to the intra-day market. The decision variables for the transmitted power and the loading of storages and use of heat pumps are defined accordingly. Further the model is defined as a multiregional model. Each country is subdivided into different regions, and the regions are further subdivided into different areas. Thus, regional concentrations of installed wind power capacity, regions with comparable low demand and occurring bottlenecks between the model regions can be considered. The subdivision into areas allows considering individual district heating grids.

19.4.1 Objective Function

The objective function (19.1) minimizes the total operation costs $Vobj$ in the whole system considered and for the current planning loop in question (cf. below, for the symbols used see also the list in Sect. 8). The first summand of the objective function describes the fuel costs. The following three summands in the same line consider additional operation and maintenance costs of electricity and heat production. The summands in the second line determine the costs due to starting additional capacity and due to transmitted energy. Further the totals of fuel taxes, of electricity taxes for heat pumps and of emission taxes are determined in the third line. Possible subsidies for individual fuels (e.g., biomass) are considered by the sum in the fourth line. The totals of the value of power plant units being online, of the value of stored water in hydro storages and of the value of the content of electricity and heat storages at the last time step T of a planning loop (i.e., the last time step of a scenario tree) reduce the total operation costs (summands in the fifth and sixth line). The values of unit groups being online and of electricity and heat storages are determined by the shadow values of (19.13) and the corresponding equations for storages. The values of the content of hydro storages are derived with a further model that optimizes the fill level of hydro storages over a year [15].

$$\begin{aligned}
\min Vobj = & \sum_{i \in I^{USING_FUEL}} \sum_{s \in S} \sum_{t \in T} \pi_s F_{r,s,t} f_{F,r}^{PRICE} + \sum_{i \in I^{ELEC}} \sum_{s \in S} \sum_{t \in T} \pi_s o_i P_{i,s,t} \\
& + \sum_{i \in I^{CHP}} \sum_{s \in S} \sum_{t \in T} \pi_s o_i \gamma_i Q_{i,s,t} + \sum_{i \in I^{HEATONLY}} \sum_{s \in S} \sum_{t \in T} \pi_s o_i Q_{i,s,t} \\
& + \sum_{i \in I} \sum_{s \in S} \sum_{t \in T} \pi_s c_i^{STARTUP} P_{i,s,t}^{STARTUP} \\
& + \sum_{r,\bar{r}} \sum_{s \in S} \sum_{t \in T} \pi_s l_{r,\bar{r}}^{TRANS,COST} P_{r,\bar{r},s,t}^{TRANS} \\
& + \sum_{i \in I^{USING_FUEL}} \sum_{s \in S} \sum_{t \in T} \pi_s F_{r,s,t} f_{F,r}^{TAX} \\
& + \sum_{i \in I^{HEATPUMP}} \sum_{s \in S} \sum_{t \in T} \pi_s f_{HEATPUMP,r}^{TAX} W_{i,s,t} \\
& + \sum_{i \in I^{USING_FUEL}} \sum_{s \in S} \sum_{t \in T} \pi_s F_{r,s,t} f_{F}^{EMISSION} f_{EMISSION}^{TAX} \\
& + \sum_{i \in I^{USING_FUEL}} \sum_{s \in S} \sum_{t \in T} \pi_s f_{F,r}^{SUBSIDY} P_{i,s,t}
\end{aligned}$$

$$
\begin{aligned}
&- \sum_{i \in I^{ONLINE}} \sum_{s \in S} \sum_{T} \pi_s Sp^{ONLINE}_{i \in I^{ONLINE},s,T} P^{ONLINE}_{i \in I^{ONLINE},s,T} \\
&- \sum_{i \in I^{HYDRO}} \sum_{s \in S} \sum_{T} \pi_s Sp^{HYDRO}_{i \in I^{HYDRO},s,T} V^{HYDRO}_{i \in I^{HYDRO},s,T} \\
&- \sum_{i \in I^{STORAGE}} \sum_{s \in S} \sum_{T} \pi_s Sp^{STORAGE}_{i \in I^{STORAGE},s,T} \\
&(V^{ELECSTORAGE}_{i \in I^{STORAGE},s,T} + V^{HEATSTORAGE}_{i \in I^{STORAGE},s,T}) \qquad (19.1)
\end{aligned}
$$

19.4.2 Market Restrictions for the Balance of Supply and Demand

The electricity demand constraint is split up into two constraints: one balance equation for the power sold at the day-ahead market and one balance equation for the power sold at the intra-day market. The constraint for the time steps, where the day-ahead market is optimized (i.e., at noon), is defined in (19.3). The equation requires that the sum of the power produced including the expected wind power production minus the planned wind power shedding plus the imported power equals the sum of the exported power to third countries that are not included in the model plus the power used for loading electricity storages and for electric heat pumps plus the exported power to other regions plus the electricity demand.

$$
\begin{aligned}
&\sum_{i \in I_r^{ELEC}} P^{DAY\_AHEAD}_{i,t} + i^{RUNRIVER}_{i,t} + i^{SOLAR}_{i,t} \\
&+ p^{BID\_WIND}_{r,t} - p^{DAY\_AHEAD,WIND\_SHED}_{r,t} \\
&+ \sum_{r} (1 - XLOSS) \cdot P^{TRANS,DAY-AHEAD}_{\bar{r},r,t} = \\
&\sum_{r} d^{ELEC,EXPORT}_{r,t} + \sum_{i \in I_r^{ELECSTORAGE} \cup I_r^{HEATPUMP}} W^{DAY\_AHEAD}_{i,t} \\
&+ \sum_{\bar{r} \in R_r^{NEIGHBOUR}} P^{TRANS}_{r,\bar{r},t} + d^{ELEC}_{r,t} \qquad (19.2) \\
&\forall t \in T^{NOT\_FIXED}, \forall r \in R
\end{aligned}
$$

If the expected wind power production is higher than the actual wind power production, a demand for up regulation arises. Conversely, there exists a demand for down regulation if the expected wind power production is lower than the actual one. The balance equation for the balancing market is described by (19.3). The up and down regulation of the unit groups and the up and down regulation of the loading of electricity storages and the use of heat pumps as well as the up and down regulation by increased/decreased import have to be equal to the difference between the expected wind power production at the bidding hour of the day-ahead market (thereby the possible wind shedding at the day-ahead market has to be considered) and the actual wind power production minus the decreased/increased export. As the model allows wind shedding also during the trade at the intra-day market, the term $P_{r,s,t}^{WIND,-}$ is added to the equation.

$$\begin{aligned}
&\sum_{i \in I_r^{ELEC}} (P_{i,s,t}^{+} - P_{i,s,t}^{-}) + \sum_{i \in I_r^{ELECSTORAGE} \cup I_r^{HEATPUMP}} (W_{i,s,t}^{+} - W_{i,s,t}^{-}) \\
&+ \sum_{r,\bar{r}} (1 - XLOSS)(P_{\bar{r},r,t}^{TRANS,+} - P_{\bar{r},r,t}^{TRANS,-}) - P_{r,s,t}^{WIND,-} \\
&= p_{r,t}^{BID_WIND} - P_{r,t}^{DAY_AHEAD,WIND_SHED} \\
&- p_{r,s,t}^{ACTUAL_WIND} + \sum_{\bar{r},r} (P_{r,\bar{r},t}^{TRANS,-} - P_{r,\bar{r},t}^{TRANS,+}) \\
&\forall r \in R, \forall s \in S, \forall t \in T
\end{aligned} \tag{19.3}$$

The heat markets are represented through an exogenously given demand for each area (19.4):

$$\sum_{i \in I_a^{HEAT}} Q_{i,s,t} = d_{a,t}^{HEAT} \forall a \in A, \forall s \in S, \forall t \in T \tag{19.4}$$

19.4.3 Demand for Ancillary and Nonspinning Secondary Reserves

The market for ancillary services is described by demand restrictions for up (19.5) and down regulation (19.6). The exogenously given demand for up regulation can be supplied either by increased power production of the power producing unit groups or by reduced loading of electricity storages or use of heat pumps, whereas the exogenously given demand for down regulation can be met by decreasing the power production or by increasing the loading of electricity storages or the use of heat pumps. Thereby it is ensured that only spinning reserves can provide primary reserves.

$$\begin{aligned}
&\sum_{i \in I_r^{ELEC}} P_{i,t}^{ANC,+} + \sum_{i \in I^{ELECSTORAGE} \cup I^{HEATPUMP}} W_{i,t}^{ANC,+} \geq d_{r,t}^{ANC,UP} \\
&\forall r \in R, \forall t \in T
\end{aligned} \tag{19.5}$$

$$\begin{aligned}
&\sum_{i \in I_r^{ELEC}} P_{i,t}^{ANC,-} + \sum_{i \in I^{ELECSTORAGE} \cup I^{HEATPUMP}} W_{i,t}^{ANC,-} \geq d_{r,t}^{ANC,DOWN} \\
&\forall r \in R, \forall t \in T
\end{aligned} \tag{19.6}$$

The market for nonspinning secondary reserves is described by demand restrictions for up regulation (19.7). The exogenously given demand for up regulation in a region is calculated as the sum of the demand for secondary reserve due to the N–1 criteria, i.e., the ability to cope with an outage situation involving the fall-out of the largest power plant, and the largest wind power forecast error causing up regulation. The two contributions are added together as two independent stochastic parameters. The secondary reserve demand can be supplied either by increased power production of the power producing unit groups, by reduced loading of electricity storages and heat pumps or imported from other regions involving reservation of the necessary transmission capacity. Some of the demand for secondary reserve is taken care of in the model by the capacity reserved for providing up regulation on the intra-day market. Therefore the demand for secondary reserve in (19.7) is reduced with this amount corresponding to the difference between the expected wind power production and the wind power forecast in each node for wind power forecasts that are lower than expected production.

$$\begin{aligned}
&\sum_{i \in I_r^{ELEC}} P_{i,t}^{NONSP,ANC,+} + \sum_{i \in I^{ELECSTORAGE} \cup I^{HEATPUMP}} W_{i,t}^{NONSP,ANC,+} \\
&+ \sum_r (1 - XLOSS) \cdot P_{r,\bar{r},s,t}^{TRANS,NONSP,ANC,+} \geq \\
&d_{r,t}^{NONSP,ANC,UP} - \max\{0, p_{r,t}^{EXP_WIND} - p_{r,s,t}^{ACT_WIND}\} + \\
&\sum_{\bar{r}} (1 - XLOSS) \cdot P_{\bar{r},r,s,t}^{TRANS,NONSP,ANC,+} \\
&\forall r \in R, \forall t \in T, \forall s \in S
\end{aligned} \tag{19.7}$$

Generally, the contribution of the individual power sources to the down regulation cannot be larger than the actual committed production or transmission. Hence, corresponding restrictions applied to power plants, electricity storages and heat pumps as well as transmission lines have to be considered. Further the possible wind shedding has to be lower than the wind power production expected when the day-ahead is cleared (i.e., at 12 O'clock).

19.4.4 Capacity Restrictions

In typical unit commitment models, the restrictions for start-up costs, reduced part-load efficiency and lead times include integer variables. However, this is hardly feasible for a model representing a national market. Therefore [20] proposes an approximation to model the restrictions in a linear way, which makes it necessary to introduce the additional decision variable $P_{i,s,t}^{ONLINE}$. The idea is illustrated in Fig. 19.1.

The capacity restrictions for the unit groups generating electricity are defined in the following equations for maximum and minimum power output. The power which is committed to the day-ahead market plus the production sold at the balancing market plus the contribution to the ancillary and nonspinning secondary reserve has to be lower than $P_{i,s,t}^{ONLINE}$, the capacity online (19.8):

$$P_{i,s,t} + P_{i,t}^{ANC,+} + P_{i,s,t}^{NONSP,ANC,+} \leq P_{i,s,t}^{ONLINE}, \forall s \in S, \forall t \in T \tag{19.8}$$

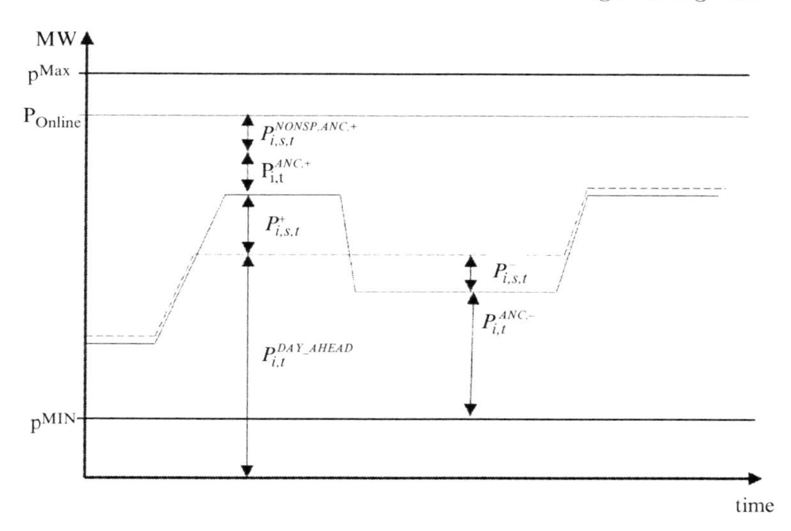

Fig. 19.1. Illustration of the contribution of a power generating turbine to the different markets and the capacity online

Thereby it is ensured that the power which is committed to the day-ahead market does not exceed the installed capacity of the considered unit group. The capacity online multiplied with the minimum output factor $p_i^{MIN\_PROD}$ forms a lower bound to the possible power output (19.10):

$$P_{i,s,t} - P_{i,t}^{ANC,-} \geq p_i^{MIN\_PROD} \cdot P_{i,s,t}^{ONLINE} \qquad (19.9)$$
$$\forall i \in I^{ELEC}, \forall s \in S, \forall t \in T$$

The value of the decision variable $P_{i,s,t}^{ONLINE}$ itself has to be lower than the maximum capacity of the unit group i including the outage factor $p_i^{GKDRATE}$ (19.10):

$$P_{i,s,t}^{ONLINE} \leq (1 - p_i^{GKDERATE}) p_i^{MAX\_PROD}, \forall i \in I^{ELEC}, \forall s \in S, \forall t \in T \qquad (19.10)$$

CHP unit groups are distinguished into extraction condensing unit groups and backpressure unit groups. To represent the possible operation modes of combined power and heat production, additional equations to match the technical restrictions are used. As the model is defined as a multiregion model, the capacity restrictions of the transmission lines have to be considered. The transmitted power plus the reservation of transmission capacity for nonspinning reserves has to be lower than the installed transmission capacity (19.12):

$$P_{r,\bar{r},t}^{TRANS,DAY-AHEAD} + P_{r,\bar{r},s,t}^{TRANS,+} - P_{r,\bar{r},s,t}^{TRANS,-} +$$
$$P_{r,\bar{r},s,t}^{TRANS,NONSP,ANC,+} \leq l_{r,\bar{r}}^{TRANS,MAX} \qquad (19.11)$$
$$\forall r, \bar{r} \in R, \forall s \in S, \forall t \in T$$

Further it is ensured that the transmission planned at the day-ahead market does not exceed the transmission capacity.

19.4.5 Fuel Consumption

Equation (19.13) determines the fuel used by conventional power plants for producing power and heat. In order to avoid that unit groups are always kept online, a fuel consumption proportional to the online capacity $P_{i,s,t}^{ONLINE}$ is included. Thereby part load production achieves lower fuel efficiency than full load production.

$$\begin{aligned}\sum_{r} F_{s,t} = e_i \cdot P_{i \in I^{USING\_FUEL},s,t}^{ONLINE} + g_i \cdot (P_{i \in I^{USING\_FUEL},s,t} \\ + \gamma_i Q_{i \in I^{CHP},s,t} + Q_{i \in I^{HEATONLY},s,t}), \\ \forall i \in I^{USING\_FUEL}, \forall s \in S, \forall t \in T\end{aligned} \tag{19.12}$$

Where e_i is the fuel consumption parameter when unit group i is online, f_i the fuel consumption parameter according to the full load efficiency when unit group i produces power. The increased fuel consumption caused by heat production of CHP plants is considered by the parameter γ , the electric power reduction due to heat production.

19.4.6 Started Capacity

Additional costs due to power plant start-ups influence considerably the unit commitment decisions of plant operators. Therefore the started capacity has to be determined (19.13):

$$P_{i,s,t}^{STARTUP} \geq P_{i,s,t}^{ONLINE} - P_{i,s,t-1}^{ONLINE}, \forall i \in I^{ONLINE}, \forall s \in S, \forall t \in T \tag{19.13}$$

19.4.7 Additional Equations

Further equations describe hydro reservoir plants, pumped hydro storage plants (and other electricity storage devices such as compressed-air storage), heat storages and electric heat pumps. Furthermore lead-times are implemented, which describe the needed time to change the capacity online of a unit group i.

19.4.8 Nonanticipativity Constraints

As a multistage scenario tree is used to model uncertainty, it has to be ensured that the decisions taken at time t must be the same if two scenarios are indistinguishable until time t by nonanticipativity constraints. A detailed formulation of the constraints can be found, e.g., in [4].

19.5 Key Model Features

19.5.1 Rolling Planning

It is not possible and reasonable to cover the whole simulated time period of for instance 2 weeks with only one single scenario tree. Therefore the model uses the multistage recursion approach with rolling planning [5]. In stochastic multistage recourse models, there exist two types of decisions: decisions that have to be taken immediately and decisions that can be postponed. The first kind of decisions are called "root decisions," as they have to be decided "here and now" and before the uncertain future is known. The second kind of decisions is called "recourse decisions." They are taken after some of the uncertain parameters become known. These "recourse decisions" can start actions which might possibly revise the first decisions. In the case of a power system with wind power, the power generators have to decide on the amount of electricity they want to sell at the day-ahead market before the precise wind power production is known (root decision). In most European countries this decision has to be taken at least 12–36 h before the delivery period. And as the wind power prediction is not very accurate, recourse actions are necessary in most cases when the delivery period is in the near future and the wind power forecast becomes more and more accurate (recourse decisions). In general, new information arrives on a continuous basis and provides updated information about wind power production and forecasts, the operational status of other production and storage units, the operational status of the transmission grid, heat and electricity demand and updated information about day-ahead and regulating power market prices. Hence, an hourly basis for updating information would be most adequate. However, stochastic optimization models quickly become intractable, since the total number of scenarios has a double exponential dependency in the sense that a model with $k+1$ stages, m stochastic parameters, and n scenarios for each parameter (at each stage) leads to a scenario tree with a total of $s = n^{mk}$ scenarios (assuming that scenario reduction techniques are not applied). It is therefore necessary to simplify the information arrival and decision structure in a stochastic model. Hence, the model steps forward in time using rolling planning with a 3 h step holding the individual hours. This decision structure is illustrated in Fig. 19.2, showing the scenario tree for four planning periods covering half a day. For each planning period a three-stage, stochastic optimization problem is solved having a deterministic first stage covering 3 h, a stochastic second stage with five scenarios covering 3 h, and a stochastic third stage with ten scenarios covering a variable number of hours according to the rolling planning period in question (in this way the determination of the shadow values is eased). In the planning period 1 the amount of power sold or bought from the day-ahead market is determined. In the subsequent replanning periods the variables standing for the amounts of power sold or bought on the day-ahead market are fixed to the values found in planning period 1, such that the obligations on the day-ahead market are taking into account when the optimization of the intra-day trading takes place.

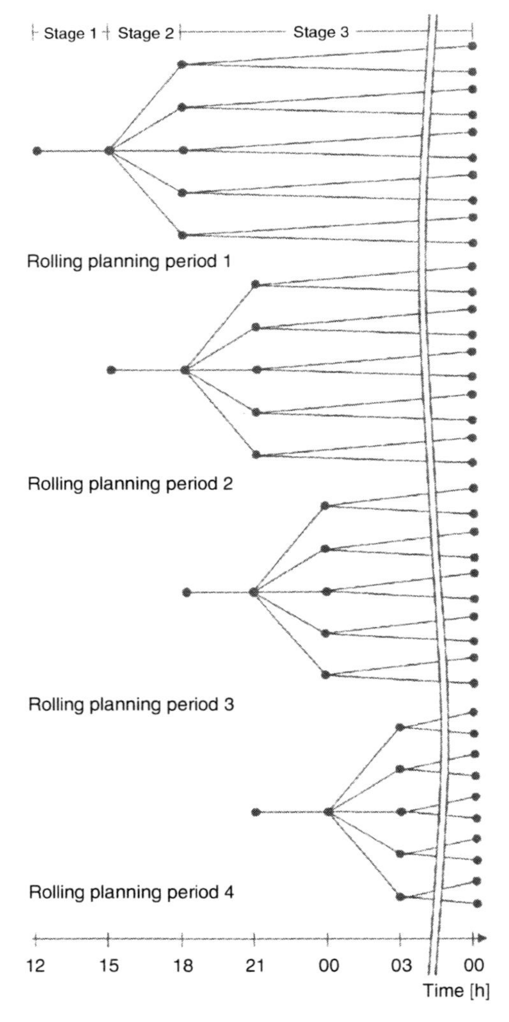

Fig. 19.2. Illustration of the rolling planning and the decision structure in each planning period within half a day

19.5.2 Scenario Creation and Scenario Reduction

The inclusion of the uncertainty about the future wind power production in the optimization model is considered by using a scenario tree. The scenario tree represents wind power production forecasts with different forecast horizons corresponding to each hour in the optimization period. For a given forecast horizon the scenarios of wind power production forecasts in the scenario tree is represented as a number of wind power production outcomes with associated probabilities, i.e., as a discrete distribution of future wind power production levels. The construction of this scenario tree is carried out in two steps:

1. Modeling of the wind speed forecast error and the simulation of the distributed wind power forecast scenarios, whose first values of the root node are identical.
2. Reduction of the wind power forecast scenarios to the scenario tree with three stages.

In the following these steps are described in more detail.

19.5.3 Modeling the Wind Power Forecast Data Process

The generation of wind power forecast scenarios is based on time-series of measured wind speed and of historical forecast errors of wind speed predictions. The increasing trend of the wind speed forecast error with rising forecast horizon is reproduced using multidimensional Auto Regressive Moving Average (ARMA) time-series:

$$X_{WF}(k) = \alpha_{WF} X_{WF}(k-1) + Z_{WF}(k) + \beta_{WF} Z_{WF}(k-1) \tag{19.14}$$

where $X_{WF}(k)$ is the wind speed error and $Z_{WF}(k)$ is the random variable with given standard deviation in the forecast hour k for the wind power farm $WF(X_{WF}(0) = 0$ and $Z_{WF}(0) = 0$, given α_{WF} and $\beta_{WF})$. The random variables $Z_{WF}(k)$ are normally distributed and created by Monte Carlo simulations resulting in a predefined large number of scenarios of the wind speed forecast error. Thereby the correlation between the forecast errors at spatial distributed wind power farms is considered following the approach of [18]. For example, data analysis from Sweden shows that the closer the stations, the higher are the correlations between forecast errors and that the correlation between different stations increases with forecast lengths.

19.5.4 Scenario Reduction

In order to keep computation times small for models representing a transnational market with a huge number of generating units, only significantly less scenarios than the scenarios created before by the Monte Carlo simulations can be used. Simply generating a very small number of scenarios by Monte Carlo simulations is not wanted since less scenarios cannot represent the distribution of wind speed forecast errors adequately. Hence, the aim is to loose only a minimum of information by the reduction process applied to the whole set of scenarios. As the current version of the scenario reduction algorithm reduces the standard deviation of the original generated scenarios, the most extreme wind power forecast scenarios have to be considered by the nonspinning secondary reserve power market. Two steps are necessary for the scenario reduction: first, the pure number of scenarios has to be reduced.

Afterwards, based on the remaining scenarios that still form a one-stage tree, a multistage scenario tree is constructed by deleting inner nodes and creating branching within the scenario tree. Therefore a stepwise backward scenario reduction algorithm based on the approach of [7] is used: the original scenario tree is modified through bundling similar scenarios or part of scenarios. Bundling two scenarios or parts of scenarios means deleting the one (or the part of the scenario) with the lower probability and adding its probabilities to the remaining one (Fig. 19.3). As a measure for the similarity of different scenarios, the Kantorovich distance between two scenarios is used.

19.5.5 Overall Tool Architecture

The overall architecture of the model is depicted in Fig. 19.4. Besides the core optimization model, the *JointMarketModel*, the tool encompasses the

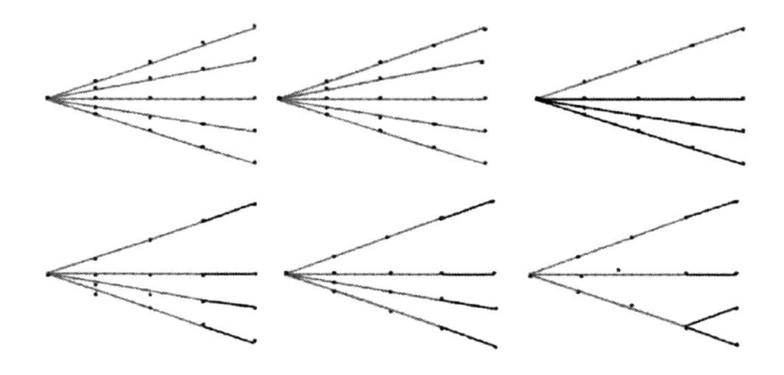

Fig. 19.3. Example for the backward scenario reduction heuristic. Source: modified figure from [11]

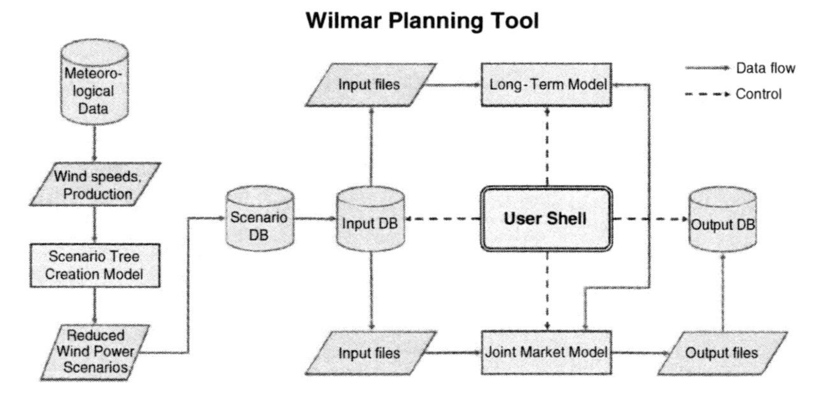

Fig. 19.4. Overall structure of the WILMAR planning tool

ScenarioTreeTool for scenario creation, the *LongTermModel* for calculation of water values as well as several data bases and control facilities.

19.6 Application

Several application studies have been done using the WILMAR model (cf. [2, 3, 15]). In the following, a few exemplary results are reported The analysis has mainly focused on wind power integration issues in a power system covering Germany, Denmark, Finland, Norway and Sweden in year 2010. A base scenario for the development of the power system configuration excluding the development in wind power capacity has been agreed upon among the project partners. The 2010 base power system configuration is a projection of the present power system configuration in Germany and the Nordic countries to 2010 by introducing investments in power plants and transmission lines that are already decided today and scheduled to be online in 2010, and by removing power plants that have been announced to be decommissioned before 2010. Likewise scenarios for fuel prices, electricity and heat demand and the other parameters in the WILMAR Planning tool have been defined. The assumptions are described in [16]. This 2010 base power system configuration is supplemented by three scenarios for the installed wind power capacity in 2010, cf. Fig. 19.5:

1. A base wind power capacity scenario consisting of a "most likely to happen" projection of wind power capacity according to a review of public information provided by the WILMAR consortium.

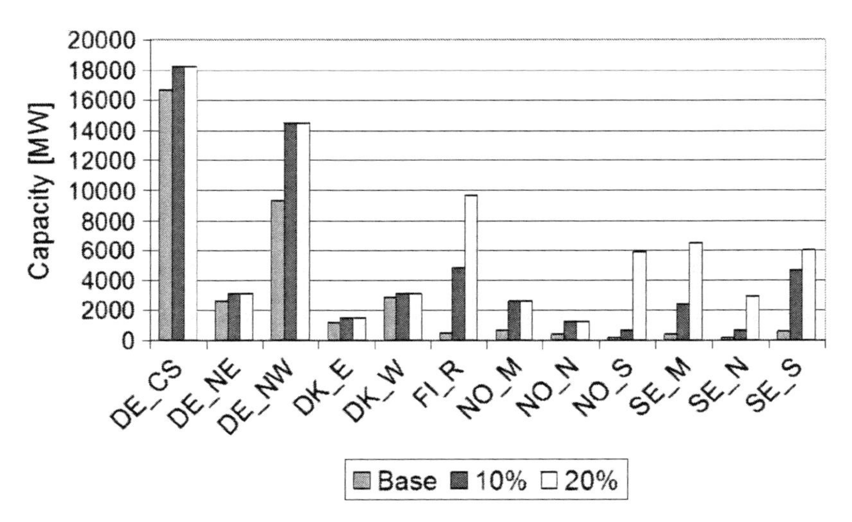

Fig. 19.5. Installed wind power capacity in each region in the three wind power capacity development scenarios

2. A 10% wind power capacity scenario consisting of installed wind power capacity in Denmark and Germany corresponding to a "most likely to happen" 2015 projection, i.e., a stronger growth than in the base scenario. A unrealistic strong growth of wind power capacity in Finland, Norway and Sweden corresponding to installed wind power producing 10% of the electricity consumption in these countries in 2010 is furthermore assumed.
3. A 20% wind power capacity scenario with the same assumption for Denmark and Germany as in the 10% scenario, but a stronger growth in Finland, Norway and Sweden with wind power production covering 20% of electricity consumption in 2010.

The reason for supplementing the base wind scenario with two unrealistic high growth wind power scenarios for especially Finland, Norway and Sweden is that we wanted scenarios where wind power production had a significant effect on the operation of the rest of the power system, and these amounts will most likely not be present in the Nordic power system (except Denmark) in 2010.

The three wind power cases have been simulated with the WILMAR Planning tool for 2010. During this year the wind power production constitutes 5.8, 11.1, and 15.1% of the total electricity consumption for respectively the wind cases: base, 10 and 20%.

The integration of wind power leads to a reduction of the total system operation costs if investment costs are disregarded, because the wind power production replaces more expensive power production. The total system operation costs consists of the sum of fuel costs, variable operation and maintenance costs, start-up costs, CO_2 emission costs and taxes and tariffs, cf. above. By comparing the system operation costs in different wind power cases, the value of different amounts of wind power production in the power system can be evaluated. As increased amounts of wind power production also lead to a decrease in the usage of hydropower, i.e., an increase in the amount of water in hydropower reservoirs, the system costs have to take into account the value of regulated hydropower not used. The avoided costs of wind power production are €35/MWh wind power production when comparing the 10% case to the Base case, i.e., adding 52 TWh wind power production in the 10% case relatively to the base wind case reduces the system costs with on

Table 19.1. Avoided system operation costs with increasing amounts of wind power production

Case name	Change system costs [M€]	Value saved water [M€]	Change windpower production [TWh]	Avoided costs per MWh extra wind [€/MWh]
10%/Base	1,614	188	52	35
20%/Base	2,275	301	90	29
20%/10%	662	195	38	22

average €35 per MWh wind power production added. A reduction in the avoided costs is expected, because as more wind power production is added, thermal production with lower and lower marginal production costs will be replaced. This is also the case, if thermal production was added to the power system. Due to this effect the value of adding wind power decreases with increasing amounts of wind power, which can be observed by the avoided costs being only €22 per MWh wind power production when comparing the 20% case with the 10% case. Focusing on the economic consequences for wind power producers when adding more wind power production Table 19.2 shows the average day-ahead power price achieved by wind power producers and the average penalties paid due to forecast errors, and Table 19.3 shows the revenue of wind power producers relatively to the penalties paid.

Table 19.2 shows that the revenue of wind power producers are reduced significantly when more wind power is added to the power system, mainly due to a reduction in the average day-ahead power price received by wind power producers, but also due to an increase in the average penalty of being in imbalance due to forecast errors. Conventional power producers also experience a reduction in average power prices, but the reduction is less than for the wind power producers, as there still will remain high-price periods when the wind power production is low. The average penalties of being in imbalance are nearly constant in the 10% to the 20% case. This is due to the extra wind power production in the 20% relatively to the 20% case being added in the hydropower dominated systems in Norway, Sweden and Finland,

Table 19.2. Average prices achieved by wind power producers and penalties paid due to wind power forecast errors

	Average day-ahead price [€/MWh Wind]	Average penalty up regulation [€/MWh forecast error]	Average penalty Down regulation [€/MWh forecast error]
Base	40.7	1.6	1.6
10%	29.7	3.3	3.2
20%	20.7	3.4	3.2

Table 19.3. Total revenue and penalties paid by wind power producers in the three wind power cases. All figures in M€except for the last column

	Revenue day-ahead	Sold intraday	Bought intraday	Total revenue	Up regulation penalty	Down regulation penalty	Penalty/ Revenue
Base	2,448	304	477	2,275	18	12	1.3%
10%	3,412	478	759	3,130	73	48	3.8%
20%	3,195	460	765	2,891	101	70	5.9%

where the hydropower with reservoir units are able to provide large amounts of balancing power without increased operational costs.

The operational integration costs of wind power production disregarding investments have also been analyzed using the three wind power case mentioned above. The integration costs have been divided into two groups:

- System operation costs due to forecast errors, which is analyzed by comparing the system operations costs in the stochastic simulation with the system costs in a WILMAR Planning tool simulation with perfect foresight, i.e., perfectly predictable wind power production.
- System operation costs due to variability, which is analyzed by comparing the system operations costs in the perfect foresight simulation with the system costs in a WILMAR Planning tool simulation with constant wind power production within each week.

Figure 19.6 shows the results for the three wind power cases. Disregarding the base wind case with low integration costs, the results show that the costs of wind being variable is larger than the costs connected to being partially unpredictable. So the time periods with low loads and large amounts of wind power production generate more costs than the balancing costs. One reason for this is that the regulating hydropower production has very low balancing costs, and that the balancing market modeled is extremely efficient. In reality balancing costs would be higher due to transaction costs and in some cases market power.

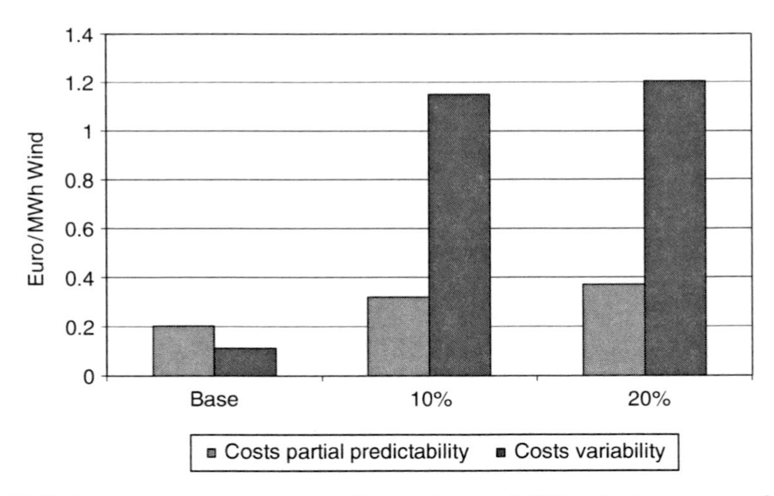

Fig. 19.6. Increase in system operation costs per MWh wind power production when comparing system cost in a model run with constant weekly average wind power production with a perfect foresight model run (Costs variability), and when comparing system costs in a stochastic model run with a perfect foresight run (Costs partial predictability) Source: [17]

19.7 Final Remarks

The WILMAR planning tool has a rather high degree of technical detail whilst including at the same stochastic programming approaches as well as a detailed modeling of the relevant markets To our knowledge no existing electricity system modeling tool except the WILMAR Planning tool handles endogenously both stochastic wind power production and water stored in hydro reservoirs, such that the WILMAR Planning tool has and will in the future contribute significantly to the technical progress within the field of modeling of electricity systems. Analysis of the technical and economic consequences of extending the wind power capacity in a power system covering Denmark, Germany, Finland, Norway and Sweden have shown the applicability of the developed tool. Notably the tool allows quantifying the impact of increased wind generation on power plant operation and on so-called integration costs. Currently the tool is applied in a detailed study for one grid operator, including several extensions such as mixed-integer unit commitment. Furthermore the tool is developed further at a European scale within the EU funded SUPWIND project. One issue to be analyzed is the impact of market design on the power prices. Another key point is the integration of the model into the operative and strategic planning processes of transmission system operators.

References

1. H. Auer et al. Cost and Technical Constraints of RES-E Grid Integration. Technical report, EU-project GreenNet, 2004. `http://www.greennet.at/downloads/WP2%20Report%20GreenNet.pdf`.
2. R. Barth, H. Brand, and C. Weber. Transmission restrictions and wind power extension – case studies for Germany using stochastic modelling. In *Proceedings of the European Wind Energy Conference, London*, 2004.
3. R. Barth, H. Brand, D. J. Swider, C. Weber, and P. Meibom. Regional electricity price differences due to intermittent wind power in Germany – Impact of extended transmission and storage capacities. *International Journal of Global Energy Issues*, Special issue "Integrating intermittent renewable energy technologies, limits to growth?":276–297, 2006.
4. J. Birge and F. Louveaux. *Introduction to stochastic programming.* Springer, New York, 2000.
5. C. S. Buchanan, K. I. M. McKinnon, and G. K. Skondras. The recourse definition of stochastic linear programming problems within an algebraic modelling language. *Annals of operations research*, 104:15–32, 2001.
6. J. Coughlan, P. Smith, A. Mullane, and M. O'Malley. Wind turbine modelling for power system stability analysis. A system operator perspective. *IEEE Transactions on Power Systems*, 22:929–936, 2007.
7. J. Dupacova, N. Gröwe-Kuska, and W. Römisch. Scenario reduction in stochastic programming – an approach using probability metrics. *Mathematical Programming Series A*, 95:493–511, 2003.

8. R. Elsässer. Kosten der Windenergienutzung in Deutschland. In *Präsentation im Rahmen der Sitzung des Wirtschafts-beirates der Union*, 2002. Berlin, 23 July 2002.
9. ETSO (ed.). European Wind Integration Study (EWIS) - towards a successful integration of wind power into European electricity grids. Technical report, Final Report Phase I, 2007. available from http://www.etso-net.org/upload/documents/Final-report-EWIS-phase-I-approved.pdf.
10. M. Fuchs. Windpower in Germany. Present situation and outlook. In *Presentation Brussels*, 2003. 23 January 2003.
11. N. Gröwe-Kuska, H. Heitsch, and W. Römisch. Scenario reduction and scenario tree construction for power management problems. In *IEEE Bologna Power Tech Proceedings*, Bologna, 2001. Downloadable at http://www.mathematik.hu-berlin.de/~heitsch/ieee03ghr.pdf.
12. M. J. Grubb. Value of variable sources on power systems. *IEE Proceedings-C*, 138:149165, 1991.
13. IEA, editor. *Integration of wind power into electricity grids. economic and reliability impacts.* Workshop Paris, 2004. 25 May 2004.
14. ILEX Energy Consulting. Quantifying the system costs of additional renewables in 2020. A report of ILEX Energy Consulting in association with Manchester Centre for Electrical Energy (UMIST) for the Department of Trade and Industry (DTI), 2002.
15. P. Meibom, R. Barth, H. Brand, and C. Weber. Impacts of wind power in the Nordic electricity system in 2010. In *Technologies for sustainable energy development in the long term. Proceedings Risø international energy conference*, 2005. Risø (DK), 23–25 May 2005.
16. P. Meibom, J. Kiviluoma, R. Barth, H. Brand, C. Weber, and H. Larsen. Value of electrical heat boilers and heat pumps for wind power integration. *Wind Energy*, 10:321–337, 2007.
17. P. Meibom, C. Weber, R. Barth, and H. Brand. Operational costs induced by fluctuating wind power production in Germany and Scandinavia. Working Paper, 2008.
18. L. Söder. Simulation of wind speed forecast errors for operation planning of multi-area power systems. In *8th International Conference on Probabilistic Methods Applied to Power Systems (PMAPS)*, Iowa, 2004.
19. D. J. Swider and C. Weber. The Costs of Wind's Intermittency in Germany: Application of a Stochastic Electricity Market Model. *European Transactions on Electrical Power*, 17:151–172, 2007.
20. C. Weber. *Uncertainty in the power industry: methods and models for decision support.* Springer, New York, 2005.

Appendix: Symbols Used

Parameters

$c^{STARTUP}$	Start-up costs
d	Demand
e	Fuel consumption for capacity being online
f	Fuel related parameter
g	Fuel consumption for electricity produced
$l^{TRANS,COST}$	Transmission cost
$l^{TRANS,MAX}$	Transmission capacity
o	Other variable cost
p^{MAX_PROD}	Max. capacity
p^{MIN_PROD}	Min. capacity
p^{BID_WIND}	Expected wind power for day-ahead scheduling
p^{EXP_WIND}	Expected wind power
p^{ACTUAL_WIND}	Actual wind power
$p^{GKDERATE}$	Availability factor
Sp	Cost dependent on capacity only
X	Forecast deviation
$XLOSS$	Transmission loss
Z	Momentaneous random forecast error
α	Autoregression coefficient
β	Moving average coefficient
γ	Electric power reduction due to heat production
π	Probability

Variables

$Vobj$	Total operation costs
F	Fuel consumption
P	Power
Q	Heat
V	Storage content
W	Electricity for loading storages

Indices, Superscripts

i, I	Unit group
k	Forecast hour
F	Fuel
$r, \bar{r}, R$	Region
s, S	Scenario
t, T	Time step
WF	Wind forecast error
$+/-$	Up/down deviation
ANC	Ancillary reserve
CHP	Combined heat and power plants
DAY_AHEAD	Valid for day-ahead scheduling
$ELEC$	Electricity

Indices, Superscripts (cont.)

$ELECSTORAGE$	"Electricity storage" plant (pumped hydro or compressed-air)
$EXPORT$	Exports to neighboring countries
$HEAT$	Heat
$HEATSTORAGE$	Heat storage
$HEATPUMP$	Heat pumps and other flexible electricity usages
$HYDRO$	Hydro power plant
$NEIGHBOUR$	neighboring countries
NOT_FIXED	Value is not fixed
$NONSP, ANC$	Nonspinning reserve
$ONLINE$	Unit online
$RUNRIVER$	Run-of-river power plants
$SOLAR$	Photovoltaics and other solar power plants
$STARTUP$	Started capacity
$STORAGE$	Hydro storage plant
$SUBSIDY$	subsidy values
TAX	Tax values
$TRANS$	Transmission
$UP, DOWN$	Up and down regulation
$USING\_FUEL$	Unit using fuel
$WIND$	Wind power plant
$WIND\_SHED$	Wind curtailment

Part IV

Stochastic Programming in Pricing

20

Clean Valuation with Regard to EU Emission Trading

Karl Frauendorfer and Jens Güssow

Summary. In the electric power industry the observed increases of electricity price dynamics combined with the characteristic periodicity of related decision processes have motivated the use of multistage stochastic programming in recent years to provide flexible models for practical applications in the sector. Specifically in power generation and trading the planning process must obey highly complex interrelations between manifold influences. They range from short term price fluctuations as observed in spot markets to long term changes of fundamental influences. Not only changes in the electric supply system itself must be considered, but also the related availability and costs of required fuels. For example, the prices and usability of natural gas in power generation also depend on the existence of respective deployment and distribution systems. Furthermore the electric power sector is exposed to manifold regulatory uncertainties related to the rules imposed by the responsible authorities. Recently environmental issues have become very popular due to the ongoing discussion on climate change. In January 2005 the European Emissions Trading Scheme (EU ETS) has been launched which by many is considered a new key element in efficient electricity market operations. In this paper we will introduce a modeling framework that considers the influence of emission trading on portfolio problems in the electric power sector by applying clean valuation schemes that particularly take fuel costs, emission efficiency in combination with investment possibilities and generation flexibility into account. Sensitivity analysis is performed with respect to changes in technology, volatilities and price scenarios.

20.1 Introduction

Since the beginning of the year 2005 the electricity markets in Europe have undergone accelerated changes in view of the *EU Emissions Trading Scheme (EU ETS)* which currently comprises a mandatory cap and trade scheme for CO_2 emission rights. It concerns large parts of the carbon emitting industry with installations above certain capacity thresholds. The electric power industry herein has a market share of more than 50%. In our work we specifically focus on this sector.

Basically the EU ETS restricts the overall amount of CO_2 emissions in EU countries by allocating a limited number of so called EU emission allowances (EUAs). One EUA gives the right to emit one ton of carbon dioxide equivalents (tCO_2e). These allowances can be freely traded between the participants. Next to OTC trading, several centralized market places have been established all over Europe with the ECX (European Climate Exchange) by far the most important.

Two trading periods are currently in focus. The first allocation interval started in January 2005 and will end in December 2007. It is basically intended to give the markets some room for settlement in order to reach minimum levels of experience and liquidity. The second 5-year interval from 2008 to 2012 coincides with the compliance period of the Kyoto Protocol with its 8% greenhouse gas (GHG) reduction targets compared to 1990. It is characterized by tighter restrictions compared to the current compliance period. The EU *linking directive* (2004/101/EC) allows participants to acquire emission rights by means of the Kyoto mechanisms CDM (Clean Development Mechanism) and JI (Joint Implementation) by converting a certain share of respective emission rights into EUAs. This basically allows market participants to acquire emission certificates by engaging in international projects approved by the UN that contribute to reductions in greenhouse gas emissions. After some initial reluctancy, this mechanism is now actively used.

Each country's national legislation is required to adopt national allocation plans (NAPs) to distribute predetermined numbers of allowances between participants which must be approved by the EU Commission. While large shares are required to be allocated for free, each country is also allowed to perform a national auction for a limited number of EUAs.

Participants submit their verified emission data to national registries and finally must surrender a sufficient number of allowances to avoid penalties. The records of real emissions from national registries for phase I (2005–2007) are used for a refined planning of phase II (2008–2012).

The trading scheme in general as well as the specific implementations of the national allocation are subject to ongoing controversial discussions. However, it is expected that carbon trading will stay a reality in the EU and that the existing market will be continued or replaced by equivalent mechanisms in the years that follow 2012.

The fast rising importance of carbon trading requires new approaches for analysis and management problems in the electric power sector. We apply multistage stochastic programming to investigate the consequences of clean valuations in dynamic investment and trading problems that are exposed to emission trading.

20.2 Market Developments and Observations

20.2.1 General Observations of the EU ETS

With the introduction of emission trading in January 2005, an important integrating market mechanism has been added to electricity markets in Europe. Economically emission allowances can be treated as freely tradable commodities that due to basically unlimited storability and free transferability have identical values in all supply regions where the rules apply. No trading limits exist apart from overall limited liquidity. According to [22] the traded market volumes in brokered OTC and exchange trading have more than tripled between 2005 (262 $MtCO_2e$) and 2006 (817 $MtCO_2e$) with an estimate of 1,750 $MtCO_2e$ for 2007. An unknown amount is traded bilaterally with estimates between 100 and 200 $MtCO_2e$ per year. Overall estimates in [2] are similar. The EU ETS currently comprises 52% of all carbon emissions in Europe, which accounts for 45% of the greenhouse gas emissions. This includes combustion installations with thermal input of more than 20 MW_{therm}, oil refineries, coke ovens as well as industrial production units for ferrous metals, cement and paper. With more than two billion allowances allocated and assuming a value of €20/tCO_2e, the overall market value is more than €40 billion each year. Trading activities have substantially increased recently indicating that the EU ETS is now generally accepted by a majority of market participants.

One crucial element for the emission market to work is that the overall contingent of EUAs is properly set on a level that is lower than typical requirements in a business-as-usual situation. Too generous allocation for 2005–2007 has led to a price crash from highs of more than €30/tCO_2e to almost zero as illustrated in Fig. 20.1. This observation has been used by opponents of the trading scheme to argue that the market mechanism does not work. However, period I was basically intended to serve as a learning phase and has been used to collect relevant data through the newly established obligatory emission registries. Presumably tighter caps based on significantly improved data and experience have been defined for the upcoming commitment period. Carry-over of emission rights from phase I to phase II is prohibited by current regulation, but is likely to be allowed for some new phase III starting in 2013. EUA prices of around €20/tCO_2e for 2008 and the following years indicate that participants consider the market situation to be competitive for the next compliance period.

Several internal and external factors determine the price of emission rights. The overallocation for phase I has led to an estimated excess of EUAs of currently about 160 $MtCO_2e$ [17] resulting in basically zero emission costs. For phase II starting in 2008 estimated fair values vary between €10 and 25/tCO_2e according to several studies. Current market observations also fall into this range. However, the estimates are exposed to a considerable

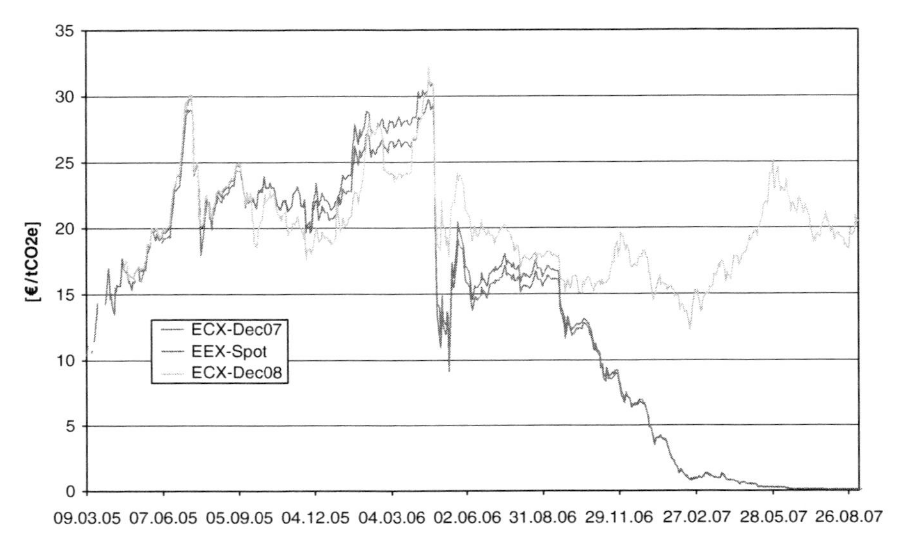

Fig. 20.1. EUA price developments for phase I and II (spot and future contracts)

degree of uncertainty. Consequences from changes and extensions in the EU ETS market design are difficult to predict. Based on the linking directive 2003/87/EC, for example, a limited share of carbon credits from the Kyoto mechanisms CDM or JI can be used to offset some of the emissions produced by replacing respective EUAs through CERs (Certified Emission Reductions) or ERUs (Emission Reduction Units). Especially CDM has become a popular option. German attempts to increase the allowable shares of CERs/ERUs from 12 to about 22% may also have considerable affects on the liquidity of the market. Another important extension for phase II refers to the integration of the aviation sector from 2011 with an estimate of 160 $MtCO_2e$ per year.

Overall for phase II an operational market can be expected with aggregate reduction requirements of 10%. The particular market situations in different countries must be individually observed. While some countries are well on track to achieve their reduction targets (e.g., Germany, Spain, or France) others are still being far off (e.g., Ireland, UK, Austria) [22].

20.2.2 Consequences for the Electric Power Sector

The importance and complexity of electricity trading has risen considerably in recent years as reflected by an increased liquidity and diversity of financial trading instruments like futures and other derivatives. According to [22, 23] most market participants expect further relevant changes to their business from the introduction of the carbon trading scheme. CO_2 trading can be considered as a completely new field of study that adds to the manifold complexity of electricity markets. In general the internalization of environmental costs by means of the EU ETS followed by some type of cost-passthrough leads to increased electricity prices. In an individual context abatement costs will be balanced with the costs to emit by holding respective emission permits. Incentives to invest into more expensive technologies with improved fuel efficiencies in turn depend on the costs of CO_2 and the risks involved.

Neglecting technical restrictions and other peculiarities in electric power trading, wholesale electricity prices are determined by their marginal costs at the time of generation. In principle they can be derived from fuel costs which in turn determine production costs taking electric efficiencies of power plants into account. Carbon trading has intensified the interrelations between electricity supply regions. Considerable repercussions on electricity prices following the news of April 2006 on lower than anticipated carbon consumption in 2005 have illustrated the importance of CO_2 prices for the energy sector.

A considerable share of free allocations is currently progressed through a mechanism called *Grandfathering* that relates the number of granted certificates to the existing generation capacities of a company. This procedure recognizes the fact that many existing power plants are long-term investments that were planned before any knowledge of an emission trading scheme. One undesired consequence of this procedure are so called *Windfall profits* which arise from the possibility to generally consider emission prices as opportunity costs even if originally allocated for free [27]. Observable EUA prices can therefore be balanced in a company's account and passed on to consumers through increased electricity prices. Even though the share of accountable emission rights has been limited by the EU, this phenomenon has raised questions about the free allocation mechanism.

On a business level the emissions resulting from electricity production must be analyzed with respect to internal abatement costs, EUA prices and the possibility to obtain emission rights from CDM/JI projects. Power generators and utilities are therefore obliged to analyze and incorporate the consequences of the EU ETS into their business planning and to analyze the risks and opportunities of different options. In the future the efficient management of carbon rights can be expected to become a crucial element for successful competition.

Consequently in face of carbon trading it becomes increasingly important for participants in the electric power industry to understand the particularities of global electricity markets in a superordinate context, since the trading of carbon not only tightens the links between different supply regions within Europe but also raises the importance of global developments in climate and energy policy. It is even likely that the EU ETS will be linked to similar trading schemes worldwide as soon as they are established.

20.3 Clean Valuation in a Multicommodity Context

20.3.1 Price Modeling in Liberalized Electricity Markets

The liberalization of electricity markets has been a field of intensive research for many years. The interdisciplinary character of the topic can be summarized as a complex combination of financial, socio-economic and engineering sciences. In business applications especially the evolvement of electricity prices and the consequences for appropriate risk management measures have been in focus for many years.

Physically electricity is a homogeneous standardized commodity that can be specified by type, voltage and frequency. However, in market terms it is far from being homogeneous due to its network bound transmission and distribution coupled with minimum requirements concerning security of supply. The load patterns of electric systems dictate certain degrees of flexibility that involve different sources of generation conditioned to network restrictions. Typically a differentiation between base- and peak-load requirements is made which implicates some sort of quality with respect to daily load patterns. Furthermore unforeseen fluctuations resulting from failures in the generation and transmission system require additional regulation and reserve power capacities that can be called when needed to maintain system stability. Technically this can be achieved through a diversified infrastructure that comprises an appropriate mix of technologies, which after actual implementation remains exposed to manifold uncertainties spanning from short-term electricity and fuel price changes to long-term regulatory and technological uncertainty. In summary electricity can be seen as a highly complex derivative with varying underlyings whose number and relevance change over time.

Influential publications in the beginning of the liberalization process have consistently stressed these unique properties of electricity even without the recent market extension. The applicability of methodologies from mathematical finance is complicated despite some obvious analogies [4, 7, 15]. Compared to financial and commodity markets the implicit requirements for load coverage lead to fundamentally different valuations even for plain standard derivatives like futures. Some contracts have been specifically designed for the special conditions in physical electricity trading. Swing options, for example, reflect the need for flexible power in predefined ranges and require special valuation methodologies [10–12]. Considering network constraints, even in generally interconnected electricity markets the buyers and sellers must handle price spreads between different regions resulting from physical restrictions in the electric power grid. The market-splitting mechanism for casual congestion occurrences in nordic countries illustrates the varying interrelation of prices in one interconnected region. These conditions often violate assumptions known from mathematical finance which basically assume free money flow and unlimited liquidity.

The modeling of electricity prices has therefore always been of considerable concern and most approaches that exist stress particular characteristics while neglecting others. Especially spot prices have been in focus of econometric models which are quite well observable in most liberalized electricity markets. They are commonly modeled through a seasonal combined with different stochastic components [3, 19, 21]. The characteristic jumps and spikes which are also typically observed are addressed in [6]. The modeling of discontinuous returns can be traced back to Merton [20]. Because of frequent structural changes in liberalized electricity markets, e.g., from regulatory changes, some more recent approaches focus on regime-switching models [13, 14]. In any case, model calibrations are difficult to perform due to insufficient availability and/or significance of historical data in a market that is characterized by high volatility, frequent structural changes and a considerable degree of seasonality with complicated daily, weekly, and yearly patterns.

Our approach to price modeling in principal follows the proposals of Pilipovic [21]. Neglecting seasonality the stochastic spot prices S_t at times t can be represented through a level model

$$S_t = S_0 + X_t \tag{20.1}$$

where X_t follows an arbitrary stochastic distribution. However, the characteristic spot price behavior is better reflected through a return model that also accounts for nonnegativity requirements. We can therefore use the log-normal of spot prices:

$$S_t = S_0 \exp X_t \tag{20.2}$$

Due to the long-term characteristic of investments in the electric power industry, we can assume a static midterm composition of installed capacities in a power system. As a consequence mean-reverting models perform well when applied to generation planning in electricity markets. Assuming a one factor mean-reverting process the stochastic component X_t is represented by the following SDE

$$dX = \kappa(\theta - X_t)dt + \sigma_x dz_x \tag{20.3}$$

where σ_x represents the diffusion factor of X and dz_x follows a Brownian motion. The long-term mean is given through θ where κ reflects the strength of mean-reversion. This model can be ultimately extended to a two-factor Pilipovic model

$$dX = \kappa(Y_t - X_t)dt + \sigma_x dz_x \tag{20.4}$$

$$dY = \sigma_y dz_y \tag{20.5}$$

where the first factor X_t models short-term fluctuations in the spot market and the second factor Y_t stands for an uncertain long-term equilibrium with dz_x and dz_y as independent Brownian motions. To account for seasonality, we simply add a time-dependent factor l_t in (20.2):

$$S_t = S_0 \exp[l_t + X_t] \tag{20.6}$$

Assuming equivalence between spot and forward prices we obtain the following relationship:

$$F_{t,T} = E_t[S_T] \tag{20.7}$$

$$F_{T,T} = S_T \tag{20.8}$$

where $F_{t,T}$ represents the forward price of electricity at t with maturity T. The forward price at any time t therefore represents the expected spot price at maturity T. Following this assumption we can use price forward curves to define seasonal patterns l_t to consistently simulate electricity prices. For the determination of arbitrage free forward curves we apply an approach by Fleten [8]. In a quadratic program discrepancies between characteristic shapes (representing short-term fluctuations of desired periodicity) and observed future prices are minimized.

20.3.2 Carbon Prices and Cost Pass-through

Carbon price observations until today suggest that the understanding of the markets is still insufficient. As a result of the overallocation in phase I the values of the certificates have basically dropped to zero. The remaining marginal prices of the certificates in principle reflect their marginal time value as an option for the unlikely event of still falling short in the end of 2007.

The experiences from phase I have been taken into account for phase II. The caps have been presumably set to much tighter limits. Market expectations on the effectiveness of the second trading period are reflected in future prices with delivery in 2008–2012. They currently range around €20/tCO_2 which clearly indicates that the market expects to be short.

The possible continuation of the European trading scheme following the Kyoto period after 2012 is still under investigation and will also have to take experiences from the second trading period into account. Longer time frames for future compliance periods are commonly considered necessary by participants and regulators. At least 10 years are recommended to achieve a desired level of confidence that actually could trigger large-scale investments into new clean technologies.

The interpretation of carbon prices in a functioning market requires to take several overlapping and interrelated influences into consideration. Due to the diverse and dynamic nature of these factors the identification of singular effects appears mostly impossible. Regulatory uncertainties and limited liquidity also allow speculators to possibly manipulate the market. Power providers are tempted to keep the book-values of their emission rights as high as possible. By including these costs in electricity prices, additional profits can be realized without additional costs which ultimately must be paid by the consumers.

An important question that arises is to what extent carbon prices are passed through to electric power prices. An empirical analysis in [27] estimates different levels ranging from 60 to 100% that not only depend on whether off-peak- or on-peak-demand is considered but also varies by region. However, the model estimates reveal considerable degrees of uncertainty. According to estimates in the influential view of PointCarbon [22] CO_2 prices are now fully reflected in power prices – at least in the important markets of Germany, UK and due to close interrelations to Germany also in the Nordic markets. Following the steep fall of carbon prices in April 2005, a statistical analysis reveals a marginal correlation of about 0.7 [€/MWh]/[€/tCO_2] in the EEX. As a theoretical consequence the average value loss of German electricity would amount to € 7 billion or even € 30 billion for all of Europe. The interdependencies are determined by the existing fuel mix of the electric supply infrastructure and the fuel price composition and emission efficiencies of the applicable power generation technologies. The increasing complexity of relevant global risk factors like oil, gas and coal price developments as well as uncertain outcomes from international climate policy make estimates on future carbon price developments even more difficult.

20.3.3 Fuel Switching following Clean Valuation

The electricity sector has by far the highest share of emission rights in the EU ETS. Therefore the analysis of carbon prices is often focussed on technical aspects of power plant operation with related fuel costs and emission efficiencies. In market terms the price developments for carbon depend on the number of available certificates compared to their consumption through facility operation, which in turn depend on external factors like economic growth or weather conditions. The new influence of emission costs also has repercussions on the system as a whole including operations and investments. The market driven emission costs imposed through EUAs are intended to provide an incentive for emission abatement which for short-term operations aim at the clean use of available capacities like switching from coal to gas and in the long-run at the development of new technologies like carbon capture and storage. However, due to the long-term nature of investments into new technologies and prevailing regulatory uncertainties undesired delays have been observed in practice.

Due to the key role of gas and coal in electric power generation and their importance in the evolving emission trading scheme we have focussed our current work on these two technologies. We highlight investment characteristics, operational flexibility and switching possibilities to analyze the consequences of the EU ETS in power plant capacity planning and operation.

Prices for Gas and Coal

Gas and coal account for about 60% of the global electricity production. Coal requires higher investment costs with usually lower variable production costs compared to gas. Global estimates on the most important generation technologies and their variations can be found in [16]. Better emission efficiencies of gas-fired power plants make them less sensitive to EUA prices. Combined with short investment cycles and low fixed costs in face of EUA prices and considerable regulatory uncertainty in the upcoming years gas fired power generation has become an attractive alternative. Depending on fuel and emission prices gas and coal may even change places in the merit order of electricity production. Several studies are currently dealing with the consequences of fuel switching and its repercussions on emissions, fuel and electricity prices.

Even though from a regulatory point of view competition should already be in full effect, gas markets in most countries of Europe are still in their initial phase towards liberalization. Gas markets in UK are the most liquid with close interrelations to prices observed in Zeebrugge and the virtual TTF (Title Transfer Facility) in Netherlands which has rapidly gained in importance recently. Due to ongoing market liberalization further reference areas are progressively developed. In the EEX, for example, day-ahead and future trading of natural gas for the market areas of BEB and E.ON GT is now possible. The significance of most historic market data in continental Europe, however, is still limited. Historic ownership of transmission and distribution infrastructures by large utilities and long-term agreements that dictate a linkage of gas prices on oil prices, for example, still prevent fast progress. Therefore we have taken the NBP (National Balancing Point) prices in the UK for reference in our analysis. The basic characteristics of gas prices are well reflected which include typical yearly price fluctuations and sensitivity to global fuel prices from substitute competition in heat and electricity production.

Coal prices can be distinguished with respect to their type (e.g., steam coal), quality (e.g., energy content) and sales location and condition (e.g., import coal for Germany including cost, insurance and freight (cif)). For Europe the API#2 index (cif ARA (Amsterdam, Rotterdam, Antwerpen)) is commonly used for reference which states future prices in [$/t]. Coal prices in general are more stable than gas prices. The main substitute for coal is gas due to its importance for electric power generation. Therefore revenues in coal power plants are commonly assumed to be dictated by gas prices. In the simulation based analysis on greenhouse gas reductions in [5] coal prices are even assumed to be constant.

Due to the recent steep rise in coal prices in the EEX based on the API#2 index we also assume coal prices to be stochastic for our modeling purposes. Figure 20.2 illustrates variable generation cost in €/MWh based on gas and coal prices observed in different markets. We have assumed typical electric efficiencies of $\eta_{el,coal} = 0.38$ for coal and $\eta_{el,gas} = 0.50$ for natural gas. The quotations have been corrected for daily interbank exchange rates. Assuming emission efficiencies $\eta_{CO2,coal} = 0.9$ and $\eta_{CO2,gas} = 0.35$ [tCO_2/MWh] we have added EUA prices at which switching from coal to gas based on variable costs would occur.

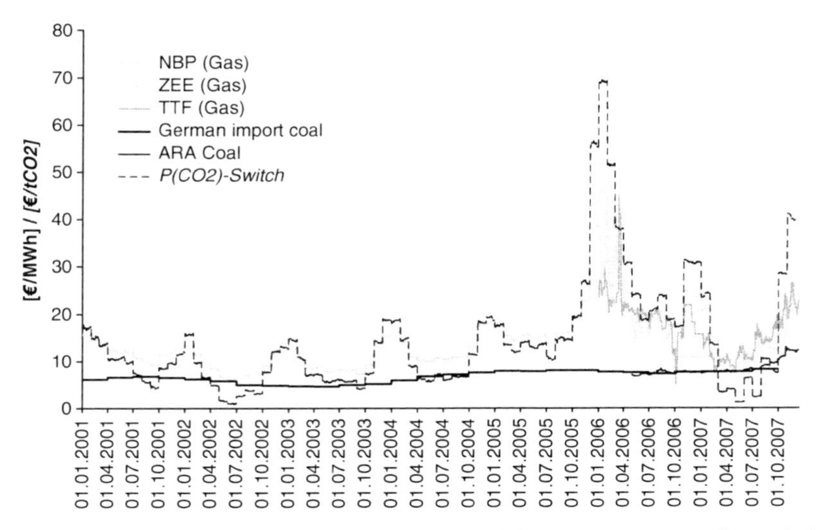

Fig. 20.2. Variable coal/gas generation costs and CO_2 trigger prices for switching

Clean Valuation

The mechanisms of clean valuation and the resulting fuel switching possibilities can be easily demonstrated considering power generation from gas and coal. For our initial illustration we neglect other factors that support operation and investment decisions in either of these technologies. Simply considering marginal generation costs from existing facilities adjusted by CO_2 prices allows us for a clean valuation of marginal power generation. Traditionally spark-spreads and dark-spreads are used to determine the margins from electricity. They basically depend on the price of fuels and electricity combined with the electric efficiencies $\eta_{el,i}$ of generation technology i at times of generation. These factors are complemented by the prices for emissions that can be observed in the EU ETS, resulting in so called *clean* spark or dark spreads. The interdependencies are complex, however, since natural gas generally has better emission efficiencies $\eta_{CO2,gas}$ than coal $\eta_{CO2,coal}$ switching between these fuels because of changed places in the merit order may occur depending on emission costs as illustrated in Fig. 20.3.

On average variable costs for gas fueled generation are higher than for coal. Gas is also generally exposed to higher fuel price volatilities. On the other hand the sensitivity of gas with respect to EUA price fluctuations is lower than for coal due to higher emission efficiencies as illustrated in Fig. 20.4. Also the risks involved in gas power plant investments are lower than for coal due to shorter planning periods and low fixed costs. In times of regulatory uncertainty this proves to be a major advantage. Gas is also superior compared to coal when it comes to generation flexibility – a profitable advantage especially during peak hour operation.

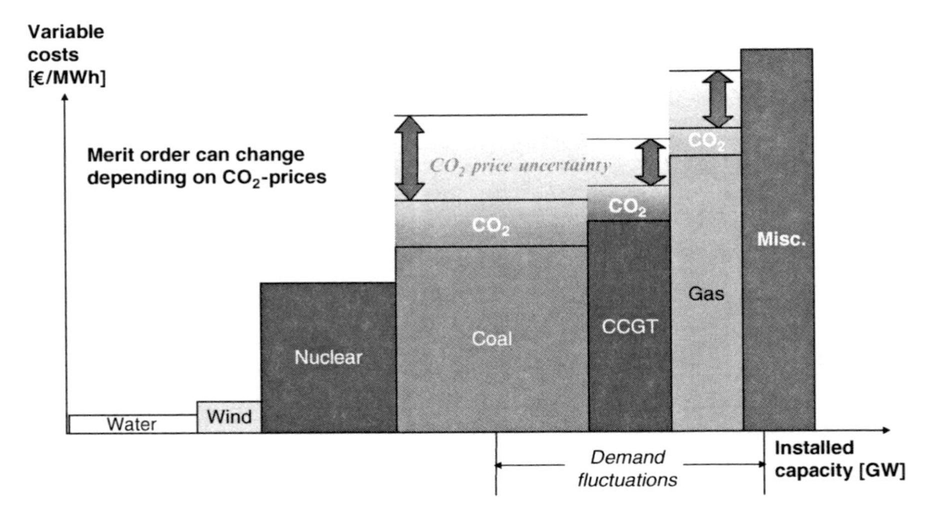

Fig. 20.3. Change in the merit order due to emission costs

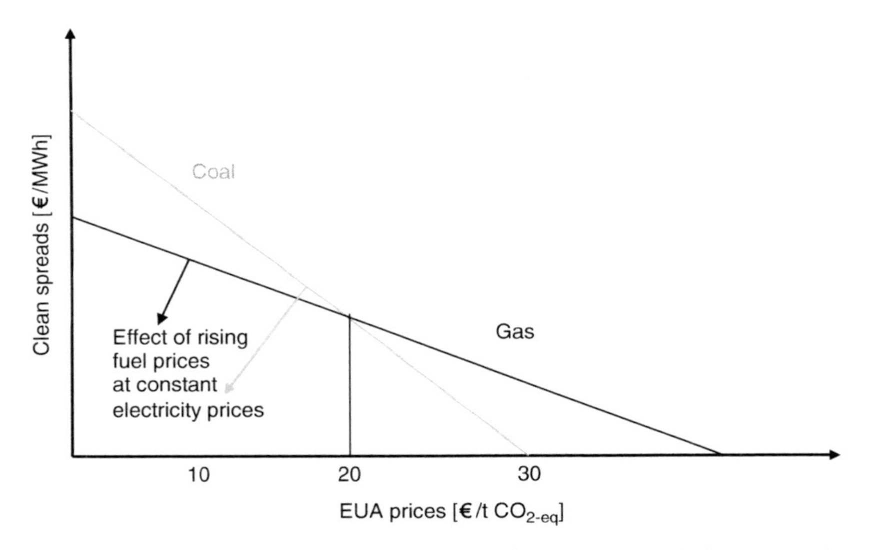

Fig. 20.4. Different sensitivity of coal vs gas with respect to emission costs triggers switching

20.4 Modeling Investment Planning and Power Generation

Quantitative tools are popular for use in the electric power industry due to the complex interdependent influence of technical and economic factors. We highlight the issues presented in the previous sections in a multistage stochastic programming framework where investment and generation flexibility are valued in a portfolio management problem with respect to prices of electricity, CO_2 and fossil fuels. The importance of fossil fueled power plants in clean valuation as investigated in [1,5], for example, has already been stressed. Therefore we only consider gas and coal fired power generation which can be used to cover different stochastic demand patterns. We perform different sensitivity analysis with respect to changes in technology, volatility and price scenarios.

20.4.1 Multistage Stochastic Program

We consider a valuation and decision problem with respect to the new challenges that stem from the introduction of carbon trading in the European electricity markets. We optimize investment decisions and power production from different technologies based on fossil fuels to cover load requirements over one planning cycle. The objective is to minimize expected costs when choosing the optimal generation mix by also taking rebalancing decisions concerning investments and power generation into account. At the end of one compliance period T which is covered by the model missing allowances can be obtained at a stochastic price that implicitly represents penalty costs for not having complied with the cap as initially defined by some regulatory measure. This also allows for a sensitivity analysis with respect to expected emission reductions and their distributions based on different price levels of emission certificates.

Our multistage stochastic program minimizes expected costs from investments, power plant operation and carbon trading over one planning cycle $t = 0, \dots, T$ to cover a stochastic load pattern. The objective function is given through:

$$\min E\left[\left(\sum_{t=0}^{T}\sum_{i\in I_t}(\eta^v_{i,t}u^v_{i,t}) + \sum_{h=0}^{H}(\eta^s_{t,h}u^s_{t,h} + \sum_{j\in J_t}\eta^g_{j,t,h}u^g_{j,t,h})\right) + \eta^e_T u^e_T\right] \tag{20.9}$$

where $u^v_{i,t}$ represents investments in technology i at cost $\eta^v_{i,t}$. Decisions $u^g_{j,t,h}$ model the generation from technology j at variable cost $\eta^g_{j,t,h}$ in each subperiod h of period t. $h = 0, \dots, H$ represents typical (e.g., hourly) operation patterns in power generation. Furthermore electricity $u^s_{t,h}$ can be bought at price $\eta^s_{t,h}$ in the spot market. u^e_t is the number of emission allowances that must be purchased at the end of one planning cycle T at cost η^e_T to fulfill requirements from capped emissions. Selling of EUAs and electricity to generate additional profits is not allowed in our setting and the spot prices for electricity are generally assumed to be unfavorable compared to the costs of own generation.

The basic stochastic requirement is to cover an uncertain load pattern $\xi^l_{t,h}$ for each subperiod h in period t through own generation $u^g_{j,t,h}$ from technology j and possibly spot market acquisitions $u^s_{t,h}$:

$$u^g_{j,t,h} + u^s_{t,h} \geq \xi^l_{t,h} \qquad j \in J_t \quad h = 0, \dots, H \quad t = 0, \dots, T. \tag{20.10}$$

The generation flexibility for each technology is limited from one subperiod to the next defined through incremental limits in the following sense

$$|u^g_{j,t,h-1} - u^g_{j,t,h}| \leq inc^{\max} \qquad j \in J_t \quad h = 1, \ldots, H \quad t = 0, \ldots, T \quad (20.11)$$

$$|u^g_{j,t-1,H} - u^g_{j,t,0}| \leq inc^{\max} \qquad j \in J_t \cap J_{t-1} \quad t = 1, \ldots, T \quad (20.12)$$

assuming constant up and down limits $inc^{\max}$ [MW] for the ramping constraints.

Investment flexibility is limited through different subsets I_t in each $t = 0, \ldots, T$. Decisions to invest into technology $i \in I_t$ in period t result in additional generation capacity from technology i for the actual and all future periods $t, \ldots, T$. Generation in each period t from available technologies j is then limited through the following capacity constraints

$$u^g_{j,t,h} \leq u^{\max}_{j,t} \qquad j \in J_t \quad h = 0, \ldots, H \quad t = 0, \ldots, T \quad (20.13)$$

that are determined through investments $u^v_{i,t}$ over time

$$u^{\max}_{j,t} = u^{\max}_{j,t-1} + u^v_{i,t} \qquad j \in J_t \quad i \in I_t \quad t = 0, \ldots, T \quad u^{\max}_{j,-1} = 0. \quad (20.14)$$

In our model calculations it is assumed that not all technologies are available for investment in all periods taking into account that unacceptable risks may arise from long-term investments with lasting effects that exceed the planning horizon T.

Generation from technology $j \in J_t$ causes emissions at intensity e^j. Furthermore we assume that missing emission allowances can be obtained from the market at the end of the compliance period T at cost η^e_T. Therefore the emission cap is recognized through the following constraint with overall limited emissions $em^{\max}$ [tCO_2]:

$$\sum_{t=0}^{T} \sum_{h=0}^{H} \sum_{j \in J_t} (e^j \cdot u^g_{j,t,h}) - u^e_T \leq em^{\max} \quad (20.15)$$

Further differentiation of this modeling framework is possible but has not been applied in our first analysis. For example, the constraints concerning generation flexibility can be further refined by distinguishing between up and down increments. Emission caps can be set for more than one compliance period within the overall planning horizon T. Also the number of subsequent periods affected through capacity investments can be limited within one planning cycle for more complex investment patterns. However, for now we have restricted the model to the above basic calibration to facilitate the analysis of significant cause–effect relations in practical examples. Otherwise the results could be blurred from a number of unclear influences in a complex setting with too many interdependent influences and decisions.

20.4.2 Solution Methodology

The modeling framework developed in the previous section represents a linear multistage stochastic program with fixed recourse exposed to manifold correlated uncertainties. In a general formulation decisions $u := (u_0, \ldots, u_T)$ with $u_t \in \Re^{n_t}$ are considered over a planning horizon of T periods, corresponding to $T+1$ stages. The decisions $u_0 \in \Re^{n_0}$ in stage $t = 0$ follow the here and now principle, i.e., assumptions on future evolvements of relevant uncertainties are not allowed. In each stage the decisions can be rebalanced with respect to realizations of the stochastic variables ω_t and decisions made up to that point. Therefore in each stage $t = 1, \ldots, T$ decisions u_t are made with respect to former decisions $u_0, \ldots, u_{t-1}$ and observations $\omega_1, \ldots, \omega_t$. In our notation $\omega^t := (\omega_1, \ldots, \omega_t)$ describes observations up to stage t, $\Omega^t := (\Omega_1 \times \cdots \times \Omega_t)$ the range of values of ω^t and $u^t := (u_0, \ldots, u_t)$ the decisions up to stage t. Furthermore we split the stochastic influences ω_t into components η_t representing influences in the objective function and ξ_t as stochastic right-hand-side component.

It follows that based on the recourse function

$$\phi_t(u^{t-1}, \omega^t) = \phi_t(u^{t-1}, \omega^{t-1}, \omega_t) \tag{20.16}$$

we can write the stochastic multistage program in its recursive backward definition

$$\phi_t(u^{t-1}, \omega^{t-1}, \omega_t) := \min \rho_t(u_t, \eta_t) + \int_{\Omega_{t+1}} \phi_{t+1}(u^t, \omega^t, \omega_{t+1})\, dP_{t+1}(\omega_{t+1}|\omega^t)$$

$$\text{s.t.} \quad f_t(u^{t-1}, u_t) \leq h_t(\xi^{t-1}, \xi_t) \tag{20.17}$$

where $\phi_{T+1}(\cdot) := 0$ and $\phi_0(u^{-1}, \omega^{-1}, \omega_0)$ represents the overall solution of the problem.

In our modeling framework decisions u correspond to the investment, generation and trading decisions that are exposed to uncertain price developments η and uncertainties in the load represented through ξ. It is well known that under assumptions formulated in [24] the value functions expose a saddle structure. In our notation this means that in (20.17) all recourse functions $\phi_t(u^{t-1}, \omega^{t-1}, \cdot)$ have a concave shape with respect to the stochastic influences in η_t and a convex shape in ξ_t. This structural property allows us to apply barycentric approximation as described in [9] for estimating scenario trees of limited dimension that yield computationally tractable stochastic multistage programs for our modeling framework. Furthermore we have applied multinomial distributions for scenario tree generation as described in [26]. Applying either scenario generation technique, the resulting large-scale linear programs have been solved using the LP solver capabilities of ILOG CPLEX 9.1.

20.4.3 Results

Investment patterns for coal and gas fired power plants are different. Low investment costs, high electric efficiencies and a considerable degree of generation flexibility in electricity markets that are exposed to extreme price volatilities and regulatory uncertainty have led to a boom of gas power investments and generation in recent years. While modern coal fired plants have investment costs of about €700/kW, the costs for Combined Cycle Gas Turbines (CCGT) is only €350/kW [25]. Average planning cycles from project initialization to power plant start-up are about 1 year for gas fired generation compared to 4–5 years for coal. Estimated lifecycles fall in a range of decades for both technologies, however, twice as many full load hours are usually assumed for coal compared to gas with load factors of more than 80%. Historically coal fired generation has been exposed to lower fuel costs and volatilities compared to gas, but because of its inferior emission efficiencies this could be at least partly offset in the future depending on levels and volatilities of CO_2 prices. As a logical consequence new technologies for coal fired generation are currently under investigation which strive for general efficiency improvements through coal gasification, for example, and also include fundamentally new approaches like carbon capture and storage.

Our model considers the effects of carbon trading on an efficiently managed electric power system including capacity investments, generation and trading. For our basic calibration summarized in Table 20.1 we assume market conditions as observed in Germany and apply a dynamic standardized load profile H0 as proposed by the VDEW and illustrated in Fig. 20.5. It represents a load pattern with characteristic high-noon and evening peaks as typically observed in Central Europe.

In our optimizations we allow for full investments into adequate generation mixes in period $t = 0$ and do not restrict the initial ramping constraint for startup of coal fired generation. Fixed costs in our model follow estimates from annual depreciation based on [18, 25]. We first analyze sensitivities with respect to CO_2 prices and emission caps. Then we provide additional results that assume variations in the load profile and generation flexibility. We conclude with an outlook on further analysis and model extensions.

Table 20.1. Basic generation technologies

Merit order	Capacity[a] [GW]	Energy cost [€/MWh]	Cap. cost [€/MW]	Lifecycle [years]	η_{el} [-]	η_{em} [tCO_2/MWh]
Baseload	26.94	12	200–1500	10–40	0.35–0.99	0
Coal	0–80	18	700	40	0.38	0.9
Gas	0–80	30	350	35	0.55	0.4

[a] Capacity investments into coal and/or gas are subject to the optimization

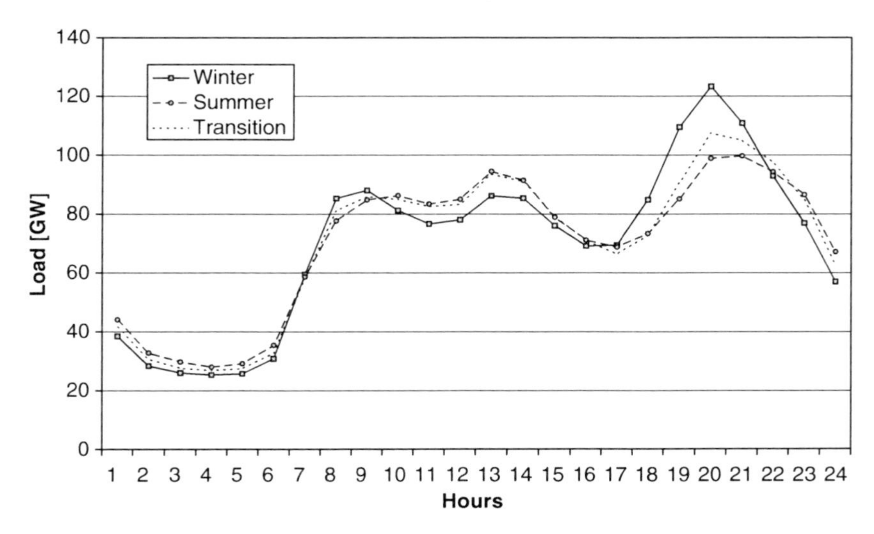

Fig. 20.5. Load patterns based on VDEW H0 weekday profile for households

CO_2 Prices and Emission Caps

The cap and trade mechanism of the EU ETS strives to provide market driven incentives for CO_2 reductions by scarcity of emission rights. In our model abatements can basically be achieved through switches in technology. We also provide the possibility to buy emission rights at the end of a planning period in case of noncompliance in certain scenarios. This coincides with the trading and penalty mechanism as foreseen in the EU ETS. As a consequence of our methodology based on multistage stochastic programming, the amount of emissions released at the end of the compliance period T is given as a probabilistic distribution rather than specific numbers. We use expected values and different risk numbers like Value-at-Risk or standard deviation for specific comparisons.

In practice scarcity of CO_2 allowances is realized through a limited initial allocation of emission rights to different industries based on national legislation which must be approved by the EU commission. Difficulties arise from the adequate estimation of efficient caps. As can be seen from the market collapse during the first trading period of the EU ETS this has already proven to be a difficult task. The current prices for the second phase, however, indicate that market participants expect the market to be short for the upcoming trading period that starts in 2008. In the latest national allocation plan for Germany a 15% reduction target has been imposed on the energy sector. This situation has also been assumed in our model calculations summarized in Table 20.2 where we analyze the influence of EUA costs on installed capacities, generation and annual emissions. The sensitivity of CO_2 reductions on emission costs is illustrated in Fig. 20.6. In our analysis a saturation effect

Table 20.2. Influence of CO_2 prices

EUA [€/tCO_2]	Emissions [$MtCO_2$][a]	Gas generation [TWh][a]	Coal generation [TWh][a]	Gas capacity [GW]	Coal capacity [GW]
0	317.44	26.81 (*7.3%*)	342.3 (*92.7%*)	17.9 (*24.4%*)	55.3 (*75.6%*)
1	314.41	32.40 (*8.8%*)	336.8 (*91.2%*)	19.3 (*26.4%*)	53.9 (*73.6%*)
5	293.39	70.62 (*19.1%*)	298.5 (*80.9%*)	27.9 (*38.1%*)	45.3 (*61.9%*)
7.5	279.91	95.11 (*25.8%*)	254.2 (*74.2%*)	32.5 (*44.4%*)	40.7 (*55.6%*)
10	272.18	109.2 (*29.6%*)	260.0 (*70.4%*)	34.9 (*47.7%*)	38.3 (*52.3%*)
15	269.72	113.6 (*30.8%*)	255.5 (*69.2%*)	35.6 (*48.7%*)	37.6 (*51.3%*)
20	269.33	114.3 (*31.0%*)	254.8 (*69.0%*)	35.6 (*48.7%*)	37.6 (*51.3%*)
25	268.98	115.0 (*31.1%*)	254.2 (*68.9%*)	35.6 (*48.7%*)	37.6 (*51.3%*)
30	268.69	115.1 (*31.2%*)	254.0 (*68.8%*)	35.6 (*48.7%*)	37.6 (*51.3%*)

[a] Annual averages

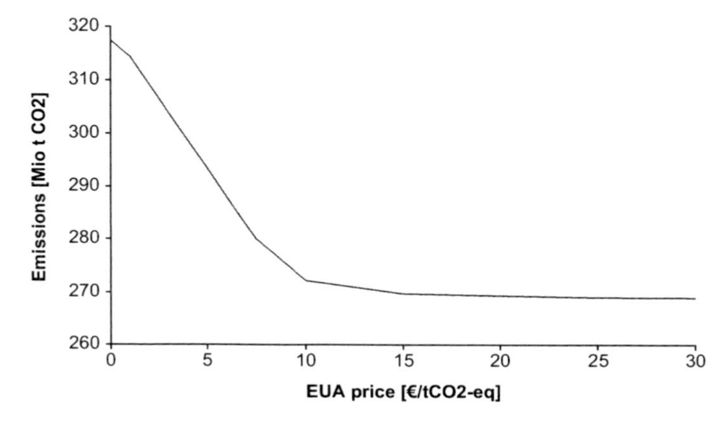

Fig. 20.6. Sensitivity of emissions vs. costs (cap set to −15%)

concerning the influence of carbon prices on emission reduction has been observed that starts at levels of about €15/tCO_2. At this price level the assumed cap of −15% has been basically achieved. The highest sensitivity in carbon reductions is observed at price levels up to 10€/tCO_2.

In a second approach we use our model to analyze effects from changing cap levels in a closed power and emission trading system which includes the influence of extra costs for noncompliance at the end of a trading period. Currently penalties of €40 and 100/tCO_2 are foreseen in case of noncompliance at the end of phase I and II respectively. These costs are in addition to any acquisition costs for the respective emission rights in subsequent periods since the obligation to deliver the missing EUAs is nevertheless sustained.

For the results in Table 20.3 we have assumed penalty costs of €40/tCO_2. It can be seen that emission targets are achieved through a switch from coal to gas without taking the penalties. Starting at about 55% reduction targets the load requirements can not be fulfilled for all scenarios without violating

Table 20.3. Influence of emission targets (penalty €40/tCO_2)

Target	Emisstion ions [MtCO_2][a]	Gas generation [TWh][a]	Coal generation [TWh][a]	Gas capacity [GW]	Coal capacity [GW]	Variable costs [€/MWh][a]
0%	312.5	35.9 (*9.7%*)	333.2 (*90.3%*)	20.2 (*27.5%*)	53.0 (*72.5%*)	43.1
5%	299.6	59.3 (*16.1%*)	309.8 (*83.9%*)	25.4 (*34.7%*)	47.8 (*65.3%*)	43.3
10%	283.9	87.8 (*23.8%*)	281.3 (*76.2%*)	31.1 (*42.5%*)	42.1 (*57.5%*)	43.5
15%	269.0	115.0 (*31.2%*)	254.1 (*68.8%*)	35.6 (*48.6%*)	37.6 (*51.4%*)	43.8
20%	253.1	143.8 (*39.0%*)	225.3 (*61.0%*)	40.2 (*55.0%*)	32.9 (*45.0%*)	44.2
25%	237.3	172.6 (*46.8%*)	196.5 (*53.2%*)	44.8 (*61.2%*)	28.4 (*38.8%*)	44.6
30%	221.4	201.5 (*54.6%*)	167.7 (*45.4%*)	49.2 (*67.2%*)	24.0 (*32.8%*)	45.0
35%	205.5	230.3 (*62.4%*)	138.8 (*37.6%*)	53.5 (*73.2%*)	19.6 (*26.8%*)	45.4
40%	189.7	259.2 (*70.2%*)	110.0 (*29.8%*)	57.9 (*79.1%*)	15.3 (*20.9%*)	45.8
45%	173.8	288.0 (*78.0%*)	81.1 (*22.0%*)	62.1 (*84.8%*)	11.1 (*15.2%*)	46.3
50%	157.9	316.9 (*85.8%*)	52.3 (*14.2%*)	66.2 (*90.5%*)	7.0 (*9.5%*)	46.7
55%	142.1	345.7 (*93.7%*)	23.4 (*6.3%*)	70.1 (*95.8%*)	3.1 (*4.2%*)	47.2
60%	129.2	369.1 (*100%*)	0.0 (*0.0%*)	73.2 (*100%*)	0.0 (*0.0%*)	47.8
75%	129.2	369.1 (*100%*)	0.0 (*0.0%*)	73.2 (*100%*)	0.0 (*0.0%*)	53.0

[a] Annual averages

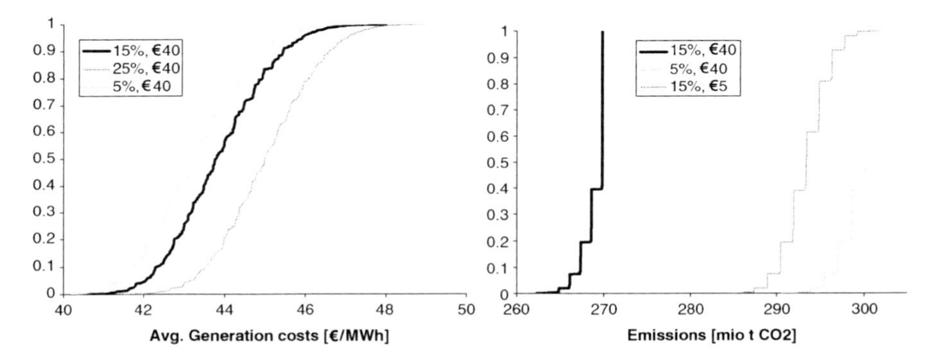

Fig. 20.7. Distributions of average generation costs and annualized emissions

emission limits. Therefore penalties are imposed as reflected in a rising slope for the variable generation costs and at some point emission levels remain constant due to technical limitations of the generation portfolio. Annualized distributions of average costs and emissions over one planning cycle are illustrated in Fig. 20.7 assuming different reduction targets and penalties.

Generation Flexibility and Volatility

In our model we also highlight issues concerning generation flexibility and volatility. Different flexibility levels have been defined in terms of hourly ramping constraints defined as a percentage of installed capacities. We assume a basic scenario with −15% reduction target and €5/tCO_2 emission costs. The results are summarized in Table 20.4.

Table 20.4. Influence of generation flexibility (−15% reduction target, €5/tCO_2)

Flexibility	Emissions [MtCO_2][a]	Gas generation [TWh][a]	Coal generation [TWh][a]	Gas capacity [GW]	Coal capacity [GW]	Acqu. [MtCO_2][a]
1%	234.9	25.0 (*55.5%*)	20.1 (*44.5%*)	51.3 (*70.1%*)	21.9 (*29.9%*)	0.0
2%	255.6	21.1 (*46.6%*)	24.2 (*53.4%*)	45.3 (*62.0%*)	27.8 (*38.0%*)	0.0
3%	267.1	18.6 (*41.2%*)	26.6 (*58.8%*)	41.3 (*56.5%*)	31.8 (*43.5%*)	0.0
5%	267.8	16.9 (*38.1%*)	27.4 (*61.9%*)	37.4 (*51.2%*)	35.7 (*48.8%*)	0.1
8%	268.6	15.2 (*35.0%*)	28.2 (*65.0%*)	34.4 (*47.0%*)	38.8 (*53.0%*)	0.2
10%	269.5	14.1 (*32.9%*)	28.7 (*67.1%*)	33.3 (*45.5%*)	39.9 (*54.5%*)	0.3
15%	270.9	12.9 (*30.6%*)	29.3 (*69.4%*)	32.3 (*44.2%*)	40.8 (*55.8%*)	1.2
25%	283.3	10.3 (*24.3%*)	31.9 (*75.7%*)	30.1 (*41.2%*)	43.1 (*58.8%*)	13.5
50%	293.3	8.1 (*19.2%*)	34.1 (*80.8%*)	27.9 (*38.1%*)	45.3 (*61.9%*)	23.5
100%	293.4	8.1 (*19.1%*)	34.1 (*80.9%*)	27.9 (*38.1%*)	45.3 (*61.9%*)	23.6

[a] Annual averages

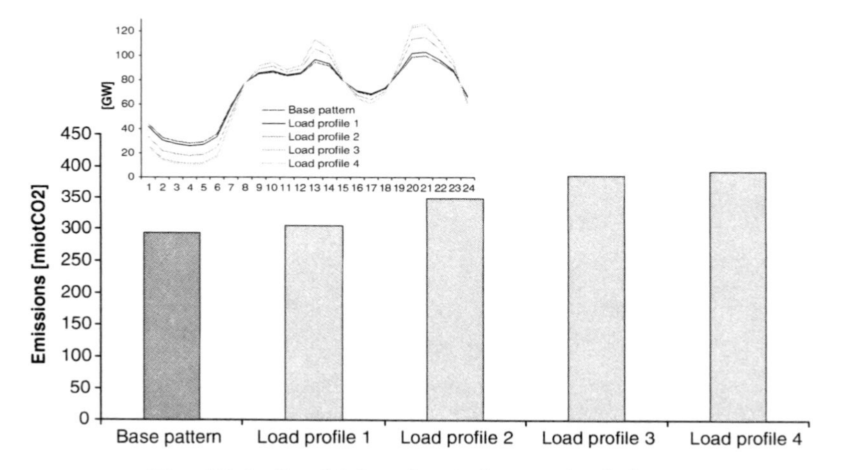

Fig. 20.8. Sensitivity of emissions on load shapes

Below 3% hourly generation flexibility for coal we see a strong incentive to switch to gas. At higher flexibility levels the effects become much weaker but are still observable. Above the 3% level we also see a willingness to buy certificates in the market as a result of high shares of coal fired power generation rising from 44.5% at the 1% flexibility level to 80.8% assuming 50% ramping constraints.

In Fig. 20.8 we illustrate the effect of increased volatility based on different load patterns. The emission levels rise with increasing volatility reflecting the growing system requirements for flexible power. However, the share of installed gas capacities compared to coal during on-peak hours also rises from 38.1 to 50.8% with generation shares rising from 19.1 to 28.1%. The resulting generation mix therefore exhibits an improved average emission efficiency which dampens the effect of overall rising emissions.

20.5 Conclusions

We have analyzed consequences of CO_2 trading following clean valuations in a multistage stochastic programming approach. The model covers important aspects of actual concern including investment, generation and emission characteristics for coal and gas. We have exposed the generation portfolio in question to stochastic fuel prices and allow for trading in energy and carbon markets to cover uncertain load patterns with varying volatility.

The achievable emission reductions have been analyzed for different market scenarios. We have varied the emission caps as well as penalties in case of noncompliance at the end of the planning period. A strong dependence of the emission levels depending on EUA prices up to levels of €15/tCO_2 for the assumed cap of −15% has been observed. This result roughly corresponds to current market prices in the EU ETS for the second compliance period. According to our analysis higher prices would only yield minor reductions. However, the strongest influence has been observed below €10/tCO_2. Up to this price level the incentive to switch from coal to gas appears to rise continuously at high rates.

In our analysis on different cap levels at penalties of €40/tCO_2 the emission targets were basically fully adapted from the system. Starting at about 60% reduction targets coal is completely substituted through gas. We conclude that at current coal–gas price spreads no effects can be expected from small variations at the chosen penalty level unless prices for gas rise much faster than coal in the future.

Further analysis has been performed with respect to generation flexibility and load volatility. The importance of generation flexibility depends on the underlying load pattern. Therefore we have increased the volatility of the basic load pattern which results in higher flexibility requirements and therefore gas powered generation. The overall rising emission levels are dampened through increased shares of gas for on-peak coverage. Nevertheless, overall our results support the idea for better demand side management which according to our analysis would also reduce emissions.

We have focussed on some important aspects of CO_2 trading in the electric power industry. There are many other issues of concern which can be included in future work. An important aspect is the continuous extension of the trading scheme by including new industries like aviation, for example. Also the consequences from a quickly expanding market especially for CDM projects may prove to have significant repercussions on the EU ETS even if allowed shares of CERs convertible into EUAs are restricted by regulation [2, 22]. For adequate evaluations the inclusion of such projects into the modeling framework must also consider the risk profiles of respective projects and the influence on the overall liquidity of the EU ETS.

Acknowledgement. We gratefully acknowledge the financial support by the Grundlagenforschungsfonds of the University of St. Gallen (Grant number G12151104).

References

1. S. Bode. Multi-period emissions trading in the electricity sector - winners and losers. Technical report, 2004. Hamburgisches Welt-Wirtschafts-Archiv (HWWA), Discussion Paper No. 268.
2. K. Capoor and P. Ambrosi. State and trends of the carbon market 2007. Technical report, World Bank Institute in cooperation with the International Emissions Trading Association, May 2007.
3. M. A. Carnero, S. J. Koopman, and M. Ooms. Periodic heteroskedastic RegARFIMA models for daily electricity spot prices. Technical report, Tinbergen Institute, Discussion Paper, TI 2003-071/4, 2003.
4. Les Clewlow and C. Strickland. *Energy Derivatives: Pricing and Risk Management.* Lacima, London, 2000.
5. E. Delarue, H. Lamberts, and W. Dhaeseleer. Simulating greenhouse gas (GHG) allowance cost and GHG emission reduction in Western Europe. *Energy*, 32:1299–1309, 2007.
6. S. Deng. Stochastic models of energy commodity prices and their applications: mean-reversion with jumps and spikes. Technical report, Industrial Engineering and Operations Research, University of California at Berkeley, February 1999. Presented at Uncertainty Workshop, Palo Alto (July 1999).
7. A. Eydeland and K. Wolyniec. *Energy and Power Risk Management.* Wiley, Hoboken, New Jersey, 2003.
8. S. -E. Fleten. Power scheduling with forward contracts. Technical report, Norwegian University of Science and Technology (NTNU), August 1998.
9. K. Frauendorfer. Multistage stochastic programming: Error analysis for the convex case. *Z. Oper. Res.*, 39(1):93–122, 1994.
10. K. Frauendorfer, D. Kuhn, and J. Güssow. Stochastische Optimierungskonzepte zur Bewertung von Swing-Optionen. Working-Paper 03-02-1, Institute for Operations Research, University of St. Gallen, Switzerland, 2003.
11. K. Frauendorfer, G. Haarbrücker, K. Kiske, and D. Kuhn. Swing-Optionen im Elektrizitätsmarkt – Bewertung und optimale Ausübung komplexer Stromderivate. *e|m|w Zeitschrift für Energie, Markt, Wettbewerb*, 2005. Heft 5.
12. G. Haarbrücker and D. Kuhn; ior/cf-HSG (Hrsg.): Valuation of Electricity Swing Options by Multistage Stochastic Programming, 2006.- URL http://www.alexandria.unisg.ch/Publikationen/29493 (2008-12-02)
13. N. Haldrup and M. Ø. Nielsen. A regime switching long memory model for electricity prices. Technical report, Working Paper, University of Aarhus, 2004.
14. R. Huisman and R. Mahieu. Regime jumps in electricity prices. Technical report, Erasmus University, 2001.
15. J. Hull. *Options, Futures and Other Derivative Securities.* Prentice Hall, Englewood Hills, NJ, 1994.
16. International-Energy-Agency. *World Energy Outlook 2006*, volume 2006. OECD – Organisation for Economic Co-operation and Development, complete edition, November 2006. IEA, Washington DC.
17. J.P. Morgan Securities Ltd. All you ever wanted to know about carbon trading 5.0, August 2007.
18. E. Lehrmann. *Informationsmanagement im Handel Strom - Eine ökonomische Analyse des Informationseinsatzes aus Sicht deutscher Verbundunternehmen.* PhD thesis, Universität – Essen, 2001.

19. E. S. Lucia, J. J. Schwartz. Electricity prices and power derivatives: Evidence from the nordic power exchange. *Rev. Derivatives Res.*, (5):5–50, 2002.
20. Robert C. Merton. Option pricing when underlying stock returns are discontinuous. *J. Finan. Econ.*, (3):125–144, 1976.
21. Dragana Pilipovic. *Energy Risk: Valuing and Managing Energy Derivatives.* McGraw-Hill, New york, 2nd edition, 2007.
22. Point Carbon. Carbon 2007 – A new climate for carbon trading, 2007.
23. PricewaterhouseCoopers. Emission critical, 2004. Connecting carbon and value strategies in utilities. `https://www.pwc.com/at/pdf/publikationen/Emissioncritical.pdf`. Cited 8 Sept 2007.
24. R. T. Rockafellar and R. J. -B. Wets. Nonanticipativity and L^1-martingales in stochastic optimization problems. In *Stoch. Syst.: Model., Identif., Optim. II; Math. Program. Study 6*, pages 170–187, 1976.
25. V. Scherer. Kombinierte Gas- und Dampfturbinenkraftwerke: Bausteine einer effizienten Stromversorgung. In Walter Blum, editor, *Moderne Wege der Energieversorgung, Leipziger Tagung 2002*, pages 93–108. DPG Deutsche Physikalische Gesellschaft, 2002.
26. H. Siede. *Multi-Period Portfolio Optimization.* PhD thesis, IfU-HSG, Universität St. Gallen, 2000.
27. J. Sijm, K. Neuhoff, and Y. Chen. CO2 cost pass through and windfall profits in the power sector. *Climate Policy (Special Issue: Emissions Allocation and Competitiveness in the EU ETS)*, 6:49–72, 2006.

21

Efficient Stochastic Programming Techniques for Electricity Swing Options

Marc C. Steinbach and Hans-Joachim Vollbrecht

Summary. We consider the valuation of contracts of electrical energy supply with optionalities. After discussing appropriate stochastic programming models and presenting especially suited solution algorithms, a set of price scenarios is simulated based on a probabilistic model of the electricity spot market price at the EEX. We determine empirically upper and lower bounds for the stochastic optimization over any scenario tree obtained by reduction techniques. Furthermore, we introduce constraints restricting all scenarios to have identical contract exercise amounts cumulated over various fixed subperiods. Calculation of the losses of the optimal value of the objective function caused by these constraints shows that, for subperiods of 1 month, no substantial loss is encountered. This suggests a temporal decoupling heuristic where the depth of scenario trees is reduced to a suitable subperiod, yielding a good approximation to the valuation problem with substantially reduced complexity.

21.1 Introduction

Due to the liberalization of energy markets, the trading of energy plays an increasingly important role in planning tasks of energy producers and distributors. In the case of electric energy, liquid spot markets have come into existence, so that the economic value of a decision can be reduced to the stock exchange value of a corresponding product. Decision makers are thus faced with the necessity of developing and solving stochastic programming models, with the stochastic process of electricity prices as a key component. Being discrete in time and value, this process can be modeled by means of scenario trees–however, trees of astronomical size. This raises the important and difficult question how to approximate the process by scenario trees of tractable size. In one type of approach [13], the tree is constructed from analytical price models, such as an adapted Pilipovic model [24, 25], possibly combined with bounding techniques to give upper and lower estimates of the optimal value. Another type of approach constructs trees from simulated scenarios by a recursive reduction technique that controls the approximation quality using suitable metrics of distributions and filtrations [16]. In the study presented here, an initial scenario fan has been generated by simulation from a price model based on [26], which includes deterministic influences (trend and seasonal periodicities) and stochastic influences (outliers and residuals, using an ARMA model), combined by multiplication.

The particular optimization problem addressed in this study is the valuation of certain swing options [4] in which the option holder may exercise, within given limits, variable energy amounts from a fixed total contract volume during a given period. Flexible contracts of this kind originate in the natural gas industry [1]; nowadays they are commonly used as hedging instruments in power and energy markets and are even being considered for pricing IT resources [7].

Valuation schemes for swing options have been developed systematically during the last decade. Most of them are based on or closely related to stochastic dynamic programming (SDP), often combined with special techniques that exploit simplifications resulting from the structural properties of specific price models or specific classes of swing options. A widely known method is the Least Squares Monte Carlo algorithm [22], which approximates conditional expectations within the SDP scheme by least squares estimates. Related Monte Carlo simulation approaches are proposed in [18] and [23]. A quantization approach (optimal discretization) is proposed in [2] and combined with a decomposition of the payoff function of certain swing options for highly accurate valuation under multifactor Gaussian price processes. In [9], a specific Poisson price model is developed deriving the probability for spikes from the electricity supply/demand ratio, and analytical price formulas are given for very simple weekly swings. In [20], electricity swing options are modeled as portfolios of forwards and call options, and a lower bound on their value is derived by linear optimization under the assumption of a Markov consumption process. In [5,6], the valuation problem is formulated and analyzed as an optimal multiple stopping problem for price processes in the form of geometric Brownian motion and general linear regular diffusion, respectively. A related formulation as stochastic impulse control problem is analyzed in [8], together with a numerical algorithm for the corresponding Hamilton–Jacobi–Bellman quasi-variational inequality. Other valuation approaches employ direct discretizations of the price process in space and time within an SDP scheme: grid discretizations are used in [32] and [15], whereas forests are used in [21] and [19]. Finally, a scenario tree model is proposed in [14], and special aggregation and reparameterization techniques are developed to obtain a multistage stochastic linear program tractable by state-of-the art optimization software like CPLEX. The multistage stochastic programming approach is the most flexible and general one as it does not impose special structural requirements on price processes or constraints of the swing option.

Many of the above references address the peculiarities of price processes in energy (specifically electricity) markets, which are characterized by mean-reversion, jumps, and spikes. For highly realistic forward price models involving these phenomena, regime switching, and stochastic volatilities, see also [10, 11].

We consider a stochastic programming model for valuating general swing options, and demonstrate that our existing optimization algorithms perform very well on that problem class. This holds regardless of a specific price model or scenario tree. Next we aim at constructing scenario trees by reduction in a way that is especially suited for the valuation problem, starting from a fan of price scenarios. We wish to develop techniques that help deciding on when to branch and how many branches to use. The approach taken here is based on the idea of using the *value of the stochastic solution* (VSS). The VSS provides information on any scenario tree obtainable from the given fan, since the fan (where branching occurs only at $t = 0$) represents near-perfect information: the complete process of future prices is already determined at $t = 1$. In the valuation problems to be considered, the VSS depends mainly on

1. The total option volume
2. The contract period (planning horizon)

The first dependence is nonmonotonic since the VSS measures the value of additional flexibility of a stochastic exercise strategy over a deterministic one. This flexibility decreases when the total option volume approaches the maximal amount within exercise limits. We evaluate the VSS numerically over the range of possible option volumes.

The second dependence is clearly monotonic: the VSS will increase with the planning horizon. As swing options usually have long contract periods, we investigate the effects of a temporal decomposition into subperiods where the cumulated energy amounts are required to be identical for all scenarios. The extra constraints will reduce the flexibility (hence the VSS), but for the price model considered we expect a moderate loss only. This is confirmed by our study, even if we distribute the total energy a priori over suitably chosen subperiods.

The paper is organized as follows. We formulate the general valuation problem in Sect. 21.2, discuss some basic properties and present our solution technology. The concrete problem and price model are given in Sect. 21.3. Three computational experiments investigating the VSS are described in Sect. 21.4, with results reported in Sect. 21.5. In Sect. 21.6 finally, we discuss the results and outline an approximation heuristic to determine the timescale and structure of tree branchings under suitable assumptions. This heuristic reduces the complexity of the valuation problem without significant loss of the approximation quality.

21.2 General Valuation Problem

In setting up the mathematical formulation of the valuation problem, we follow [14] where more details can be found. We consider a planning horizon $[0, T]$ divided into periods $t = 1, \ldots, T$ of unit length, where $[0, T]$ represents the (remaining) contract period and each subinterval represents an exercise period. During period t, the option holder may exercise any power p_t within an agreed range $\mathcal{P}_t = [p_t^-, p_t^+]$. Often there are additional ramp constraints (or *ratchets*), limiting the power difference $r_t = p_t - p_{t-1}$ to some range $\mathcal{R}_t = [r_t^-, r_t^+]$. Finally, the cumulated energy e_t up to time t may vary within $\mathcal{E}_t = [e_t^-, e_t^+]$. In practice, the bounds $p_t^\pm$ will typically be constant or change just a few times over $[0, T]$, ratchets will usually have the form $r_t^\pm = \pm\rho$ with ρ fixed, and energy limits $e_t^\pm$ will only be specified at T and possibly a few more time instances.

Further, let K denote the strike price of the option, s_t the spot price during period t, $k_t := K - s_t$, and e_0, p_0 the initial energy and power values at $t = 0$. Here p_0 will only be relevant in the presence of ratchet constraints, and we have $e_0 = 0$ if the planning period coincides with the entire contract period.

21.2.1 Stochastic Programming Model

Our probabilistic model is based on a finite number of scenarios with associated probabilities, organized as a scenario tree. Let V denote the set of nodes (vertices) of the tree, τ_j the probability of node j, $L_t \subseteq V$ the level set of nodes at depth t, and L the set of leaves; further $1 \in L_1$ the root, $j \in L_t$ the "current" node, $S(j)$ its set of successors, $i \equiv \pi(j)$ its unique predecessor (if $t > 1$), and $\Pi(j) = \{1, \ldots, i, j\}$ the unique path from the root to j. Finally define $V^* := V \setminus \{1\}$. The subtree rooted in j has respective vertex set, level sets and leaves $V(j)$, $L_t(j)$, and $L(j)$. Below, the vertex set is often taken to be $V = \{1, \ldots, N\}$ where nodes are numbered in any ascending order. (In the deterministic case we have $V = \{1, \ldots, T\}$.)

More concretely, the stochastic price process (s_t) has discrete realizations s_j defining the scenario tree and inducing realizations e_j, p_j, r_j, k_j of e_t, p_t, r_t, k_t. We define $\pi(1) = 0$ with e_0, p_0 as above, and variable vectors $r, p, e \in \mathbf{R}^{|V|}$ with associated feasible sets $\mathcal{R}_V, \mathcal{P}_V, \mathcal{E}_V$, where $r = (r_j)_{j \in V}$ and $\mathcal{R}_V = \prod_{j \in V} \mathcal{R}_j$, etc.

The general valuation problem then becomes a stochastic LP,

$$\underset{r,p,e}{\text{Minimize}} \sum_{j \in V} \tau_j k_j p_j \tag{21.1}$$

$$\text{subject to } p_j = p_{\pi(j)} + r_j, \qquad j \in V, \tag{21.2}$$

$$e_j = e_{\pi(j)} + p_j, \qquad j \in V, \tag{21.3}$$

$$(r, p, e) \in \mathcal{R}_V \times \mathcal{P}_V \times \mathcal{E}_V. \tag{21.4}$$

This is a multistage model in control form, with "incoming control" in the terminology of [31]. This means that the state variables p_j, e_j depend on the control variable r_j of the *same* node, reflecting the assumption that the decision for period t is made *after* observing the actual price s_j for period t, at time $t-1$. Concretely, we control the exercise process via the power differences r_t; the exercised power p_t is the sum of these differences (first integral), and the cumulated energy e_t is the sum of the powers (second integral).

In the absence of ratchets, one obtains a simplified stochastic LP model,

$$\underset{p,e}{\text{Minimize}} \quad \sum_{j \in V} \tau_j k_j p_j \tag{21.5}$$

$$\text{subject to} \quad e_j = e_{\pi(j)} + p_j, \qquad j \in V, \tag{21.6}$$

$$(p, e) \in \mathcal{P}_V \times \mathcal{E}_V. \tag{21.7}$$

Again we have incoming control form, now with states e_j and controls p_j. Thus we control the exercise process with the power p_t, and obtain the cumulated energy e_t as first integral.

Throughout the paper we let $K := 0$; hence minimizing $\sum \tau_j k_j p_j$ is the same as maximizing $\sum \tau_j s_j p_j$, the expected spot market value of the contract.

21.2.2 Aggregated Formulation and Critical Prices

The two models above represent the most straightforward and most flexible formulations, in that branchings of the scenario tree and bounds on all variables (including cumulated energy) are allowed at each node. In practice, energy bounds will only be present at a few points. Moreover, due to the large number of decision periods, branchings of the scenario trees will occur at relatively few time instances only. In this case every tree node j may represent several periods, $t_{\pi(j)} + 1, \ldots, t_j$, where $t_{\pi(1)} = 0$ and $t_j = T$ for $j \in L$. Each vector of node variables holds only a single energy value (associated with t_j): $(r_j, p_j, e_j) \in \mathbf{R}^{2d_j+1}$ or $(p_j, e_j) \in \mathbf{R}^{d_j+1}$, where $d_j := t_j - t_{\pi(j)} - 1$. Letting $\mathbf{1}_j := (1, \ldots, 1) \in \mathbf{R}^{d_j}$, the stochastic LP models can now be reformulated with aggregated periods:

$$\underset{r,p,e}{\text{Minimize}} \quad \sum_{j \in V} \tau_j k_j^T p_j \tag{21.8}$$

$$\text{subject to} \quad (I - N_j) p_j = M_j p_{\pi(j)} + r_j, \qquad j \in V, \tag{21.9}$$

$$e_j = e_{\pi(j)} + \mathbf{1}_j^T p_j, \qquad j \in V, \tag{21.10}$$

$$(r, p, e) \in \mathcal{R}_V \times \mathcal{P}_V \times \mathcal{E}_V \tag{21.11}$$

and

$$\underset{p,e}{\text{Minimize}} \quad \sum_{j \in V} \tau_j k_j^T p_j$$

$$\text{subject to} \quad e_j = e_{\pi(j)} + \mathbf{1}_j^T p_j, \qquad j \in V,$$

$$(p, e) \in \mathcal{P}_V \times \mathcal{E}_V.$$

Here $N_j \in \mathbf{R}^{d_j \times d_j}$ contains unit entries on the lower secondary diagonal and zeros elsewhere, while $M_j = \mathbf{1}_j \mathbf{1}_{\pi(j)}^T \in \mathbf{R}^{d_j \times d_{\pi(j)}}$, containing a single unit entry in the upper right corner. These formulations reduce the total numbers of variables to roughly two thirds and one half, respectively.

It is well-known that the LP without ratchets has a simple solution in the deterministic case: there exists a critical price s^* such that $p_t = p_t^-$ is optimal whenever $s_t < s^*$, and $p_t = p_t^+$ is optimal whenever $s_t > s^*$. (The optimal power for $s_t = s^*$ depends on the actual energy bounds.) In fact, the critical price is given by the optimal dual variables, $s^* = K - y = K - w^- - w^+$. The aggregated formulation just introduced yields a direct generalization of this result to the stochastic case. From the Lagrangian

$$
\begin{aligned}
L(p, e, y, v^\pm, w^\pm) = \sum_{j \in V} \tau_j \big[k_j^T p_j - y_j(e_{\pi(j)} + \mathbf{1}_j^T p_j - e_j) \\
- (v_j^-)^T (p_j - p_j^-) - w_j^- (e_j - e_j^-) \\
- (v_j^+)^T (p_j^+ - p_j) - w_j^+ (e_j^+ - e_j)\big]
\end{aligned}
$$

we obtain the dual LP

$$
\begin{aligned}
\underset{y, v^\pm, w^\pm}{\text{Maximize}} \quad & -e_0 y_1 + \sum_{j \in V} \tau_j \left[(p_j^-)^T v_j^- + e_j^- w_j^- - (p_j^+)^T v_j^+ - e_j^+ w_j^+\right] \\
\text{subject to} \quad & \mathbf{1}_j^T y_j = k_j + v_j^+ - v_j^-, \qquad j \in V, \\
& y_j = w_j^- - w_j^+, \qquad j \in V, \\
& \tau_j y_j = \sum_{k \in S(j)} \tau_k y_k, \qquad j \in V \setminus L, \\
& v^\pm, w^\pm \geq 0.
\end{aligned}
$$

Starting with $y_j = w_j^- - w_j^+$ for $j \in L$, the remaining multipliers y_j are thus determined recursively as expectations of their successor variables,

$$
y_j = \sum_{k \in S(j)} \frac{\tau_k}{\tau_j} y_k = \sum_{k \in L(j)} \frac{\tau_k}{\tau_j} (w_k^- - w_k^+).
$$

This defines a martingale process on the scenario tree, which induces a *martingale process of critical prices* $s_j^* = K - y_j$. In other words: along every branch of the scenario tree, the optimal exercise strategy is as in the deterministic case, just with different critical prices. Of course, the same holds in the nonaggregated model for every node sequence without branching.

Note finally that one can rewrite the *implicit* transition equation (21.9) of the aggregated problem with ratchets as an *explicit* one, $p_j = G_j p_{\pi(j)} + E_j r_j$, with $E_j := (I - N_j)^{-1}$ containing unit entries in the lower triangle and zeros else, and $G_j := (I - N_j)^{-1} M_j = E_j M_j$ containing unit entries in the last column and zeros else. All the LP variants above can thus be written in the following incoming control form, which is a special case of the tree-sparse problems considered in [31]:

$$\begin{aligned}
\underset{u,x}{\text{Minimize}} \quad & \sum_{j \in V} \tau_j (d_j^T u_j + f_j^T x_j) \\
\text{subject to} \quad & x_j = G_j x_{\pi(j)} + E_j u_j + h_j, \qquad j \in V, \\
& (u, x) \in [u^-, u^+] \times [x^-, x^+].
\end{aligned}$$

21.2.3 Stochastic Programming Algorithms

In solving the valuation problems above, the main difficulty is the potentially excessive size of scenario trees. Our tree-sparse solution approach is based on the general idea of tackling the problem by iterative optimization methods that preserve the overall sparse structure in the linear subproblems, so that the latter are solvable by highly efficient linear algebra. The idea goes back to trajectory optimization problems [27, 28] and has been extended to stochastic optimization in [29–31]. Specifically, we use a primal–dual interior point method in connection with a factorization of the large tree-sparse KKT system defining the Newton step in each iteration. The KKT system can be interpreted as a linear–quadratic regulator problem defined over the scenario tree, with additional local and global constraints arranged into a generic hierarchical scheme. The associated factorization can be interpreted as a hierarchical sequence of projections in each node, combined with a dynamic programming recursion over the tree. Assuming that the number of variables in all nodes is comparable, this algorithm achieves optimal complexity $O(|V|)$ with respect to both memory and runtime. To exploit application-specific structural properties, a software tool has been developed [17] that analyzes the specific KKT structure and generates a custom factorization as source code in C++. Data structures and operations of the custom implementation are designed with the aim of minimal storage requirements and moderate operation counts. This approach is particularly efficient on regularly structured models like the one under consideration. For instance, the constraints matrix contains only unit entries (± 1) in fixed positions. Consequently, the custom implementation can represent the matrix entirely in code. It needs no memory to store entries or even entry positions. Similarly, the KKT matrix factorization has many such entries which are detected and represented in code; storage will only be allocated for the remaining entries, along with code that computes their values. Table 21.1 provides computational statistics for the solution of the concrete valuation problem with and without ratchets, using a scenario fan with 1,000 scenarios (see below). Observe that each problem is solved in less than 5 min on a 2 GB machine, with total memory (code and data) less than 12 times the size of one KKT vector. Note also that the per-iteration effort depends only on the size of the scenario tree but not on its topology.

Table 21.1. Computational statistics for concrete valuation problem on one processor of a 2.67 GHz Intel Core 2 E6700; relative accuracy 10^{-10} (duality gap+residuals)

Stochastic LP	Without ratchets	With ratchets
Number of rows	4,415,001	8,830,002
Number of columns	8,830,002	13,245,003
Number of nonzeros	13,245,002	30,905,004
Percentage of nonzeros	3.4×10^{-5}	2.6×10^{-5}
Order of KKT system	13,245,003	22,075,005
KKT vector memory	101.1 MB	168.4 MB
KKT factor memory	101.1 MB	202.1 MB
Total process memory	1,181 MB	1,859 MB
Number of iterations	89	77
Total solution time	219 s	288 s
Time per iteration	2.46 s	3.74 s

Table 21.2. Computational comparison for concrete valuation problem on one CPU of a 2 GHz Intel Xeon 5130 server with 12 GB RAM: IPM/TreeKKT vs. CPLEX10.1.1

Solution time in seconds: (read data) + presolve + solve

Scenarios	IPM/TreeKKT	CPLEXdual simplex	CPLEXbarrier
100	(0.3) + 0 + 30.2	(8.3) + 2.1 + 13.3	(8.3) + 2.3 + 39.3
200	(0.6) + 0 + 62.1	(51.6) + 4.2 + 34.6	(70.6) + 4.6 + 73.6
300	(0.8) + 0 + 97.0	(92.5) + 6.3 + 63.9	(109.8) + 7.0 + 121.1
500	(1.4) + 0 + 171.7	(66.0) + 10.9 + 127.4	(66.8) + 11.6 + 218.7
1 000	(2.7) + 0 + 401.1	(243.4) + 21.4 + 453.1	(237.9) + 23.5 + 529.9

Total memory in MB: (read data) + solve

Scenarios	IPM/TreeKKT	CPLEXdual simplex	CPLEXbarrier
100	(0) + 213	(170) + 581	(170) + 705
200	(0) + 397	(319) + 1192	(319) + 1410
300	(0) + 580	(481) + 1791	(481) + 2130
500	(0) + 946	(700) + 3240	(700) + 3810
1 000	(0) + 1862	(1940) + 6007	(1940) + 7180

Table 21.2 gives a comparison of runtime and storage requirements of our solver IPM/TreeKKT with ILOG CPLEX10.1.1 (dual simplex and barrier) on the valuation problem with ratchets for various numbers of scenarios. Both solvers run on the same machine but under different operating systems: CPLEX under Windows XP Professional, and IPM/TreeKKT under GNU/Linux (openSuSE 10.3). CPLEX reads an LP file and then solves the problem after some preprocessing; IPM/TreeKKT only reads the problem dimensions and price scenarios and then solves the problem without any preprocessing. The additional storage and memory requirements for reading the LP files into CPLEX (in parentheses)

are mainly shown for completeness: they can probably be reduced to the corresponding values of IPM/TreeKKT by using the CPLEX callable library interface, and will be disregarded in the following discussion.

The runtime comparison shows the typical behavior of interior point methods and (dual) simplex for an LP. The simplex algorithm clearly outperforms our interior point method on smaller problem instances. For 100 scenarios it is nearly twice as fast, with 15.4 s versus 30.2 s. Conversely, our interior point method performs better on large instances, with 401.1 s versus 474.5 s on the 1,000 scenario problem. The CPLEX interior point method (barrier) behaves similar in principle but is between 26% and 38% slower than our algorithm.

The main advantage of IPM/TreeKKT becomes obvious in the memory comparison. A linear interpolation for $n = 500$ and $n = 1{,}000$ scenarios yields as memory requirements $30 + 1.832n$ for IPM/TreeKKT, $473 + 5.534n$ for CPLEX dual simplex, and $440 + 6.74n$ for CPLEX barrier. This is exact for IPM/TreeKKT (up to rounding errors) while CPLEX actually needs less memory on the smaller instances. In any case, on the large instances CPLEX dual simplex and CPLEXbarrier require 3.0 times and 3.7 times more memory per scenario than IPM/TreeKKT, respectively. On the 12 GB server, the largest solvable instances thus have roughly 1,650 and 2,000 scenarios, versus 6,300 with IPM/TreeKKT.

21.3 Concrete Valuation Problem

21.3.1 Evaluated Contract

A bilateral contract with optionalities for obtaining electrical energy from an energy supplier has to be valuated. This valuation is based on its optimal exercise at the European Energy Exchange (EEX) spot market: the value of an exercise of the contract at a particular hour is interpreted as the corresponding value at the EEX spot market. The contract defines exercise rights for obtaining variable amounts of energy:

- Contract period in h: $T = 4416$ (2 quarters Q1, Q2, 1.7.2006 – 31.12.2006)
- Contract volume in GWh: $e^{\pm}_{2208} = 50$ (Q1), $e^{\pm}_{4416} = 240$ (Q2)
- Power limits in MW: $\mathcal{P}_t = [0, 90]$ (Q1), $\mathcal{P}_t = [25, 145]$ (Q2)
- Ratchets in MW: $r^{\pm}_t = \pm 60$

This type of contract can be viewed as a swing option [4]. With a swing option, the owner has the possibility to hedge his risks at the spot market with flexible exercise profiles of the option, which other hedging instruments such as futures are not capable of because of their rigid exercise structure allowing just to hedge with a fixed exercise that may differ only from peak to off-peak period. General swing options may require also temporal constraints such as a minimum period between two swings (changes in exercise). In our example, this is not required. For an example, Fig. 21.1 shows part of an optimal exercise strategy for a simulated spot price scenario.

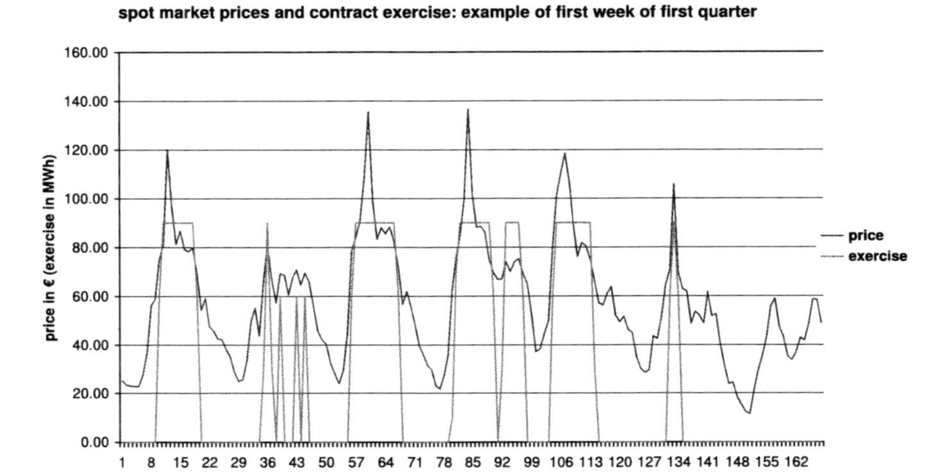

Fig. 21.1. Example of an optimal exercise for a single scenario (partial view)

21.3.2 Price Model

From the numerous electricity price models published so far, we have chosen the one described in [26]. Details and their justification can be found there. In this section we just sketch this model and outline where our model differs from it.

The spot price s_t is decomposed into four components: two deterministic ones, the trend s_t^{tr} and the seasonal part s_t^{seas}, and two stochastic ones, the outlier part s_t^{out} and the residual part s_t^{res}. The price s_t without outliers is modeled as

$$s_t = s_t^{\text{tr}} s_t^{\text{seas}} s_t^{\text{res}}. \tag{21.12}$$

The trend follows an exponential model; the seasonal part models daily profiles and the yearly seasons. The latter are modeled by a trigonometric polynomial with a basic oscillation of 1 year plus the first harmonic of half a year. For the daily profiles, we define five categories of days: Monday or day after or between holidays, Tuesday–Thursday, Friday or day before holiday, Saturday, Sunday or holiday. For each of these categories and for every hour (1–24), we estimate an independent model based on the trigonometric polynomial. The residuals are assumed to follow an ARMA process with time lags of $t-1$, $t-24$ and $t-25$. For estimating the parameters, we transform the historical residuals to obtain a normal distribution. Outliers are then added to the model by a random process that creates an outlier at t with probability p_t^{out}, depending on whether there has been an outlier at certain preceding points of time. The values of outliers are modeled by the gamma distribution. All probabilistic models are calibrated on historical data, specifically, EEX spot prices of the years 2003–2006.

Scenario Simulation

We generate n price scenarios by Monte-Carlo simulation using the price model just presented. The scenarios are assumed to have identical probability $\tau_j = 1/n$. This yields a scenario fan as in Fig. 21.2, which represents the approximation of the price process that our experiments are based on (see Sect. 21.4). Note that the construction of a scenario tree according to [16] starts from such a scenario fan, to which certain reduction steps are applied. Note further that, with n in the range of tractability ($n = 1,000$ in our experiments), the probability of obtaining a scenario fan rather than a more general tree is close to one since the probability of sampling two identical prices at $t = 1$ is almost zero. Figure 21.3 shows 30 simulated scenarios for a period of 1-week.

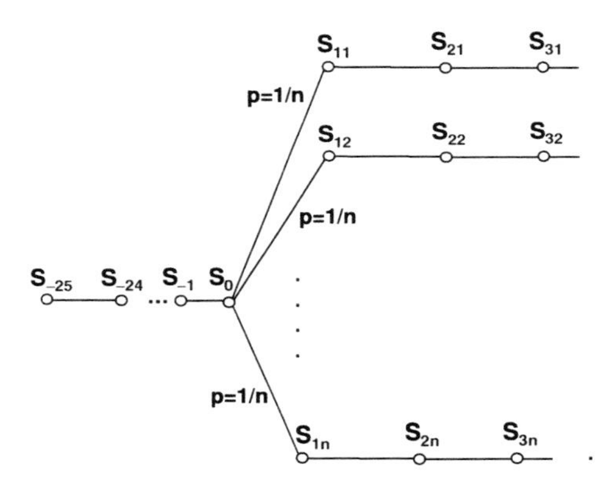

Fig. 21.2. The simulated scenario fan with spot market prices

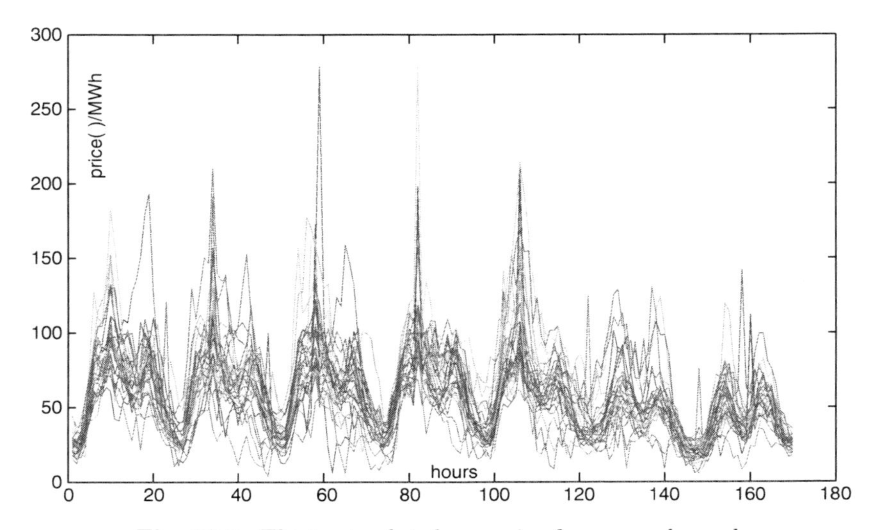

Fig. 21.3. Thirty simulated scenarios for a sample week

Approximation with a Tree or a Fan?

This question motivates the experiments to be presented in the next section. In the following, we briefly discuss this item.

There are strong arguments for preferring general scenario trees (branching at any point of time) over a scenario fan (branching only at $t = 0$) [3, 4, 12]: in a stochastic process, chance unfolds all the time and not only at the very beginning. Optimization based on a fan works with an erroneous process model, since all decisions after the first step are deterministic (from $t = 1$ on, the future price process is completely known until the end of the contract period). This should lead to a result that is too optimistic since the algorithm has perfect information about the future.

On the other hand, the theory of scenario tree reduction [16] tells us that a tree constructed by reduction from a fan approximates the distribution and filtration given by that fan, where the reduction method guarantees stability in the sense that convergence of reduced scenario trees to the reference tree (fan) under a particular metric leads to convergence of the optimal values to the optimal value for the reference tree (fan).

It seems that the question whether a sampled fan (of moderate size) or a tree constructed by reduction from such a fan gives a better approximation of the optimal value cannot be answered in general, notwithstanding the strong structural argument given at the beginning of this paragraph in favor of a tree. For this reason, we perform three experiment series, to quantitatively evaluate the impact of the structural constraints that a tree introduces to the valuation problem.

21.4 Computational Experiments

21.4.1 Experiment 1

Using the 1,000 scenarios just described, we solve the valuation problem (21.1) to (21.4) for the contract of Sect. 21.3.1 over the following range of contract volumes (in GWh):

$$e^{\pm}_{2208} = 50\alpha \in [0, 95], \quad e^{\pm}_{4416} = 55.2 + (240 - 55.2)\alpha. \tag{21.13}$$

Thus, α is a common scaling factor for the nominal exercise volumes of quarters Q1 and Q2 in excess of the base load $\sum_{\tau=1}^{t} p^{-}_{\tau}$. Three cases are distinguished: (1) *expected value:* deterministic optimization of the single scenario of expected prices, $\bar{s}_t = \sum_{j \in L_t} \tau_j s_j$; (2) *wait-and-see:* separate optimization of each scenario; (3) *here-and-now:* joint optimization over the scenario fan. The difference of the mean optimal value of (2) and the optimal value of (3) is called the *expected value of perfect information* (EVPI). For the given fan it turns out to be almost zero, indicating (as expected) that the nonanticipativity constraint at $t = 0$ (decisions being identical for all scenarios) has a negligible impact. The difference of the optimal values of (3) and (1) is called the *value of the stochastic solution* (VSS). This value is more interesting in our case. Results will be discussed in Sect. 21.5.1.

21.4.2 Experiment 2

Here we valuate the contract over individual scenarios with its nominal volume, $e^{\pm}_{2208} = 50$, $e^{\pm}_{4416} = 240$, but with additional interscenario volume constraints

$$e_j = e_k \quad \forall j, k \in L_t, \quad \forall t \in \mathcal{T}, \tag{21.14}$$

where $\mathcal{T} \in \{\mathcal{T}_d, \mathcal{T}_w, \mathcal{T}_m\}$ selects the set of final periods either of each day, week, or month within the contract period. Thus we require that the cumulated energy exercised during each of these subperiods agrees for all scenarios, while the respective amounts are subject to optimization.

Compared to experiment 1 we use a relatively small number of 30 scenarios. This is due to the fact that we do not have a suitable modeling environment for stochastic optimization with interscenario constraints, so we have to set up the deterministic equivalent manually. Consequently, the results of experiment 2 are not statistically significant. However, these results just serve for motivating experiment 3 which runs again on 1,000 scenarios.

21.4.3 Experiment 3

Next we valuate the contract again over individual scenarios and with its nominal volume, but now with additional per-scenario volume constraints

$$e_j = \bar{e}_t \quad \forall j \in L_t, \quad \forall t \in \mathcal{T}, \qquad \mathcal{T} \in \{\mathcal{T}_d, \mathcal{T}_w, \mathcal{T}_m\}, \tag{21.15}$$

where $\bar{e}_t$ denotes the optimal cumulated volume from experiment 1, case 1. Thus we require again that the cumulated energy exercised during each subperiod agrees for all scenarios, but now with prescribed amounts.

21.5 Computational Results

21.5.1 Experiment 1

Figure 21.4 shows the optimal contract value for the three cases of experiment 1, in dependence of the contract volume of quarter Q1. It turns out that the factor α is bounded by the freely exercisable energy, $\sum_\tau (p^+_\tau - p^-_\tau)$, during quarter Q2: the valuation problem becomes infeasible at

$$\hat{\alpha} = \frac{240 - 50 - 2208 \times 0.025}{2208 \times (0.145 - 0.025)} \gtrapprox 1.9656, \quad \text{or} \quad \hat{e}^{\pm}_{2208} \gtrapprox 98.2789. \tag{21.16}$$

We call $100 e^{\pm}_{2208} / \hat{e}^{\pm}_{2208}$ the *coverage*; it measures the percentage of hours of Q2 necessary to exhaust the contract volume when exercising p^+_t during

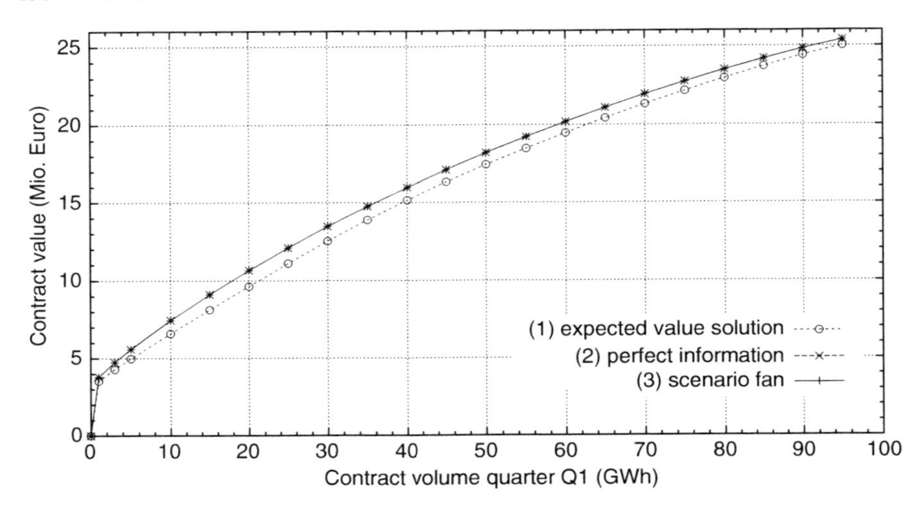

Fig. 21.4. Optimal values for Experiment 1

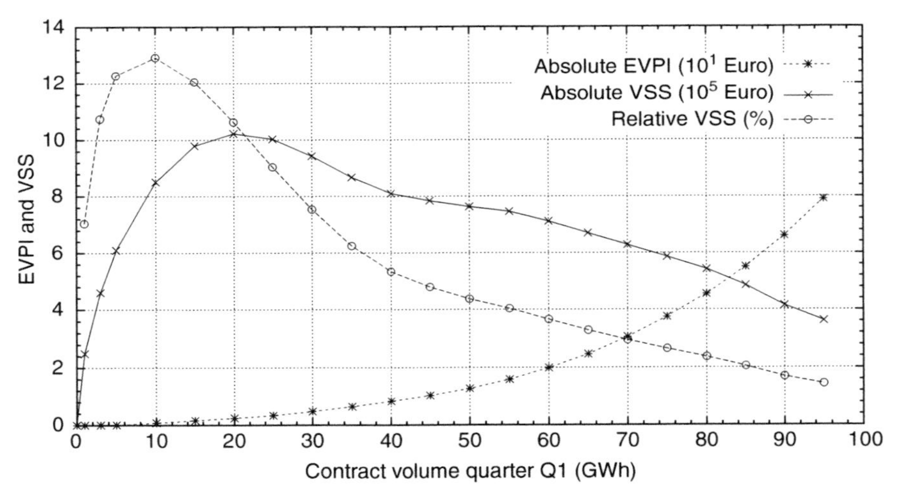

Fig. 21.5. EVPI (absolute) and VSS (absolute and relative) for the scenario fan

these hours, thus indicating a lack of flexibility. We will use $e^{\pm}_{2208}$ as a good approximation of the coverage.

Turning to Figs. 21.4–21.5, we confirm that the EVPI (difference of cases 2 and 3) is indeed very small (<80€ throughout). Hence, we compare case 1 ("deterministic optimization") with cases 2, 3 ("stochastic optimization"). As can be seen from Fig. 21.5, the gain by stochastic optimization (VSS) increases rapidly for small coverage reaching an early maximum, and then decreases moderately toward large coverage (i. e., little flexibility in exercising the contract). The absolute VSS varies between 0.35 and 1 Million €, having its maximum at 20% coverage. The relative VSS reaches a maximum of 13% at 10% coverage, then decreases to 4% at about 50% coverage and to 2% at about 85% coverage.

Note that the optimal contract value for any scenario tree based on the same scenarios must lie between the values of cases 1 and 2; thus the VSS studied in Fig. 21.5 (plus the negligible EVPI) is an upper bound for the given set of scenarios.

21.5.2 Experiment 2

Next we consider the modified problem where interscenario constraints require the cumulated energy exercised during certain subperiods (day, week, or month) to agree for all scenarios. Subperiods of an hour would be equivalent to case 1 above (deterministic optimization), yielding a lower bound on the optimal value. The given subperiods must result in contract values between cases 1 and 2.

Table 21.3 gives the relative losses of the optimal contract value with respect to case 3 for the first quarter. The monthly constraint has very little influence on the value, and also the weekly constraint causes only a moderate loss. Therefore the variability that is exploited by perfect information hardly extends over more than a week. This qualitative interpretation is not surprising, however. It reflects the fact that, in our price model, the long-term behavior is modeled mainly by the *deterministic* components, trend and season. The stochastic component of outliers has little influence, and the residuals generate volatility only over a short range because of the small autoregression time lag (25 h). Thus we can only expect a very moderate accuracy of the long-term approximation from the sampled scenarios.

Figure 21.6 shows the optimal daily exercise volumes (not the hourly volumes, for the sake of better visibility) under the weekly constraint for all 30 scenarios. These values vary considerably both within single scenarios during a week and between different scenarios at a given hour. The difference between workdays and weekend is quite apparent.

Figures 21.7–21.8 show the weekly profits under the daily and the monthly constraint, respectively. It can be seen for the first quarter (which has a low

Table 21.3. Mean loss (in %, over first quarter) due to subperiod constraints

0.71	Identical monthly amounts
1.77	Identical weekly amounts
4.59	Identical daily amounts
7.70	Identical hourly amounts (deterministic optimization: case 1)

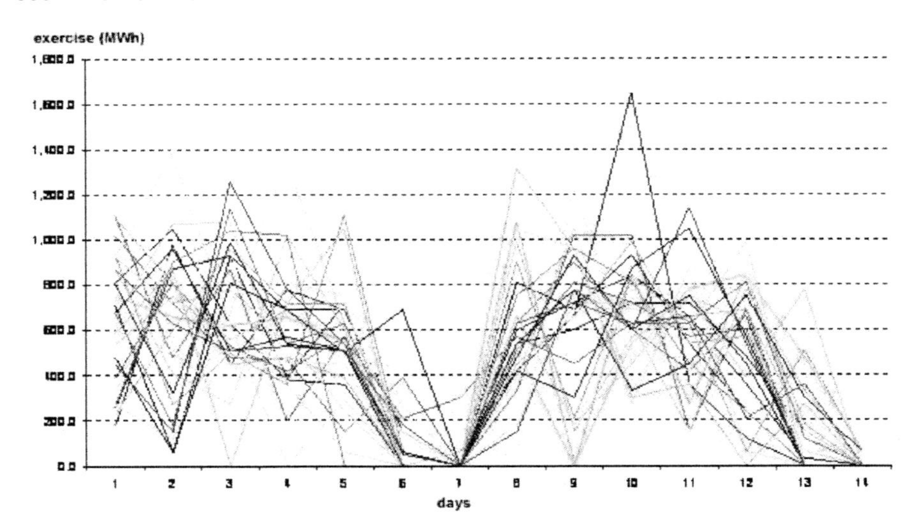

Fig. 21.6. Daily exercises under the weekly constraint

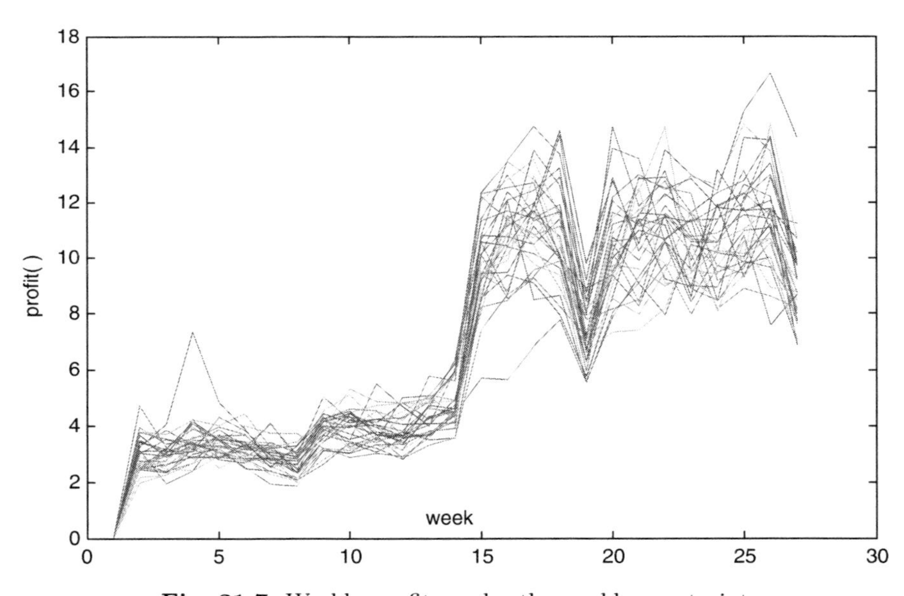

Fig. 21.7. Weekly profits under the weekly constraint

coverage) how the monthly constraint results in a higher variance of the profit than the daily one, due to potentially higher flexibility in exercising the contract.

Although these results are not statistically significant, the qualitative conclusions are supported by the quantitative results of experiment 3.

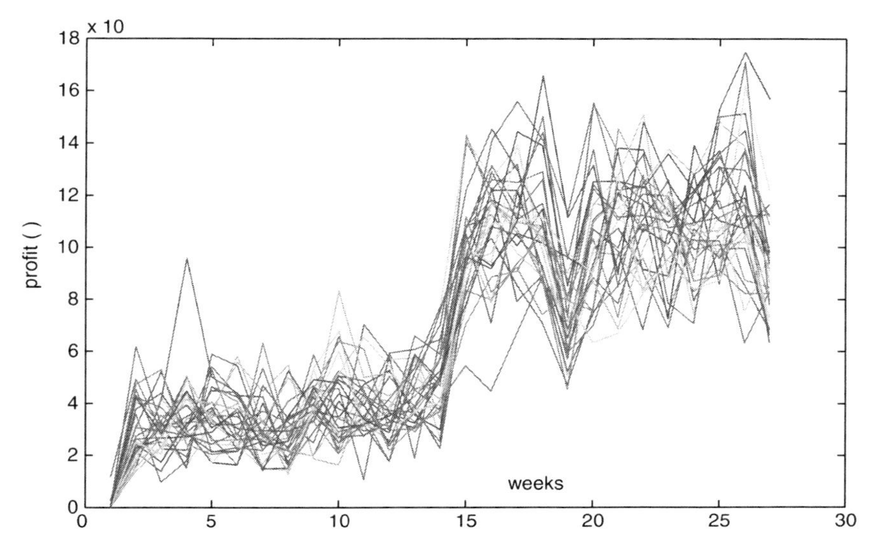

Fig. 21.8. Weekly profits under the monthly constraint

Table 21.4. Mean loss (in %, over both quarters) due to subperiod constraints

0.4	Fixed monthly amounts (deterministically optimal)
1.0	Fixed weekly amounts (deterministically optimal)
4.2	Identical hourly amounts (deterministic optimization: case 1)

21.5.3 Experiment 3

In this experiment, the cumulated energy of every scenario on every subperiod (hour, week, or month) is fixed, matching the corresponding optimal amount of the deterministic optimization (experiment 1, case 1).

Table 21.4 presents the results for the mean loss with respect to the stochastic optimization on the scenario fan, i. e., case 3. The difference to Table 21.3 results from the fact that in experiment 2 we consider only the first quarter but here both quarters. Again, as in experiment 2, we see that the monthly constraint results in just 10% loss of the VSS (relative VSS = 4.4% at 50 MWh), and the weekly constraint results in less than 25%.

21.6 Discussion

The motivation for the investigations in this work was to analyze the gain of a stochastic optimization over the deterministic optimization (VSS) for a concrete valuation problem, in dependence on the flexibility for exercising the contract in its original form (experiment 1) or with additional subperiod constraints on the cumulated energy (experiments 2 and 3). The results reported in the last section encourage us to sketch in this section a heuristic for dimensioning and structuring the scenario trees for the specific problem of valuating swing options. This heuristic consists of the following steps:

1. Determine the bounds on the optimal contract value, the lower bound given by the optimal value of the mean price scenario (expected value solution), and the upper bound given by the mean optimal value over all individual scenarios (wait-and-see solution).
2. For various uniform subdivisions of the contract period with different values of the subperiod length, determine for each subinterval of every subdivision the cumulated exercise amount of the expected value solution.
3. For every subperiod length, determine the mean loss of the optimal value with respect to the upper bound of step 1 if the subperiod exercise amounts of all scenarios are fixed to the values calculated in step 2.
4. For a given approximation tolerance, select on the basis of step 3 the largest subperiod length whose mean loss remains within the tolerance. Now define approximate valuation problems using the selected subperiod length:
   (a) For optimizing the exercise strategy over some initial period within the first subperiod (just the first day, say), use the valuation problem for the first subperiod with the cumulated exercise volume determined in step 2. This reduces the planning horizon; hence we can generate a scenario tree of substantially reduced complexity (by the methods cited in Sect. 1, for instance). See Fig. 21.9a for an illustration of the principle.
   (b) For valuating the contract over the entire planning horizon, the problem can be temporally decomposed to obtain independent subproblems over the selected subdivision. For each subperiod, a separate scenario tree can be generated, having as initial spot price the mean price over all scenarios and restricting the cumulated energy to the amount determined in step 2. Again, this will yield a substantial reduction of the problem complexity. See Fig. 21.9b for an illustration of the principle.

Next we discuss how the subproblems created by the temporal decomposition or reduction (as a special case) are related to the original problem and to each other. Of course, the selected approximation tolerance will not yield any hard bounds on how much one loses by the decomposition; rather, it is a measure how well the contract volume has been distributed over the subperiods. Given that the price model and the limited number of sampled scenarios will reflect the long-term behavior rather inaccurately in any case, we feel that the valuation error induced by the decomposition should be tolerable. It remains to discuss feasibility issues. Intertemporal dependencies of the decisions (exercises) in the original valuation problem (Sections 21.2–21.3) are introduced by

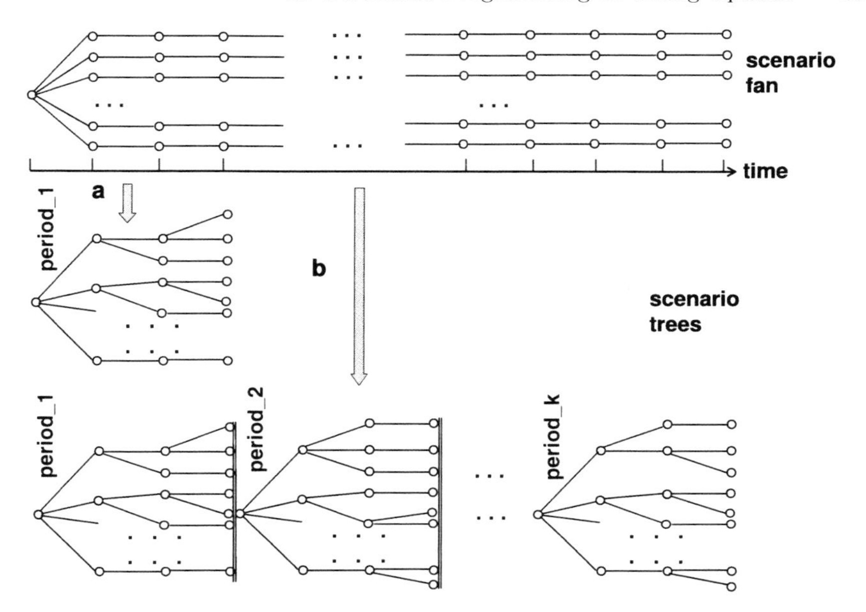

Fig. 21.9. Steps 4a and b: reducing the scenario trees

- The bounds on the cumulated exercise volume
- The ratchets

The first dependence is a long-term dependence, the second a short-term one. By introducing the subperiod constraints on cumulated exercise amounts as in experiment 3, we break the first dependence while keeping the total volume, thereby guaranteeing feasibility with respect to the cumulated energy of the original problem. The ramp constraints (ratchets) across subperiod boundaries, however, are dropped in step 4 of the temporal decomposition.

Now, when optimizing over the first subperiod (case 4a), the dropped ratchets belong to the following subperiod, and feasibility is still guaranteed with respect to all constraints of the original problem.

In case of step 4b, dropped ratchets belong to the beginning of every subperiod after the first, and will generally be violated. This can be prevented by additional constraints if desired. For instance, we could optimize the subproblems backward in time, recursively tightening exercise bounds in the last hour of the previous subperiod to force consistence with the ratchets relative to the optimal exercise in the first hour of the current subperiod. However, we are primarily interested in a good approximation of the contract volume, and in generating a good exercise strategy for some initial period. In practice the model will be reoptimized after that initial period, and the contract provides enough flexibility to generate a feasible exercise strategy this way. Therefore it is not necessary to enforce full feasibility in the decomposed problem.

In conclusion, the suggested heuristic promises the advantage of obtaining a good approximation of the contract value with substantially reduced effort and with easily parallelizable subproblems.

21.7 Conclusion

We have determined optimal values of an energy supply contract of swing option type for several stochastic and deterministic optimization models.

We have also calculated, with respect to a scenario fan, the expected value solution (given by deterministic optimization based on the mean price scenario), and the mean contract value under perfect information (given by averaging the optimal values of all individual scenarios). The two solutions provide lower and upper bounds, respectively, for the contract value obtained with any scenario tree constructed from the given fan by reduction techniques. The difference between the bounds equals the value of the stochastic solution (VSS) plus the negligible expected value of perfect information (EVPI).

By introducing additional subperiod constraints (thus fixing the exercise amounts of all scenarios to the corresponding optimal amounts of the deterministic optimization), we found that a smaller optimization horizon can be chosen without significant loss of the resulting contract value. This also allows for temporal decomposition of the valuation problem into independently solvable subperiod problems, yielding drastic reductions of the size of scenario trees. In combination with the fast solution algorithms presented in Sect. 21.2.3, we are thus able to solve good approximations of the valuation problem efficiently.

References

1. Angelo Barbieri and Mark B. Garman. Understanding the valuation of swing contracts. Energy and Power Risk Management, 1996.
2. Olivier Bardou, Sandrine Bouthemy, and Gilles Pagès. Optimal quantization for the pricing of swing options. Technical report, Gaz de France, 6 April 2007. eprint arXiv: 0705.0466, 2007 – `arxiv.org`.
3. John R. Birge and François Louveaux. *Introduction to Stochastic Programming.* Springer, Berlin, 1997.
4. Lea Blöchlinger, Karl Frauendorfer, and Gido Haarbrücker. Vertragsbewertung in der Stromwirtschaft unter Anwendung der stochastischen Optimierung. Working Paper ior/cf-HSG 06-04-01, Universität St. Gallen, Switzerland, 2006.
5. René Carmona and Savas Dayanik. Optimal multiple-stopping of linear diffusions and swing options. Technical report, Princeton University, New Jersey, 2003.
6. René Carmona and Nizar Touzi. Optimal multiple-stopping and valuation of swing options. Technical report, Princeton University, New Jersey, October 2004.

7. Scott H. Clearwater and Bernardo A Huberman. Swing options: A mechanism for pricing IT peak demand, 2005.
8. M. Dahlgren. A continuous time model to price commodity-based swing options. *Rev. Derivat. Res.*, 8(1):27–47, 2005.
9. Matt Davison and Lindsay Anderson. Approximate recursive valuation of electricity swing options. Technical report, The University of Western Ontario, Ontario, 29 October 2003.
10. Shijie Deng. Stochastic models of energy commodity prices and their applications: Mean-reversion with jumps and spikes. POWER Working Paper PWP-073, University of California Energy Institute, Berkeley, February 2000.
11. Stein-Erik Fleten and Jacob Lemming. Constructing forward price curves in electricity markets. *Energy Econ.*, 25(5):409–424, 2003.
12. Nicole Gröwe-Kuska, M. Lucht, Werner Römisch, G. Spangardt, and Isabel Wegener. *Mittelfristige risikoorientierte Optimierung von Strombezugsportfolios kleiner Marktteilnehmer.* VDI-Bericht 1792. VDI Verlag, Düsseldorf, 2003.
13. Jens Güssow. *Power Systems Operation and Trading in Competitive Energy Markets.* PhD thesis, Universität St. Gallen, Switzerland, 2001.
14. Gido Haarbrücker and Daniel Kuhn. Valuation of electricity swing options by multistage stochastic programming. Working Paper Series in Finance 45, Universität St. Gallen, Switzerland, December 2006.
15. Ben M. Hambly, Sam Howison, and Tino Kluge. Modelling spikes and pricing swing options in electricity markets. Technical report, University of Oxford, Oxford, 24 April 2007.
16. Holger Heitsch and Werner Römisch. Scenario tree modelling for multistage stochastic programs. Preprint 296, 2005.
17. Andrei Huţanu. Code generator for sparse linear algebra in stochastic optimization. Diploma thesis, "Politecnica" University of Bucharest, 2002.
18. Alfredo Ibáñez. Valuation by simulation of contingent claims with multiple early exercise opportunities. *Math. Finan.*, 14(2):223–248, 2004.
19. Patrick Jaillet, Ehud I. Ronn, and Stathis Tompaidis. Valuation of commodity-based swing options. *Manage. Sci.*, 50(7):909–921, 2004.
20. Jussi Keppo. Pricing of electricity swing options. *J. Derivat.*, 2(11):26–43, 2004.
21. Ali Lari-Lavassani, Mohamadreza Simchi, and Antony Ware. A discrete valuation of swing options. *Can. Appl. Math. Quart.*, 9(1):35–74, 2001.
22. Francis A. Longstaff and Eduardo S. Schwartz. Valuing American options by simulation: A simple least squares approach. *Rev. Finan. Stud.*, 14(1):113–147, 2001.
23. N. Meinshausen and Ben M. Hambly. Monte Carlo methods for the valuation of multiple exercise options. *Math. Finan.*, 14(4):557–583, 2004.
24. Dragana Pilipovic. *Energy Risk: Valuing and Managing Energy Derivatives.* McGraw-Hill, New York, 1998.
25. Dragana Pilipovic and John Wengler. Getting into the swing. *Energy and Power Risk Management*, 2(10), 1998.
26. Hagen Klaus Schmöller. *Modellierung von Unsicherheiten bei der mittelfristigen Stromerzeugungs- und Handelsplanung*, volume 103 of *Aachener Beiträge zur Energieversorgung.* Klinkenberg Verlag, 2005.
27. Marc C. Steinbach. A structured interior point SQP method for nonlinear optimal control problems. In Roland Bulirsch and Dieter Kraft, editors, *Computational Optimal Control*, volume 115 of *International Series of Numerical Mathematics*, pages 213–222. Birkhäuser Verlag, Basel, Switzerland, 1994.

28. Marc C. Steinbach. *Fast Recursive SQP Methods for Large-Scale Optimal Control Problems.* Ph.d. dissertation, Universität Heidelberg, 1995.
29. Marc C. Steinbach. Recursive direct algorithms for multistage stochastic programs in financial engineering. In Peter Kall and H.-J. Lüthi, editors, *Operations Research Proceedings 1998*, pages 241–250, Berlin, Springer, 1999.
30. Marc C. Steinbach. Hierarchical sparsity in multistage convex stochastic programs. In Stanislav P. Uryasev and Panos M. Pardalos, editors, *Stochastic Optimization: Algorithms and Applications*, pages 385–410, Kluwer, Dordrecht, The Netherlands, 2001.
31. Marc C. Steinbach. Tree-sparse convex programs. *Math. Methods Oper. Res.*, 56(3):347–376, 2002.
32. Andrew C. Thompson. Valuation of path-dependent contingent caims with multiple exercise decisions over time: The case of take-or-pay. *J. Finan. Quantit. Anal.*, 30(2):271–293, 1995.

22

Delta-Hedging a Hydropower Plant Using Stochastic Programming

Stein-Erik Fleten and Stein W. Wallace

Summary. An important challenge for hydropower producers is to optimize reservoir discharges, which is subject to uncertainty in inflow and electricity prices. Furthermore, the producers want to hedge the risk in the operating profit. This article demonstrates how stochastic programming can be used to solve a multireservoir hydro scheduling case for a price-taking producer, and how such a model can be employed in subsequent delta-hedging of the electricity portfolio.

22.1 Introduction

The main challenge for a hydro producer with reservoir capacity is deciding on how much electricity to produce today versus future periods. To obtain the best possible balance between immediate and future costs of using the water, uncertain factors (inflow and electricity prices) must be considered. Stochastic optimization models for generation planning are in regular use in hydro-dominated systems [11].

An additional challenge for a hydropower generator is to reduce the risk of low profit from its entire operation. Risk management adds value by reducing the expected cost associated with financial distress.

In this article, electricity prices and inflow are modeled as stochastic processes. The price submodel is calibrated to future- and swap prices that are observed in the market. Price and inflow are assumed to be negatively correlated, and the price process is exogenous to the production optimization model, consistent with a price-taker assumption. The negative correlation is due to the positive relationship between local inflow and inflow for the whole system, and the negative relationship between system inflow and price. The price-taker assumption makes the analysis valid for a small hydropower producer operating in a well-functioning electricity market.

For the purpose of illustrating our modeling approach, relatively simple statistical models are fitted based on historical spot prices and inflow. Monte Carlo simulation is employed to generate an initial set of scenarios for price and inflow. Based on this high number of scenarios, a scenario tree is generated using the approach of [1]. In a real situation, more care should be taken modeling the scenarios in terms of analyzing historical data, incorporating expert judgments and existing forecasting models, and in the construction of event trees, making sure that the underlying data generating processes are well represented (including the information/stage structure) so that in turn the hydropower plants are operated efficiently. The corresponding stochastic program [12] is set up and solved as a large deterministic equivalent LP [23]. This creates acceptable solution times for a problem with 14 reservoirs and ten power stations. The optimal objective function value converges as the discretization of the stochastic processes is refined.

Profit risk is sought reduced using so-called delta-hedging. Delta-hedging of a portfolio means buying and selling contracts so that the total hedged cash flows are insensitive to short-term movements in the contract prices. This is a standard method for risk hedging as explained in textbooks, e.g., [17]. In contrast to the approach of [10], or of [3,6,14,19], with delta-hedging there is no need to expand the power optimization model with contract trading and a risk averse objective function. This means that less effort needs to be spent in model development and maintenance, and that the computing time will be shorter. We discuss how delta-hedging can be implemented in the context of an electricity portfolio and provide initial calculations.

The remainder of the article is structured as follows: The relevant markets are outlined in Sect. 22.2, while hedging is described in Sect. 22.3. Section 22.4 is devoted to modeling the stochastic processes as well as the decision problem itself. Results are given in Sect. 22.5, while we sum up in Sects. 22.6 and 22.7.

22.2 The Nordic Power Market

In the aftermath of the 1991 deregulation of the Norwegian power system, a Nordic power exchange was formed, Nord Pool. As the other Nordic countries joined Norway in the deregulation process, the scope of the exchange's activities steadily widened. Today the exchange is responsible for a number of markets, of which the day-ahead market for physical delivery is central. There are also financial markets (Eltermin and Eloption) which enable future power trading.

22.2.1 The Physical Market

Nord Pool offers exchange of electricity through the Elspot market, which is a day-ahead market with an hourly resolution. Participation in the market is voluntary. Participants submit their bids electronically on bidding forms, and a market clearing calculation is performed, determining the price for each of the 24 h of the following day. A so-called system price is also calculated, as an average over the 24 h, under an assumption of unlimited transmission capacity within the whole system. Price areas and counter-trading is used to handle congestion problems.

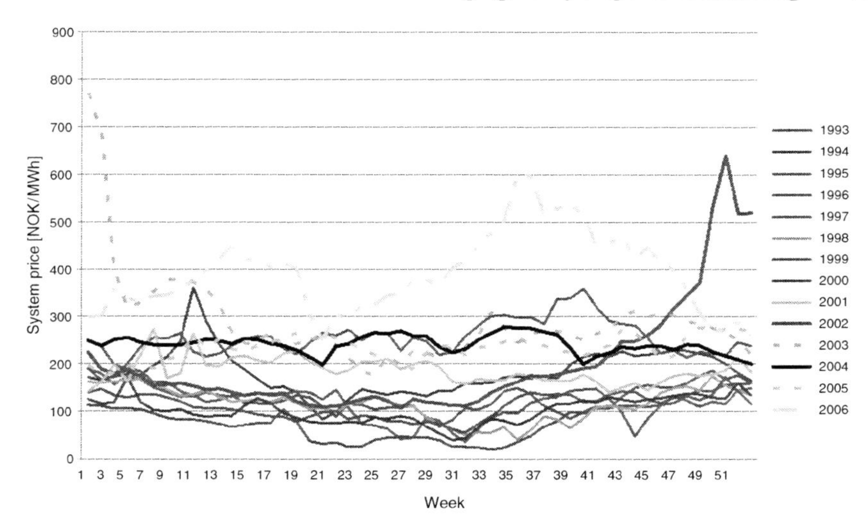

Fig. 22.1. Weekly system price in NOK/MWh from 1993 to 2006, source: Nord Pool. 1 € ≈ 8 NOK

The system price typically shows seasonal and diurnal patterns [16]. More than 95% of Norwegian production comes from hydro plants. Consequently the system price will be very much dependent on reservoir levels and inflow. As electricity-based residential heating is the norm in Norway, load usually increases in periods with cold spells (and correspondingly low inflow). This may induce spikes in the spot price.

22.2.2 The Financial Market

In the financial markets Nord Pool offers futures, swaps, options, electricity certificates, as well as emission allowance and certified emission reduction contracts. In the Eltermin market futures and swaps are traded. Future contracts are traded on a daily and weekly basis, swaps for months, quarters, and years. All contracts are standardized and have a size of 1 MW during the delivery period. All these contracts have financial settlement, and the system price is used as a reference. The swap contracts are termed "forwards" on the exchange, and in this article the two terms will be used interchangeably. The Eloption market offers European options, or "swaptions," with quarters and year forward contracts as the underlying.

For a hydro producer the information available in the financial markets of Nord Pool is useful when planning the optimal use of the water resources, as is investigated empirically by [7].

22.2.3 Electricity Price Characteristics

Several studies on the characteristics of electricity prices have been made, e.g., [16]. Typical observed characteristics for the system price are price spikes, mean-reversion, seasonality and excess kurtosis and skewness in price changes and log returns.

22.3 Hedging of Power Production

The idea of hedging electricity portfolios via stochastic programming was introduced by [9, 10]. The model was a "traditional" multistage stochastic program with a focus on the integration of production and financial trading. In this article our goal is to use a hydropower case to discuss a different approach to hedging, closer to what is learned by students in business schools and universities. A different but related approach is demonstrated in [21].

The key insight in modern option pricing theory is that it is possible to construct financial portfolios with exactly the same payoff structure as the underlying derivative. Hedging hydropower production means a search for a set of products that do exactly this – replicate the cash flows generated by hydro production. There are several possible reasons why a producer would prefer to hedge production, typically based on capital market imperfections (in a wide sense) that imply that risk averse behavior increases the value of the firm.

[22] argue that hedging should not affect the actual production plan, given an efficient derivative market for hedging price risk. According to standard financial theory the market value of a financial contract is zero when first entered into and will therefore not change the market value of production. A change in the production plan, will, however, change the market value of the production.

Risk is in this context typically related to price and inflow. It is not possible to hedge all risk related to the production, as no market for inflow risk exists. In addition, financial products needed to hedge high resolution price risk, e.g., spikes, are not available. Financial contracts that are liquid and available for price hedging have an increasing swap term (length of the delivery period) as the time to maturity increases. If the only available instrument is a 1-year contract, weekly price risk cannot be hedged, as the contract only reflects average risk over the entire period of 1 year. Furthermore it would be necessary to make some sort of assumption about the reservoir levels at the end of the period. Despite these shortcomings, we hereafter assume that the market for hedging price risk is complete.

The fact that it is nearly impossible to hedge all risk related to hydropower production complicates the methods used for risk-neutral pricing and thus also the hedging process itself. [18] also encountered this problem when trying to find a price for options when underlying prices can jump. We are going to use the same idea as he introduced, namely that the additional risk over price risk, is assumed not systemic and can be diversified away.

If inflow risk is assumed to be nonsystemic, investors can hedge against it by holding a well-diversified portfolio. In a sufficiently liquid market a risk premium on inflow risk will represent an opportunity for excess profit. This window of opportunity would not last. In such a case the risk premium of inflow risk should be zero.

Even though it is nearly impossible to perfectly hedge hydropower production, it is possible to reduce the risk significantly. For a producer going short on futures and swaps is an effective way to lower the risk to an acceptable level. To achieve a consistent result the producer needs to plan and price production in such a way that it is possible to estimate how sensitive the production value is to changes in value of the available future and forward contracts.

Delta-hedging is usually explained in terms of hedging an option that has been sold. The hedger should try to maintain a position of $delta = \Delta$ number of shares so that the risk in the total position is close to zero. Delta is simply the derivative of the option price with respect to the stock price. As the stock price changes, so does the delta, and the hedger must buy or sell to maintain a total position of zero risk. In theory, the position must be rebalanced continuously, but in practice a delta-hedger will wait for the position to become somewhat unhedged before trading. With F as the price of the underlying, and V as the value of the option (or portfolio) to be hedged, the delta is

$$\frac{\partial V}{\partial F} = \Delta\,. \tag{22.1}$$

If V depends nonlinearly on F, any change in the value of the underlying leads to a change in delta. [24] describes a central-difference-estimator to find the approximate change in option value when the value of the underlying increases or decreases. If we assume that optimal expected cash flows from hydropower production can be seen as an option with the forward curve as the underlying, the delta can be expressed in the following way:

$$\frac{\partial V}{\partial F} = \Delta \approx \frac{V^{\Delta +} - V^{\Delta -}}{2}\,, \tag{22.2}$$

where $V^{\Delta *}$ is the electricity portfolio value resulting from a unit shift in the forward curve.

The portfolio value V depends in principle on all futures and swaps that together constitute the forward curve. One possibility is to define a vector of deltas, one for each traded product. However, the producer wants to be hedged against price risk, and not all moves in the forward curve are equally likely. It is natural to start with looking at the risk of a general shift in prices.

It is possible to go beyond the delta to consider other greeks such as the gamma, for changes in the delta, and the vega, for changes in the volatility. This is left for future work.

22.4 Production Models – Theory and Implementation

In this section we present the most important assumptions and explain the price and inflow models used for representing the stochastic processes. We then give a mathematical description of the medium to long term planning problem. Finally we give a short presentation of the actual power plant system from which input data has been extracted.

We assume that the hydro producer is a price-taker. Decisions made by the producer will not influence electricity prices. For most hydropower producers in the Nordic region this is a reasonably valid assumption. For large-scale producers such as Statkraft, it is more dubious. Such actors could employ their market power to manipulate prices. This is not a subject in this analysis, however.

22.4.1 Stochastic Models

Stochastic variables in the model are electricity spot prices and inflow. This is the norm for long term production planning models [11]. Furthermore a modest negative correlation between price and inflow is assumed. Again this makes sense, as load levels in Norway typically increase in periods with low inflow (during the winter), and on longer term, draughts leads to increased prices.

Price Model

Many price models have been suggested for the dynamics of electricity prices. Electricity companies tend to replace their models from time to time. The method of scenario generation we have chosen is not affected by the choice of price model, so we opt for a simple one-factor mean-reverting process. The use of such a model enables us to capture some of the more important properties of electricity prices, in particular the tendency to revert to a long-run level. The price process is a variant of the Ornstein–Uhlenbeck process with time-dependent expectation and is expressed as:

$$\frac{\mathrm{d}\Pi_t}{\Pi_t} = \kappa(\theta_t - \ln \Pi_t)\mathrm{d}t + \sigma \mathrm{d}Z_t \ . \tag{22.3}$$

Here $\mathrm{d}Z_t$ is a Wiener process. Using Ito's Lemma and the log of the electricity price with the transformation $g_t = \ln \Pi_t$ we get the discrete model:

$$\Delta g_t = \kappa\left(\theta_t - g_t\right)\Delta t + \sigma\sqrt{\Delta t}\varepsilon_t \ , \tag{22.4}$$

where ε_t is a standard normal random variable. To find the risk-adjusted process market prices of derivatives are used. According to [2], the relationship between market prices and the parameters of (22.3) is:

$$\theta_t = \frac{1}{\kappa}\frac{\partial \ln F_{0,T}}{\partial t} + \kappa \ln F_{0,T} + \frac{\sigma^2}{4\kappa}\left(1 - e^{-2\kappa T}\right) \ . \tag{22.5}$$

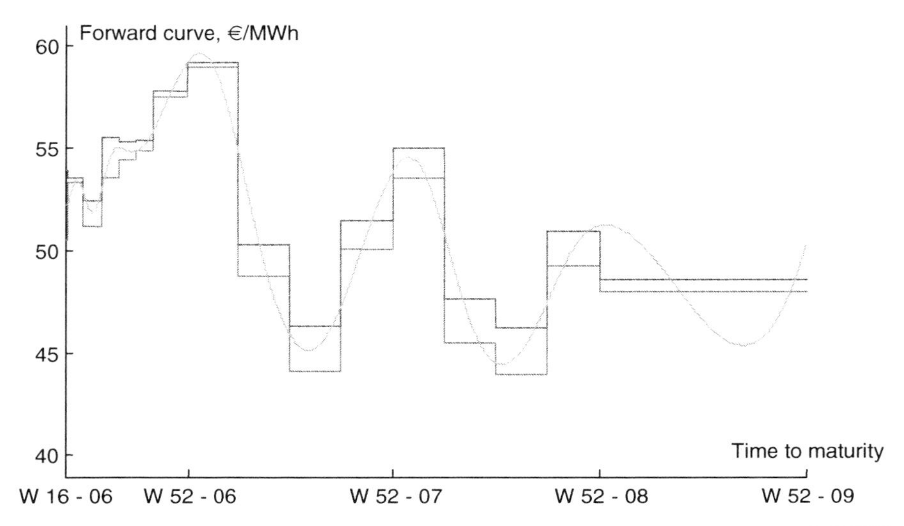

Fig. 22.2. Smoothed forward curve with bid-ask rectangles. Prices are in € /MWh

Here $F_{0,T}$ is the current electricity forward curve, for forwards with delivery at time T. The electricity price model was calibrated using information from the term structure on 17 April 2006 for swap and future contracts from `www.nordpool.no`. The term structure was smoothed, see, e.g., [8]. If the term structure is displayed 2-dimensionally, with the bid-ask spread visible, as in Fig. 22.2, the smoothed curve will pass through these bid-ask rectangles. Some adjustments were also made to the end of the term structure to achieve a more realistic seasonal effect. The parameters σ and κ in (22.5) were found using weekly spot prices from Trondheim in the period from 1996 to 2005.

Inflow Model

The inflow process is in general multidimensional and has strong seasonal components. The main bulk of inflow to reservoirs comes during spring, whereas in winter the precipitation accumulates as snow. Forecasting the inflows and capturing the structure of the processes and their degree of predictability is of vital importance to hydro scheduling models. This issue is discussed by [20].

The model for inflow has the same structure as for price (22.3). The model used also has seasonal expectation and variance. Autocorrelation is often present in inflow series. The one-factor mean-reverting process corresponds by discretization to an AR(1)-process, autocorrelated at lag 1. For the inflow the parameters have been estimated using weekly historical inflow data for the period 1951–2001 from the Nea–Nidelva river system in Norway. Weekly volatilities and mean-reversion coefficients were estimated using OLS.

22.4.2 Event Tree Modeling

One might consider using stochastic dynamic programming to solve this problem. However, this is a multireservoir case, and price and inflow would need to be states as well. Due to the curse of dimensionality, this approach is not practical. Instead, we employ linear multistage stochastic programming [12].

To model how uncertainty (represented by stochastic inflow and electricity prices) unfolds over time, event trees are generated. The first node in the tree is called the root node. Each node in the tree represents a decision point, or equivalently a state, corresponding to a realization of the random variables up to the stage of state n, denoted by $t(n)$. Every state except the root node has a predecessor node, denoted $a(n)$.

Creating event trees which provide a satisfactory description of the stochastic processes is a considerable challenge. An overview up to about 2000 can be found in [4]. A reasonable update of later work is part of [13].

Starting from fan scenarios, i.e., scenarios without a stage structure, has some advantages since it makes it easy for the problem owners to replace the price and/or inflow model with whatever they prefer, in particular simulators or "black boxes" they may have available. [1,5] describe methods to construct event trees based on fans. Since our scenarios have been generated this way, we apply their approach. After all, the way we generate scenarios does not effect the main purpose of this article: To illustrate the use of delta-hedging. However, in a real setting, we would approach the scenario generation in a more careful way, as discussed by [13], since the reduction technique may have weaknesses when starting from a fan (it should ideally be used to reduce a too large scenario tree with appropriate stage structure to a smaller one with the same structure.)

22.4.3 Deterministic Equivalent of the Stochastic Problem

In this section we present the mathematical program used for the production planning problem. Uncertainty for inflow and price is taken into account via joint discrete distributions, and are represented by an event tree with n nodes which represent different states in the stochastic process. This stochastic model is formulated as a deterministic equivalent linear program.

Data

t	Index for periods. Let $t(n)$ be the period belonging to node n.
$i, j \in \mathcal{I}$	Indices for reservoirs in set $\mathcal{I}$
$\mathcal{U}_i$	Set of reservoirs upstream of reservoir i whose outflow will go to reservoir i
$\mathcal{R}_i$	Set of reservoirs upstream of reservoir i whose spill will go to reservoir i

$n, \mathcal{N}$	Index, set for nodes in the event tree
S_t	Set of nodes in period t
$a(n)$	Index of predecessor to node n
P_n	Unconditional probability of the state in node n
π_n	Electricity price in node n
D_t	Discount factor for period t
K_i	Water-to-energy coefficient for reservoir i
$\nu_{i,n}$	Inflow in node n for reservoir i
$L_{max,i,t}$	Upper bound in period t for reservoir level in reservoir i
$L_{min,i,t}$	Lower bound in period t for reservoir level in reservoir i
$L_{end,i,n}$	End level for reservoir i in node n
$Q_{max,i,t}$	Upper bound in period t for discharge through the station for reservoir i
$Q_{min,i,t}$	Lower bound in period t for discharge through the station for reservoir i
r	Risk free interest rate

Decision Variables

V	Value of production for the whole planning period
$l_{i,n}$	Reservoir level in node n at the start of period $t(n)$ for reservoir i
$r_{i,n}$	Spill in node n during period $t(n)$ for reservoir i
$w_{i,n}$	Hydropower generation in node n during period $t(n)$, $w_{i,n} = K_i q_{i,n}$
$q_{i,n}$	Production discharge in n from reservoir i

Objective Function

$$\max V = \sum_{i \in \mathcal{I}} \sum_{n \in \mathcal{N}} P_n \pi_n D_{t(n)} K_i q_{i,n} \tag{22.6}$$

Constraints

$$l_{i,n} - l_{i,a(n)} + q_{i,n} + r_{i,n} - \sum_{j \in \mathcal{U}_i} q_{j,n} - \sum_{j \in \mathcal{R}_i} r_{i,n} = \nu_{i,n}, \quad n \in \mathcal{N}, i \in \mathcal{I} \tag{22.7}$$

$$l_{i,n} = L_{end,i,n}, \quad i \in \mathcal{I}, n \in S_T \tag{22.8}$$

$$L_{min,i,t} \leq l_{i,n} \leq L_{max,i,t}, \quad n \in \mathcal{N}, i \in \mathcal{I} \tag{22.9}$$

$$Q_{min,i,t} \leq q_{i,n} \leq Q_{max,i,t}, \quad n \in \mathcal{I}, i \in \mathcal{I} \tag{22.10}$$

$$q_{i,n}, r_{i,n} \geq 0, \quad n \in \mathcal{N}, i \in \mathcal{I} \tag{22.11}$$

The objective function (22.6) is the sum of the discounted expected future revenues from each period. There are no direct variable costs of hydropower generation; all costs are fixed. We use the risk free interest rate for discounting, because risk is already adjusted for in the stochastic process for the cash flows, in that it is calibrated to the forward curve. In (22.7) the reservoir in each node n is dependent on the reservoir level in the predecessor node $a(n)$. The initial storage levels are $l_{i,0} = L_{init,i}$ for all reservoirs i. End reservoir levels are fixed by (22.8). The constraints on the flow of water in (22.10) have time indices since the time intervals have different lengths. With *discharge* we mean water being used for electricity production. *Spill* is the amount of water that is not utilized. This could typically occur in situations where the reservoir is full. Time-varying bounds on reservoir levels (22.9) and discharges (22.10) reflect physical, technical and environmental concerns.

22.4.4 Model Implementation

The deterministic equivalent described in the previous section has been used to solve the hydropower production problem for a number of plants and reservoirs in Mid-Norway, in the Nea–Nidelva waterway. The optimization itself was done on a 2.4 GHz Intel Celeron CPU with 3.71 GB RAM. Scenarios were generated as described in Sect. 22.4.2.

Period of Analysis

The typical horizon for hydro scheduling is a few months to a few years. A typical length of the first time step ranges from 1-week to 1-month. The hydro scheduling model gives signals to hydro unit commitment via marginal values of stored water in the reservoirs and/or via total generation during the first week.

In our case the planning horizon is divided into 14 periods and spans April 2006 to October 2007. The first six periods are weeks, the next four periods are months, and the final four periods are quarters. This corresponds to the swap term of the products traded at Nord Pool at the beginning of the first period. The stage structure is illustrated in Fig. 22.3.

The production facilities in Nea–Nidelva currently consists of a catchment area of 3,100 km^2, ten reservoirs and 14 plants with a total installed production capacity of 614 MW. The waterway has its origin in Sweden and ends in

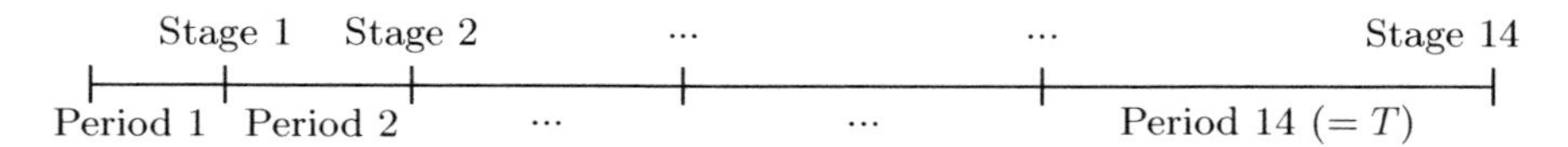

Fig. 22.3. Stage structure

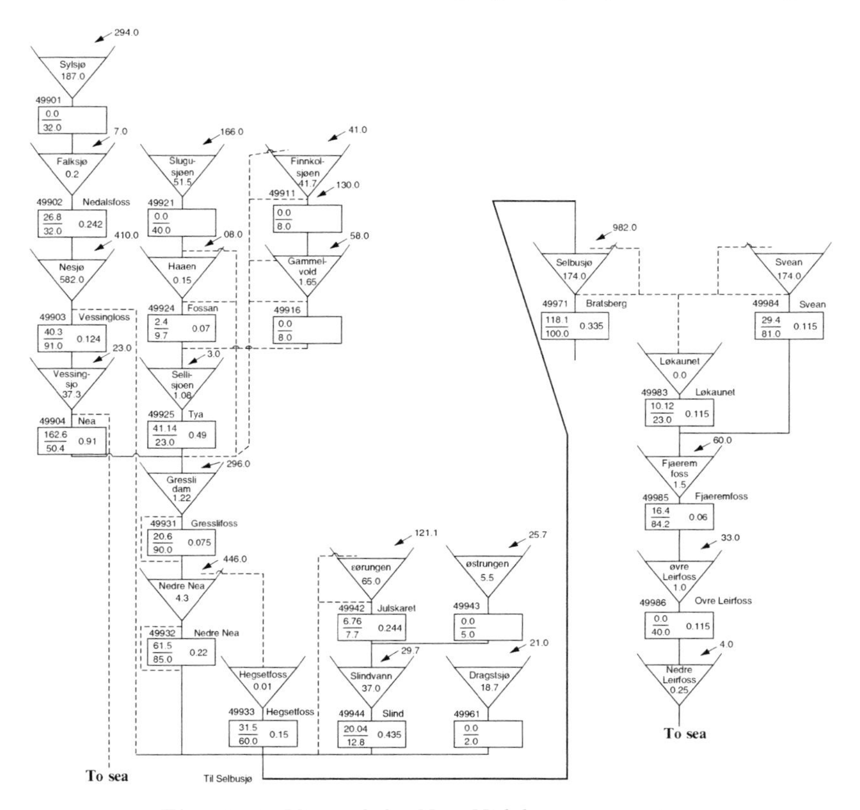

Fig. 22.4. View of the Nea–Nidelva water system

Table 22.1. Input data for convergence analysis

f_T	S_{fan}	Correlation
0.725	500	−0.2

the city of Trondheim, a distance of 160 km. A general view of parts of the waterway is shown in Fig. 22.4. The ratio of aggregate reservoir capacity to annual inflow is relatively high (64%), which makes this system of reservoirs well suited for a production planning/risk management analysis. To simplify the problem somewhat we have used fixed water flow to energy conversion coefficients for the power stations. This means that the energy efficiency is not affected by the actual reservoir level. The topology of this system is such that the real efficiency does not vary much with weekly flow and reservoir levels, so dealing with this issue in more detail is left for future work.

The value of water at the end of the planning horizon depends on the time of year, the reservoir levels and price levels. This value function is hard to estimate, however, and instead of using such a function we have chosen to set target levels for the reservoir at the end of the planning horizon. End reservoir levels have been set according to:

$$L_{end,i,n} = f_T L_{max,i}, i \in \mathcal{I}, n \in S_T \ , \tag{22.12}$$

where $f_T \in [0, 1]$ is a parameter governing the relative end levels. Starting levels were set to the actual historical levels in week 16 2006, at approximately half full.

22.5 Results

22.5.1 Optimization Analysis

The optimization is performed using the dual simplex algorithm in Mosel XPRESS. The risk free rate of interest is set at 3.5% in all analysis. The analysis of convergence is done with the parameter set in Table 22.1.

S_{fan} represents the number paths constructed via Monte Carlo sampling of (22.4) and the corresponding inflow equation, before the construction of the event trees. The optimal value of expected production during the planning period seems to converge with an increasing number of nodes in the event tree, as shown in Fig. 22.5.

22.5.2 Value of Production

The subsequent analysis is done using the parameters presented in Table 22.2.

The reduction level in Table 22.2 is a parameter used in the scenario generation process. The number of nodes in the resulting event tree decreases with an increasing reduction level. Table 22.3 shows the value of operating revenues for various reservoir levels at the end of the planning period. As expected the value is higher for lower end reservoir levels.

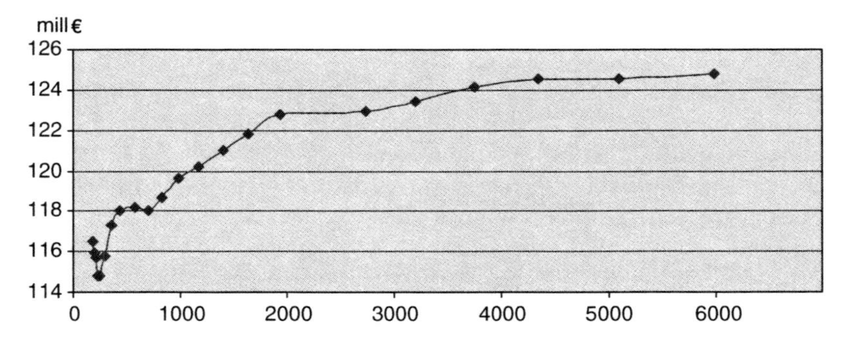

Fig. 22.5. Optimal value of production (vertical axis) for specified number of nodes in the event tree

Table 22.2. Scenario generation parameters for analysis of production value

S_{fan}	Reduction level	S_{tree}	Nodes
1,000	0.35	771	3,701

Table 22.3. Value of operating revenues for different reservoir end levels

f_T	Correlation	Value[M€]
0.750	–0.2	123.7
0.725	–0.2	126.0
0.700	–0.2	128.3
0.675	–0.2	130.5

Table 22.4. Value of production for different reservoir end levels, no correlation between price and inflow

f_T	Correlation	Value[M€]
0.750	0.0	124.1
0.725	0.0	126.4
0.700	0.0	128.6
0.675	0.0	130.9

By assuming no correlation between inflow and prices (Table 22.4) the resulting values do not differ much from the case with assumed negative correlation. This could indicate that the low negative correlation does not necessarily have a significant effect on the value of production[1]. It can not be ruled out that it may have an effect on hedging.

22.5.3 Expected Production Strategy

For the analysis below we used the following parameters:

Figures 22.6 and 22.7 present examples of expected production and reservoir level respectively. Expected production displays a clear seasonal variation in addition to a downward trend. The seasonality is partly a result of the seasonal variation in the term structure of futures prices.

22.5.4 Expected Cash Flow

The expected operating revenue in each period is presented in Fig. 22.8. From this figure it is clear that there is a certain resemblance between the cash flow and forward curves. Intuitively, this makes sense, as the forward curve is the main source of information on future spot price levels.

[1] An absolute value of correlation of 0.2 may be too low to draw conclusions from. Further analysis is left for future work.

Table 22.5. Parameters for analysis of the production strategy

f_T	Correlation	S_{fan}	Reduction level	Nodes	S_{tree}
0.7	–0.2	1,000	0.12	7,395	937

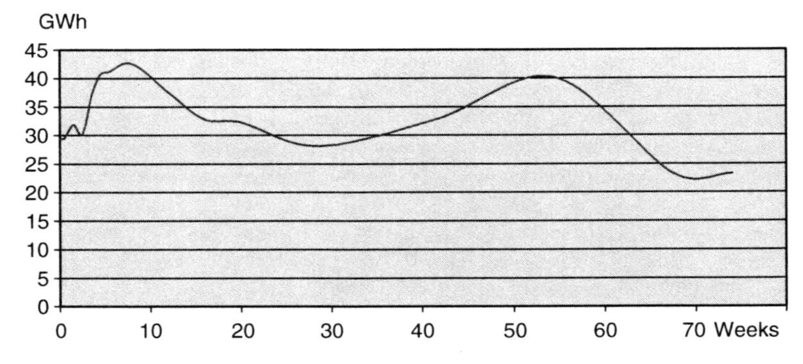

Fig. 22.6. Expected weekly production for case in Table 22.5

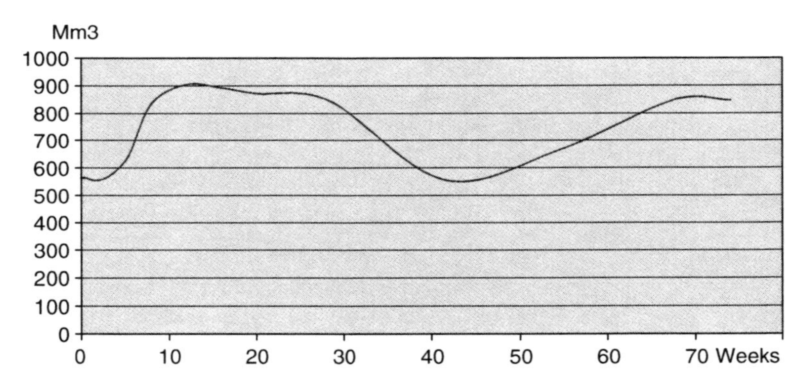

Fig. 22.7. Expected reservoir levels for case in Table 22.5

22.5.5 Delta-Hedging

By adding and subtracting one unit for all contracts in the forward curve it is possible to obtain an expression for the sensitivity of the cash flows and the total production value in the subperiods of the planning problem. This can be achieved by using (22.2).

Here the value of the cash flows V is in €and delta is in MWh. The delta is calculated from model instances created using the parameters in Table 22.6.

The results from the sensitivity analysis for each interval in the planning period are shown in Table 22.7. For each period the delta can be observed as a quantitative discrepancy in the expected cash flows. The conclusion from the

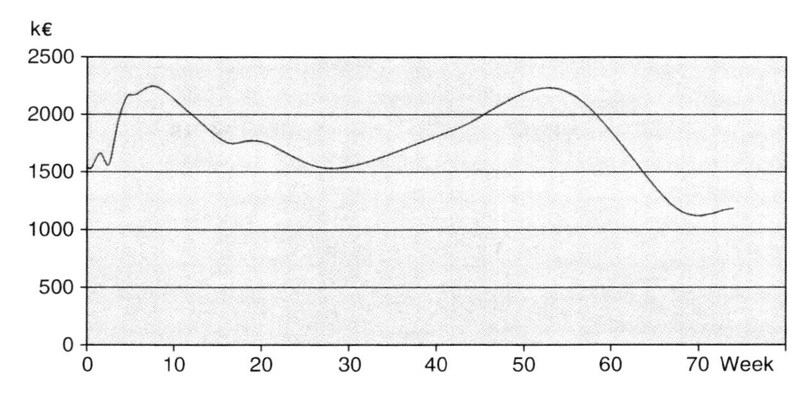

Fig. 22.8. Expected weekly cash flows for case in Table 22.5

Table 22.6. Parameters for delta-hedging

f_T	Correlation	S_{fan}	Reduction level	Nodes	S_{tree}
0.7	-0.2	500	0.12	4,058	472

Table 22.7. Cash flow sensitivities for parallel shifts in the forward curve

Period[t]	1	2	3	4	5	6	7
V^{Δ^+}[k€]	1,382	1,223	1,241	1,640	1,990	1,970	8,197
V^{Δ^-}[k€]	1,780	2,009	1,741	1,896	2,123	1,995	8,431
$\Delta[GWh]$	-199	-393	-250	-128	-66	-12	-117

8	9	10	11	12	13	14	Whole period
7,777	7,668	7,619	20,299	26,036	30,412	15,760	133,214
9,047	7,362	7,224	17,904	24,393	27,202	14,228	127,336
-635	153	197	1,198	821	1,605	766	2,939

table is that the producer is recommended to go short in the swap and future products that spans the next 3.5 months, and to go long in the products that span the rest of the planning horizon.

22.6 Discussion

Much has been left for future work, since this is a first attempt at delta-hedging of a portfolio with hydropower. Further sensitivity analysis with respect to correlation, end reservoir level and number of fan scenarios before scenario reduction is a natural next step. An interesting future possibility is to compare integrated risk management such as in [10] with delta-hedging.

One could also measure the performance of the delta-hedging over time, and compare the performance with, e.g., delta hedges that come from deterministic generation scheduling, or scheduling heuristics. Furthermore, it would be interesting to investigate the effects of the long-term negative relationship between local inflow and local prices, that leads to a natural hedge meaning the risk-minimizing position is less than expected electricity sales.

The electricity price model takes many of the typical empirical characteristics of electricity prices into account. However, since we only need weekly average prices, the need for short term characteristics, such as spikes, is reduced.

Any real use of this approach would have to be more careful about how scenarios are generated from the stochastic models of price and inflow.

The current status of the development of this model is that it remains a case study. The owner of the power plant, Trondheim Energi, has been taken over by Statkraft, who may be less interested in hedging cash flows.

A hedging strategy should be based on second-order market information, in the form of a term structure of volatility. The model in (22.3) has a simple volatility structure that does not fully reflect real market dynamics. It would also be preferable to have a model providing a better representation of the correlation between prices and inflow. It is possible to update and upgrade the models for prices and inflow used in this analysis, albeit possibly at the cost of keeping the disadvantages associated with creating reduced multistage event trees from two-stage scenario fans. Using more sophisticated models for the stochastic processes will also allow for more realistic and more effective hedging strategies. For example, there is just one random factor driving (22.3). In reality, a model with many factors is needed to capture a large part of the variance [15].

A hedging strategy involving frequent trading could lead to large transaction costs. On the other hand, if too much time pass between each time the portfolio is updated, it could lead to unnecessary losses. The optimal trading frequency must be found in future work.

22.7 Conclusion

The model gives reasonable results. For an increasing number of nodes in the event tree the optimal value of the production converged towards a stable level. This production value increased when the fixed reservoir level at the end of the planning horizon was lowered. An assumed negative correlation between price and inflow did not seem to have a significant effect on the expected production value.

Even with 6,000 nodes in the event tree the model did not need more than a couple of minutes to solve. This suggests that there could be a considerable potential for using such models to plan the hydro production and risk management.

Acknowledgement. We thank Martin G. Pedersen and Ståle Skrede for research assistance, and Trondheim Energi, esp. Erling Kylling, for providing the case data. Fleten acknowledges financial support from the Research Council of Norway through project 178374/S30, and Wallace through project 171007/V30.

References

1. H. Heitsch and W. Römisch. Scenario reduction algorithms in stochastic programming. *Computational Optimization and Applications*, 24(2):187–206,2003.
2. L. Clewlow and C. Strickland. *Energy Derivatives - Pricing and Risk Management.* Lacima, London, UK, 2000.
3. J. Doege, P. Schiltknecht, and H. J. Lüthi. Risk management of power portfolios and valuation of flexibility. *OR Spectrum*, 28(2):267–287, 2006.
4. J. Dupačová, G. Consigli, and S. W. Wallace. Scenarios for multistage stochastic programs. *Annals of Operations Research*, 100:25–53, 2000.
5. J. Dupačová, N. Gröwe-Kuska, and W. Römisch. Scenario reduction in stochastic programming. *Mathematical Programming*, 95(3):493–511, 2003.
6. A. Eichhorn, H. Heitsch, and W. Römisch. Scenario tree approximation and risk aversion strategies for stochastic optimization of electricity production and trading. In J. Kallrath and P. Pardalos, editors, *Optimization in the Energy Industry.* Springer, Berlin, 2008.
7. S. -E. Fleten and J. Keppo. Empirical analysis of hydroelectric scheduling. *Working paper, Norwegian University of Science and Technology*, 2008.
8. S. -E. Fleten and J. Lemming. Constructing forward price curves in electricity markets. *Energy Economics*, 25(5):409–424, 2003.
9. S. -E. Fleten, S. W. Wallace, and W. T. Ziemba. Portfolio management in a deregulated hydropower-based electricity market. In E. Broch, D.K. Lysne, N. Flatabø, and E. Helland-Hansen, editors, *Proceedings of the 3rd international conference on hydropower (Hydropower'97)*, pages 197–204, Trondheim, Norway, July 1997. Balkema, Rotterdam, 1997.
10. S. -E. Fleten, S. W. Wallace, and W. T. Ziemba. Hedging electricity portfolios using stochastic programming. In C. Greengard and A. Ruszczyński, editors, *Decision Making under Uncertainty: Energy and Power*, volume 128 of *IMA Volumes on Mathematics and Its Applications*, pages 71–93. Springer, New York, 2002.
11. O. B. Fosso, A. Gjelsvik, A. Haugstad, B. Mo, and I. Wangensteen. Generation scheduling in a deregulated system. The Norwegian case. *IEEE Transactions on Power Systems*, 14(1):75–81, 1999.
12. P. Kall and S. W. Wallace. *Stochastic Programming.* Wiley, Chichester, 1994.
13. M. Kaut and S. W. Wallace. Evaluation of scenario generation methods for stochastic programming. *Pacific Journal of Optimization*, 3(2):257–271, 2007.
14. J. Kettunen, A. Salo, and D.W. Bunn. Dynamic risk management of electricity contracts with contingent portfolio programming. *Submitted manuscript,* `http://www.sal.hut.fi/Publications/m-index.html`, 2007.
15. S. Koekebakker and F. Ollmar. Forward curve dynamics in the Nordic electricity market. *Financial Management*, 31(6):74–95, 2005.
16. J. J. Lucia and E. S. Schwartz. Electricity prices and power derivatives. Evidence from the Nordic Power Exchange. *Review of Derivatives Research*, 5(1): 5–50, 2000.

17. R. L. McDonald. *Derivatives Markets.* Addison Wesley, Boston, MA, 2003.
18. R. C. Merton. Option pricing when underlying stock returns are discontinouos. *Journal of Financial Economics*, 3(1):125–144, 1976.
19. S. Sen, L. Yu, and T. Genc. A stochastic programming approach to power portfolio optimization. *Operations Research*, 54(1):55–72, 2006.
20. Tejada-Guibert, S. A. Johnson, and J. R. Stedinger. The value of hydrologic information in stochastic dynamic programming models of a multireservoir system. *Water Resources Research*, 31:2571–2579, 1995.
21. I. Vehviläinen and J. Keppo. Managing electricity market price risk. *European Journal of Operational Research*, 145(1):136–147, 2003.
22. S. W. Wallace and S. -E. Fleten. Stochastic programming models in energy. In A. Ruszczynski and A. Shapiro, editors, *Stochastic programming*, pages 637–677. Vol. 10 of *Handbooks in Operations Research and Management Science*. Elsevier, Amsterdam, 2003.
23. R. J. -B. Wets. Stochastic programs with fixed recourse: The equivalent deterministic program. *SIAM Review*, 16(3):309–339, 1974.
24. P. Glassermann. *Monte Carlo Methods in Financial Engineering.* Springer, New York, 2003.

Index